HUMAN FORM AND FUNCTION

Human Form and Function
A Basic Approach

Marvin R. Barnum

*St. Louis Community College
at Florissant Valley*

Goodyear Publishing Company, Inc.
Santa Monica, California

Library of Congress Cataloging in Publication Data

Barnum, Marvin R
  Human form and function.

  Bibliography: p.
  Includes index.
  1. Human physiology.   2. Anatomy, Human.
I. Title
QP34.5B35   612   78-23374
ISBN 0-87620-385-3

Copyright © 1979 by Marvin R. Barnum
Published by Goodyear Publishing Company, Inc.
Santa Monica, California 90401

All rights reserved. No part of this book may be
reproduced in any form or by any means without
permission in writing from the publisher.

ISBN: 0-87620-385-3
Y-3853-2

Current printing (last digit):
10 9 8 7 6 5 4 3 2 1

Editing: Thomas Belina
Illustrations: Basil Wood
Special color plates: Dick Oden
Cover illustrations: Dick Oden
Typesetting: Computer Typesetting Services, Inc.
Editing and production management: Brian K. Williams

Printed in the United States of America

Dedicated to the memory of my father,
ELDEN KENNETH BARNUM

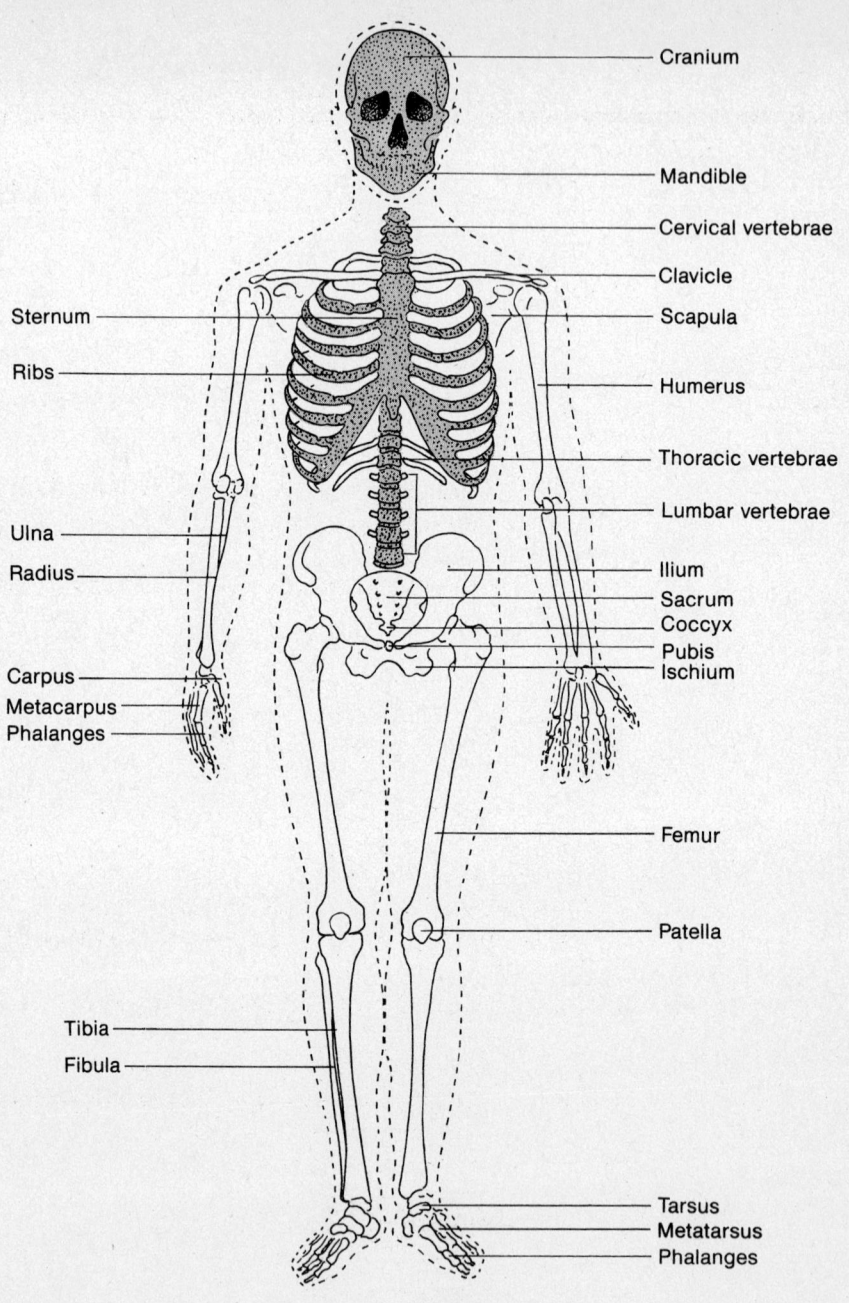

Contents

PREFACE   xvii

1   INTRODUCTION   1

       Homeostasis   2
       The Scientific Method   3
       Scientific Measurement: The Metric System   3
       Summary and Review   10

2   OVERVIEW OF ANATOMY AND PHYSIOLOGY   11

       Body Organization   12
       Terms of Reference   12
       Preliminary Biological Concepts   17
       Summary and Review   21

(For chemistry and physics review, see the appendixes at the end of book.)

APPENDIX A   BASIC CHEMISTRY   315
APPENDIX B   SOME BASIC PHYSICAL CONCEPTS   344

3   CELL STRUCTURE AND FUNCTION   23

       The Cell   24
       Characteristics of Cells and Cellular Organelles   24
       Cellular Activities   30
       Summary and Review   32

4   TISSUES, MEMBRANES, SKIN   33

       Tissues   34
       Membranes   39
       Skin   39
       Summary and Review   45

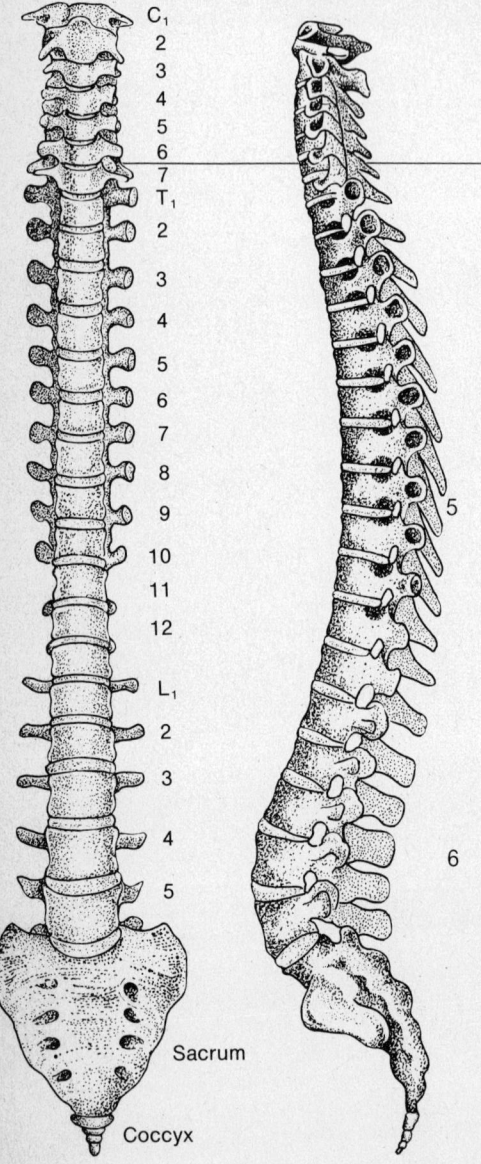

Plates 1-4 follow page 46.
PLATE 1 Muscles of the Head and Neck
PLATE 2 Principal Superficial Muscles of the Anterior Trunk
PLATE 3 Principal Superficial Muscles of the Posterior Trunk
PLATE 4 A Section of Skin and the Musculature of the Arm and Leg

5 THE SKELETAL SYSTEM   47

Cartilage   49
Bone   49
Skeletal Components   54
Fractures   72
Bone Articulations   72
Diseases of the Skeletal System   73
Calcium Level Influences in the Aged   74
Summary and Review   74

6 THE MUSCULAR SYSTEM   76

Types of Muscle Tissue   77
Microscopic Structure of Muscle Tissue   77
Neural Control of Muscle Contraction   79
Chemistry of Muscular Contraction   80
Classification of Muscles   85
Major Muscles of the Body   85
Diseases of Muscles   92
Muscle Reduction with Aging   96
Summary and Review   96

7   THE NERVOUS SYSTEM   98

   The Nerve Impulse   102
   Communicating Pathways   103
   Divisions of the Nervous System   107
   Cranial Nerves   116
   The Autonomic Nervous System   118
   Diseases and Disorders of the Nervous System   121
   The Aging Nervous System   122
   Summary and Review   122

8   COORDINATION: THE SENSES   124

   The Eye   125
   The Ear   131
   Sense of Smell   136
   Sense of Taste   136
   Aging of the Senses   136
   Summary and Review   137

9   THE HEART AND BLOOD VESSELS   138

   The Heart   139
   Arteries, Veins, and Capillaries   148
   Circulation Patterns   151
   Diseases of the Circulatory System   161
   Hypertension in the Elderly   163
   Summary and Review   164

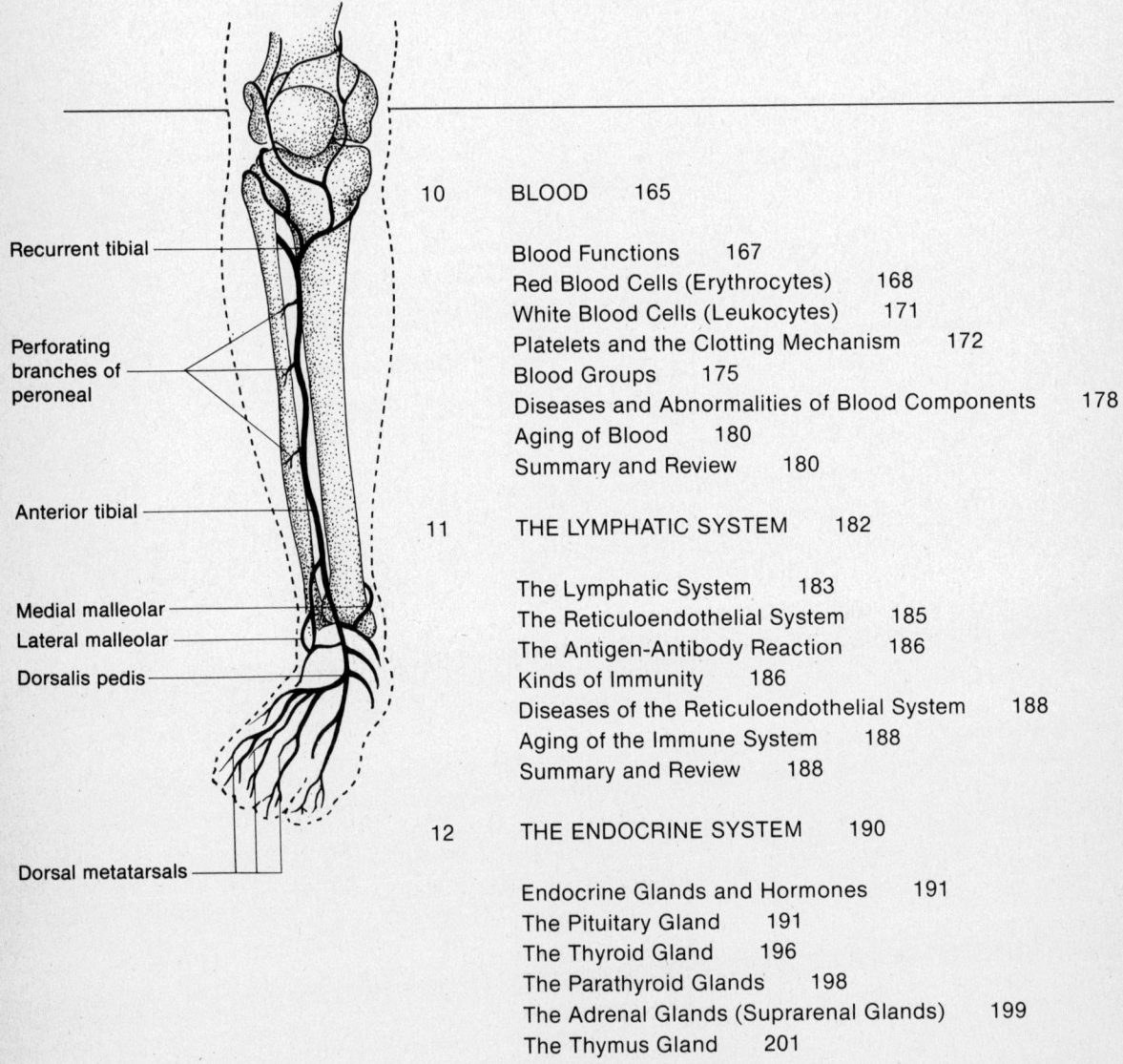

10 BLOOD 165

    Blood Functions 167
    Red Blood Cells (Erythrocytes) 168
    White Blood Cells (Leukocytes) 171
    Platelets and the Clotting Mechanism 172
    Blood Groups 175
    Diseases and Abnormalities of Blood Components 178
    Aging of Blood 180
    Summary and Review 180

11 THE LYMPHATIC SYSTEM 182

    The Lymphatic System 183
    The Reticuloendothelial System 185
    The Antigen-Antibody Reaction 186
    Kinds of Immunity 186
    Diseases of the Reticuloendothelial System 188
    Aging of the Immune System 188
    Summary and Review 188

12 THE ENDOCRINE SYSTEM 190

    Endocrine Glands and Hormones 191
    The Pituitary Gland 191
    The Thyroid Gland 196
    The Parathyroid Glands 198
    The Adrenal Glands (Suprarenal Glands) 199
    The Thymus Gland 201
    The Pancreas 202
    Aging and Endocrine Function 206
    Summary and Review 206

Plates 5-9 follow page 206.
PLATE 5   Position of Organs of the Trunk
PLATE 6   Spinal Cord, Nerve Pathways, and Associated Tissues
PLATE 7   Circulation—the Heart and Blood Vessels
PLATE 8   Digestive System
PLATE 9   Kidney

13   THE DIGESTIVE SYSTEM   207

Teeth   208
Anatomy of the Mouth and Throat   209
The Esophagus   212
Anatomy of the Stomach   212
Anatomy of the Small Intestine   217
Anatomy of the Large Intestine   223
The Liver   225
Physical Disorders of the Gastrointestinal Tract   227
Infectious Diseases of the Digestive System   229
Aging and the Gastrointestinal Tract   230
Summary and Review   230

14   NUTRITION AND METABOLISM   232

Carbohydrate Metabolism   233
Caloric Value of Foods   235
Fat Metabolism   236
Protein Metabolism   237
Enzymes   237
Vitamins   237
Minerals   238
Water   239
Medical Problems Related to Nutrition   239
Summary and Review   241

15   THE EXCRETORY SYSTEM   242

       Basic Anatomy of the Urinary System   243
       Structure of the Nephron   245
       How Materials Get into Bowman's Capsule   246
       Formation of Urine   247
       Diuretics   251
       Maintaining Acid-Base Balances   251
       Maintaining Fluid Balances   252
       Composition of Normal Urine   253
       Composition of Abnormal Urine   253
       Micturation   254
       Renal Diseases   255
       The Aging Kidney   258
       Summary and Review   258

16   BODY FLUIDS AND ELECTROLYTES   259

       Functions of Body Fluids   260
       Fluids and Body Weight   260
       Organs that Regulate Fluids   260
       Electrolytes   261
       Acid-Base Balance   266
       Respiratory and Metabolic Acidosis   266
       Respiratory and Metabolic Alkalosis   266
       Summary and Review   267

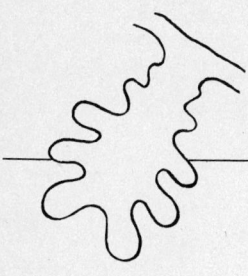

Normal alveoli

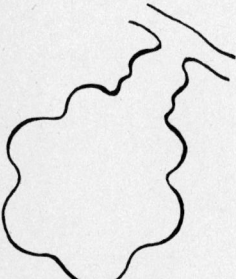

Emphysema (alveolar septa missing)

Pneumonia (alveoli filled with fluids)

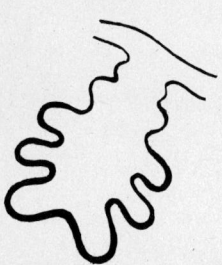

Fibrosis (alveolar walls thickened)

17 THE RESPIRATORY SYSTEM 268

   Anatomy of the Breathing Apparatus 269
   The Mechanisms of Breathing 270

   Plates 10-13 follow page 270.
   PLATE 10 Respiratory System
   PLATE 11 Male Reproductive System
   PLATE 12 Female Reproductive System
   PLATE 13 Conception and Implantation

   The Alveoli 273
   Regulation of Breathing 274
   Types of Breathing 277
   Physical Obstruction and Resistance to Air Flow 278
   Gaseous Transport 278
   Microbial Diseases of the Respiratory System 282
   Respiratory Changes in the Aged 283
   Summary and Review 283

18 REPRODUCTION 286

   The Male Reproductive System 286
   The Female Reproductive System 291
   Pregnancy 295
   Fetal Development 295
   Diseases of the Reproductive System 296
   Nonmicrobial Diseases of the Reproductive System 297
   The Aging Reproductive System 297
   Summary and Review 297

19 HUMAN GENETICS 299

    Chromosomes and Karyotypes 300
    Homologous Pairs of Chromosomes 300
    Single-Gene Defects 303
    Sex Determination 304
    Sex-Linked Traits 304
    Multifactorial Inheritance 305
    Chromosomal Abnormalities 305
    Sex Chromosome Defects 307
    Tests for Genetic Abnormalities 310
    Summary and Review 313

APPENDIX A BASIC CHEMISTRY 315

    Organized Matter 316
    Elements 316
    Compounds 316
    Solutions 317
    Energy 317
    Atoms and Atomic Structure 318
    Ions 322
    Acids, Bases, and Salts 324
    Organic Chemistry 325
    Biologically Important Organic Substances 331
    Summary and Review 342

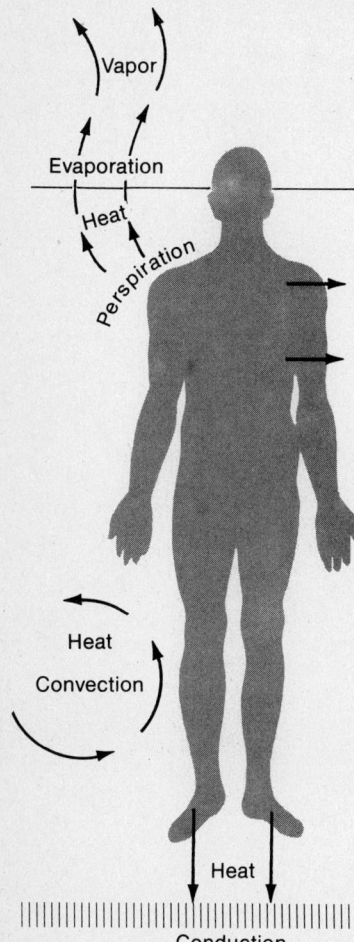

APPENDIX B   SOME BASIC PHYSICAL CONCEPTS   344

    Gases   345
    Liquids   351
    Energy   354
    Heat   355
    Light   356
    Electricity   357
    Summary and Review   359

APPENDIX C   DISCUSSION OF COLOR PLATES   361

GLOSSARY   368

BIBLIOGRAPHY   390

INDEX   395

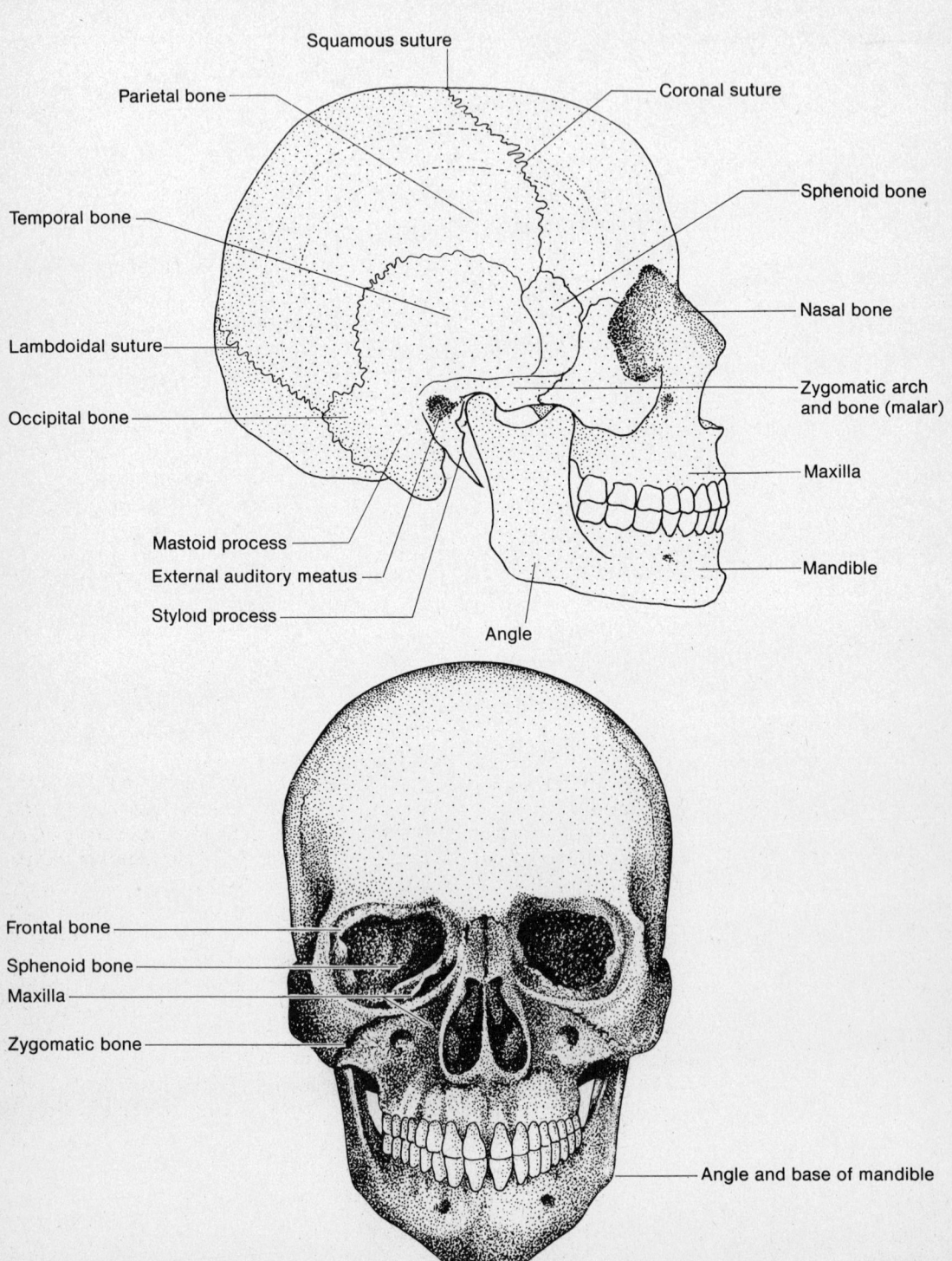

Preface

Why yet another anatomy-physiology textbook? This question to my mind has several answers, and I hope you will find agreement with at least some of them.

The decade of the 1970s has demanded that all of us in the academic community educate increasingly more people in an increasingly more efficient way. Without arguing the merits of this issue, it is a fact that the college student today is less prepared, in terms of prerequisite background, to enter into the kind of burgeoning scientific study our discipline now describes.

Traditional anatomy and physiology texts, which were certainly appropriate to the students and curriculum prior to 1970, have indeed kept pace with the current growth of our scientific knowledge. Unfortunately, many of our students today are not prepared to handle the level of chemistry, biology, and physics which today's sciences require for understanding and learning. Part of this problem stems from the changing purpose of academic preparation. Prior to 1970, most students were enrolled in four-year degree programs that required chemistry and biology prerequisites to the anatomy and physiology course. Today about half of all students enrolled in anatomy and physiology courses are in two-year technical programs that may not have chemistry or biology prerequisites. It is obvious that a text appropriate to one program may not be appropriate to another, and this is my major justification for writing yet another text—to produce one that will fill the specific need of students of limited background, where the anatomy and physiology course is an orientation, rather than the area of major interest.

By addressing myself to this specific application, I have necessarily affected the depth of coverage by selective discussion of basic concepts, but I have maintained the current breadth of coverage. There should be no surprises in the information I am dispensing because, above all else, I have made accuracy paramount—primary sources and reviewers have assured me that I've achieved this goal. Rather than being different in *what kind* of anatomy and physiology I have written, I have purposely been different in *how* it is presented, keeping the information accurate, simple, uncluttered, and interesting to learn. I have made a serious attempt to "underwhelm" the student.

This text is oriented toward health science curricula, including nursing, allied health technologies, and physical education. In the chapters on body systems, I first pursue the healthy state of the body, then proceed to discuss both microbial and nonmicrobial diseases, and conclude with discussion of the effects of aging upon each body system.

The appendix on basic chemistry and the appendix on basic physical concepts can serve either to review these topics or to help teach the material to students who have no prior background in these areas. The concepts discussed in these appendixes are those most commonly associated with the chemical and physical processes of human structure and function.

I have included a comprehensive self-pronouncing glossary, a feature I believe will be valuable to the students. With over 700 entries, it is more extensive than the glossaries found in other anatomy and physiology texts.

Having completed the writing of this book, I would like to be able to say "I did it my way." However, this book could not have been successfully completed without the aid of many individuals. I want to extend a sincere thank you to the reviewers who provided expert editing and sound advice on scientific concepts and who, in general, helped build my confidence in the writing task. The reviewers are:

Robert Agnew, El Centro College
Lee Armstrong, Diablo Valley College
Gerald Bessey, Los Angeles Valley College
Robert Catlett, University of Colorado
Charles Levy, Boston University
Marion Kretzer, St. Louis Community College
Irene Newby, Santa Ana College

The typists, who succeeded in turning out a professional-looking manuscript from handwritten copy, are Lois Phillips, Gail Warmann, and Betty Barnum.

Special acknowledgement must be given-to the Goodyear Publishing Company staff, particularly to my editor, Clay Stratton. His firm guidance, friendly persuasion, and positive attitude kept me motivated and enthusiastic about writing.

Marvin R. Barnum

1 Introduction

MAJOR CONCEPTS
HOMEOSTASIS
THE SCIENTIFIC METHOD
SCIENTIFIC MEASUREMENT: THE METRIC
  SYSTEM
  Length
  Weight
  Volume
  Temperature
  Density and Specific Gravity

The concept of homeostasis is central to an understanding of how the body adapts to changing environments. When the body fails to adapt, the result may be sickness, disease, or death.
  Most of this chapter deals with the scientific method and scientific measurement.

The major objective of medicine is to preserve life and maintain health. Health, however, has many definitions. The most widely used is that stated in the preamble to the constitution of the World Health Organization: "Health is a state of complete physical, mental and social well-being and not merely the absence of disease or infirmity." Although this definition encompasses all the conditions normally associated with what we commonly call health, it lacks precision. What, exactly, do the words "social well-being," "disease," and "infirmity" mean?

Another definition, also somewhat imprecise, is: "Health is the continuing adjustment of an organism to its environment." (Disease, therefore, could be defined as the reverse—"the imperfect continuing adjustment of an organism to its environment.") Thus, the healthy person is one whose body is adaptable—that is, capable of subtle physiological changes in response to an ever-changing external environment.

External and internal environmental changes may be as simple as small variances in temperature, humidity, and light or as extreme as a bacterial invasion of the body.

HOMEOSTASIS

French physiologist *Claude Bernard* (1813–1878), the first to use scientific experimentation in physiological research, was also the first to demonstrate how the external environment, that which constantly surrounds the living organism, differed in nature from the internal environment which influences body functions. As the external environment changes, he realized, so must the body internally adjust to meet the crisis, and failure to make such an adjustment resulted in illness or death.

Internal self-adjusting mechanisms are generally referred to as *systems*. Systems stabilize such delicate functions as cellular chemical balances, cell division, and cellular growth and repair. But even in an unchanging external environment an organism's chemical balances would not remain unchanged. Development, growth, and aging bring about changing internal environments. The healthy body is capable of adapting to these internal changes as it is to external changes.

We do not know with complete confidence why certain internal changes, such as aging, occur. We recognize the features associated with aging, such as hardening of the arteries, wrinkling of the skin, and loss of certain physical and mental faculties, but the underlying causes of these changes continue to be perplexing problems.

An all-encompassing word, *"homeostasis"* (coined by American physiologist Walter B. Cannon), a word derived from the Greek words meaning "stay the same," is used to summarize this concept of discrete checks and balances made by internal mechanisms in response to external and internal environmental forces.

Changing external and internal environments would have drastic, even fatal, consequences for the body were it not for self-regulating mechanisms that aid in restoring a function to its normal state once it is disturbed by stress. *Stress* is defined as a stimulus that disrupts a stable internal environment.

For example, if an artery is cut, nerve impulses sent to the smooth muscle cause the muscle to contract, thus reducing the flow of blood out of the body. The blood-clotting process is also initiated to further reduce blood loss. If the blood loss significantly reduces the total volume, other processes are initiated to restore the proper volume.

In summary, the body is capable of instituting controls that aid in maintaining normal conditions. Body homeostasis is the sum of many homeostatic mechanisms operating on smaller scales.

A special type of control called a *feedback control* serves to maintain homeostatic mechanisms by initiating a process, then halting the process when the normal state is achieved. For example, carbon dioxide (CO_2), a product of metabolism, must be removed by the respiratory system. The concentration of CO_2 in *extracellular fluids* (fluid outside cells—for example, blood) must not exceed certain levels or other body systems may be adversely affected. In order to maintain constant CO_2 levels, the breathing rate fluctuates as body activity fluctuates. Increased activity produces more CO_2. This higher concentration of blood CO_2 stimulates nerves that control the breathing rate. As a consequence, the breathing rate is accelerated, thereby increasing the rate of CO_2 loss from the body. When the activity is reduced or stopped, the blood CO_2 level returns to normal and consequently the breathing rate returns to normal. This represents a self-regulating, feedback control.

Homeostasis in more general terms refers to the precisely adaptable nature of an organism or its parts which allows the organism to function most effectively.

Throughout this text, homeostatic relationships will be stressed.

THE SCIENTIFIC METHOD

The scientific method is an efficient, common-sense problem-solving method used in scientific discovery. In most scientific discovery, there are identifiable steps that lead to the discovery. These steps constitute the scientific method and are as follows:

1. A problem is recognized to exist.
2. Facts and relevant information are collected through observation and/or experimentation.
3. Collected facts and relevant information are classified into some specified order and generalizations are drawn in view of known laws of nature.
4. A working hypothesis is formulated.
5. The working hypothesis is tested.
6. The hypothesis is accepted, rejected, or modified.

Individuals trained to solve problems by using the scientific method commonly follow these six steps. They are, however, also aware that "unscientific" factors, such as trial and error or happy accidents, may aid them in finding a more rapid solution to a problem.

In addition, they should be aware that certain precautions must be observed if a solution is to be found. For example, the problem must be carefully delineated and narrowly defined. Many problems go unsolved because their scope has not been limited.

Another precaution is to be on guard for prejudice and bias. The significance of the outcome of an experiment is often missed because the "undesired" result is not produced. The "desired" results are those which are thought to be necessary to bring about a specific order that the experimenter thinks is correct. Natural order does not always follow a researcher's assumed order.

SCIENTIFIC MEASUREMENT: THE METRIC SYSTEM

A scientist needs tools to accurately record observations. As an experiment is often an essential part of the scientific method, the design of the experiment must be such that it can be reproduced again and again by other scientists. The recorded observations must precisely indicate such things as how much, how far, how hot, how high, and so on. In order to make these observations with accuracy, a system of measurement is necessary.

There are two measuring systems employed in the world. The *English system* employs the linear measurements of the inch, foot, and yard; the volume measurements of the pint, quart, and gallon; the weight measurements of the ounce, pound, ton; and so on.

The *metric system* is used by most other countries except the United States and is used for scientific work in all countries, including the United States.

The metric system is based on the factor of 10. Each unit is larger or smaller than other units by a factor of 10. This permits easy manipulation of measurement numbers. If we look at the English system—12 inches to a foot, 3 feet to a yard, 1,760 yards to a mile and so on—we can see no systemic relationship between the units. The metric system, once an understanding of its basic design is gained, is much easier to use.

Length

The unit of length in the metric system is the *meter*. A meter, as indicated in Figure 1.1, is 39.37 inches long. The meter is an arbitrary length which, by international agreement in 1889, was defined as the distance between two

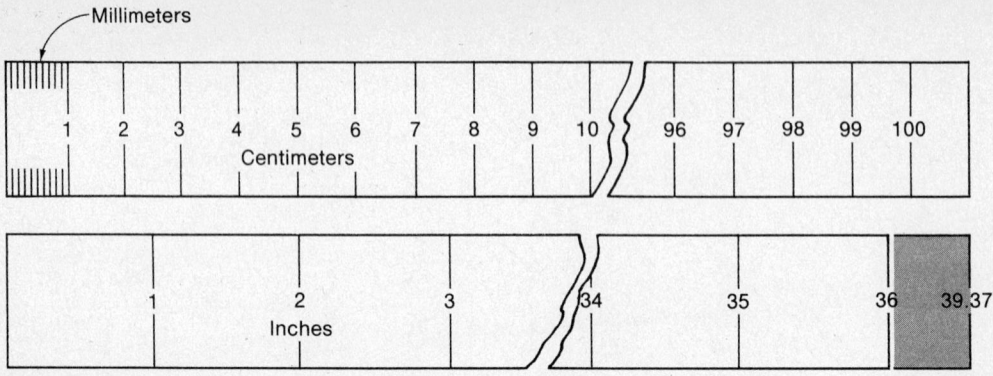

Fig. 1.1 Comparison of metric and English units.

marks inscribed on a metal bar made of an alloy of platinum and iridium. This particular bar is kept at the International Bureau of Weights and Measurements near Paris, France. The Bureau of Standards in Washington, D.C. has an exact copy of this bar that is used as a standard in this country.

A more recent international standard of length has been accepted based on the specific wavelength of light. This new definition of the meter does not appreciably change the measurement of length, however, and it does not change the relationship between the English and metric systems.

In order to measure lengths shorter than 1 meter, the meter can be divided into 10 equal parts. One of these divisions (0.1 meter) is called a *decimeter*, abbreviated *dm*. If a meter is divided into 100 equal divisions, each of the smaller divisions is called a *centimeter*, abbreviated *cm*. This is the equivalent of dividing each decimeter into 10 smaller divisons. Ten centimeters equal 1 decimeter and 10 decimeters equal 1 meter. Therefore, there are 10 × 10 (100) centimeters in 1 meter.

If we wish to have even smaller units, a centimeter may be divided into 10 divisions. One-tenth of a centimeter is called a *millimeter*, abbreviated *mm*. How many millimeters are in a meter? Ten millimeters times 10 centimeters times 10 decimeters gives the answer—1,000 millimeters. One millimeter is equal to 1/1000 or 0.001 meter.

In metric measurement, the same set of prefixes is attached to each base unit. Each prefix designates some multiple of or division by 10. The prefixes are as follows:

Milli: thousandths (0.001) (10^{-3})
Centi: hundredths (0.01) (10^{-2})
Deci: tenths (0.10) (10^{-1})
Base Unit (Example–Meter)
Deka: tens (10) (10^{1})
Hecto: hundreds (100) (10^{2})
Kilo: thousands (1,000) (10^{3})

In linear measurement, the base unit is the meter. When the preceding prefixes are added to the term meter the unit of length is specified. For example, 1/10 meter is called a *decimeter*. One-hundredth meter is a *centimeter*. Ten meters is called a *dekameter*. Figure 1.2 compares metric linear units.

Weight

The metric base unit for weight measurement is the *gram*, abbreviated g. This is a considerably smaller unit than the pound or ounce. As shown in the comparison chart, Table 1.1, one pound equals 435.6 grams; one ounce equals 28.35 grams.

The same metric prefixes are used for weight measurement as are used for linear measurement. A *decigram* is 1/10 g, a *centigram* is 1/100 g, and a *milligram* is 1/1000 g. Ten grams equals a *dekagram*, 100 grams equals a *hectogram*, and 1,000 grams equals a *kilogram*. It is not always necessary to know the names of all the different units, as they can also be described as fractions or multiples of the base unit. For example,

4 HUMAN FORM AND FUNCTION

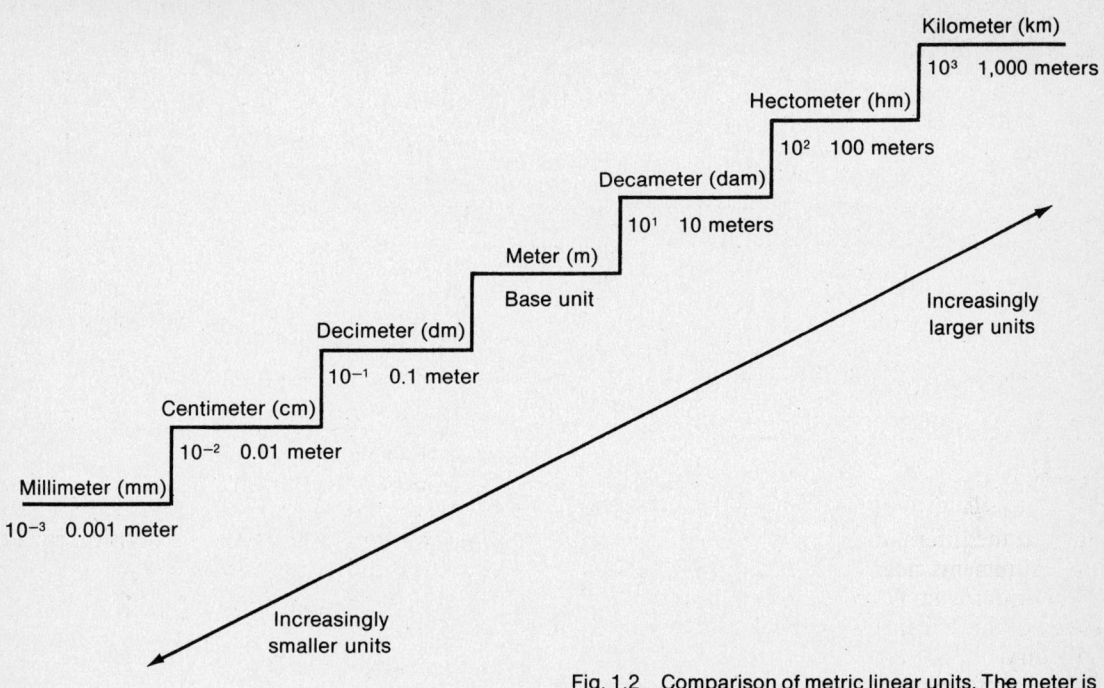

Fig. 1.2 Comparison of metric linear units. The meter is the base unit. Units up the stairway increase as multiples of 10 meters. Units down the stairway decrease by 1/10 meter.

Table 1.1 English and Metric Systems Compared

QUANTITY	ENGLISH SYSTEM	METRIC SYSTEM	EQUIVALENTS
Length	FOOT	CENTIMETER	
	1 foot = 12 inches	1 centimeter = 10 millimeters	1 in. = 2.54 cm
	1 yard = 3 feet	1 decimeter = 10 centimeters	1 ft. = 30.5 cm
	1 mile = 5,280 feet	1 meter = 100 centimeters	1 meter = 39.37 in.
		1 kilometer = 1,000 meters	1 km = 0.62 mile
Weight (Mass)	POUND	GRAM	
	1 pound = 16 ounces	1 gram = 1,000 milligrams	1 ounce = 28.35 g
	1 ton = 2,000 pounds	1 kilogram = 1,000 grams	1 pound = 453.6 g
			1 kg = 2.2 pounds
Volume		MILLILITER	
	2 pints = 1 quart	1,000 milliliters = 1 liter	1 liter = 1.06 qt
	4 quarts = 1 gallon		

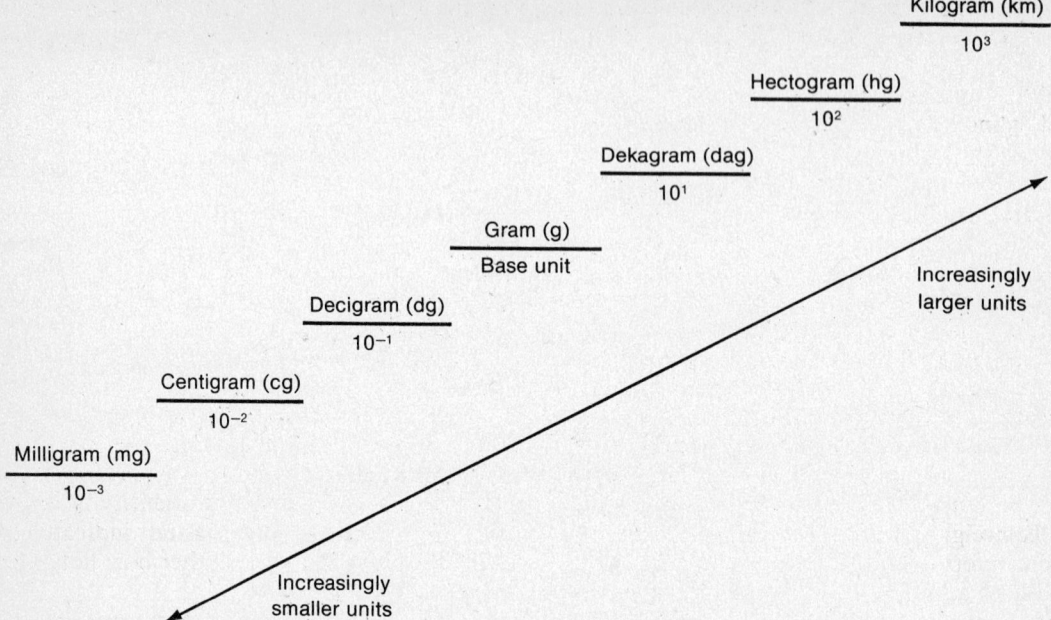

Fig. 1.3 Comparison of units for metric weight (mass). The same prefixes used here were previously used in linear measurement. This consistency of unit prefixes from one type of measurement to another is the advantage of the metric system.

something might weigh 1/10 of a gram (0.1 g), or 1/100 of a gram (0.01 g), or 1/1000 of a gram (0.001 g), and so on. Figure 1.3 illustrates the comparison of units of weight.

Volume

The volume of a liquid, gas, or solid is a measurement of the space it occupies. The volume of a solid—a cube, for example—is determined by multiplying the length times the width times the height. If the dimensions of the cube are measured in centimeters, the answer would be in units called *cubic centimeters,* abbreviated *cc.* If the cube measured exactly 1 centimeter long, 1 centimeter wide, and 1 centimeter high, the volume would be 1 cubic centimeter (1 cc). Figure 1.4 illustrates the formula for determining the the volume of a cube or rectangular solid.

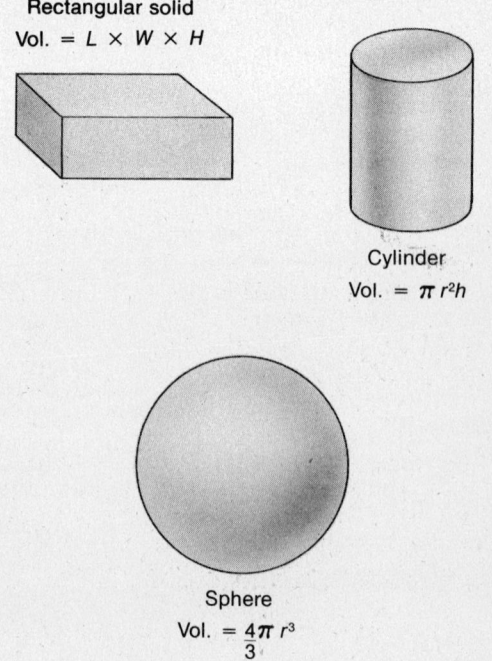

Fig. 1.4 Formulas for deriving volumes of solid objects. Volume is measured in cubic centimeters or milliliters (1 cubic centimeter = 1 milliliter, or 1 cc = 1 ml).

6 HUMAN FORM AND FUNCTION

Note also the formulas for deriving volumes of other solids—for example, a cylinder and a sphere.

When measuring liquid volumes in graduated cylinders, the surface of the liquid will be curved. The bottom of this curve (the *meniscus*) should be read as the volume. In order to eliminate error in reading the volume, the viewer's eye must be at the level of the meniscus, as shown in Figure 1.5.

When the standardizing procedures were first being developed by international agreement, it was intended that one kilogram (1,000 grams) would be the weight of 1,000 cubic centimeters of water at 4 degrees Celsius. This volume of water (1,000 cubic centimeters) is referred to as a *liter*, abbreviated *l*. If 1,000 cubic centimeters of water weigh 1,000 grams, then 1 cubic centimeter (cc) weighs 1 gram. The prefix *milli*, as noted before, refers to 1/1000 of something. One thousandth of a liter is a *milliliter*, abbreviated *ml*.

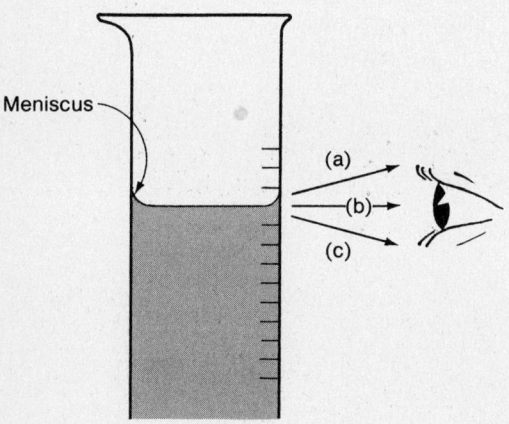

Fig. 1.5 Location of the eye in reading the volume in a graduated cylinder. (a) If eye is above meniscus, too small a volume is observed. (b) If eye is on the same level as meniscus, the correct volume is observed. (c) If eye is below meniscus, too large a volume is observed.

(Figure 1.6 compares metric volume units.) As 1/1000 of a liter is equal to 1 cubic centimeter, we now have two equivalent units—milliliter and centimeter. In other words, 1,000 milliliters equal 1 liter; 1,000 cubic centimeters also equal 1 liter. Quantities equal to the same thing are equal to each other. Therefore, 1 cubic centimeter equals 1 milliliter. One milliliter of water at 4 degrees Celsius weighs 1 gram.

One ml is approximately 15 drops from an eyedropper. The term approximate is necessary since the size of eyedroppers is variable.

Temperature

The measurement of temperature is similar to other forms of measurement in that some arbitrary units are established to indicate different intensities of heat.

Common thermometers are based on the principle that liquids or solids expand when heated. If a liquid is contained within a glass tube, it will expand when heated and fill the tube more completely. The liquid will expand a precise amount when a precise intensity of heat is applied. A thermometer does not indicate how much heat is present, but rather how hot something is.

Degrees are merely equal markings on a glass tube. The Fahrenheit scale has 180 such markings between the freezing and boiling points (32 and 212 degrees).

Twenty years after the Fahrenheit scale was developed, Anders Celsius of Sweden suggested that the scale should have 100 equal markings with 0 degrees indicating the freezing point of water and 100 degrees indicating the boiling point. (The two scales are compared in Figure 1.7.) The Celsius scale is much less cumbersome and is used by the scientific world. The system fits in well with the metric system of weights and measurements.

Since the Fahrenheit scale is more commonly used in the United States than the Celsius, it is often necessary to determine equivalents between the two scales. The equivalents appear in many texts, but these references may not always be accessible. Therefore, there is a need to know how to determine temperature equivalents.

Note in Figure 1.7 that most of the degree markings on the two scales are different for different temperatures. There is, however, one temperature at which both Fahrenheit and Cel-

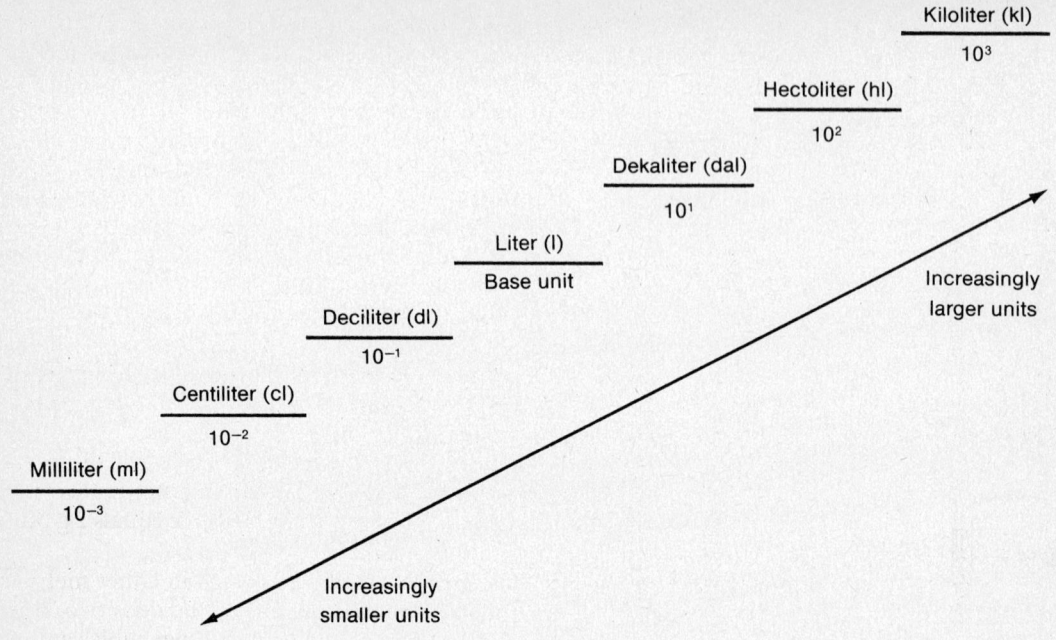

Fig. 1.6 Comparison of metric units for volume. Once again the same metric prefixes used for length and mass are used for volumetric measurement.

sius scales indicate the same numerical value: –40 degrees. This provides a rather simple method for determining equivalents on the two scales. If you wish to convert Fahrenheit degrees to Celsius, merely add the number 40 to the Fahrenheit reading, multiply the sum by 5/9, then subtract the number 40.

Example: To change 212 degrees Fahrenheit to degrees Celsius:

$$212 + 40 = 252$$
$$252 \times 5/9 = 140$$
$$140 - 40 = 100 \text{ degrees Celsius}$$

If you wish to convert Celsius degrees to Fahrenheit degrees, the same procedure is used except that the fraction is 9/5 rather than 5/9.

Example: To change 100 degrees Celsius to degrees Fahrenheit:

$$100 + 40 = 140$$
$$140 \times 9/5 = 252$$
$$252 - 40 = 212 \text{ degrees Fahrenheit}$$

You will have no trouble in remembering which fraction (5/9 or 9/5) you need to use if you reason that the numerical Celsius will always (except for –40) be smaller; therefore, you need to multiply by the fraction which will give you a number smaller than the one you started with. In the first example, the fraction is 5/9.

When you change from Celsius to Fahrenheit, since Fahrenheit will be numerically larger than Celsius, multiply by 9/5. The fractions 5/9 and 9/5 are derived from the fact that there are 100 degree marks between boiling and freezing on the Celsius scale and 180 degree marks between these two extremes on the Fahrenheit scale. Five Celsius degrees equal 9 Fahrenheit degrees. Therefore, 5/9 degree Fahrenheit equals 1 degree Celsius and 9/5 degree Celsius equals 1 degree Fahrenheit.

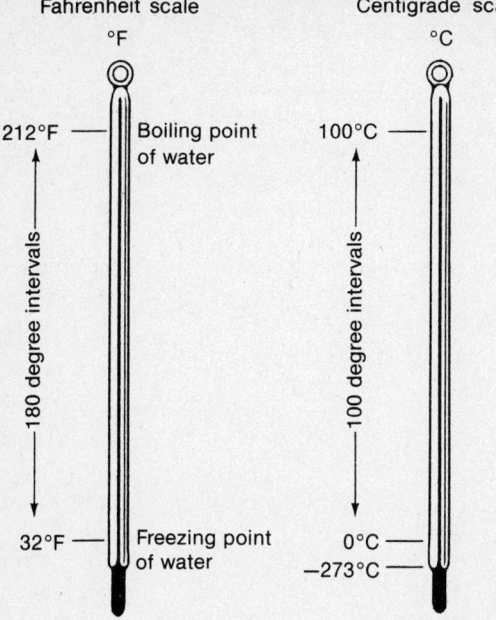

	°F	°C
Average room temperature	77	25
A very cold day	−20	−29
A hot day	100	38
Normal body temperature	98.6	37

Fig. 1.7 Fahrenheit and Centigrade temperature scales: comparison of common temperature reference points.

Another method of converting one temperature scale to another uses the following formulas:

$$\text{Degrees Celsius} = \frac{5(F-32)}{9}$$

$$\text{Degrees Fahrenheit} = \frac{9}{5}C + 32$$

Calories are units expressing quantities of heat. A *small calorie* (cal) is the amount of heat required to raise the temperature of 1 gram of water 1 degree Celsius. The *large calorie* (Kcal or Cal) consists of 1,000 small calories. The caloric value of foods is expressed in large calories.

Density and Specific Gravity

As noted previously, 1 cc of pure water weighs 1 g. If you weigh 1 cc of alcohol, milk, or urine, you would find that these do not weigh 1 g. From common experience we know that some objects are heavier than others even when they are the same size. For instance, a block of iron is heavier than a block of wood of equal volume.

The scientific term for the weight of a specific volume of a material is *density*. The density of a substance is derived by dividing its weight by its volume, or, stated as a formula, $D = W/V$. Suppose we want to determine the density of a block of wood that weighs 16 g and has a volume of 8 cc. Sixteen g divided by 8 cc equals 2 g per cc (2 g/cc).

Density may be expressed in either metric or English units. For example, the density of water in the metric system is 1 gram per cubic centimeter. In the English system it may be expressed as 62.4 pounds per cubic foot. It may also be expressed as 16.41 grams per cubic inch. Thus, when expressing density of any object, it is important to determine and indicate which units are being used.

The term *specific gravity* is used to compare densities of different substances. Specific gravity is the ratio of the weight of a substance compared to the weight of an equal volume of water. If a substance has a specific gravity of 2, it is twice as heavy as a comparable volume of water. A substance that is lighter than water will have a specific gravity less than 1 and will float if it does not readily mix with water. The specific gravity of gasoline is 0.7, or 7/10 times as heavy as water. Since gasoline has a specific gravity less than 1 and does not mix with water, it will float.

In order to determine the specific gravity of a substance, it is necessary to measure out a precise volume of the substance, weigh it, then divide the weight of an equivalent volume of water into this number. *Example:* Find the specific gravity of olive oil if given a sample of 2 cc.

$$\text{Weight of 2 cc of olive oil} = 1.84 \text{ g}$$

$$\text{Weight of 2 cc of water} = 2 \text{ g}$$

$$\frac{\text{Specific gravity}}{\text{of olive oil}} = \frac{\text{weight of substance}}{\text{weight of equal vol. of water}}$$

$$= \frac{1.84 \text{ g}}{2 \text{ g}} = 0.92$$

Since the specific gravity of olive oil is less than 1 it will float on water.

Notice that when referring to specific gravity no units are designated, since the specific gravity of a substance is numerically the same as the density. Specific gravity is a ratio, comparing densities of substances with the density of water. Table 1.2 shows the specific gravity of some common substances.

Note that the specific gravity of milk and urine are shown as a range rather than a definite unit. These substances are mixtures in which the components may vary by weight. Seawater is also a mixture. The components, however, are fairly uniform.

The concept of specific gravity in the health sciences is important in determining the concentration of impurities that may be found in body fluids. Pure substances such as water, alcohol, or glycerine, will always have the same specific gravity (at a given temperature). The specific gravity of urine and other body fluids, being mixtures, have a variable specific gravity depending to some degree upon the health of the individual. It has been clinically determined that a narrow range of specific determinations may be called normal. The deviation of the specific gravity of body fluids from the clinical norm might be an indication of some body malfunction. Thus, specific gravity determinations are a useful diagnostic tool.

SUMMARY AND REVIEW

A. The term homeostasis relates to the uniformity that is maintained by the body's internal environment. External environmental factors that influence the internal environment include:
   1. temperature
   2. light
   3. humidity
   4. bacterial invasion
B. The scientific method is a problem-solving process which involves the following steps:
   1. Recognition of a problem.
   2. Collection of facts and relevant information through observation and/or experimentation.
   3. The drawing of generalizations from the collected data.
   4. Formulation of a working hypothesis.
   5. Testing the hypothesis.
   6. The acceptance, modification, or rejection of the hypothesis.
C. The metric system is used in all scientific weights and measures. The base units are as follows:
   1. length—meter
   2. weight—gram
   3. volume—liter
   4. temperature—Celsius scale
D. Quantities of heat are expressed in calories. A calorie is defined as the amount of heat required to raise the temperature of 1 g of pure water 1 degree Celsius.
E. Density refers to the weight of a specific volume of a substance.
F. Specific gravity is the weight of a specific volume of a substance divided by the weight of an equivalent volume of pure water.

Table 1.2 Specific Gravity of Common Substances

SUBSTANCE	SPECIFIC GRAVITY
Acetone	.791
Alcohol, ethyl	.791
Alcohol, methyl	.810
Ether	.736
Glycerine	1.26
Milk	1.028–1.035
Sea water	1.025
Olive oil	.918
Urine	1.010–1.025

2
Overview of Anatomy and Physiology

MAJOR CONCEPTS
  BODY ORGANIZATION
  TERMS OF REFERENCE
    Surgical Regions of the Abdomen
    Body Cavities
  PRELIMINARY BIOLOGICAL CONCEPTS
    Diffusion and Osmosis
    Active Transport
    Filtration
    Bioelectricity
    Enzymes

The study of anatomy and physiology requires a bit of prerequisite knowledge. This perhaps has been gained already in basic biology courses. Students unfamiliar with basic science concepts should see Appendixes A and B. These two appendixes plus the present chapter should set the stage for all the chapters in between.

Anatomy is the study of bodily structure; *physiology* is the study of bodily function. These two terms are usually used in reference to the structure and function of normal individuals, not too young and not too old, and, unless otherwise stated, free of disease or organ malfunction.

Until the concept was established that the cell and its functions are of primary importance in disease, medical research made slow progress. The external appearance of what constitutes a disease is often actually an overall manifestation of some chemical or physical disruption of cells. Thus, an understanding of cell functions is necessary to an understanding of both health and disease.

Another facet of anatomy and physiology is scientific terminology and its application to what may be called body geography. Review the new terms introduced in each chapter. This will help secure your understanding of the material that follows.

BODY ORGANIZATION

A primary part of the study of anatomy and physiology is learning the organization of the body and the integrating mechanisms which hold the organization together.

The basic unit of life is the *cell*. The cell builds *tissues* that make up *organs* that collectively build *systems*. Proper function of the body depends upon the degree of efficiency with which systems integrate. When one system fails to perform in a normal manner, other systems are directly or indirectly affected.

Life is organized matter; when this organization fails, the consequence is death.

TERMS OF REFERENCE

Two initial terms of reference are *superior* and *inferior*. *Superior* means toward the head end of the body. Another term synonymous with superior is *cranial*, which means head end.

Inferior means away from the head. Another term synonymous with inferior is *caudal*, which means toward the tail.

To illustrate these terms we would say that the head is superior to the breastbone. The foot is inferior to the head.

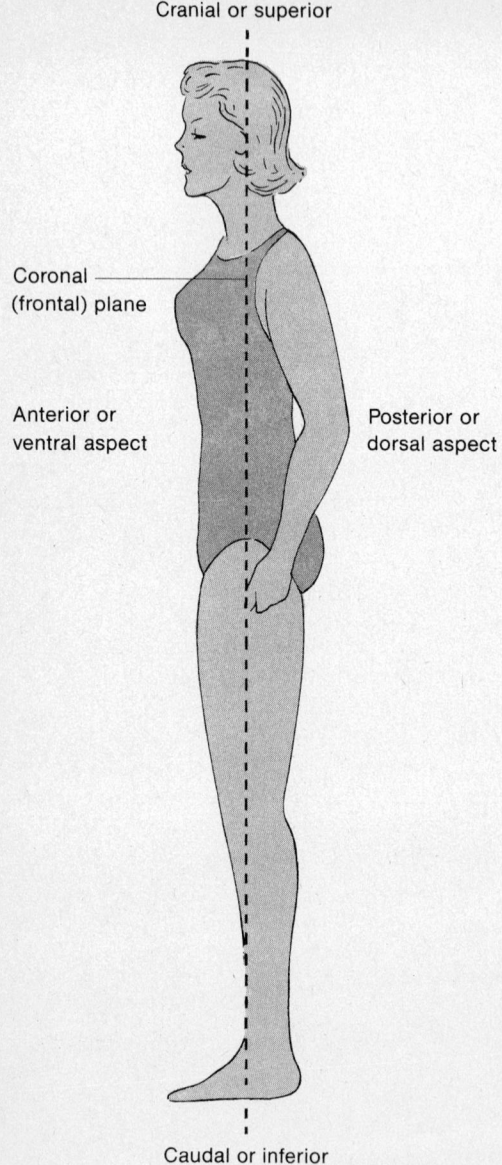

Fig. 2.1 The cranial, caudal, posterior, and anterior aspects of the body. These terms are used to describe a general location or position relative to another location or position.

Two terms used to indicate front and back locations on the body are *anterior* and *posterior*. Anterior refers to the front of the body. A term synonymous with anterior is *ventral*. The nose is on the anterior or ventral surface of the body.

Posterior refers to the back of the body. Another term synonymous with posterior is *dorsal*. The backbone is posterior or dorsal to the heart. All the terms used thus far are illustrated in Figure 2.1.

Other terms used to describe body locations include *medial, lateral, proximal,* and *distal.*

Medial refers to toward the midline of the body. If an imaginary line is drawn, as in Figure 2.2, dividing the body into right and left halves, and extended between the legs, the inside of the leg is its *medial* surface, the outside is its *lateral* surface. The breastbone is medial to the tip of the shoulder; the ear is lateral to the nose.

Proximal means nearest the trunk or point of origin of a part. For example, the knee cap is proximal to the ankle. The term *distal* means away from the trunk or point of origin of a part. The toes are distal to the ankle and the kneecap.

Body planes describe parts of the anatomy if it were sectioned or divided into segments.

The *coronal plane,* shown in Figure 2.3 is a plane through the body dividing it into anterior and posterior portions. The plane extends from the superior to the inferior ends of the body.

A *transverse plane* is a horizontal plane dividing the body into upper and lower parts. This plane is also called a *cross section.*

The *sagittal plane* divides the body into right and left portions. If a section is made exactly in the center of the body, the plane is said to be *midsagittal.* Each half of the body is a mirror image of the other.

Surgical Regions of the Abdomen

For convenience in locating abdominal parts, the abdomen is divided by imaginary lines into nine regions. These, as shown in Figure 2.4, are the *right and left hypochondriac, right and left lumbar, right and left iliac, epigastric, umbilical,* and *hypogastric.*

The top horizontal line crosses the abdomen at the level of the ninth rib cartilage. The term hypochondriac (beneath cartilage) indicates that this line is beneath the rib cartilages. The lower horizontal line crosses the abdomen at the

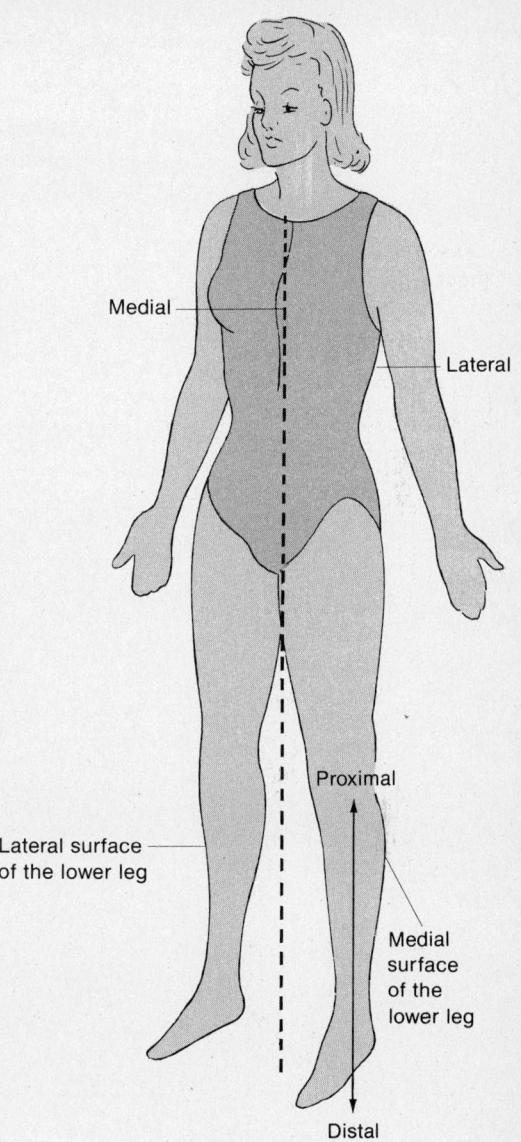

Fig. 2.2 The medial, lateral, proximal, and distal aspects of the anatomy.

OVERVIEW OF ANATOMY AND PHYSIOLOGY

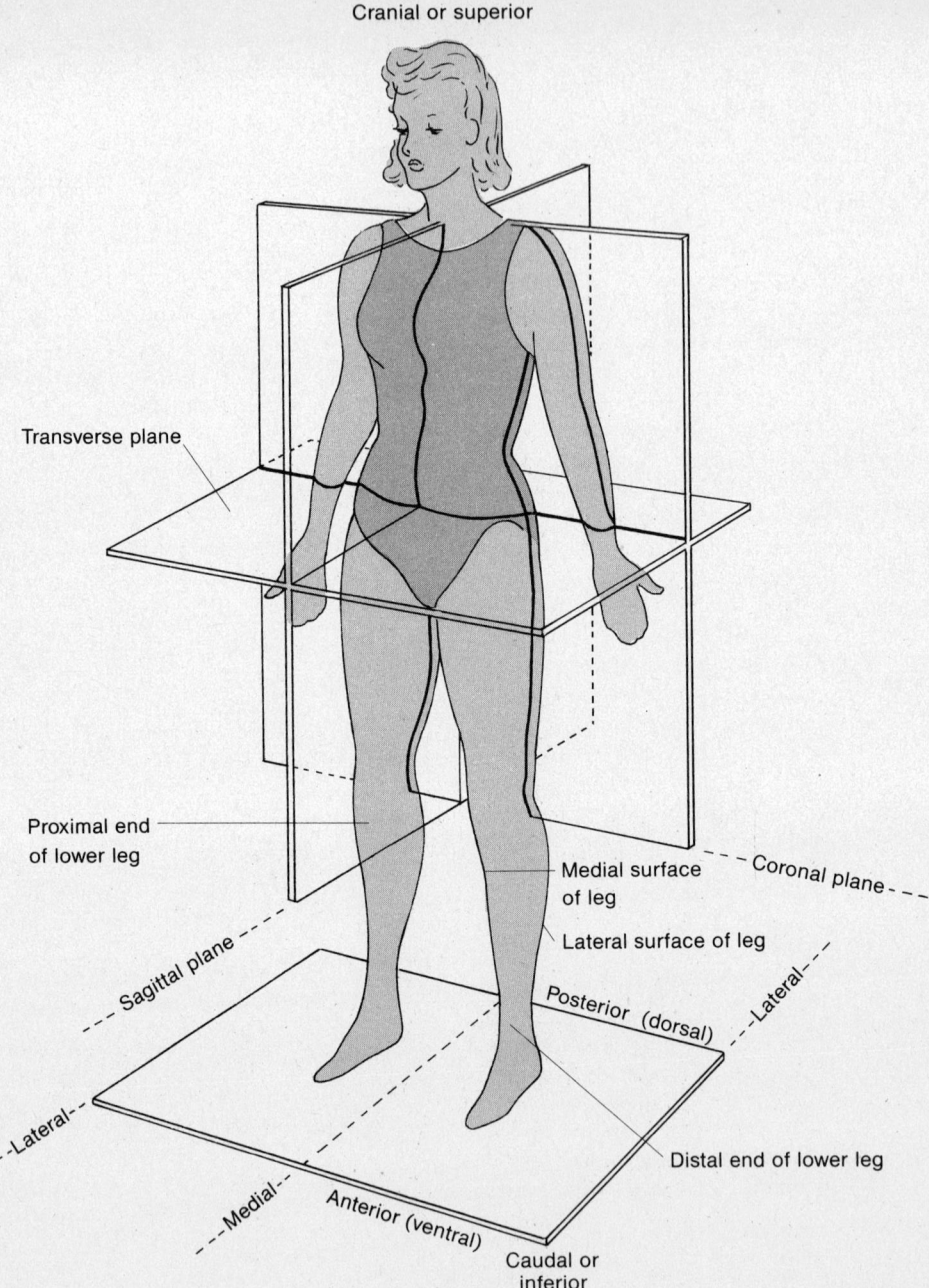

Fig. 2.3 Body planes are hypothetical sections that describe the parts of the anatomy as if it were sectioned or divided into segments.

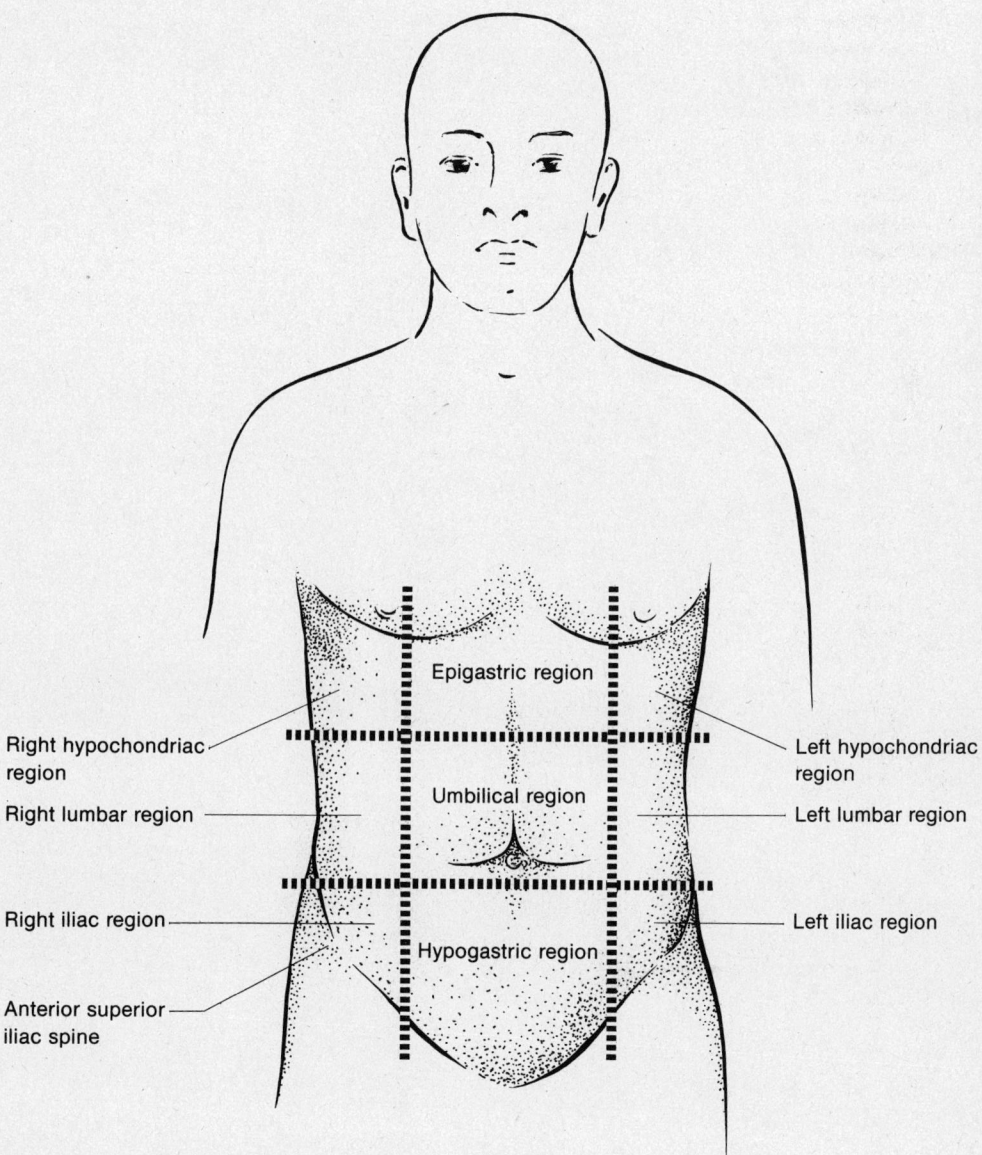

Fig. 2.4 Surgical regions of the body. For convenience in locating abdominal regions, the abdomen is divided by imaginary lines into the nine regions shown here.

OVERVIEW OF ANATOMY AND PHYSIOLOGY

level of the upper tips of the hipbone (iliac crests). This line runs through or near the navel.

The vertical lines are located midway between the lateral body surfaces and the midsagittal line.

A descriptive location of the surgical regions is as follows:

The *right hypochondriac region* is on the upper right side of the abdomen. Its superior line is the breast nipple and its inferior line is the ninth rib cartilage. It extends horizontally from the body's right lateral surface. The medial line of the right hypochondriac region is shared with the epigastric.

The *epigastric* (top of stomach) region lies above the stomach and is bounded laterally by the right and left hypochondriac regions.

The *left hypochondriac region* has the same relative locations as the right hypochondriac except it is on the opposite side of the body.

The *right lumbar region* is beneath the right hypochondriac and above the horizontal line drawn through the iliac crests. Its medial line is shared with the umbilical region.

The *left lumbar* shares the same superior and inferior lines and is on the side opposite the right lumbar region.

The *umbilical region* has the same superior and inferior boundaries as the right and left lumbar. Its lateral lines are shared with the medial right and left regions.

The *right and left iliac regions* have the iliac crest as their superior line.

The *hypogastric* (below stomach) is located beneath the navel and shares the line of the right and left regions.

Body Cavities

A body cavity is not in fact a cavity, but rather a compartment filled with organs, glands, membranes, and tissues. These compartments include the cranial, thoracic, and abdominal cavities.

The *cranial cavity*, as shown in Figure 2.5, houses the brain and is continuous with the vertebral cavity containing the spinal cord. The cranial cavity is unique in several respects. It is,

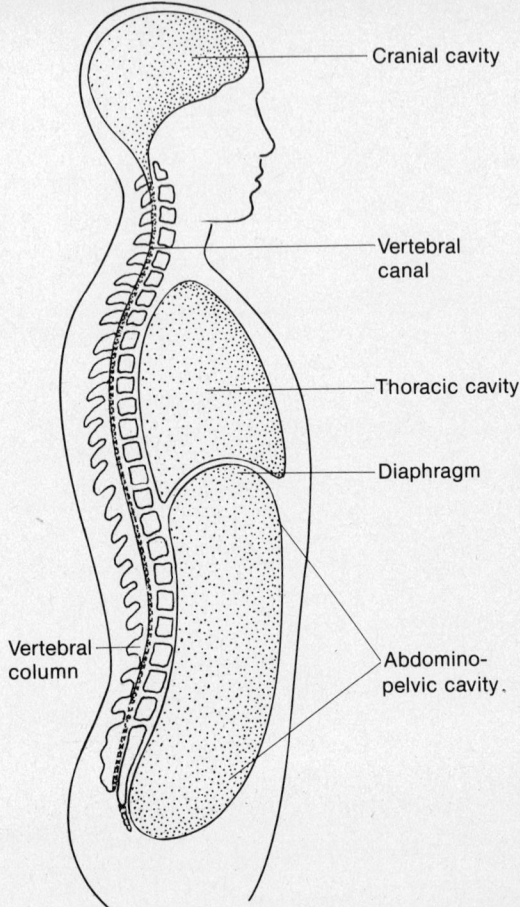

Fig. 2.5 Body cavities are compartments filled with organs, glands, membranes, and tissues. The major body cavities include the cranial, thoracic, and abdomino-pelvic compartments.

for example, the only cavity completely enclosed by bone. The skull, in addition to enclosing the brain, contains smaller cavities including those of the eye, ear, semicircular canals, and sinuses.

The *thoracic cavity* encloses the lungs, the heart, and major arteries and veins serving these structures. The thoracic cavity is bounded posteriorly by the spinal column, laterally by the ribs, anteriorly by the breastbone (sternum), and is separated from the abdominal cavity by the muscular diaphragm.

The *abdominal cavity* is the largest body compartment, housing the *viscera* (stomach, intestines, colon), reproductive, and excretory organs.

This cavity is, except at the backbone, bounded by layers of muscle. The floor of the abdominal cavity is enclosed by muscles and bones of the pelvic region. The abdominal cavity extends from the muscular diaphragm into the pelvic cavity. Although the cavity is continuous, it is often arbitrarily divided into two cavities, the abdominal and pelvic, with a single name, the *abdomino-pelvic* cavity. The pelvic portion of the cavity is below a line drawn at about the level of the upper edge of the hipbone. In the male, a fourth cavity is evident in the form of the scrotum, containing the testes. The scrotum is not a true cavity, however, since it is continuous with the abdominal cavity.

PRELIMINARY BIOLOGICAL CONCEPTS

Biological concepts will be discussed later in the text in detail. An introduction to a few concepts of major importance, however, will be useful at this point.

Diffusion and Osmosis

Diffusion is the physical process that permits molecules to become uniformly distributed. Molecules in an area of high concentration move to an area of lower concentration. Diffusion may occur across membranes, and thus becomes the mechanism by which oxygen and nutrients enter cells and wastes leave cells.

Osmosis is a special kind of diffusion pertaining only to the movement of water molecules across membranes. Water moves from where it is more concentrated to where it is less concentrated. For example, if pure water is placed on one side of a membrane and a 5 percent salt solution is placed on the other side, water moves from the pure side into the solution. The difference in water concentrations on either side of a membrane creates *osmotic pressure,* and the greater the difference the higher the osmotic pressure. A higher osmotic pressure forces water molecules across a membrane at a greater rate, as illustrated in Figure 2.6.

Active Transport

Osmosis does not explain the transfer of all substances across membranes. For example, there are many instances in which substances move in a direction opposite to osmotic pressures. These substances move against the *concentration gradient,* that is, from an area of low concentration into an area of higher concentration. This process is explained by a chemical union of the transmitted substance with a carrier molecule that aids the transfer across membranes. Certain membranes of the body maintain higher concentrations of some substances on one side of a membrane as a normal, and essential, body function. The movement of molecules across membranes against (up) a concentration gradient is called *active transport.* Because the process involves chemical bonding, it therefore uses energy. Energy is supplied in the form of ATP (adenosine triphosphate). ATP is an energy rich molecule stored throughout the body that can readily release energy for body needs.

Figure 2.7 illustrates the process of active transport. Since there are more sodium ions

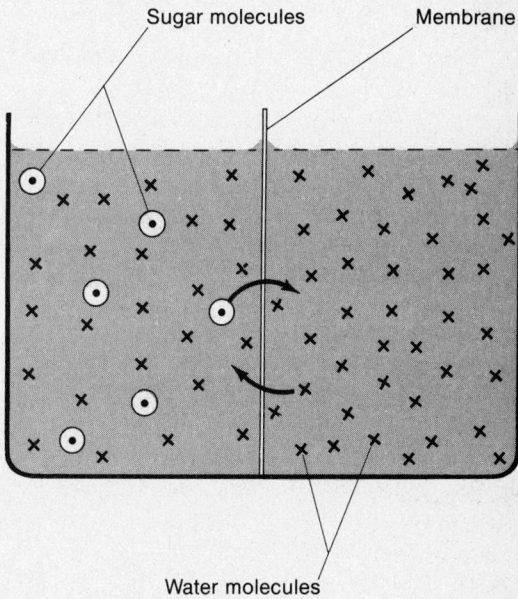

Fig. 2.6 The greater concentration of sugar molecules on the left of the membrane will create an osmotic pressure moving sugar molecules across the membrane to the water side. Water molecules on the right of the membrane will flow to the left side.

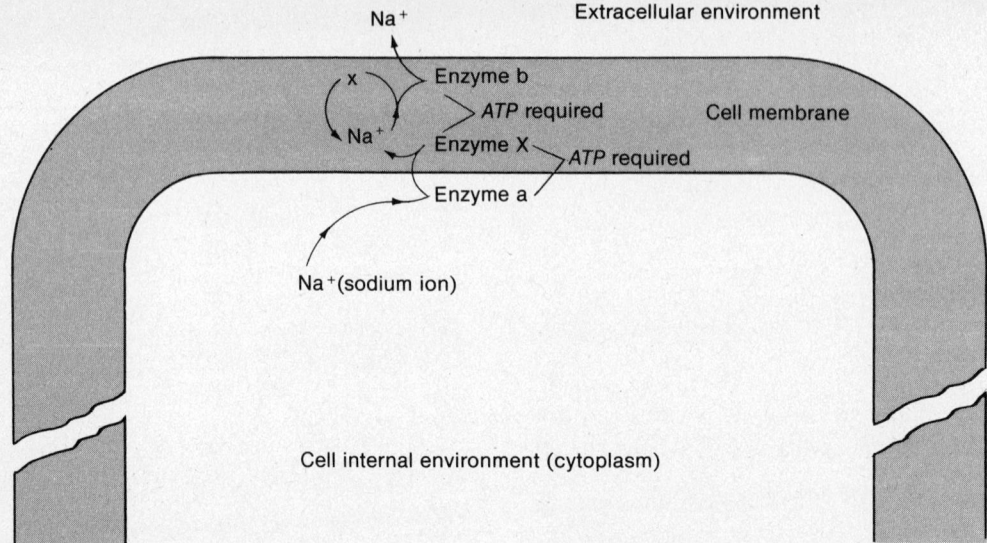

Fig. 2.7 In this theoretical model of active transport, excess sodium ions must be removed from the cell often against a concentration gradient. Sodium ions (Na^+) under the influence of enzyme a join enzyme X, forming a loose bonded molecule that carries Na^+ across the cell membrane. At the outer membrane surface, enzyme b aids in releasing the sodium ion from X enzyme. X enzyme may then be recycled to pick up more Na^+ from the inner cell membrane.

outside the membrane, it would seem logical that by osmosis the ions would move down the concentration gradient toward the lower concentration. Indeed, sodium ions do "leak" into the cell, but they are apparently pumped back out. This mechanism is called the *sodium pump*.

The exact way this pump works is unknown, but the concept helps explain how a cell is capable of maintaining the concentration difference on either side of a membrane. If sodium ions leak into the cell they have to be pumped back out against a stronger concentration gradient. In other words, the pump works opposite to the action of osmosis. Since this is a dynamic rather than passive function, energy is required to move the sodium ions out of the cell.

The movement of other substances across the cell membrane, such as potassium ions (K^+), calcium ions (Ca^{++}), sugars, and amino acids, probably involves some type of pump similar to the sodium pump. In fact, the same mechanism may be responsible for several substances.

Filtration
To filter means to separate two substances by permitting one substance to pass through minute pores of some separating structure. Human membranes contain pores that permit the passage of some small particles but resist the passage of large particles. Regardless of size, substances cannot passively cross membranes—some force is required to push them through the pores. This force is usually *hydrostatic pressure*, the pressure created and maintained by heart ventricular contraction and the elasticity of arteries and veins. This pressure literally pushes substances through membrane pores.

Bioelectricity
The membranes covering muscle and nerves maintain different concentrations of ions on either side. Since ions are charged particles (either positive or negative), they establish a polarity. A membrane is said to be polarized when the number of positive ions on one side of the membrane exceeds the number of positive ions on the other side. The side with the fewer positive ions is said to be negative.

Membranes of muscle and nerves in the resting state maintain a positive charge on the outside and a negative charge on the inside. The flow of ions across membranes, creating imbalances or balances in ionic charges, is called *bioelectricity*. This form of electricity is not the same as electricity flowing in electric wires (current electricity). There are some similarities but also many differences. For a discussion of current electricity, see Chapter 21.

We may think of bioelectricity as being a flow of ions across membranes. Processes such as active transport, diffusion, and sodium pumps help maintain specific ionic imbalances across membranes. When a membrane is stimulated by chemical or physical means, an area of the membrane becomes depolarized. This means that the membrane for some reason suddenly becomes easier to penetrate so that positively charged ions, such as sodium, rush to the inside of the membrane, thus equalizing the charge on either side. This rush of ions and the consequent depolarization is called an *action potential*.

Action potentials are measured in units of electricity called millivolts (1/1000 volt). A volt is a unit expressing the potential difference between two oppositely charged surfaces.

Enzymes

The human body is faced with complicated problems in maintaining the delicate chemical balances essential for life. Even a slight imbalance in some of the body's chemical reactions may bring about serious illness or death. Enzymes play a critical role in mediating and coordinating many of these crucial chemical reactions.

Enzymes are organic substances that accelerate and control biological chemical reactions. Specifically, an enzyme might be defined as a protein that acts as a catalyst.

Most chemical reactions occurring in the normal body are controlled by enzyme activity. This activity includes aiding the decomposition of large molecules into small molecules during digestion, and the synthesis of small molecules into large molecules for storage and tissue building or, as we will see later, for the release of energy from food substances. Other specialized functions of enzymes will be discussed with specific systems.

Computerized Medical Care Is Here

Fifty years ago, Buck Rogers fantasies envisioned men flying in space, manned space laboratories, laser guns, and much more. Today research in medicine and medical care has spawned its own fantasies for the future. Computers can be programmed to monitor body chemistry on a minute-by-minute basis. Disease, stress, and organ malfunction may change the flow of electrons across membranes, alter normal organic molecules, damage tissue, or induce certain glands to hypersecrete. Analysis of body fluids and membrane polarity, which is possible today, merely touches the tip of an iceberg of possibilities.

Computers of the future may be used to correctly diagnose a disease and indicate the proper medical care necessary to combat it. Patients "wired" to a computer will receive the proper medication by computer-controlled injections—and not simply so many times per day, but instead precisely when they need it and in the correct quantity.

The missing ingredient in computerized medicine is the doctor-patient relationship, but perhaps the computer may even solve this dilemma. Since we do not know for a fact that these personal relationships aid in health care, researchers might program a computer for answers.

This discussion seems to suggest that there is no need in the future for individuals to study medicine. Quite the contrary. This relatively new field of health care will require a higher level of expertise in medicine, chemistry, and psychology than ever before. Medical-computer technology should be directed, not followed.

Since enzymes are protein molecules, they are destroyed or deactivated, as are other proteins, by excessive heat, strong alkali, or other enzymes.

Enzymes are named according to the substance they act upon, which is called a *substrate*. The suffix "ase" is tacked onto the end of the substrate name to form the name of the enzyme. For example, the enzyme which acts upon sucrose, converting it to glucose, is called *sucrase*. An active enzyme that will destroy penicillin is called *penicillinase*.

The precise nature of the catalytic function of enzymes is not known. However, some hypothetical models explaining this activity have been designed.

In order for any chemical reaction to take place, the reacting substances must be brought into close contact with each other, and they must be supplied with a certain level of energy before they will react. For example, in order for a match to burn, heat energy must be applied in the form of friction when the match is drawn across a rough surface. A match lying quietly on the table at room temperature does not ignite. The same is true with a container of gasoline at room temperature. Energy must be added to the combination of gasoline and oxygen in order for the chemical combination to occur—sometimes producing a violent chemical reaction.

The level of energy required varies with different chemical substances. For example, it takes more energy to ignite a piece of wood than it does to ignite the same amount of gasoline.

Digestion of foods is basically a chemical reaction. Carbohydrates, fats, and proteins must be changed chemically in order for energy to be released from them. A carbohydrate such as a lump of sugar sitting at room temperature will not undergo chemical changes. This lump of sugar can be ignited with a flame and burned, releasing energy from it in the form of heat and light.

The body's metabolic processes can also release energy from sugar. However, the intense temperatures which are necessary to begin the reaction and those that result from the reaction cannot be tolerated by the human tissues. This is where enzymes come into the picture.

Enzymes are *organic catalysts*. They are capable of mediating the chemical reactions that occur within the body and still maintaining the systems at a very low temperature. The temperature at which these chemical reactions are carried out in the body is the normal body temperature of 37°C (98.6°F). The function of enzymes then is to mediate and permit chemical reactions at body temperature rather than the very high temperatures necessary outside the body.

Another function of enzymes is to regulate chemical reactions so as to release energy gradually. If all the energy were released at once, as in a gasoline explosion, or even a match being lighted, the excessive amount of energy would damage body tissues. Enzymes permit the slow release of energy from food materials. In this way, the body not only can utilize the small packets of energy but also can synthesize molecules that store the released energy for future use.

A hypothetical model can be used to help explain enzyme action. We know that molecules speed up in motion when heat is applied, which may explain why heat is necessary for a chemical reaction to take place. As molecules speed up, there is greater opportunity for two or more molecules to collide and combine, thereby producing a chemical reaction.

If some mechanism were available where two molecules stuck to a surface in close proximity to each other, the chemical reaction would then not have to await a random collision with specific molecules. Enzymes are thought to provide this mechanism, as shown in Figure 2.8.

Note that the enzyme surface configuration is such that the two reacting chemical substances, labeled A and B, are firmly attached and held together by the enzyme structure. Once the substances A and B are combined to produce a new molecule, AB, the enzyme could peel off or pull away from the new molecule and serve as a mold for other similar molecules. Energy is consumed in this process, as it is whenever substances are chemically united.

Enzymes are specific with regards to which chemical reactions they mediate. The substance shown as C in the diagram would not be able to join either A or B because the surface of the enzyme would not permit this particular configuration.

Figure 2.8 illustrates the chemical reaction in which two substances combine, or are synthesized into another substance. The reverse of this is *decomposition*, a process just the opposite of that illustrated in the figure. For example, in

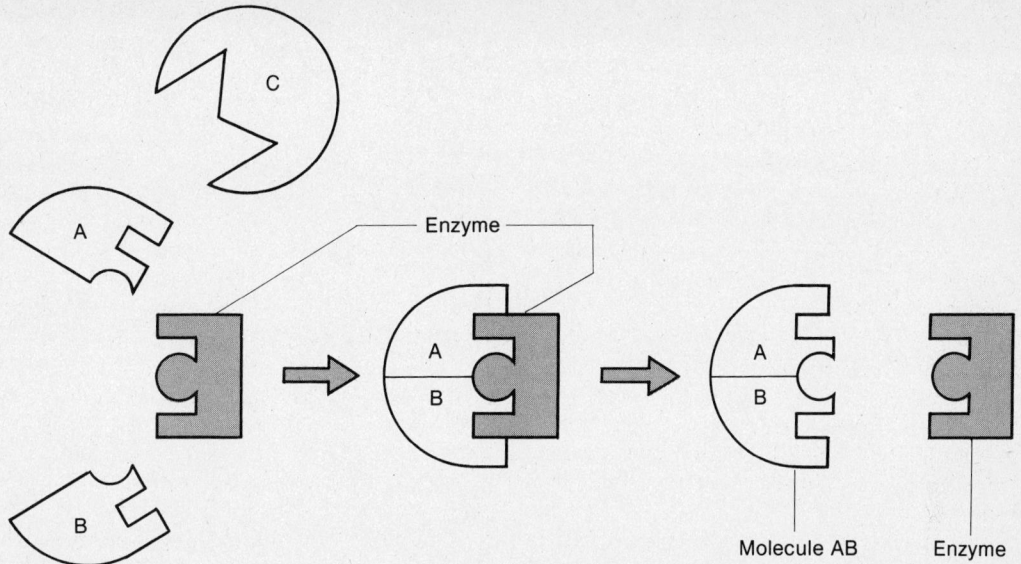

Fig. 2.8 A specific enzyme aids in the union of substances A and B forming a new molecule AB. The enzyme releases the molecule to be used again in the synthesis of another AB molecule. Substance C cannot, because of its molecular configuration, join either A or B with the enzyme shown. If substance C is to unite with either substance A or B, another specific enzyme would be required.

decomposition, the compound AB would be surrounded by an enzyme that would help break the bonds holding the molecule AB together. Once this attracting bond was severed, the separate molecules, A and B, would be released. When the bond between A and B is broken, energy is released.

This energy can then be captured by other body mechanisms and stored for future use. Both enzymes and substrates have detailed and intricate molecular shapes that allow them to fit together in a very precise position. The surface shapes of enzymes and substrates must fit together almost as a key fits a lock. In fact, this theory is called the *lock and key theory*.

The precise mechanism by which an enzyme catalyzes a specific reaction is still to be discovered. The answer seems to be in the particular configurations of enzymes and substrate molecules.

SUMMARY AND REVIEW

A. Anatomy and physiology is a study of form and function. The form inherently is dictated by the function of each body part.
B. The body is organized (arbitrarily) into a hierarchy, with cells as the basic units. Tissues are composed of cells. Organs are composed of tissues. Systems are composed of organs, and the organism is composed of systems. The hierarchy is thus exemplified: cells ⟶ tissues ⟶ organs ⟶ systems ⟶ organism.
C. Terms of reference in anatomy
   1. Superior (cranial): toward the head.
   2. Inferior (caudal): toward the tail.
      a. Anterior: front or ventral.
      b. Posterior: back or dorsal.
      c. Medial: toward the midline of the body.
      d. Lateral: toward or nearest the outside.
      e. Proximal: nearest the point of origin.
      f. Distal: away from the point of origin.

D. Body planes
 1. Coronal plane: divides the body into front and back parts (anterior, posterior parts).
 2. Transverse plane: divides the body into upper and lower parts.
 3. Sagittal plane: divides the body into right and left parts.
 4. Midsagittal plane: exact division of the body so that one segment is a mirror image of the other.
E. Surgical regions of the abdomen: The abdominal region can be arbitrarily divided for reference convenience into nine regions—the right and left hypochondriac regions, right and left lumbar, right and left iliac, epigastric, umbilical, and hypogastric regions.
F. Body cavities
 1. Cranial cavity: houses brain.
 2. Vertebral cavity: houses spinal cord.
 3. Thoracic cavity: encloses lungs, heart, and major arteries arising from the heart.
 4. Abdominal cavity: contains the viscera, reproductive, and excretory organs.
G. Biological concepts
 1. Osmosis: the process of substances diffusing across membranes from an area of high concentration to an area of lower concentration.
 2. Active transport: process whereby substances may be transported across membranes against a concentration gradient.
 3. Sodium pump: an active transport process to return sodium from the inside of a cell to the outside.
 4. Filtration: the separation of substances by permitting some substances to pass through a membrane while holding back others.
H. Bioelectricity
 1. Bioelectricity is produced by the flow of charged particles (ions) from one part of a cell to another part or from one part of the body to another.
 2. Ions of major importance include those of sodium and potassium.
I. Enzymes
Enzymes are organic catalysts essential to mediate most chemical reactions occurring in the body.
 1. A substrate is a substance an enzyme works on.
 2. Digestion involves the action of enzymes upon food substances to break down large molecules into smaller ones.
 3. Enzymes are called organic catalysts because they catalyze the chemistry of living systems.

3

Cell Structure and Function

MAJOR CONCEPTS
  THE CELL
  CHARACTERISTICS OF CELLS AND
  CELLULAR ORGANELLES
    Cell Membrane and Function
    Endoplasmic Reticulum and Ribosomes
    Nucleus
    Mitochondria
    Vacuoles
    Cilia
    Flagella
    Cytoplasm
    Nucleolus
    Lysosomes
    Golgi Body
  CELLULAR ACTIVITIES
    Phagocytosis
    Pinocytosis
    Cell Movement
    Cell Reproduction

Cells are the basic units of all living organisms. Some relatively simple organisms may be composed of only one or a few cells. The human organism is made up of billions of cells.

In this chapter a preliminary discussion examines basic cell structure. Later in the chapter detailed information concerning cellular function is presented.

"Life" is an imprecise term. Attempts to define this thing we call life probably began with the first conscious human beings on earth. From past ages to the present, these attempts have been singularly unsuccessful. When we attempt to define life, about the best we can do is to describe certain characteristics that all living things have in common. Life in these terms then becomes a dynamic process, an aggregation of functions.

For example, we assume that an organism is alive if we recognize what we call life signs. Clinically speaking, these are referred to as *vital signs* and consist of such functions as heartbeat or pulse, respiratory movements, pupillary sensitivity, muscular sensitivity to mechanical stimulus, and temperature. Except for temperature, all these vital signs involve movement. We cannot say, however, that anything that moves is alive because physical forces such as gravity may cause an inanimate object to move.

The purpose of this text is not, however, to examine the scientific or philosophical attempts to define life, but rather to examine life processes in minute detail beginning with the most elementary living unit—the cell.

THE CELL

In order to solve a complicated problem, it is essential to simplify it into the components that make up the problem, that is, into its basic units. The basic unit of living organisms is the *cell*.

We have only recently (within the last 150 years) thought of the cell as the basic unit of life. Robert Hooke (1665) first described things that he called cells. By looking at thin sections of cork with crude magnifying lenses, he was able to see the dead and empty cells of cork bark—actually the cellulose outer skeleton of cells. Much later, Schleiden and Schwann (1839) proposed the "cell theory," which in essence stated that new cells can only come from preexisting cells. This theory, quite revolutionary at the time, is a little over 150 years old.

A single cell is a living unit. Groups of cells having similar functions are classified as *tissues*, which in turn make up *organs*. Higher levels of organization are called *organ systems*, which make up the major unit called an *organism*.

Some living forms consist of only a single cell. Examples of these are found primarily in lower forms of life, such as *protozoa* (single celled animals), *algae* (single celled plants), and *bacteria*.

Characteristic cellular activities include such processes as movement, excitability, the capability to transform nutrients into usable energy, and the capability of reproduction.

Before examining these characteristics in detail, a clear understanding of the basic structure of a cell is necessary. We will use a *generalized cell*, as shown in Figure 3.1, to indicate those structures and functions found in most cells. Some cells lack some of the features of a generalized cell, while others have additional features.

The structures shown in Figure 3.1 have been intentionally kept few in number. The ones listed, however, are of great importance and will be referred to consistently throughout this text. At this time cell structure and function will be discussed only in general terms. Greater emphasis will be given to individual structures in future chapters.

CHARACTERISTICS OF CELLS AND CELLULAR ORGANELLES

This minute bit of living matter called a cell is a literal dynamo. Here, energy from food is released, stored and used, nerve impulses are generated, metabolic wastes excreted, and chemicals secreted.

Examining the activity of one cell in the body will not reveal much about how the whole body functions, nor will it shed much light on behavior patterns of organisms. However, knowledge gained through the study of a cell and its interaction with other cells often provides clues to higher level activities. For example, we know that the sensation of being ill is a cumulative effect and is the external manifestation of what is going on at the cellular level.

Cells in general have a varied range of size, all the way from those measured in microns to those measured in centimeters, but most single cells are too minute to be observed with the unaided

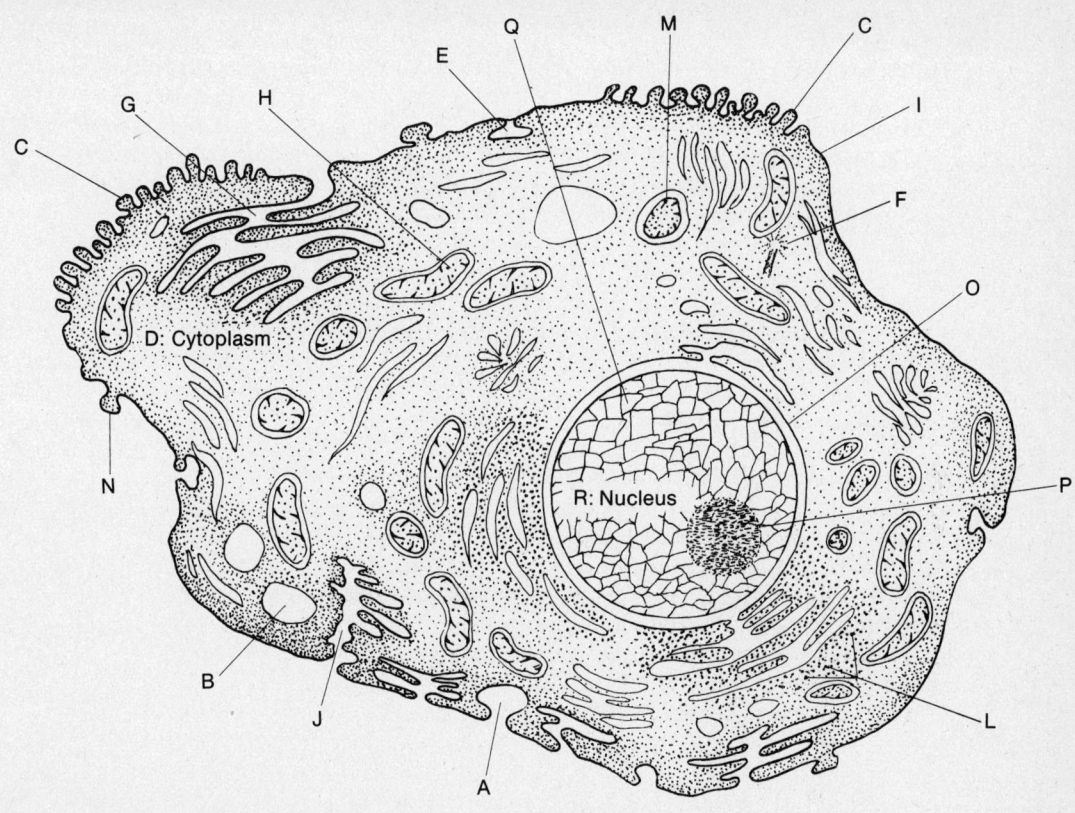

Cellular components	Function	Cellular components	Function
A. Phagocytic vesicle	Nutrient intake	J. Golgi body	Synthesizes complex organic molecules
B. Vacuole	Nutrient storage	K. Cell wall	Protection, rigidity
C. Ciliated border	Increased surface area	L. Ribosomes	Protein synthesis
D. Cytoplasm	Nutrient and gas distribution	M. Lysosome	Contains lysing enzymes
E. Pinocytic vesicle	Fluid and nutrient intake	N. Evagination	Increases surface area
F. Centriole	Cell division	O. Nuclear membrane	Boundary for nucleus
G. Endoplasmic reticulum	Nutrient and chemical distribution	P. Nucleolus	Protein synthesis
H. Mitochondria	Metabolism	Q. Chromation	Chromosome formation
I. Cell membrane	Absorption	R. Nucleus	Cell division

Fig. 3.1 Diagram of a generalized animal cell.

eye. The light microscope can increase the magnification of the cell up to approximately 1,000 diameters and is commonly used to study cells in the laboratory. The electron microscope can provide even higher magnifications, as shown in Figure 3.2.

Cells come in all shapes as well as sizes. Some of the more typical types of cells found in the human body are shown in Figure 3.3. The variable shapes of these cells, as we will see later, is somewhat determined by their functions. The subcellular particles shown in the figure are called *organelles*.

Cell Membrane Structure and Function

Surrounding all cells is a complex organelle called a *cell membrane*, or *plasma membrane*. At first glance the cell membrane might be compared to a plastic bag enclosing the cell contents. It is flexible, transparent in most cases, and seems to hold in the cell contents quite well. Upon closer scrutiny, the plastic bag analogy becomes totally unacceptable.

The cell membrane is a highly complex structure. It not only passively contains the cell components, it is also the structure through which all nutrients must enter the cell and all wastes must be secreted. It further functions in being selective of which substances enter and leave. This selective property of the cell membrane is called *differential permeability*. This permeability can be altered by several factors, including certain drugs.

There is an important relationship between the surface area of a cell and the volume of the cell that often serves as a limiting factor in the growth of a cell. The membrane surface is important in controlling the movements of materials into and out of the cell. Raw materials must pass into the cell so that metabolism may occur. Waste products must pass out.

For a moment, visualize a cell as a spherical structure, like a ball. In any given sphere, there is a certain ratio between the surface area and the volume. In a properly functioning cell, there must be sufficient ratio of surface area to volume to permit the passage of enough materials into and out of the cell so that the metabolic needs of the cell can be satisfied.

Suppose the diameter of the cell was doubled. As a cell or a sphere grows, the surface area increases at a rate based on the square of the cell radius, but the volume increases at a rate based on the cube of the cell radius. In other words, as a cell enlarges, the volume increases faster than the surface area. Table 3.1 shows the rate of increase of surface area and volume each time the cell radius is increased.

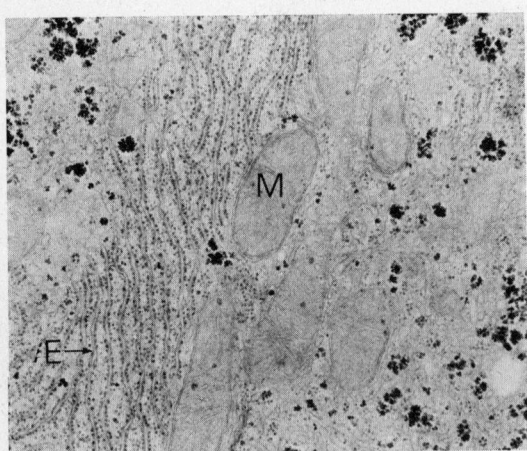

Fig. 3.2 Electron micrograph showing endoplasmic reticulum at *E*. The dark dots that line the reticulum are ribosomes. Mitochondria shown at *M* are the sites of cellular respiration. Magnification is 20,000 x. *(Courtesy Santiago Plurad.)*

Table 3.1 Cell Rate of Growth. As a cell grows, the volume increases at a faster rate than the surface area. For example, a cube 1 cm on all sides has a surface area of 6 sq. cm and a volume of 1 cc. If the dimensions are increased, the ratios are as shown.

SIDE OF CUBE IN CM	SURFACE AREA IN SQ. CM	VOLUME IN CUBIC CM	RATIO OF AREA: VOLUME
1	6	1	6:1
2	24	8	3:1
3	54	27	2:1
4	96	64	1.5:1
5	150	125	1.2:1
6	216	216	1:1

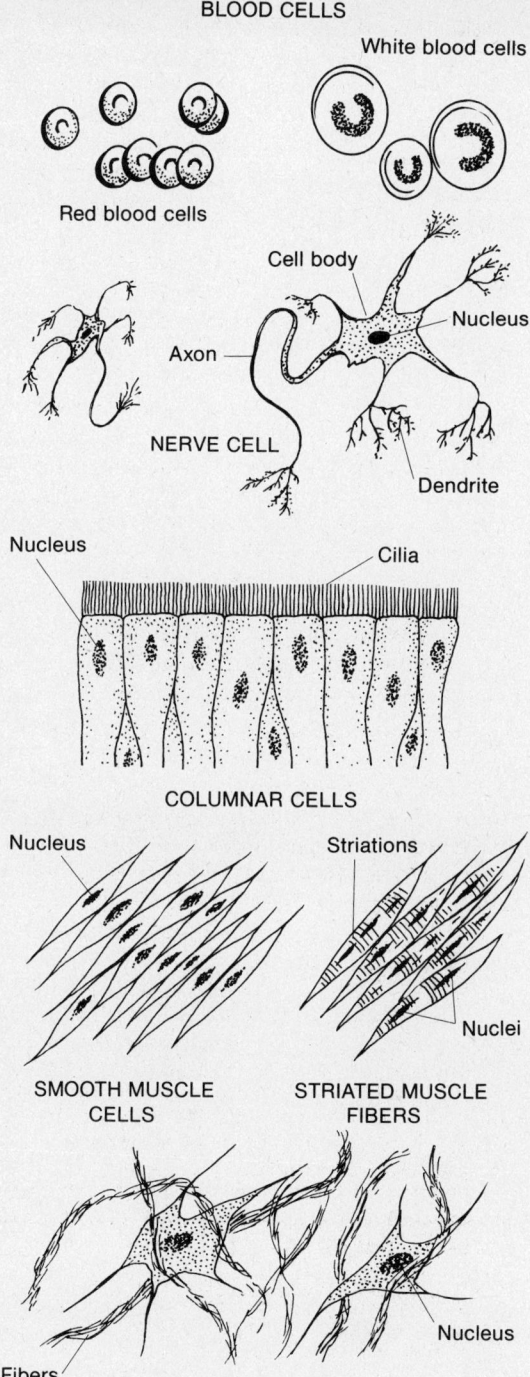

Fig. 3.3 Several types of cells found in the body are illustrated here. Cells come in many shapes and sizes. Note particularly the varied shapes. The shape of the cell is generally determined by the specific function it is designed to carry out.

A disproportionate ratio of surface area to volume would limit the metabolic activities of the cell because the surface area available would not permit the interchange of sufficient amounts of materials. Cells approaching this critical ratio may avoid the problem of insufficient surface area by dividing into two cells or by changing the surface configuration by invagination or evagination, thus increasing the amount of the surface area.

Invagination is an infolding of the cell membrane. *Evagination* is an outfolding. Figure 3.1 illustrates invagination at *A* and evagination at *N*.

The movement of substances across cell membranes depends upon several factors including osmosis, active transport, and the passive flow through membrane pores.

It seems logical to assume the water and substances soluble in water might simply diffuse through the cell membrane. Some other kinds of molecules, however, also enter and leave. The explanation currently given is that the cell membrane has minute openings or pores through which these substances can pass, but these hypothetical pores have not as yet been observed.

In summary, the cell membrane is a unique, differentially permeable membrane capable of maintaining a constant internal cellular environment. This constancy is maintained regardless of the extracellular environment in a normal functioning body. All food substances required for the body must cross the cell membrane and enter the cells. All metabolic waste materials must leave through the cell membrane.

Endoplasmic Reticulum and Ribosomes

The *endoplasmic reticulum* is a mass of fine, tubelike structures found throughout the cell, as shown in Figure 3.2. The fine tubes are bound by membranes and are believed to be a transporta-

CELL STRUCTURE AND FUNCTION

tion network through which nutrients and minerals are carried to different parts of the cell. It is believed that the endoplasmic reticulum is continuous with other cellular membranes. This would facilitate the easy flow of materials throughout the cell.

Ribosomes are dark staining bodies located along the inner surfaces of some of the endoplasmic reticulum. These structures, as we will see later, play a role in protein synthesis.

Nucleus

The *nucleus* is a spherical organelle surrounded by the *nuclear membrane*. The contents within this membrane include a fluid called *nucleoplasm* and chromosomes made up largely of *DNA* (deoxyribonucleic acid). It is through DNA that the nucleus controls many cell activities, such as protein synthesis. Some of these proteins are enzymes that catalyze chemical reactions throughout the body. Other proteins whose structure is dictated by chromosomal DNA function in muscle contraction or are instrumental in establishing differential osmotic pressures.

In the past, the nucleus was referred to as the governing organelle of a cell. It is more specific to say that DNA molecules, located within the nucleus, dictate not only the characteristics of each individual but, in addition, most of the chemical reactions occurring in the body.

Mitochondria

Mitochondria are rod-shaped sacs that contain enzymes essential for the release of energy from nutrient molecules. Mitochondria are large enough to be seen with the light microscope with proper staining. Higher magnifications are provided by the electron microscope, as in Figure 3.2. Mitochondria are called the powerhouses of cells since this is where energy is released from nutrients.

Substances such as glucose enter the mitochondria and are chemically decomposed *(hydrolyzed)* with the subsequent release of energy. Mitochondria are capable of capturing the released energy and utilizing it to create a new molecule, *adenosine triphosphate* (ATP), that can be stored in various tissues for later use.

The role of enzymes in the release of energy cannot be overemphasized. Each biological chemical reaction is controlled in some manner by enzymes, each one specialized to carry out a specific synthesis or decomposition reaction.

Vacuoles

Vaculoes are cavities, found throughout the cytoplasm, that contain fluids or particulate matter. Substances found within vacuoles may have been placed there by the processes of phagocytosis or pinocytosis. (See the next major section, "Cellular Activities.")

Cilia

Some specialized cells have fine hairlike surfaces *(cilia)* that are capable of movement. Some cells having cilia line the upper respiratory tract. The cilia, by wavelike motion, constantly move foreign particles, such as dust and bacteria, upward and outward toward external surfaces.

Flagella

Some cells, such as sperm cells, have a relatively long, fiberlike extension called a *flagellum*. This structure, through a whiplike motion, is capable of propelling the sperm cell through the reproductive tract. More will be said about sperm cell movement later, in the chapter on reproduction.

The precise mechanism by which cilia and flagella are made to move is unknown. However, it is known that the mechanism is an energy-utilizing function and that cell membrane excitability is involved in the process.

Cytoplasm

Cytoplasm is the fluid part of a cell contained between the cell membrane and the nuclear membrane. It does not include the fluid content of the nucleus, which is called nucleoplasm. Cytoplasm, within which are found the various organelles, is referred to as *intracellular fluid*.

A general overview of the relationship of cell fluids to other body fluids can be gained by considering the foods and liquids we need to stay alive. The foods we eat are chemically decomposed and, along with water absorbed into the bloodstream from the intestines. The products of digestion, water, and other essential substances are transported by the circulatory system to the vicinity of cells. These substances then diffuse out of tiny blood vessels (capillaries) into a space between the capillaries and the cells called the *interstitial space*.

The fluid accumulating in this space, called *interstitial fluid*, has approximately the same composition as blood plasma. A major difference is that blood plasma has a much higher concentration of protein. Interstitial fluid surrounds all cells and is the source of all nourishment and water for cells.

The cell membrane is particularly selective when it comes to admitting chemicals from interstitial fluid, that is, the membrane is *selectively permeable*. Note in Table 3.2 the difference in concentrations of substances found within the cell compared to the interstitial fluid and plasma. Notice in particular the differences in sodium and potassium ions. Interstitial fluids contain much more sodium and much less potassium than intracellular fluid. Other differences include the higher concentrations of phosphates, sulfates, and proteins in intracellular fluids.

A comprehensive look at the effect these concentrations produce is given in Chapter 16.

Table 3.2 Fluid Content Comparison. Comparison of fluid content between blood plasma, interstitial fluid, and cytoplasm units listed as mEq/liter. (See Chapter 16.)

	ION	PLASMA	INTER-STITIAL FLUID	CYTO-PLASM
Positive ions	Sodium	148	142	9
	Potassium	4.3	4.1	145
	Magnesium	3.0	2.7	40
	Calcium	4.0	3.8	0
Negative ions	Chloride	106	114	6
	Bicarbonate	28	30	11
	Phosphate	2.1	2.0	90
	Sulfate	1.1	1.1	20
	Protein	17.0	1.0	67
	Organic acids	3.2	3.4	0

Nucleolus

The *nucleolus* is a dark staining structure found within the nucleus. As the nucleolus lacks a membrane, it is debatable whether or not this structure is a true organelle. The nucleolus is composed of a special type of RNA and is believed to be involved in the synthesis of complex protein molecules.

Lysosomes

The *lysosome* is one of the many structures in the body about which very little is known. The prefix *lysis* means to shred, tear, or destroy. The suffix *some* means body; therefore, lysosome means a destructive body. Actually, lysosomes appear to be sacs of enzymes capable of destroying potentially harmful substances or particulate matter that may enter cells. Note in Figure 3.1 the cellular component, *A*, labeled phagocytic vesicle. Certain cells—white blood cells in particular—form these vesicles, permitting extraneous material, including bacteria, to enter the cell. This process, called *phagocytosis*, is illustrated in Figure 3.3.

The ingested matter, a single bacterium, for example, surrounded by a membrane, is transported to a lysosome. The membrane surrounding the bacterium and the lysosome membrane fuse, then dissolve, forming one body. Lysosomes contain powerful enzymes that, when coming in contact with the bacterium, digest it and thus destroy it. This activity is an important and crucial mechanism that protects the body from invasion by microorganisms.

Other activities attributed to lysosomes include the digestion and removal of cells which routinely die and are replaced and the chemical process which is set in motion at death.

Lysosomes may be involved in such processes as sperm entering into the egg membrane, the implantation of the zygote into uterine tissues, the release of ova from ovaries, and a host of other activities. The role of this organelle no doubt will be explored in greater depth in the future.

Golgi Body

The specific activities of the *Golgi body* have only recently been explored, although they were first described in 1898 by Camillio Golgi. This structure seems to be a packaging center for complex proteins. These proteins, some combined with sugars, become a part of other organelles, such as the cell membrane. Other products of the Golgi body may be packaged as lysosomes or sent out of the cell as "detective" molecules helping other cells decide what substances are potentially harmful and so aiding in eliminating them from the body. The Golgi body thus plays a distinct role in secretion.

CELLULAR ACTIVITIES

Body cells are highly diverse in their activities depending upon what specific function they serve. For example, the function of a nerve cell is to transmit signals to different parts of the body. A contracted muscle cell brings about movement. Another type of cell may have a secretory function that influences the activity of other cells. The discussion of cells thus far has been limited to those general aspects that most cells have in common. At this point we will examine some lesser known cell activities. Keep in mind, however, that not all cells—in some instances only a few—have these specialized activities.

Cells that Crawl, Devour, and Die

Some white blood cells (leukocytes) *have the ability to crawl into and out of microscopic spaces in the walls of blood vessels. These cells may migrate toward a wound site following a concentration gradient produced by substances flowing away from the injury. This process is called* diapedesis. *Once the leucocyte arrives at the wound, it can engulf bacteria in large numbers, so many in fact that the cell itself dies. Massive numbers of dead white cells, bacteria, dead tissue cells, and interstitial fluid which accumulate at the wound site is called pus.*

Phagocytosis

Phagocytosis is a process which literally means "the cell eats." White blood cells have the capability of ingesting substances that have not been previously digested, such as bacterial cells, as shown in Figure 3.4.

In phagocytosis, a bacterial cell becomes attached to the cell membrane, which invaginates and finally pinches off, forming a small vesicle inside the cell containing the particle. Once inside the cell, enzyme action decomposes the vesicle wall and the particle. The decomposed substances may then be used by the mitochondria for synthesizing energy rich compounds for cellular metabolism.

Pinocytosis

The process of *pinocytosis* is similar to phagocytosis except that it probably involves molecular size particles in solution. The term pinocytosis means "cell drinking." A vesicle is formed as in phagocytosis; once inside the cell, the vesicle is disintegrated by enzyme action, releasing the ingested substances into the cytoplasm.

Cell Movement

Although phagocytosis and pinocytosis may be considered to be cell movement, other kinds of movement are more obvious. Some cells, such as red blood cells, are passively carried about by the circulating fluids, while others, such as some white blood cells, have an intrinsic movement of their own. This particular type of movement is similar to that of the single celled protozoa, the ameba, and is thus called *ameboid* movement. By this process, these specialized cells can move about under their own power.

Cell Reproduction

New ideas are slow to be accepted by the scientific community. The cell theory, formally proposed in 1838 by Schleiden and Schwann, has undergone close examination and some revision. The currently accepted cell theory states that all living things are composed of cells and that new cells are formed by the division of preexisting cells. The theory further states that all cells are composed of similar chemical constituents, that they carry on similar metabolic activities, and

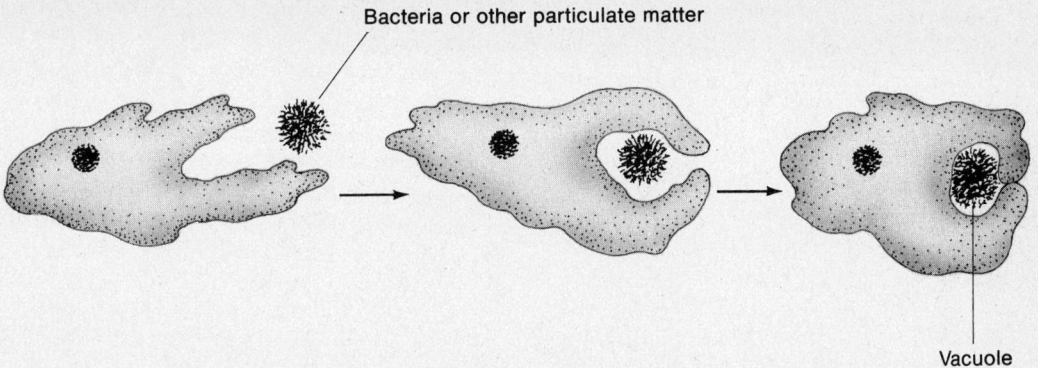

Fig. 3.4 A diagrammatic representation of phagocytosis. The cell engulfs (eats) the particulate substance.

that the overall activity of an organism is the sum of all cell activity and their interactions.

Every cell in the human body originated from a single cell—the fertilized egg. Growth is the result of cell division.

Mitosis The process by which a single cell divides is called *mitosis*. All cells of the body except sex cells (eggs and sperm) reproduce by mitotic division.

The mitotic process essentially involves the division of the nuclear contents of a cell. The material outside the nucleus, the *cytoplasm*, is also divided; this, however, is another process called *cytokinesis*. The discussion here will be limited to the division of the nuclear materials.

Mitosis begins when, in response to some initiating signal (we are not sure what), the nucleus begins a series of structural changes. Its internal structure becomes rearranged in a systematic and predictable manner. These predictable changes are the *mitotic process*.

Mitosis is usually represented as a series of steps or phases, which may lead to the impression that the process stops and starts in a pulsating manner. The process once initiated continues in a smooth, uninterrupted fashion from beginning to end. The ultimate result is the formation of two identical cells, each having exactly the same kind and number of chromosomes. This process is repeated time and again, with the number of cells doubling each time. Mitosis is the process that accounts for the billions of cells which make up the human body.

We can only speculate what causes cells to start dividing. Surface-to-volume ratios probably play a role. The length of time required for the entire mitotic process to be completed varies with the age of the individual, being most rapid during the early development stages and early childhood, and slowing down to a replacement function in adults. Cancer is a condition in which the mitotic rate is excessively high and uncontrolled.

Artificial Environment in Tissue Culture

For many years, researchers have been able to grow human cells in laboratory vessels. Artificial media designed to provide the proper nutrition have until recently depended upon the inclusion of blood serum. Precisely what blood serum contains which makes it indispensable for cell growth is still not known. Researchers at the University of California at San Diego have succeeded in assimilating an artificial medium which lacks blood serum. The new growth medium components may be altered to help identify precisely what hormones or other factors are required for cell growth. Cancer cells proliferate at the expense of normal cells. This research may be able to provide answers to questions concerning cancer cell metabolism. If requirements for tumor growth can be identified in different kinds of tissue, we will have taken giant steps toward controlling such growth.

Meiosis *Meiosis* is the name given to sex cell reproduction. Sex cells (eggs and sperm), commonly called *gametes,* normally have half the number of chromosomes *(haploid number)* that body cells have *(diploid number).* The haploid number of chromosomes in human gametes is 23.

Meiosis begins in diploid cells located in the ovaries in the female and in the testes in the male. The essential function of meiosis is to form gametes having the haploid number of chromosomes from cells having diploid number of chromosomes. The necessity of this reduction in chromosome number is to promote the constancy of chromosomes in body cells within each species.

A human egg with 23 chromosomes, if fertilized by sperm with 23 chromosomes, results in a single cell having 46 chromosomes, from which all body cells ultimately are produced by mitosis. Each chromosome in an egg can be matched with a homologous chromosome in a sperm cell; therefore, we speak of matched pairs in cells having the diploid number of chromosomes.

SUMMARY AND REVIEW

A. Life is best defined as an accumulation of processes called vital signs.
B. The cell is identified as the basic unit of life. Characteristics of some cells or cellular activities include:
    a. absorption
    b. excitability
    c. movement
    d. energy utilization
    e. reproduction
C. Characteristics of cells
    1. Cells are generally microscopic.
    2. Cells vary in shape, size, and function.
D. Cell membrane (plasma membrane)
    1. permits selected substances to flow into the cell (differential permeability).
    2. flow into and out of cells may be controlled by surface-to-volume ratios.
    3. invagination and evagination increase surface area.
E. Endoplasmic reticulum: An elaborate tubular network extending throughout the cell, thought to be involved in transporting materials from one part of the cell to another.
F. Ribosomes: Protein synthesis sites within cells.
G. Nucleus: The organelle containing DNA, the major chemical substance responsible for indirectly controlling all body functions.
H. Mitochondria: The so-called powerhouse of cells due to the activity involved in the release of energy from nutrients.
I. Vacuoles: Cavities scattered throughout cells containing enzymes, fluids, or particulate matter.
J. Cilia: Specialized appendages capable of movement to propel particulate matter.
K. Flagella: Long filaments that provide locomotion for certain kinds of cells.
L. Cytoplasm: A complicated fluid found between the cell membrane and the nuclear membrane. Consists of nutrients, water, ions, minerals, and gases essential for life.
M. Cellular activities
    1. Phagocytosis: A process by which a cell ingests particulate matter.
    2. Pinocytosis: Similar to phagocytosis except that fluids instead of particulate matter are ingested.
N. Cell movement: Certain cells are capable of various kinds of movement by extending or contracting certain parts. This type of movement may be described as creeping or ameboid.
O. Cell reproduction
    1. All cells arise from preexisting cells. Human life begins with a single cell—the fertilized egg.
    2. The process of cell division for body cells (excluding egg and sperm) is called mitosis.
    3. Cytokinesis is the process of cytoplasmic division during mitosis.
P. Meiosis: The name given to sex cell (egg and sperm) division. The chromosome number is reduced from the diploid to the haploid number.

4

Tissues, Membranes, Skin

MAJOR CONCEPTS
  TISSUES
    Types of Epithelial Tissue
    Special Glandular Epithelium
    Connective Tissue
    Contractile Tissue
    Nervous Tissue
    Blood
  MEMBRANES
    Mucous Membranes
    Serous and Synovial Membranes
    Cutaneous Membranes
  SKIN
    Sensory Nerves
    Accessory Structures
    Microbial Diseases of the Skin
    Burns
    Decubitus Ulcers
  THE AGING SKIN AND ACCESSORY ORGANS

The key concept of this chapter is the concept differentiation. This is the process of cell specialization. Specific body tissues, membranes, and the skin are each identified. Each of these tissues are discussed relative to structure and function.

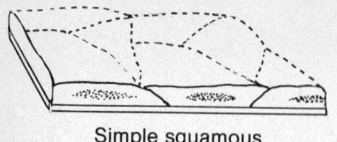
Simple squamous

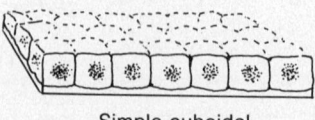
Simple cuboidal

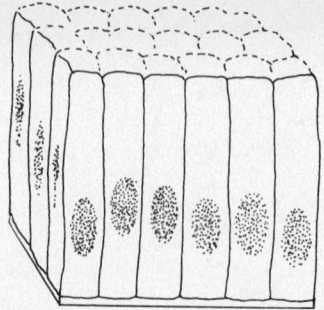
Simple columnar

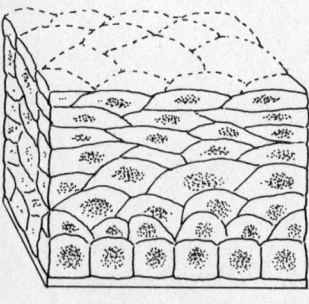
Stratified squamous

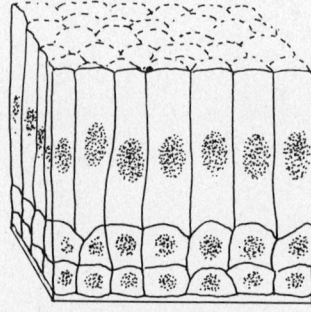
Stratified columnar

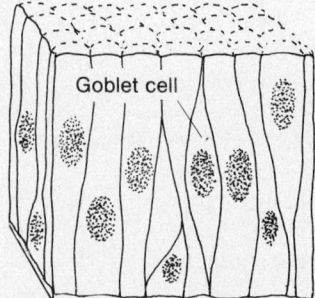
Pseudostratified columnar

Fig. 4.1 Types of epithelial tissues. These tissues have several different functions including protection of underlying cells, secretion, and the absorption of nutrients. Some epithelial tissues possess cilia on their free surfaces. An example of this epithelium is the lining of the upper trachea. The cilia maintain a wavelike motion moving particulate matter upward away from the lungs.

Single cells such as bacteria are capable of independent existence if basic needs are available. A single bacterium can take in nutrients, carry on metabolism for the release and storage of energy, reproduce itself, and in some instances, move about. Some cells of a slightly higher level than bacteria exist in colonies with each cell still carrying on a largely independent existence. Still higher forms also exist in colonies, but within these colonies there is a division of labor. Certain cells have specific functions that others lack. The task performed by several of the specialized cells is sufficient for the whole colony.

TISSUES

A *tissue* is defined as a group of specialized cells having similar shape and function. The specialization of human cells has advanced to such a degree that most of the cells are dependent upon other cells for their very existence. The following is a brief survey of general tissue types. Each of these types will be discussed at greater detail as we get involved with specific body systems.

Many tissues are formed into a definitive mass called an *organ*. An organ is defined as a group of different tissues joined structurally and cooperating functionally to perform a composite task. A group of organs that provides for a basic body need is called an *organ system*, or more commonly, a *body system*. Body systems make up the overall unit we call an *organism*.

Epithelial tissues are composed of cells which cover the external surfaces of the body and line internal surfaces. Figure 4.1 shows several different structural types of epithelial tissues.

Epithelial tissues have several functions including protection of underlying cells, secretion, and absorption of nutrients. Some epithelial tissues possess cilia on their free surfaces, such as the ciliated tissue that lines the trachea. The cilia here maintain a wavelike motion which constantly moves mucus and other debris upward. The accumulation of debris in the lower throat brings about a throat-clearing reflex. Another example of ciliated epithelium is found in the cells lining the oviducts. The cilia aid in propelling the egg along the oviduct toward the uterus.

The secretory function of epithelial cells is highly varied; some of the substances secreted by these cells are tears, oil, saliva, mucus, sweat, milk, and digestive enzymes.

Underlying all epithelial tissue is connective tissue that helps bind and hold the epithelium in place. Connective tissue also helps strengthen the epithelium since it generally exists in thin sheets with very little intercellular substance. This connective tissue layer is called the *basement membrane*. Beneath the basement membrane is a layer of muscle tissue. Thus, the basement membrane serves as an adhesive tissue between epithelium and muscle, as illustrated in Figures 4.2 and 4.3.

Epithelial tissue in general lacks a direct blood supply. Blood vessels located in the basement membrane supply nutrients and oxygen for epithelial cells.

Types of Epithelial Tissue

Epithelial tissue is named according to the shape and arrangement of the cells of which it is composed. Cells may be *flat* (squamous), *columnar* (column shaped), or *cuboid* (cube shaped), as shown in Figure 4.1.

Epithelial cells may be arranged in single flat sheets *(simple)*, or they may consist of more than one layer *(stratified)*.

Simple squamous epithelium is an absorbing tissue. It is found wherever diffusion and filtration occur. Nutrients and gases can rapidly diffuse across these thin sheetlike cells. Examples of this tissue include the lining of blood vessels, the innermost lining of the heart, kidney tubules, and respiratory tissues. The smallest blood vessels, the capillaries, are almost exclusively a single layer of simple squamous epithelium. The alveolar sacs of the lungs are also composed of this epithelium.

Stratified squamous epithelium is composed of several layers of flat cells. (Skin, a special kind of stratified squamous, is discussed later in the chapter.)

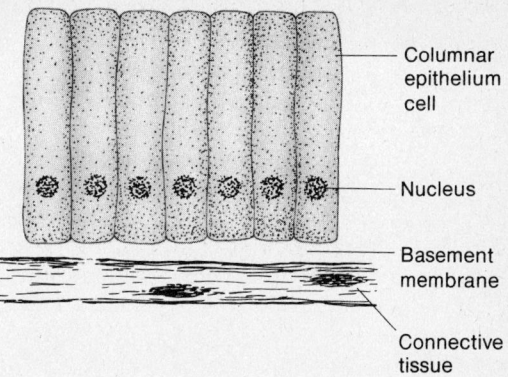

Fig. 4.2 Illustration of the relationship between epithelium, basement membrane, and connective tissue.

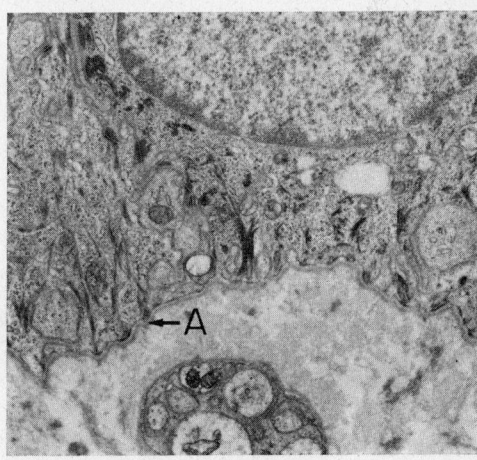

Fig. 4.3 Electron micrograph of basement membrane, shown at A. *(Courtesy Santiago Plurad.)*

Simple cuboidal epithelium is made up of a single layer of cube-shaped cells. These cells have both protective and absorbing functions and can be found in kidney tubules, the thyroid gland, and in the covering of the ovaries.

Simple columnar epithelium is a single layer of cells that are shaped like columns or tubes. These cells may be found lining the intestines, where their main function is absorption. Scattered among the columnar cells of the intestines are special secretory cells called *goblet cells* that secrete an alkaline, viscous fluid called *mucus*.

Pseudostratified columnar epithelium is a single layer of columnar cells. The cells appear to be made up of several layers owing to cilia, which extend from them into the cavity they line. Much of the respiratory tract is lined with this kind of tissue, which aids in removing bacteria and debris from the tract. The cilia move in a wavelike motion, moving foreign particulate matter toward the body exterior.

Stratified columnar epithelium is made up of multiple layers of columnar cells. The tissue has a function of protection, absorption, and secretion and is found lining the stomach and some of the respiratory tract.

Special Glandular Epithelium
Although goblet cells secrete mucus, they do not constitute a separate tissue owing to the fact that they are scattered throughout other cells.

In some organs secretory cells have multiplied and invaginated into the underlying connective and muscular tissue. This accumulation of specialized secretory cells is called a *gland*.

Glands that maintain a tubular pathway and opening into another structure are called *exocrine glands*. Examples of exocrine glands are sweat glands, oil glands, and parts of the pancreas that secrete digestive enzymes.

Glands that have lost their direct connection with other body structures are called *endocrine glands*. The secretions of these glands, *hormones*, diffuse directly into the bloodstream.

Cell Specialization: The Unanswered Question

Except for a few highly specialized cells such as gametes, red blood cells, and skeletal muscle cells, every cell in the body has the same number and kinds of chromosomes. We know that cells specialize (differentiate) to provide the multiplicity of functions required by the living body. But if each cell carries the same genetic information, how do cells become differentiated to become liver cells or brain cells or stomach cells? This is an unanswered question.

Research has provided some clues as to when cells may be expected to specialize. For example, the beginning of certain tissue specialization, such as parts of the eye, can be fairly well pinpointed as a specific time after conception. What makes cells specialize into eye tissue, however, is not known.

If a cell mass is divided into two clusters before differentiation begins, each cluster is capable of growing into a complete organism. In humans, identical twins are formed in this manner. If a single complete cell taken from any part of the body could be induced to begin cell division, theoretically a whole organism could be produced—presuming that a proper environment was provided for the developing organism. This process is called cloning.

It is interesting and somewhat frightening to contemplate that a person, through cloning, might be able to produce an identical twin, or 50 identical twins—or 10,000.

Connective Tissue
Note in Figure 4.4 that the structure of *connective tissue* cells is quite different from that of epithelial cells. The most common forms of connective tissue are bone and cartilage *(gristle)*. Other forms include tendons and ligaments, and the sheathlike layers of cells which attach organs to other body structures.

Connective tissue gives the body shape and allows organs to be attached and firmly held in specific positions. Cartilage gives shape to the ears and nose and also forms a padding between joints. This padding absorbs shocks that otherwise would be directly transmitted to bones, which are more easily shattered.

The skeleton in the early embryonic stages is largely made up of cartilage. Bone later develops

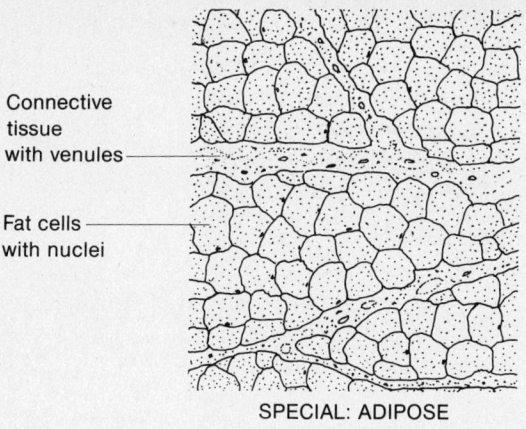
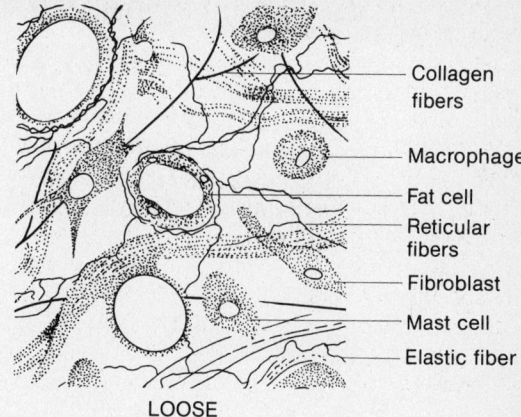
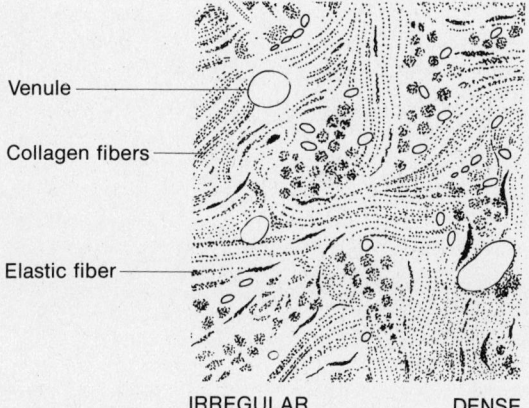
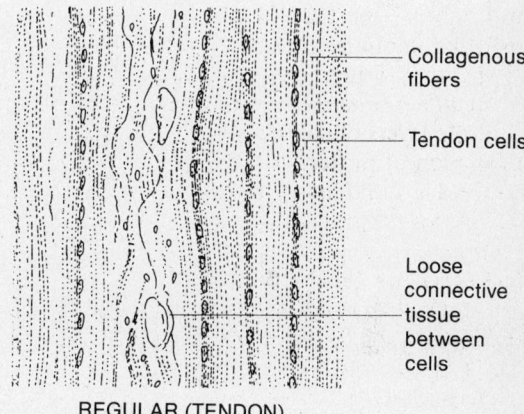

Fig. 4.4 Types of connective tissue. Connective tissue gives the body shape and allows organs to be attached and firmly held in position.

and replaces this basic cartilage formation. Calcium and phosphorus salts deposited in and around the bone cells *(osteocytes)* give bone its great rigidity. As a person ages, the ratio between inorganic salts, such as calcium and phosphorus, and organic matter is greater, making bones more brittle.

Collagenous fibers make up the protein *collagen*, a common connective tissue found mostly in muscle. High temperatures (150 to 180° C) change collagen to gelatin. Collagen fibers are shown in Figure 4.5 at *A*.

Contractile Tissue

Contractile tissue is mostly that of the musculature. There are three different types—*smooth* (visceral), *striated* (skeletal), and *cardiac*. These are shown in Figure 4.6. All three types of muscle are organized into fibers that have an inherent contractible nature so that they can shorten due to either nervous or hormonal stimulation. This contraction permits movement in different parts of the body.

Normally, the smooth type of muscle is not controlled by the voluntary senses. This muscle

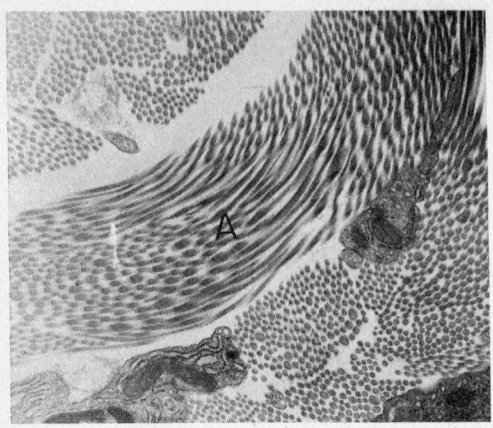

Fig. 4.5 Electron micrograph of collagen fibers, labeled A. These fibers are coiled protein having high tensile strength. Collagen fibers bind the basement membrane to epithelial tissues. *(Courtesy Santiago Plurad.)*

tissue is found in the vascular system, the blood vessels, and many of the tubes and ducts located throughout the body. Heart muscle (cardiac muscle) is a striated form, but it is not under voluntary control as is the normal striated form of muscle.

Nervous Tissue

Nerve cells are highly specialized cells which carry messages from one part of the body to another for precise coordinating activities, which must be constantly controlled. Nerve cells are so highly specialized that they have lost the capability of reproducing themselves. A complete discussion of nervous tissue can be found in Chapter 7.

Blood

Blood is classified as a tissue since it has all the properties that by definition a tissue has. It contains highly specialized cells to carry out very precise, specific body functions. More will be said about the structure and function of blood as we discuss specific organs and organ systems.

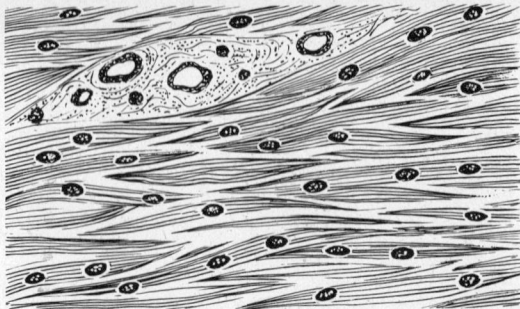

Visceral or nonstriated (smooth) involuntary muscle tissue.

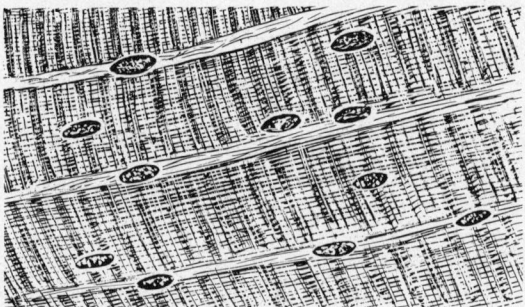

Skeletal or striated voluntary muscle tissue.

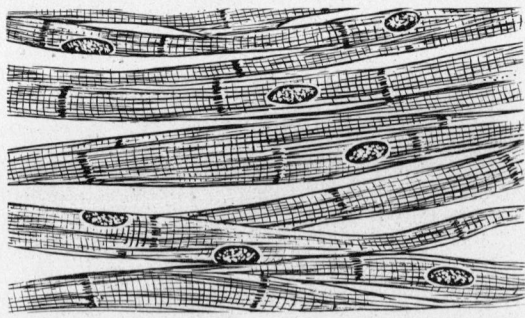

Cardiac (heart) muscle tissue

Fig. 4.6 Three types of muscle tissue. All three types of muscle have cells arranged into fibers that have the capability to shorten (contract). This contraction may produce locomotion, pump blood, close a body opening, move nutrients along the digestive route, or facilitate many other body activities.

MEMBRANES

Membranes constitute a higher level of organization than tissues. Membranes are composed of several different kinds of tissues, the variety depending upon the function required. Membranes are generally classified as sheets of tissue that cover or line body organs. Membranes may protect organs, hold organs in place, secrete fluids, or serve as boundaries. The more common membranes include *mucous, serous, synovial,* and *cutaneous* membranes.

Mucous Membranes

Mucous membranes secrete a thick, sticky substance called *mucus.* These membranes generally line body cavities that open to the exterior. These cavities are found in the digestive, respiratory, reproductive, and urinary systems.

Mucus serves a variety of functions. It may serve as a lubricant to facilitate the movement of materials through the gastrointestinal tract. It may protect tissues from enzymes and, due to its alkaline content, may neutralize acids.

Serous and Synovial Membranes

These membranes line body cavities that do not open to the exterior. Serous membranes line the abdominal and thoracic cavities and cover the organs within these cavities. Serous membranes lining the abdomino-pelvic cavity are double layered, with the outer layer, the *parietal peritoneum,* closely adhering to the wall of the abdomen. The innermost layer, the *visceral peritoneum,* adheres to the organs in the cavity. Fluid secreted between the two layers acts as a lubricating and friction-reducing substance to permit some movement of organs within the cavities.

The thoracic cavity is divided into three segments; two segments enclosing the lungs, the third enclosing the heart. All of these segments are lined with serous membranes. Those lining the lung cavities are called *pleura.* The outermost layer of the pleura is called the *parietal pleura;* the innermost layer is called the *visceral pleura.* The serous membrane covering the heart, the *pericardium,* is also composed of parietal and visceral layers. Serous fluid secreted between the layers permits the free movement of the lungs and heart in such confined space.

Synovial membranes line joint cavities. Secretions of these membranes also serve to lubricate the surfaces of articulating bones, as shown in Figure 4.7.

Cutaneous Membranes

These membranes make up the outer layers of the skin and, along with accessory structures such as hair, nails, oil and sweat glands, and nerve endings are termed the *integument.*

Since the skin is in direct contact with the external environment it is an important barrier to invasion from microorganisms. Other functions include:

1. A receptor for stimuli such as temperature, chemicals, and physical factors.
2. A regulator of body heat levels.
3. A storage compartment for fats, glucose (in the form of glycogen), protein, minerals, and ions.
4. An excretory organ to remove certain body wastes.

SKIN

Skin is composed of several kinds of tissue arranged in different layers. The outermost layer, the *epidermis,* and an inner layer, the *dermis,* make up the major portion of the structure called integument or skin, as shown in Figure 4.8. These two layers are firmly cemented together. Occasionally, when excessive friction upon the epidermis irritates the tissue, fluid accumulates between the layers causing a *blister.*

Although the epidermis varies in thickness in different parts of the body, it is generally composed of a very thin layer of stratified squamous epithelium. Since epithelial cells lack a direct blood supply, they must receive nutrition and oxygen from capillaries in the dermis. These substances diffuse out of capillaries into fluid spaces that separate epidermal cells from dermal cells.

Only the innermost layer of the epidermis has access to nutrients and it is only here that we find

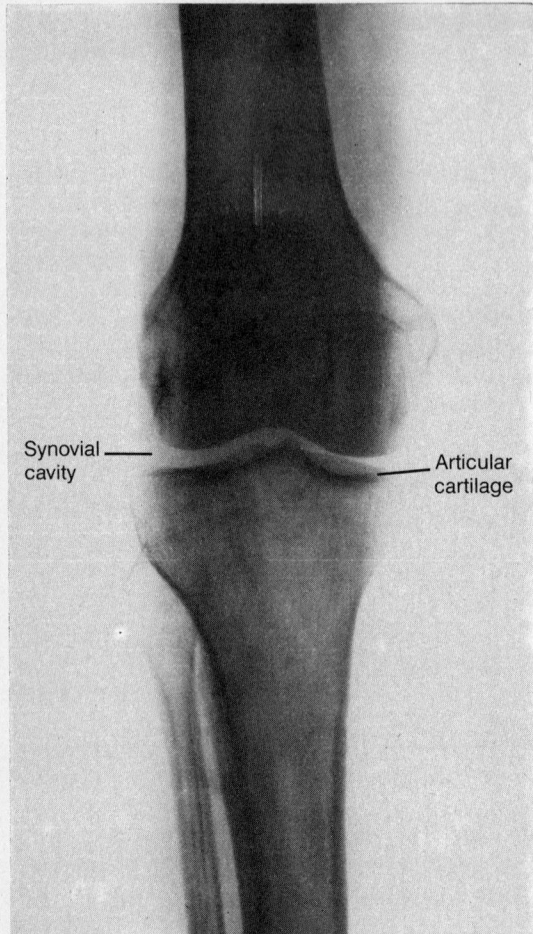

Fig. 4.7 X ray showing knee joint, a typical synovial cavity. *(Author's collection.)* Synovial membranes line joint cavities such as the knee illustrated here. Secretions of these membranes also serve to lubricate the surfaces of articulating bones.

> ### Black Skin Color Changes
>
> *Most textbooks, when they discuss skin color changes caused by body malfunction, indicate color changes evident in white skin only. White skin contains melanin granules only in the lower layers of the epidermis. Black skin contains more melanin granules, and they are found throughout all layers of the epidermis. Rashes, cyanosis, pallor, and jaundice are all more difficult to observe in the dark-skinned person, and therefore it is often necessary to look for these conditions in not so obvious places. If the skin is so heavily pigmented that cyanosis cannot be detected in lips, nail beds, and so on, with casual observation, close inspection and accurate observation of the precyanotic color is important. If direct observation fails and cyanosis is suspected, blood tests indicate higher than normal levels of reduced hemoglobin.*
>
> *Jaundice may be detected by observing the color of the hard palate in bright daylight. The sclera of the eyes of black-skinned persons often contain heavy deposits of carotene, which might be mistaken for jaundice.*
>
> *The keys to accurate observations of color changes in black skin are:*
>
> 1. *observe normal color,*
> 2. *anticipate and look for changes,*
> 3. *observe behavior changes that may indicate a disorder, then search for color changes.*

living cells. This living layer of cells, the *stratum germinativum*, is the only area where mitosis occurs in the epidermis. This layer also contains specialized cells (melanocytes) that synthesize the pigment *melanin*. This pigment, along with *carotene* (stored in fat tissue), and *hemoglobin* (blood pigment), give human skin a wide variation of colors. White skin contains fewer melanin granules than black skin. Skin tanning (more noticeable in white skin) is due to the effect of darkening of melanin granules due to ultraviolet rays of sunlight.

As new cells are produced in the stratum germinativum, older cells are pushed upward and away from the source of nutrition. Cut off from nutrition, they die. The protoplasm of these cells undergoes a chemical change and becomes hardened, or *keratinized*, forming a tough protective membrane.

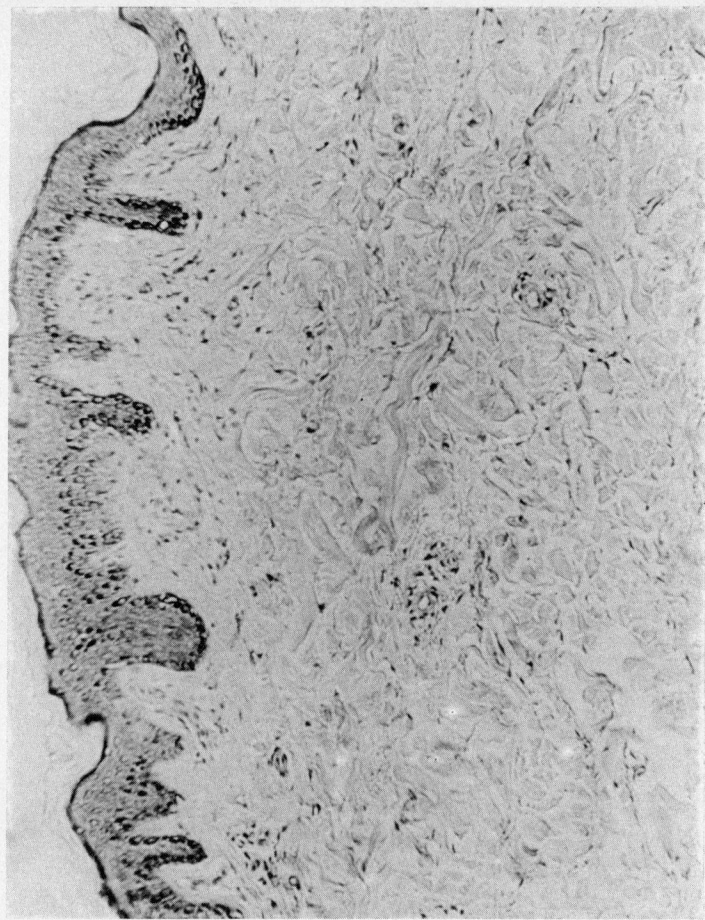

Fig. 4.8 A photomicrograph of a cross section through abdominal skin showing epidermal components and the dermis. (Author's collection.)

Located above the stratum germinativum are three other layers, the *stratum granulosum*, the *stratum lucidum*, and the outermost layer, the *stratum corneum*. The stratum corneum is the excessively keratinized layer which acts as a waterproof covering of the body. The stratum lucidum is the layer of clear cells, and the stratum granulosum is a layer containing dark staining granules, hence its name.

Sensory Nerves
Some of the sensory nerves in the skin detect cold, heat, touch, pressure, and pain.

Cold and heat receptors located in the dermis are capable of detecting minute changes in temperature. This provides the body with a protective device, particularly when the body is subjected to excess heat and cold. In addition to protection, these nerves are also responsible for helping to maintain normal body temperatures by stimulating other nerve centers located in the brain that, in turn, induce sweat-gland secretion.

The evaporation of sweat is a cooling process. In response to a warm environment, brain cen-

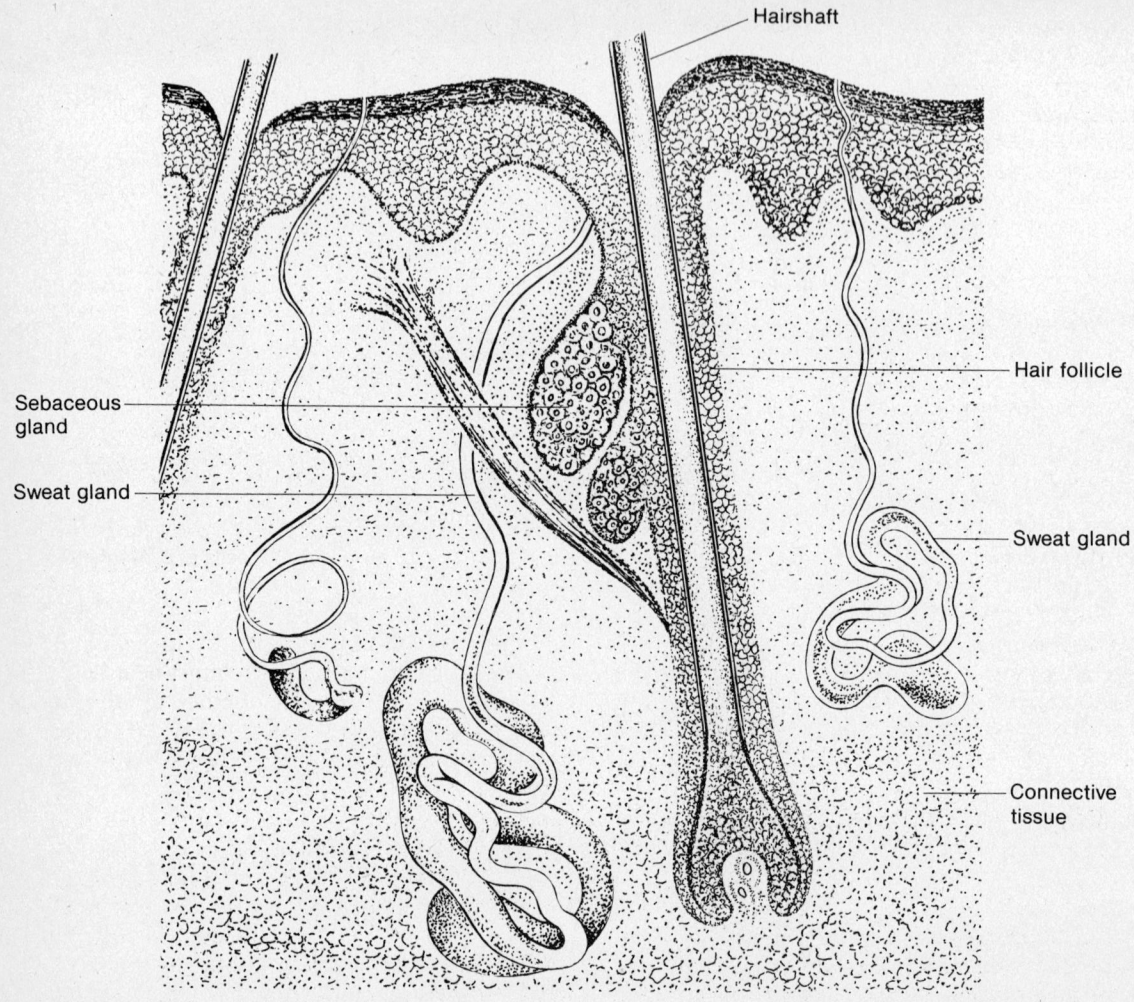

Fig. 4.9 A cross section of skin showing sebaceous glands and sweat glands. Sweat glands help rid the body of wastes and provide dampness on the skin surface. The evaporation of this moisture is a cooling process for an overheated body. Sebaceous, or oil, glands secrete substances that keep the skin pliable and flexible.

ters send impulses to capillaries located in the dermis. The capillaries are dilated *(vasodilation)*, which effectively brings the blood nearer the surface of the skin so that excess heat carried by the blood can be radiated away from the body. This explains why the skin appears to be flushed during vigorous exercise.

In a cold environment the body must conserve heat. Blood capillaries are constricted *(vasoconstriction)*, which causes the blood to be pulled away from the skin surface, preventing radiation.

Touch and pressure receptors, also located in the dermis, are most highly concentrated in the fingertips. When touch receptors are stimulated, impulses are sent to the brain where they are interpreted as light touch. The reception of touch stimuli provides the capability of identifying textures of touched substances. When you place your fingertips on an object, you immediately recognize whether it is smooth or rough, and what general shape it has. These receptors quickly lose their power to conduct impulses if the fingertips remain motionless on the object.

Pain receptors are located deeper in the dermis and are stimulated when excessive pressure is applied or when the skin is penetrated.

Accessory Structures

The skin accessory structures include *sweat glands, sebaceous glands, hair,* and *nails.*

Sweat glands located in the dermis are found in most of the body's surface. They are more numerous in the palm of the hand and the sole of the foot. Sweat provides two important functions—temperature regulation due to its evaporation from the skin surface, and excretion of waste materials such as urea, uric acid, and ammonia.

Under severe conditions, such as vigorous exercise, high temperature, and humidity, profuse sweating may remove as much as 8 to 10 liters of water from the body in a 24-hour period. Water lost in the form of sweat carries electrolytes with it (ions in solution, particularly sodium), which may lead to deficits resulting in extreme physiological disturbances. If these lost fluids are not replaced, the body becomes dehydrated.

Sebaceous glands are oil glands associated with hair follicles, as shown in Figure 4.9. They empty their secretions around the hairshaft and serve to lubricate the hair, keeping it soft and pliable.

Hair serves a protective and insulating function. For example, hair on the head cushions the head against mechanical injury and screens out harmful sun rays. The hair of eyelashes, nostrils, and the ear canal filters out debris that might otherwise enter and affect body tissues. Figure 4.11 shows a single hair at a magnification of 12,800×.

Microbial Diseases of the Skin

Intact skin and mucous membranes, acting as the first line of body defenses against invasion by *pathogenic* (disease-causing) *organisms,* are in nearly constant contact with organisms that can cause serious infections if they break through this defensive system. A few microorganisms

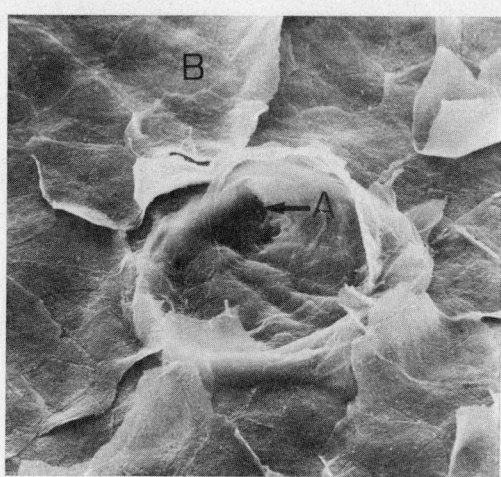

Fig. 4.10 Electron micrograph of skin showing a sweat duct opening at *A* and horny skin cells at *B*. Magnification 1300 x. *(Courtesy Santiago Plurad.)*

and small worms have the ability to penetrate intact skin, but in general there must be a break in the skin or mucous membranes to provide entrance for the invading organism. This tissue break need not be great and may actually be so small that the person may not notice it.

Bacterial Diseases All of the following skin diseases must gain entrance through skin or mucous membranes that are broken. Each of the organisms considered in this section are, once they have gained initial entrance through the broken skin, able to create a local infection. A local infection may destroy not only surface tissues, but also may provide the invading organism with a route into deeper tissues—the blood and major body organs and systems.

A bacterium of the genus *Staphylococci* causes a variety of localized skin infections including pimples, boils, and in unison with other bacteria or fungi, the infections of *acne* and *impetigo.*

Acne is an eruptive inflammation of sebaceous glands more common in adolescents than adults. Causes of acne may include increased sebaceous gland secretion, sex hormones

Viral Diseases These include *warts* and *fever blisters*. Fever blisters are caused by a virus called *Herpes simplex*. Once the virus attacks a tissue, such as the lips, nose, or genital organs, it may remain in the tissue as an inactive agent. When the body loses some of its normal resistance through illness, such as a cold or influenza, the virus can again exhibit its presence.

Rashes may be caused by either chemical action or as a response to a systemic disease.

The rash associated with poison ivy or poison oak causes a special kind of *dermatitis*, or inflammation of the skin. A list of other dermatitis-causing agents is given in the box. Many, if not most, of these agents produce what is called an *antigen-antibody response*. The body, in an attempt to neutralize the agent, produces specific substances called *antibodies*. The interaction between the agent *(antigen)* and the antibody may cause different levels of tissue damage exemplified as a rash. The immune system is discussed more thoroughly in Chapter 11.

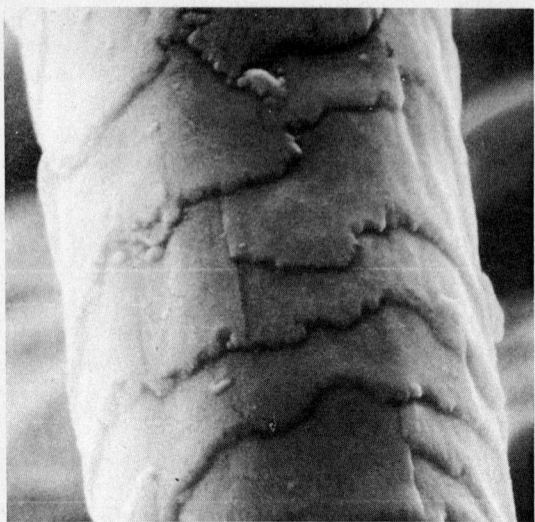

Fig. 4.11 Electron micrograph of a single hair. *(Courtesy Santiago Plurad.)*

(androgens), bacteria, yeasts, and fungi. The psychological impact that a serious case of acne may have upon an adolescent can leave emotional scars that may persist through life. Teenage years represent a time of life when a good personal appearance is of highest priority, and acne removes the capability of looking one's best.

Acne is difficult to treat, supposedly due to the variety of causes. Maintaining extreme cleanliness helps in some cases. Medications, vitamins, hormones, and specialized, restrictive diets are also recommended in some cases.

The penetration of microorganisms into skin membranes can also cause small abscesses. *Pus* is made up largely of destroyed white blood cells that have migrated to the wound site to attack an invading organism.

The skin normally maintains numerous bacteria, many of which are trapped in the thin layer of oils poured out by oil glands. Their numbers are kept in control by some skin secretions. Proper hygienic skin care also removes large numbers of bacteria.

Dermatitis-Causing Agents

Rashes are often an external manifestation (skin eruption) of some allergic reaction. Some elements you may be allergic to are:

1. *Things you breathe (dust particles, chemical vapors, animal danders, feathers, pollen, molds, cosmetics).*
2. *Things you eat or drink (foods, drinks, medicines).*
3. *Things you touch (fabrics, poison plants, metals, woods, plastics, medicines, cosmetics, soaps).*
4. *Things you are injected with (vaccines or drugs, venom from insect stings or snakebites).*
5. *Things you are infected with (bacteria, fungi, molds).*
6. *Physical factors (heat, cold, light, sun).*
7. *Emotional factors (conscious or unconscious feelings).*

Fungal Diseases These include such conditons as *ringworm*, *ringworm of the feet* (athlete's foot), and *scalp ringworm (Tinea capitis)*.

The entire group of diseases caused by fungi is called *dermatophytoses*. These diseases usually establish residence in the keratinized layer of the skin through a break or scratch. The fungi release an enzyme capable of dissolving *keratin,* the protein that makes up hair, nails, and the outer layer of the epidermis. These diseases have little influence upon deeper skin tissues. Some fungal secretions, however, may be toxic to living cells, causing the antigen-antibody reaction.

Another danger from fungal disease is the possibility of offering a site for the invasion of other microorganisms that may do even greater damage to the skin and its underlying tissues.

Burns

The skin faces few conditions more serious than burns. Burned skin is destroyed tissue. The more tissue destroyed, the more serious the consequences are for the underlying tissues—and indeed for the whole body systems.

Two major problems are created with burned skin:

1. the exposure of normally protected tissues to pathogens, and
2. the loss of body fluids.

Burns are classified according to the degree of damage done. *First degree burns* cause damage to the epidermal layer, causing redness and swelling *(edema)* due to fluid accumulation. *Second degree burns* damage both the epidermis and dermis. Blisters containing tissue fluids may form. *Third degree burns* destroy both the epidermis and dermis. The skin may be charred. Nerve endings are destroyed, leaving the burned area insensitive to touch. In third degree burns, the loss of body fluids along with electrolytes they contain becomes a critical problem. These fluids must be replaced and steps immediately taken to cover the burned area to prevent further loss of fluids. Special preparations may be used to cover the burn to prevent fluid loss. Since the skin germinating layer has been destroyed new cells are not produced, and if skin transplants are not utilized, disfiguring scar tissue forms.

Decubitus Ulcers

If skin cells fail to receive nutrition they die. A *decubitus ulcer* is the result of prolonged pressure on skin covering a bony prominence and is usually caused by a patient's inability to move. The areas of the body generally affected include the pelvic bones, ankles, wrists, and heels. If a paralyzed or unconscious patient remains in one position, the weight of the body or limb creates pressure, reducing circulation in the skin over the bony prominence. Skin cells die, hemorrhage occurs, and eventually there is a break in the skin. Microbial infection then may be a serious problem. If untreated, the ulcer becomes larger due to damage of surrounding cells.

THE AGING SKIN AND ACCESSORY ORGANS

The aging skin typically becomes thin, pale, dry, and less elastic. Dryness, caused by reduced oil gland secretion, leads to itching. The delicate skin of the elderly is susceptible to abrasion and other injury. Thus infection is a constant threat.

The hair may become dull and thin, and the nails brittle. Teeth tend to loosen and fall out.

There is much evidence that aging of the skin and accessory organs of the skin can be slowed with proper nutrition and good hygiene.

SUMMARY AND REVIEW

A. Tissues are defined as a group of specialized cells having similar shape and function. The major types of tissues include:
   1. Epithelial tissue
      a. simple squamous
      b. stratified squamous
      c. simple cuboidal
      d. simple columnar
      e. pseudostratified columnar
      f. stratified columnar
   2. Special glandular epithelium—composed of specialized cells which secrete oils, sweat, enzymes, or hormones.
   3. Connective tissues—these give the body shape, hold organs in place, and protect certain organs.
   4. Contractile tissue—this is muscle tissue:
      a. smooth
      b. striated
      c. cardiac

5. Nervous tissue—composed of specialized cells that conduct information from one part of the body to another.
6. Blood—as a tissue is responsible for transporting essential substances throughout the body. Other functions include blood clotting, and heat transfer.

B. Membranes are sheets of tissue that line or cover organs. The more common ones include:
   1. mucous
   2. serous
   3. synovial
   4. Cutaneous

C. The skin is composed of two major kinds of tissues—the epidermis and the dermis.
   1. The epidermis is composed of distinct layers:
      a. stratum corneum
      b. stratum lucidum
      c. stratum granulosum
      d. stratum germinativum
   2. Functions of the skin include protection, secretion, and reception of stimuli.
   3. Skin accessory structures include sweat glands, sebaceous (oil) glands, and hair.

D. Microbial diseases of the skin include diseases caused by bacteria, viruses, and fungi.
   1. Bacterial diseases such as boils, pimples, and acne may be caused by a variety of bacteria—among them staphylococci is usually implicated.
   2. Viral diseases include warts and fever blisters.
   3. Fungal diseases include the ringworm types (athlete's foot, scalp ringworm).

E. Burns of the skin are classified according to the degree of tissue damage, third degree burns being the most serious. Special attention must be given burn patients to prevent the loss of fluids and electrolytes from the body and to prevent infection.

F. Decubitus ulcers occur in the skin when circulation is cut off by pressure against bony prominences.

G. The skin of elderly persons is normally:
   1. thin
   2. pale
   3. dry
   4. less elastic than youthful skin
   5. more susceptible to damage

PLATE 1   MUSCLES OF THE HEAD AND NECK

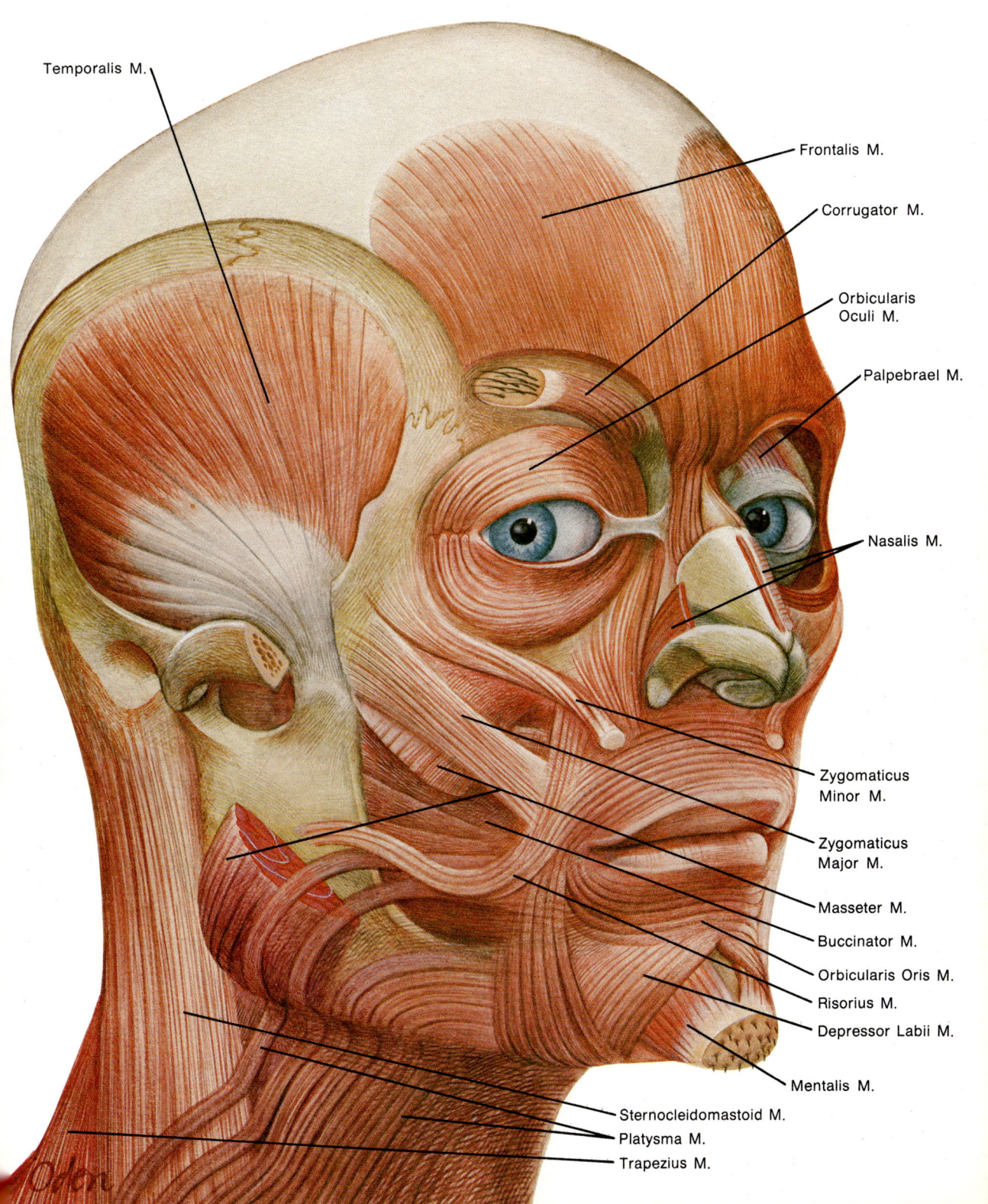

PLATE 2 PRINCIPAL SUPERFICIAL MUSCLES OF THE ANTERIOR (VENTRAL) TRUNK

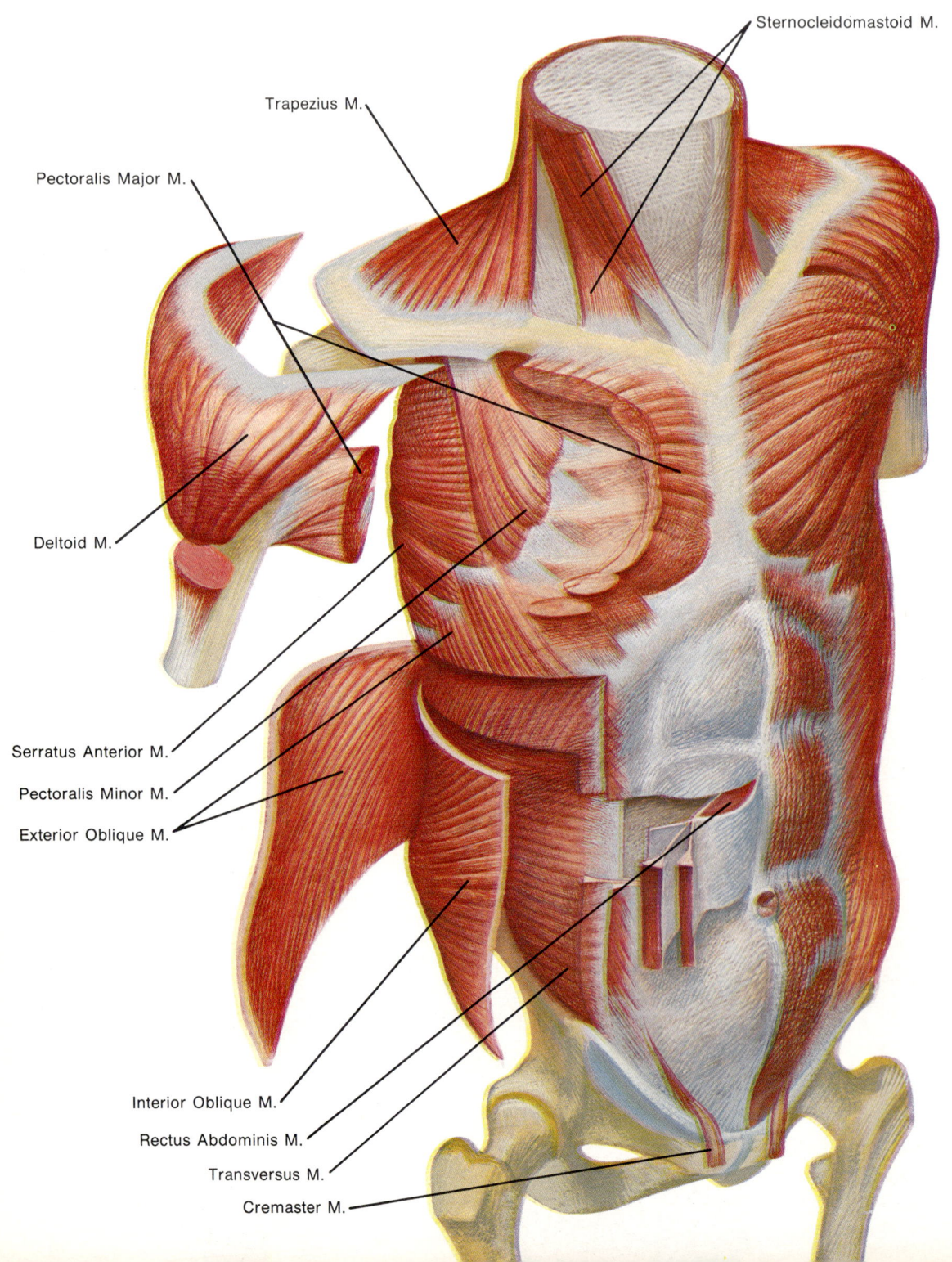

PLATE 3 PRINCIPAL SUPERFICIAL MUSCLES OF THE POSTERIOR (DORSAL) TRUNK

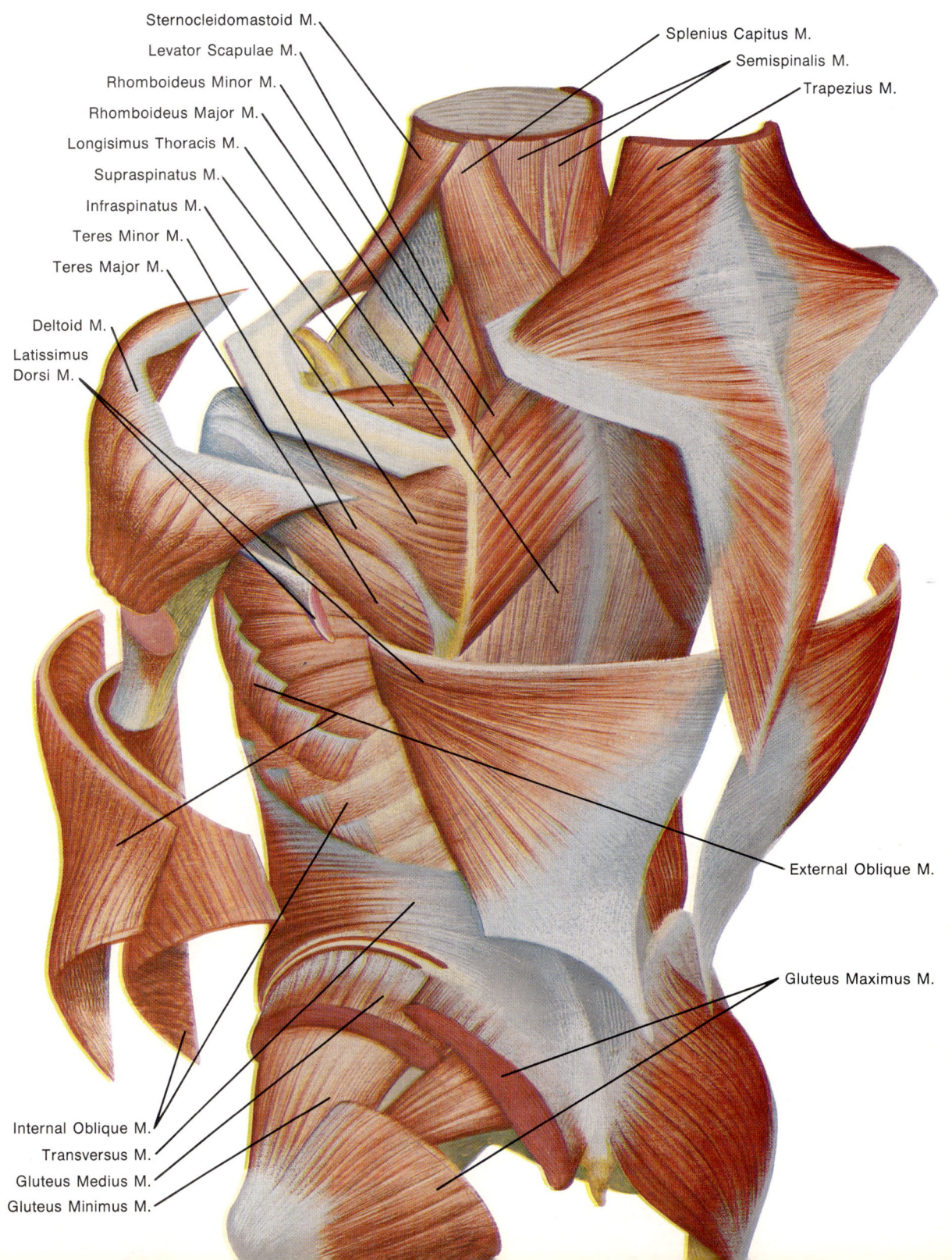

PLATE 4 A SECTION OF SKIN AND THE MUSCULATURE OF THE ARM AND LEG

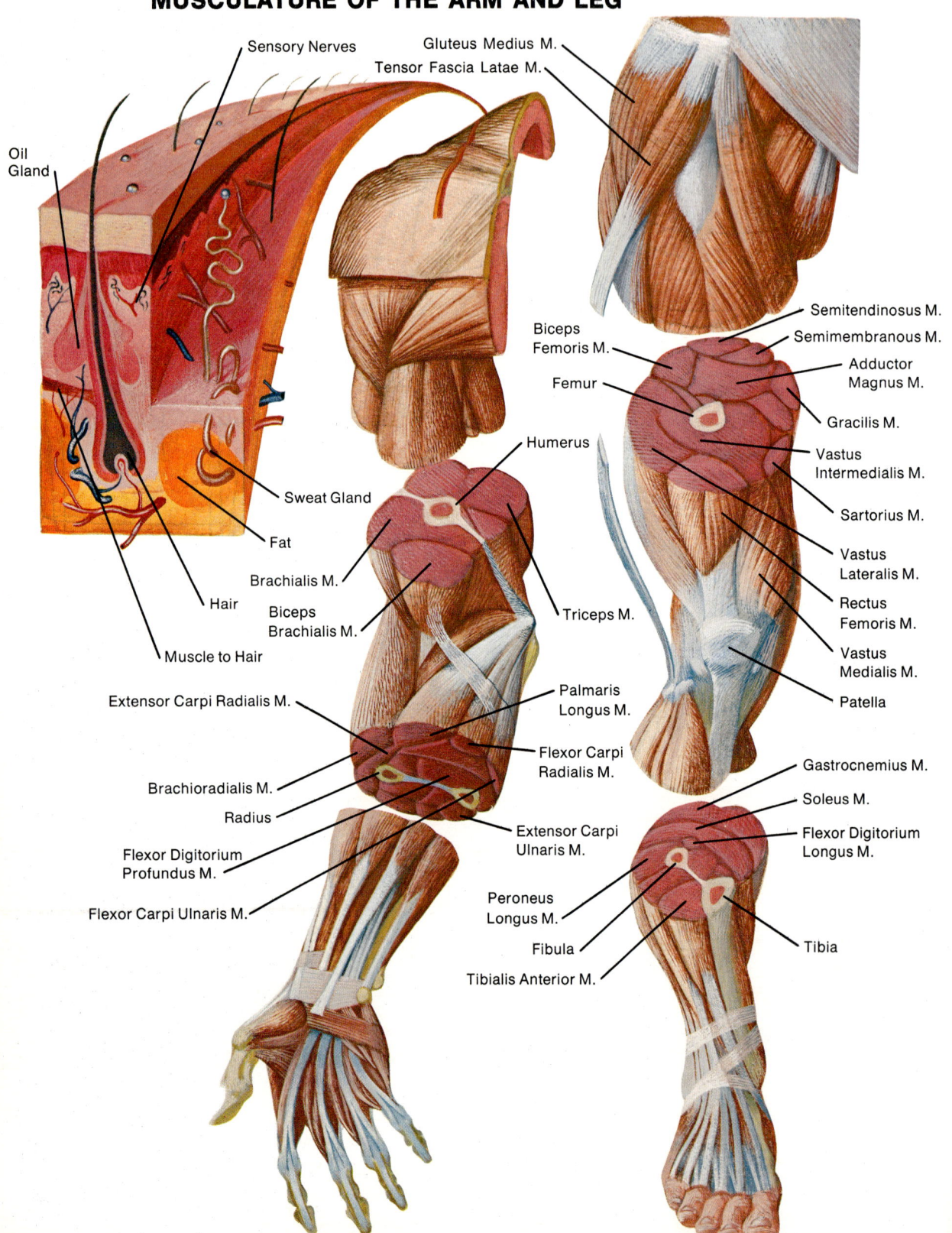

5
The Skeletal System

MAJOR CONCEPTS
  CARTILAGE
    Hyaline Cartilage
    Fibroelastic Cartilage
  BONE
    Bone Marrow
    Classification of Bones
    Bone Markings
    Morphology of Bones
    How Bones Grow
  SKELETAL COMPONENTS
    Upper Appendicular Skeleton
    Lower Appendicular Skeleton
    Axial Skeleton
    Effects of Hormones and Vitamins on Bone Growth
  FRACTURES
  BONE ARTICULATIONS
    Synarthroses
    Amphiarthroses
    Diarthroses
  DISEASES OF THE SKELETAL SYSTEM
    Nonmicrobial Diseases
    Microbial Diseases
  CALCIUM LEVEL INFLUENCES IN THE AGED

The skeletal system includes bone, cartilage, and other connective tissue. All these tissues are composed of living cells having the same nutritional needs as other cells of the body. These tissues support and protect other more delicate tissues. An additional and very important function of bone is to act as a reservoir for calcium and phosphorus.

The major part of this chapter deals with the description and location of various bones of the body.

The skeletal system is composed of bone, cartilage, and other connective tissues. Because the study of these components is usually limited to the examination of dried or preserved specimens, the term skeletal has the connotation of dried bones. But in fact, of course, the components of the skeleton are living tissues. As living tissues, they are composed of cells that, like other cells in the body, require nutrition and oxygen and, like other cells, carry on normal cell activities.

The skeletal system provides protection for

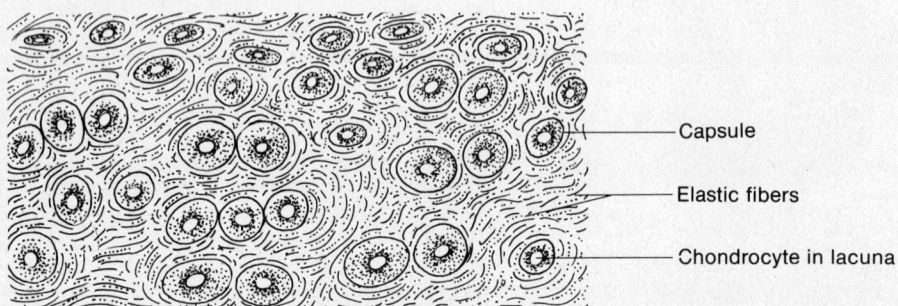

ELASTIC CARTILAGE

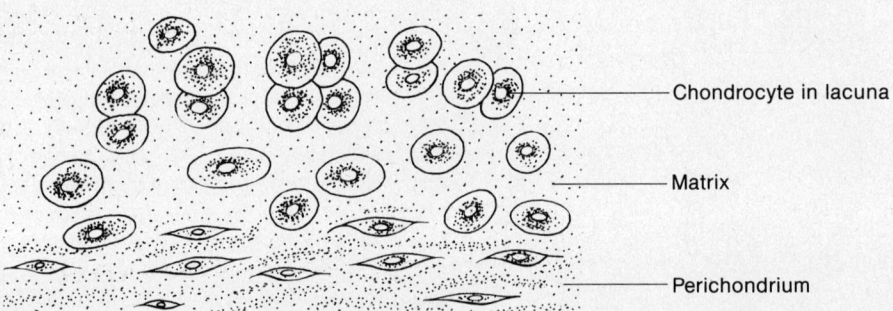

HYALINE CARTILAGE

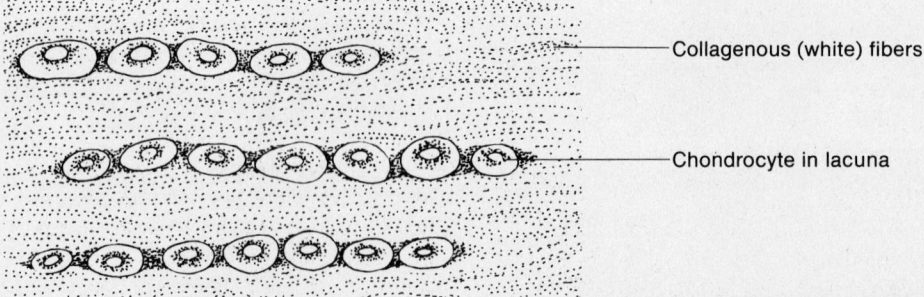

FIBROCARTILAGE

Fig. 5.1 Three types of cartilage: hyaline, fibrocartilage, and elastic cartilage. Cartilage contains relatively few cartilage cells (chondrocytes) randomly scattered throughout an intercellular substance called the matrix.

delicate tissues, gives the body shape, permits the movement of body segments, and serves as a reservoir for essential minerals such as calcium and phosphorus.

CARTILAGE

Cartilage is the initial supporting tissue of the embryo. It differs from other types of tissue in several respects. Cartilage cells, called *chondrocytes*, are located in spaces called *lacunae*, as shown in Figures 5.1 and 5.2.

Cartilage contains relatively few number of chondrocytes, randomly scattered throughout an intercellular substance called the *matrix*. The matrix is a protein and polysaccharide substance having a plasticlike consistency. The flexible, plastic nature of the matrix gives cartilage special qualities, making it ideal for cushioning joints and forming soft features such as the nose and ears.

Blood vessels bring nutrients and oxygen only to the outer surface of cartilage *(perichondrium)*. Since there are no blood vessels within the cartilage itself, cartilage cells are nourished solely by the diffusion of nutrients through the matrix. Simple diffusion through the matrix accounts for the transport of essential substances as well as the removal of cartilage cell waste products.

Hyaline Cartilage

Hyaline cartilage is the most common type of cartilage. This translucent, pearllike, plastic tissue gives shape to the nose and acts as the cushioning tissue at the ends of long bones *(epiphyseal cartilage)*.

Hyaline cartilage makes up the early fetal skeleton and is later replaced by bone. It is also found in ringlike structures around air passages of the breathing apparatus, such as the larynx, trachea, bronchial tubes, and bronchioles. Without these rings, the air tubes would collapse during inhalation, thus preventing the entry of air into the lungs.

Fibroelastic Cartilage

Fibroelastic cartilage has very few chondrocytes and has many more fibers running throughout the matrix for more rigid support. Fibroelastic cartilage is found at the *pubic symphysis,* where the pubic bones join on the ventral surface. It is also found in the joints between vertebrae, functioning as a cushion.

BONE

Bone is composed of cells called *osteocytes,* which are surrounded by organic and inorganic substances that provide rigidity for the bone tissue. Osteocytes are directly supplied with nutrients and oxygen by an extensive blood vessel and capillary system.

Bone is constantly changing its chemical composition. Short-term changes occur as calcium from the blood is deposited in bones and as calcium from bone is released back into the blood. Longer-term changes occur in both structure and chemical composition as aging progresses.

The basic unit of bone tissue is the *Haversian system*, illustrated in Figures 5.3 and 5.4. The Haversian system includes a Haversian canal, osteocytes, lacunae, and canaliculi. The *Haversian canal* runs parallel to the axis of bone and contains large blood vessels. Blood vessels and nerves enter from the outside by way of *Volkmann's canals. Osteocytes* are arranged in a concentric pattern around the Haversian canal, giving the tissue a layered appearance. Osteocytes are located in small compartments called

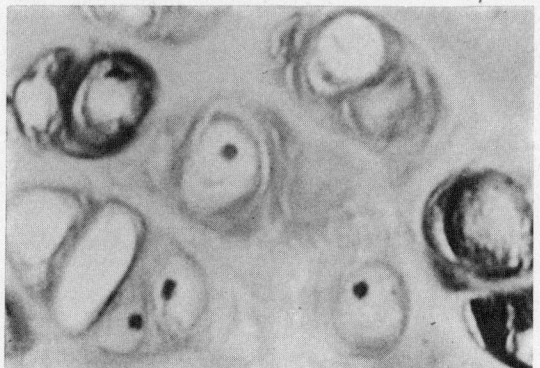

Fig. 5.2 Photomicrograph of hyaline cartilage. *(Author's collection.)*

lacunae. The areas between layers of lacuna, called *lamellae,* are composed of organic and inorganic substances that give bones their rigidity. Radiating away from each lacunae are tiny canals called *canaliculi,* which connect the lacunae of one layer to another. Canaliculi transport essential materials from the Haversian canal to the outermost layer of osteocytes within a Haversian system.

The inorganic component of bone is largely calcium and phosphorus salts. Protein and polysaccharides make up the bulk of the organic portion of bone. Note in Figure 5.5 that there are two different structural formations of bone. In the outermost layer of *compact bone,* the Haversian system and the interconnecting canals are very compact and close together. Internally, the bone is less compact, with the osteocytes more scattered. The osteocytes are not provided nutrition by the Haversian system but rather by diffusion of materials through the loosely organized bone. This type of bone is called *spongy bone* or *cancellous bone.* Spongy bone is also found in the ends of long bones and the inside of most flat and irregularly shaped bones.

Bone Marrow

As an embryo develops, cartilage begins to be replaced by bone. Another material, *bone marrow,* is formed in the shafts of long bones and between the layers of flat bones. This material, though not structurally important, is of great importance for other functions.

Two types of bone marrow are recognized, *yellow marrow* and *red marrow.*

Yellow bone marrow, located in the shafts of long bones, serves as a storage reservoir for fats. *Red bone marrow,* located in the ends of some long bones and between the layers of most flat bones, has the critical function of producing new red blood cells. In an adult, the red marrow is the major site of red blood-cell production. Other functions of red marrow include:

1. the formation of some white blood cells, particularly those having phagocytic activity;
2. the formation of *thrombocytes* and *platelets*, which are important factors in blood clotting; and
3. the reclamation of old red blood cells, using some of the reclaimed components to make new red blood cells.

Diagnosing the health of blood-cell forming sites requires a sampling of red bone marrow. Prominent bones with only skin and connective

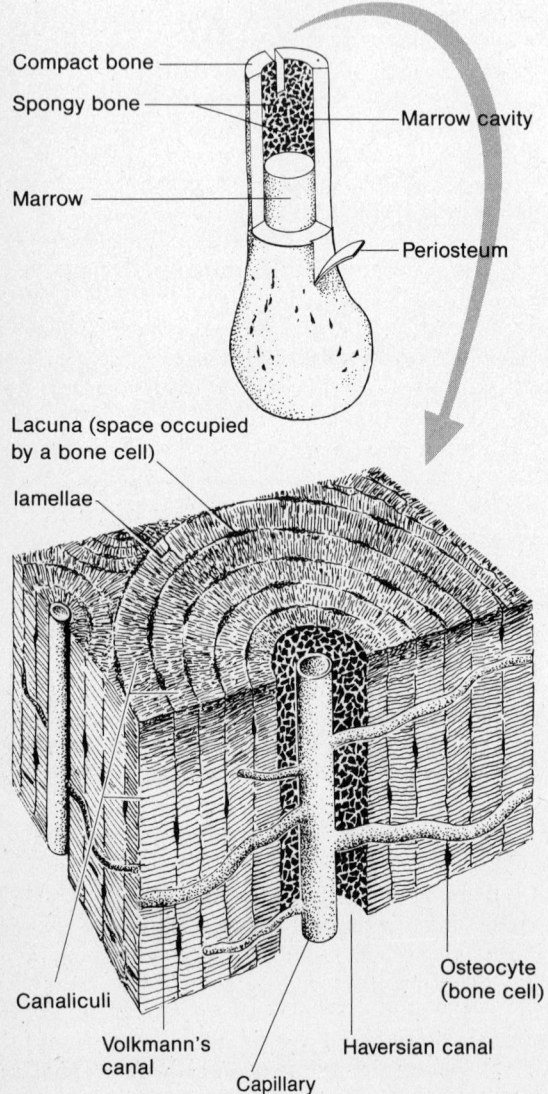

Fig. 5.3 A thin section of compact bone showing the Haversian system—osteocytes, lacunae, lamellae, Haversian and Volkmann's canals.

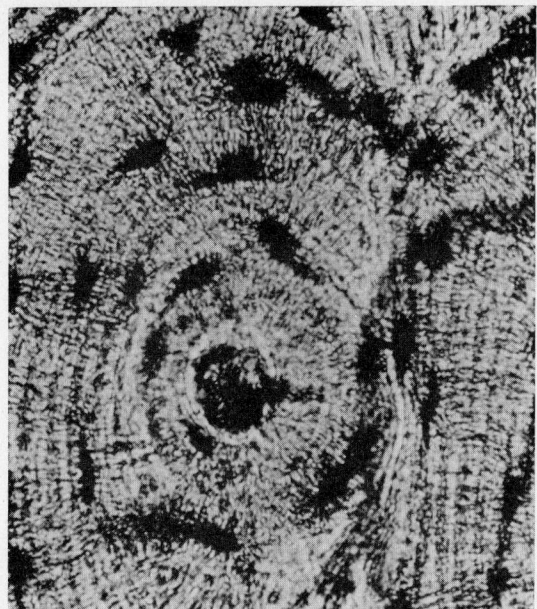

Fig. 5.4 Photomicrograph of a cross section of ground bone. *(Author's collection.)*

tissue covering them, such as the sternum, iliac crest, and the tibia are good locations. A special needle, inserted through the skin into the bone marrow cavity, is used to aspirate a small quantity of bone marrow. Microscopic examination may aid in diagnosis.

Classification of Bones

Bones are generally classified according to their shape. *Long bones* are located in the arms, legs, and fingers. *Short bones*, which are more cube shaped, are found in the wrists and ankles.

Flat bones consist of a thin layer of spongy bone sandwiched between layers of compact bone. The sheetlike structure of flat bones makes them ideal for protecting underlying tissues. For example, the cranium, composed of flat bones, protects the brain. Other flat bones serve as attachments for large muscles. These bones include the scapula, sternum, and pelvis.

Irregular bones are those that do not fit the previous three categories. Such bones usually have projections extending from them, as exemplified by the bones found in the vertebral column.

Bone Markings

Bone markings are identifiable protuberances, indentations, openings, or connecting surfaces. Although it is not essential to know all the details related to the minute markings on all bones, the following terms will help you gain a fuller understanding of the structure and function of bones.

Condyle: A rounded projection located where two bones articulate at the joint. For example, a condyle of the femur articulates with the tibia, and the occipital condyles at the base of the skull articulate with the first neck vertebra.

Crest: A ridge of bone to which muscles are attached. An example is the iliac crest of the hipbone.

Foramen: A hole in a bone that may serve as the entry for blood vessels and nerves. An example is the foramen magnum of the skull through which the spinal cord passes.

Fossa: A shallow depression in the bone. The mandibular fossa, which forms a pocket for the articulation between the skull and the mandibular condyle, is an example.

Meatus: A tube-shaped opening. For example, the external auditory meatus is a canal leading inward toward the middle ear.

Head: The expanded rounded ends of long bones. The femur head, which articulates with the pelvis, is an example.

Sinus: A cavity of spongelike air space in bone, such as the frontal air sinuses located in the eyebrow ridges.

Spine or Spinous Process: A relatively sharp and elongated projection that serves as muscle attachment. The spine of the scapula is an example.

Morphology of Bones

A typical long bone, such as the femur, has several identifiable parts, including the *epiphysis, diaphysis, medullary cavity, endosteum, periosteum,* and *articular cartilage.* These are shown in Figure 5.5. The *epiphyses* are the extreme ends of long bones and are composed of spongy bone. Their relatively large size make them ideal for muscle attachment and articulation with other bones.

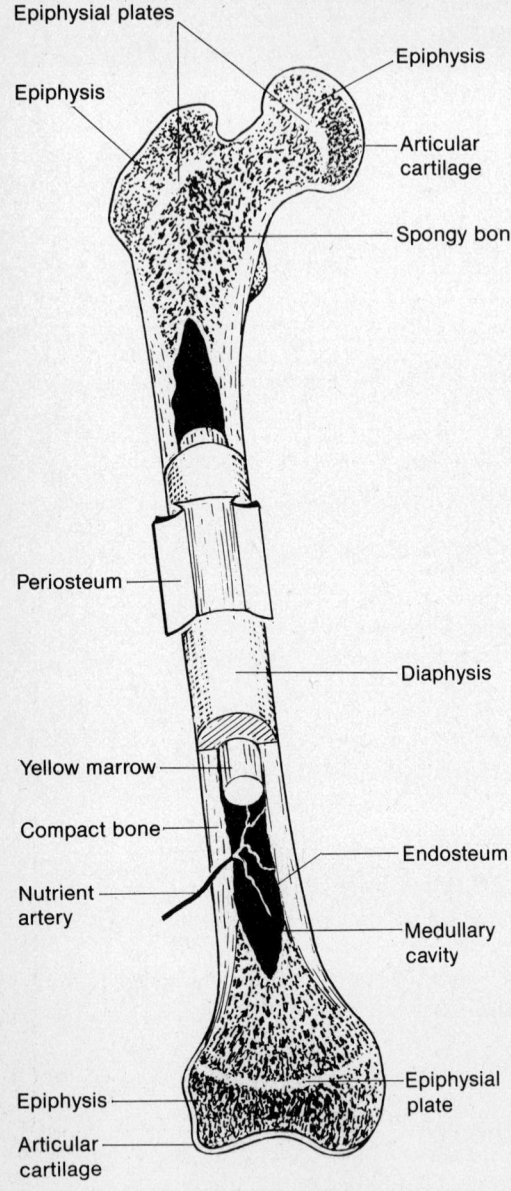

Fig. 5.5 Sagittal section through the femur. The epiphyses are mostly composed of spongy bone, whereas the diaphysis is composed of compact bone. Note the relationship of the periosteum, blood vessels, nerves, and bone tissue.

The *diaphysis* is the shaft of bone between the epiphyses. It is made up largely of compact bone and surrounds the medullary cavity. The *medullary cavity* contains the yellow bone marrow. The *endosteum* lines the walls of the medullary cavity. It contains cells called *osteoblasts*. These specialized cells are bone-forming cells. They are active during normal skeleton growth and are essential for the repair of broken bones.

Another type of specialized cell, the *osteoclast*, is found in the endosteum. Osteoclasts absorb bone, allowing the medullary cavity to enlarge as growth progresses.

The *periosteum* is a tough, fibrous, tight-fitting tissue that contains osteoblasts, which lay down new bone during growth and in the repair of broken bones.

The articular cartilage is hyaline cartilage covering the epiphysis of long bones. It acts as a cushioning surface between jointed bones.

How Bones Grow

In the embryo the first skeleton is one of cartilage. The cartilage serves as a model for later replacement by bone. This replacement process is called *ossification*. Calcium, phosphorous, and organic materials needed for bone formation in the embryo must come from the mother's blood, and during pregnancy there is a severe drain on the calcium and phosphorous deposits in the mother's bones.

As mentioned before, these minerals are constantly being deposited in bones and being reabsorbed into the bloodstream as a normal body function. During pregnancy, however, more minerals are reabsorbed from maternal bones than deposited. Hence, there is a need for the mother to increase her intake of calcium and phosphorous, as well as other nutrients.

Bone formation gets a rather late start in the embryo, beginning around the sixth week of development. Most bones of the body are formed by a gradual replacement of the cartilage model with bone cells and mineral deposits. This type of formation is called *endochondral* (within cartilage). A few of the skull and face bones are formed by the invasion of fibrous membranes by bone cells and minerals. These bones are called *intramembranous* (within membranes). Both formation processes produce bone in an identical manner.

Endochondral bone begins with a cartilage model shaped like the bone it is to become, as shown in Figure 5.6 and in the X ray in Figure 5.7. The cartilage lacunae enlarge and merge, forming larger cavities. In long bones this process begins approximately halfway along the axis of the diaphysis. Blood vessels enter the cartilage cavities bringing with them the bone-forming cells called osteoblasts. The blood also brings calcium, phosphorous, and other bone building components.

The process of producing bone is called *ossification*. The midway point along the diaphysis is called the *initial* or *primary ossification center*. Ossification continues toward each epiphysis. Other ossification centers become established at the epiphyses of long bones, with ossification continuing toward the midpoint of the diaphysis.

Bone laid down in this manner produces spongy or cancellous bone. The epiphyses of long bones remain spongy, while along the diaphysis the initial spongy bone is restructured into compact bone.

Bones lengthen owing to a mass of vigorously growing bone cells between the epiphyses and the diaphysis. This layer, called the *epiphyseal plate* (also called a disc or line), is an area of rapid multiplication of osteoblasts. As the osteoblasts multiply and move outward, cells left behind become active in forming new bone. The epiphyseal plate is present until young adulthood, when bone growth has reached its maximum.

Measure the Length of One Bone and Tell How Tall You Are

A precise relationship exists between the length of certain bones and height. In fact, even if only one bone of a prehistoric human were found, we could still determine almost exactly how tall the individual was. The following gives the living height in inches:
  Male:
    Length of humerus × 2.894 + 27.811
    Length of radius × 3.271 + 33.829
    Length of femur × 1.880 + 32.010
    Length of tibia × 2.376 + 30.970
  Female:
    Length of humerus × 2.754 + 28.140
    Length of radius × 3.343 + 31.978
    Length of femur × 1.945 + 28.679
    Length of tibia × 2.352 + 29.439
  Example: The author's measured radius (male) is 12 inches. 12 in. × 3.271 + 33.829 = 73.081 in. 73.081 ÷ 12 = 6.09 ft = author's height.

Reprinted by permission from Time, The Weekly Newsmagazine; copyright Time, Inc.

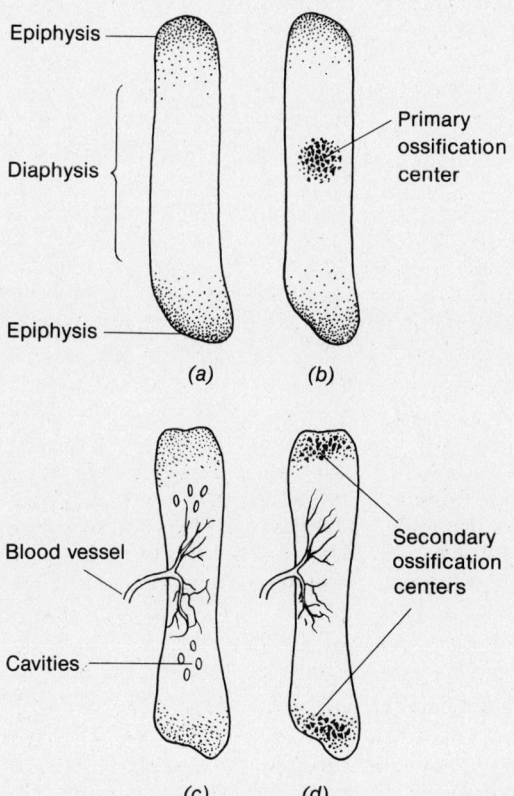

Fig. 5.6 Endochondral bone formation. *(a)* Cartilage model of bone. *(b)* Midway along the diaphysis a primary ossification center develops. *(c)* Cavities form in the new bone. *(d)* Secondary ossification centers develop.

Bones grow in diameter by removing bone deposits from the lining of the medullary cavity (endosteum) and adding new bone to the outside. The specialized cells called osteoclasts dissolve minerals from along the endosteum. These

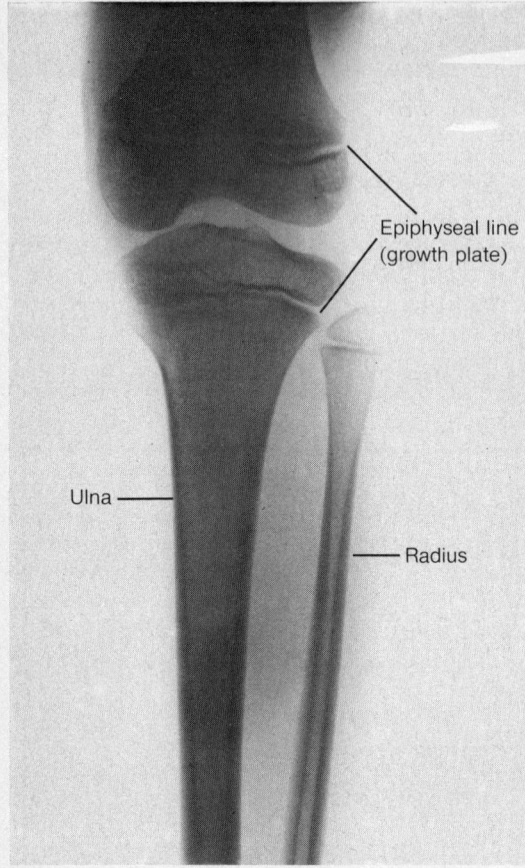

Fig. 5.7 X ray showing growth line (epiphyseal line). *(Author's collection.)*

minerals are picked up by the blood. The periosteum contains great numbers of osteoblasts that may then use reclaimed minerals from the blood to build up the outer surface of the bone. These two processes produce longer and larger diameter bones. Factors that influence the rate and extent of bone growth are discussed later in the chapter.

Intramembranous bone is formed in a manner similar to endochondral except that the ossification process begins in fibrous membranes rather than cartilage.

SKELETAL COMPONENTS

A complete skeleton, as shown in Figures 5.8 and 5.9, is made up of more than 200 bones. Learning the names of the bones in the human body can be simplified by grouping the related bones. The first major division is that of the *axial* and the *appendicular* portions of the skeleton.

The axial skeleton is that portion which forms the axis of the body. This segment includes the head, vertebral column, ribs, and sternum. These bones are shown as the shaded parts in Figure 5.8. There are 80 bones in the axial skeleton.

The appendicular skeleton is made up of all 126 other bones. These include the bones of the arms, legs, shoulder, and pelvic girdles.

Upper Appendicular Skeleton

Bones of the upper appendicular skeleton include:

1. Scapula (shoulder blade)
2. Clavicle (collarbone)
3. Humerus (upper arm bone)
4. Radius (lower arm bone on the thumb side)
5. Ulna (lower arm bone on the little finger side)
6. Carpals (wrist bones)
7. Metacarpals (bones of the palm of the hand)
8. Phalanges (finger bones)
9. Innominate bones (hip bones)

The *scapula*, shown in Figure 5.10, is a large, flat, triangular bone forming part of the shoulder girdle. Its extensive flat surfaces are well suited for attachment of neck and shoulder muscles. The posterior (dorsal) surface of the scapula is divided, by a spine, into two areas—the *supraspinous fossa* above the spine and the *infraspinous fossa* below. The spine projects laterally, forming the *acromion process*, which articulates (joins) with the clavicle (collarbone). Another process, the *coracoid process*, located medial and anterior to the acromion process, serves as an attachment for the tendon of the biceps muscle. Immediately beneath these two processes is the *glenoid fossa*, a depression that articulates with the upper arm bone (humerus).

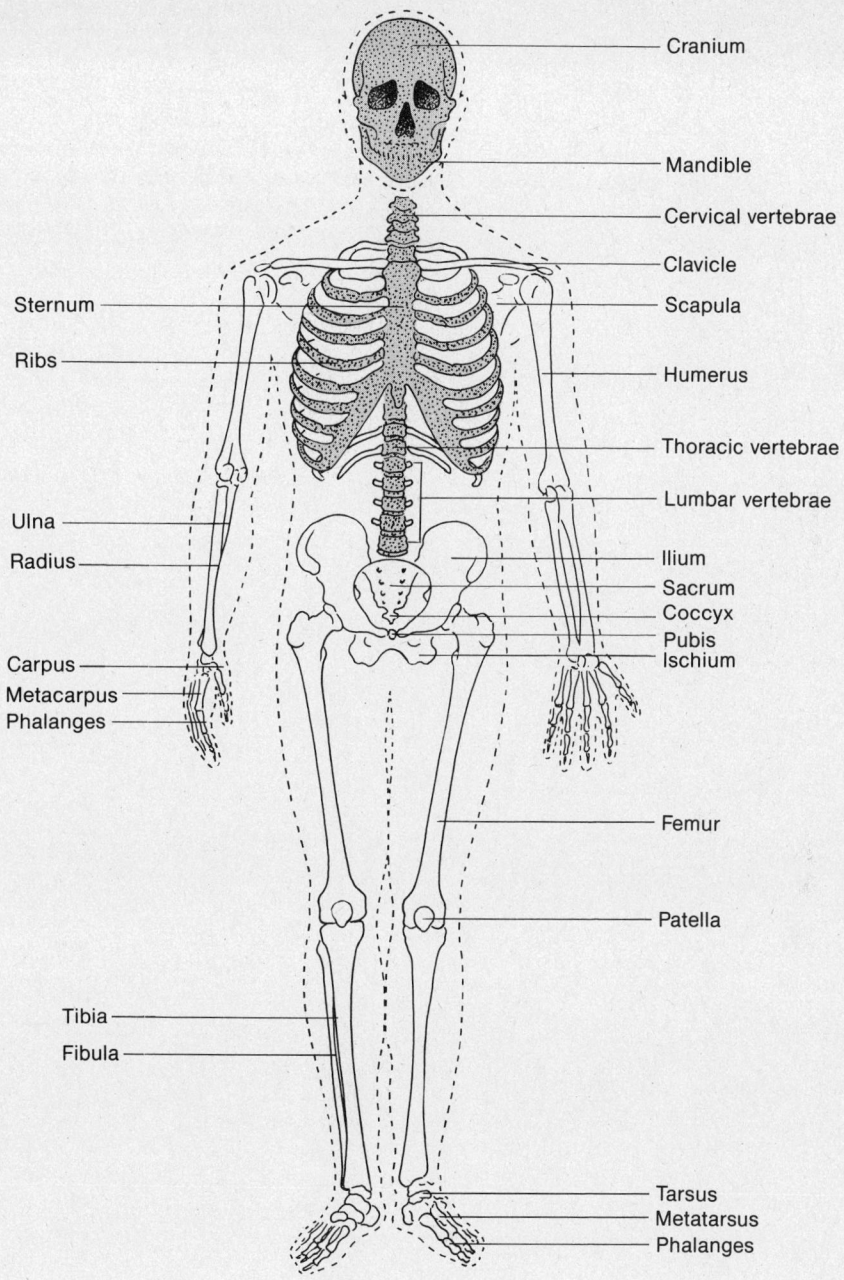

Fig. 5.8 Front view of a complete skeleton. The axial skeleton is that portion which forms the axis of the body. This segment includes the head, vertebral column, ribs, and sternum. The appendicular skeleton includes bones of the arms, legs, shoulder, and pelvic girdles.

THE SKELETAL SYSTEM

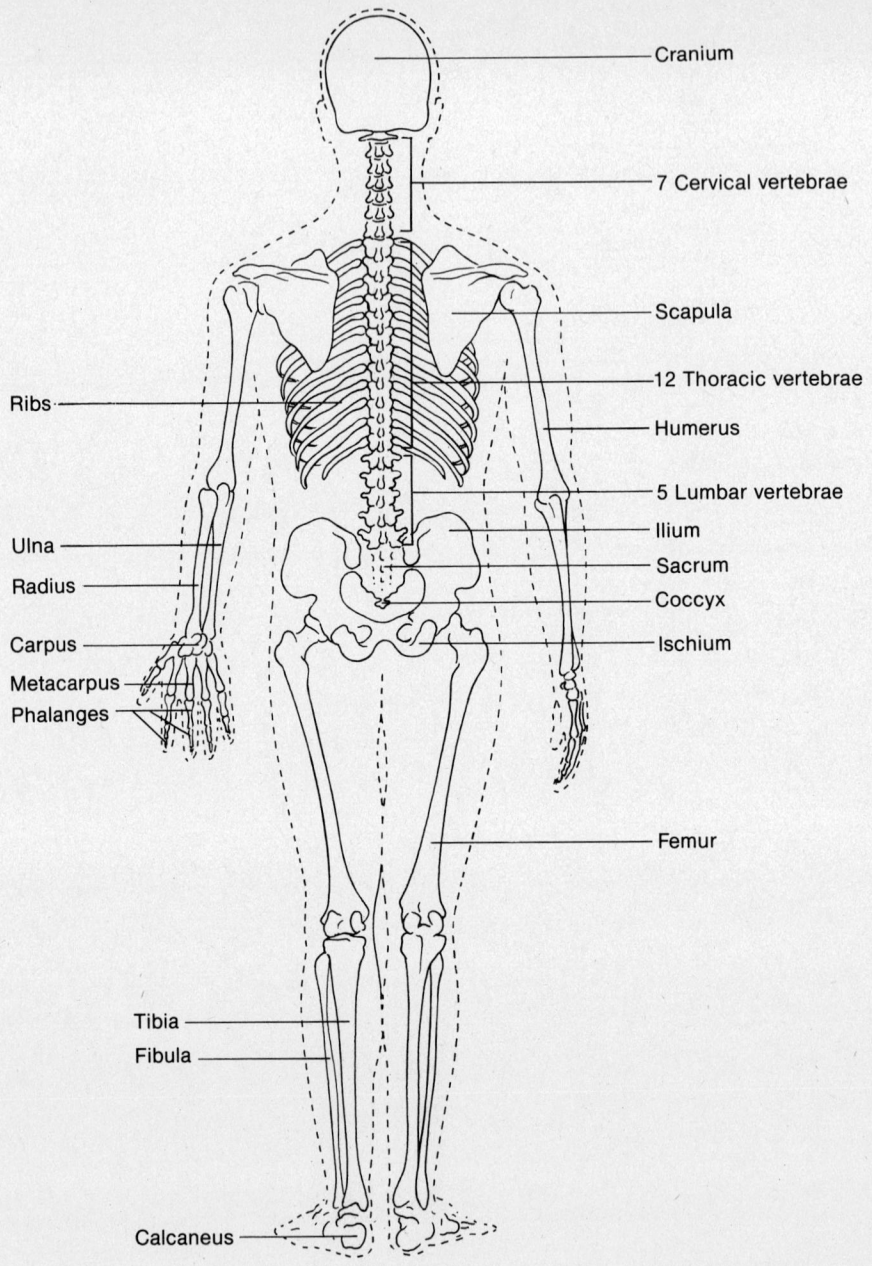

Fig. 5.9 Back view of a complete skeleton.

56 HUMAN FORM AND FUNCTION

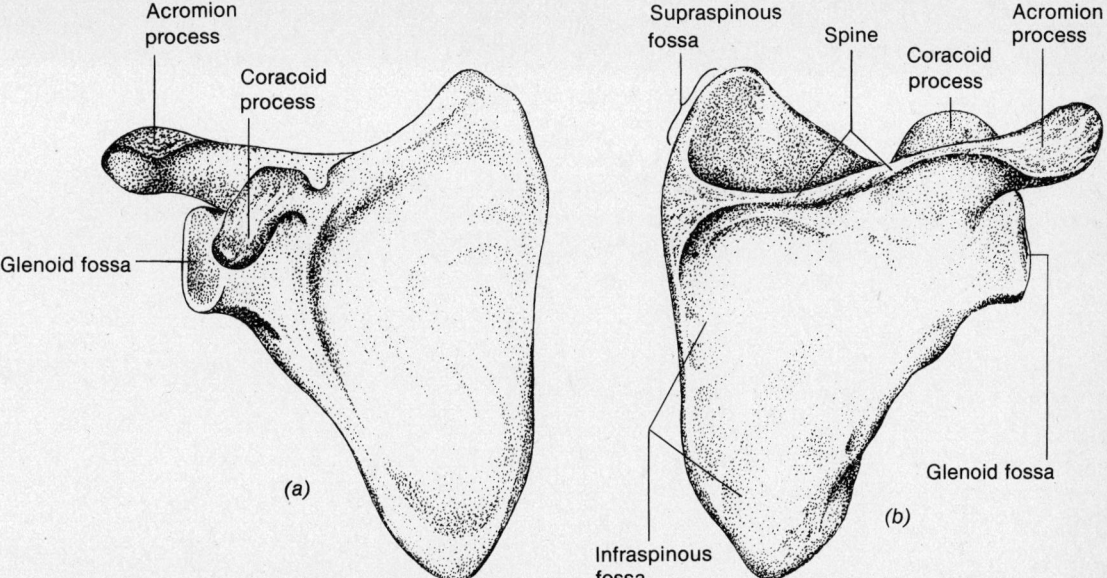

Fig. 5.10 Right scapula. *(a)* Anterior view. *(b)* Posterior view. The scapula, commonly called the shoulder blade, is a large, flat triangular bone forming part of the shoulder girdle.

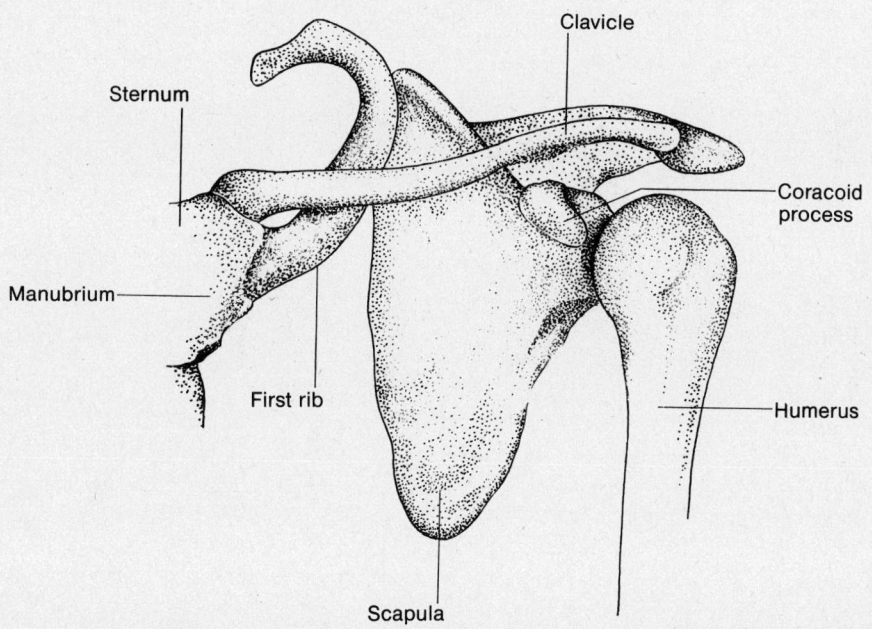

Fig. 5.11 The clavicle or collarbone, showing how the clavicle articulates with the shoulder and sternum (breastbone).

THE SKELETAL SYSTEM 57

The *clavicle* (collarbone), shown in Figure 5.11, is a short, tubular bone articulating with the acromion process of the scapula and the *manubrium* of the sternum. The function of the clavicle is to brace the shoulder. The clavicle is the only bony attachment between the shoulder bones and the axial skeleton. Fractures of the clavicle are common due to falling on the shoulder.

The *humerus,* shown in Figure 5.12, is the upper arm bone articulating with the glenoid fossa of the scapula at its proximal end and distally with the radius and ulna. The head of the humerus, shaped like a ball, forms a *ball-and-socket joint* with the scapular glenoid fossa. Immediately distal to the head is the surgical neck. This is a common fracture area. The distal end of the humerus has two prominences—the *trochlea,* which articulates with the ulna, and the *capitulum,* which articulates with the radius. The *olecranon fossa,* also on the distal end, is a depression that accepts the *olecranon process* of the ulna.

The *ulna* shown in Figure 5.13, is the larger of the two lower arm bones. It articulates with the humerus proximally and with the wrist bones (carpals) distally. The olecranon process, located on the proximal end of the ulna, articulates with the trochlea of the humerus. When the lower arm is extended, the olecranon process fits into the olecranon fossa on the posterior surface of the humerus. This articulation, shown in the X ray in Figure 5.14, prevents the overextension of the lower arm.

The *radius,* also shown in Figure 5.13, the smaller lower arm bone, forms a pivot joint with the humerus at its proximal end. Its distal end forms a slightly movable joint with the ulna. This joint is tightly bound by an *annular ligament* that holds the radius and ulna together, yet permits the hand to be rotated. Moving the hand so that the palm faces forward from the anatomical position is called *supination.* Turning the back of the hand forward is called *pronation.* The rounded proximal end of the radius is called the *head.* Immediately posterior to the head is a

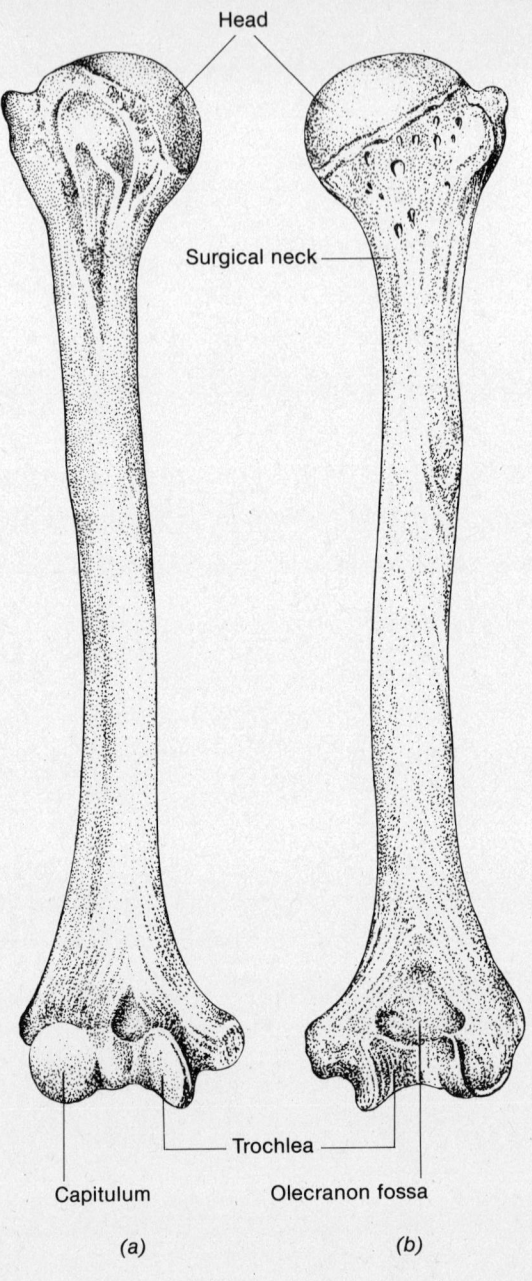

Fig. 5.12 The right humerus. *(a)* Anterior view. *(b)* Posterior view. This bone articulates with the glenoid fossa of the scapula at its proximal end. Its distal articulation is with the radius and ulna.

58 HUMAN FORM AND FUNCTION

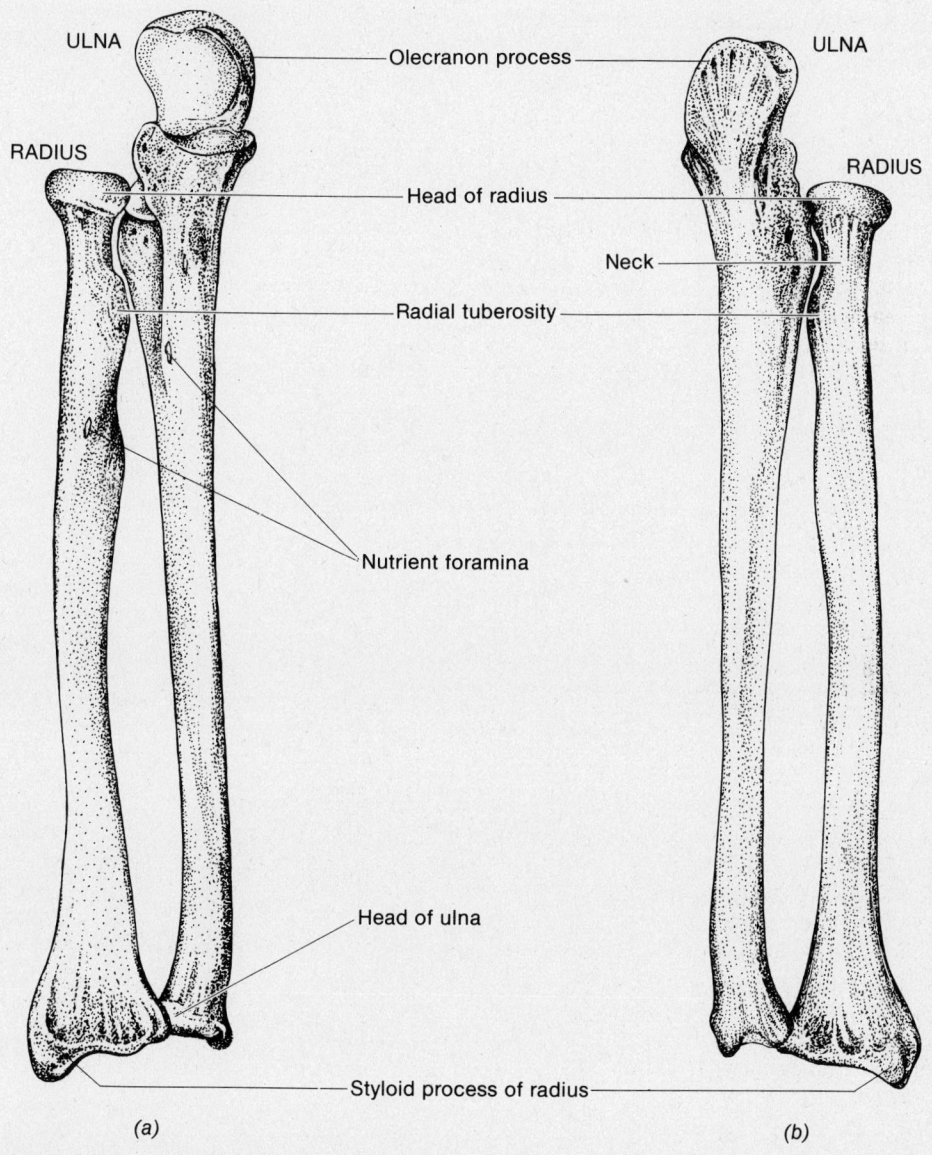

Fig. 5.13 Lower arm bones showing (a) right radius and ulna anterior surface; and (b) right radius and ulna posterior surface.

THE SKELETAL SYSTEM 59

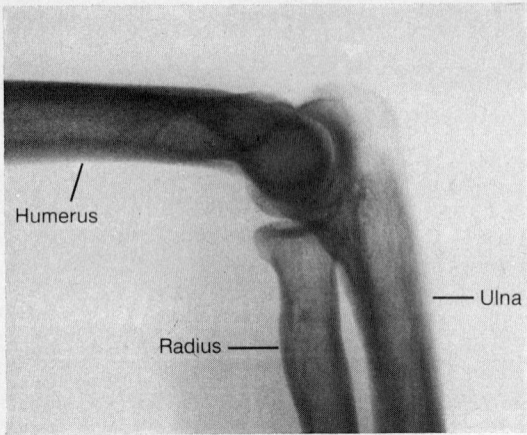

Figure 5.14 X ray illustrating the articulation of the humerus and the radius and ulna at the elbow. *(Courtesy Radiology Department, Normandy Osteopathic Hospital, St. Louis, Mo.)*

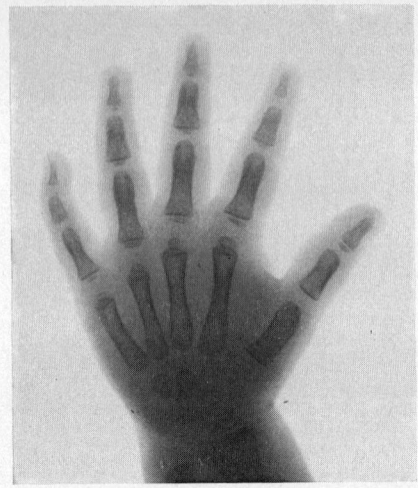

Fig. 5.16 X ray of an infant's hand. Note the incomplete ossification of the hand and wrist bones. *(Author's collection.)*

slender part, the neck. The *radial tuberosity* is a raised projection that serves for muscle attachment. Bone cells require nutrition and receive it from blood vessels, which enter the bone through openings called *nutrient foramina*. These holes also admit nerves.

Carpals (wrist bones) are a set of eight irregular-shaped bones, as shown in Figure 5.15. Figure 5.16 shows an X ray of these bones in an infant's hand. The arrangement of these bones permits great flexibility of hand movement. The wrist bones are arranged in two rows of four bones. The proximal row, beginning on the thumb side, includes the *trapezium, trapezoid, capitate,* and *hamate*. The distal row includes the *scaphoid, lunate, triangular,* and *pisiform*.

Fractures of the wrist bones in the majority of instances involve the scaphoid bone.

The *hand* bones include those of the palm *(metacarpals)* and fingers *(phalanges)*. Metacarpals are numbered one through five beginning with the thumb.

There are 14 phalanges per hand. The thumb has two phalanges, the other fingers have three each.

Lower Appendicular Skeleton

Bones of the lower appendicular skeleton include:

1. Innominate (hipbone, which helps to form the basinlike pelvic cavity)
2. Femur (the upper leg bone between the hip and the knee)
3. Patella (flattened triangular kneecap)
4. Tibia (the larger of the lower leg bones, sometimes called the shinbone)
5. Fibula (smaller of the two lower leg bones)
6. Tarsals (anklebones)
7. Metatarsals (bones of the foot and instep)
8. Phalanges (bones of the toes)

The *innominate* (hipbone) forms the pelvic girdle, shown in Figure 5.17 In early fetal life, the pelvic girdle, as shown in Figure 5.18, is composed of three separate bones—the upper *ilium,* the lower *ischium,* and the anterior *pubis.* Before birth, these three bones fuse, forming the basinlike *pelvis.* The ilium joins with the sacrum to form the *sacroiliac joint,* a fibrocartilage articulation. At the fused intersection of the three bones is a large depression, the *acetabulum,* into which the head of the femur fits to form a ball-and-socket joint.

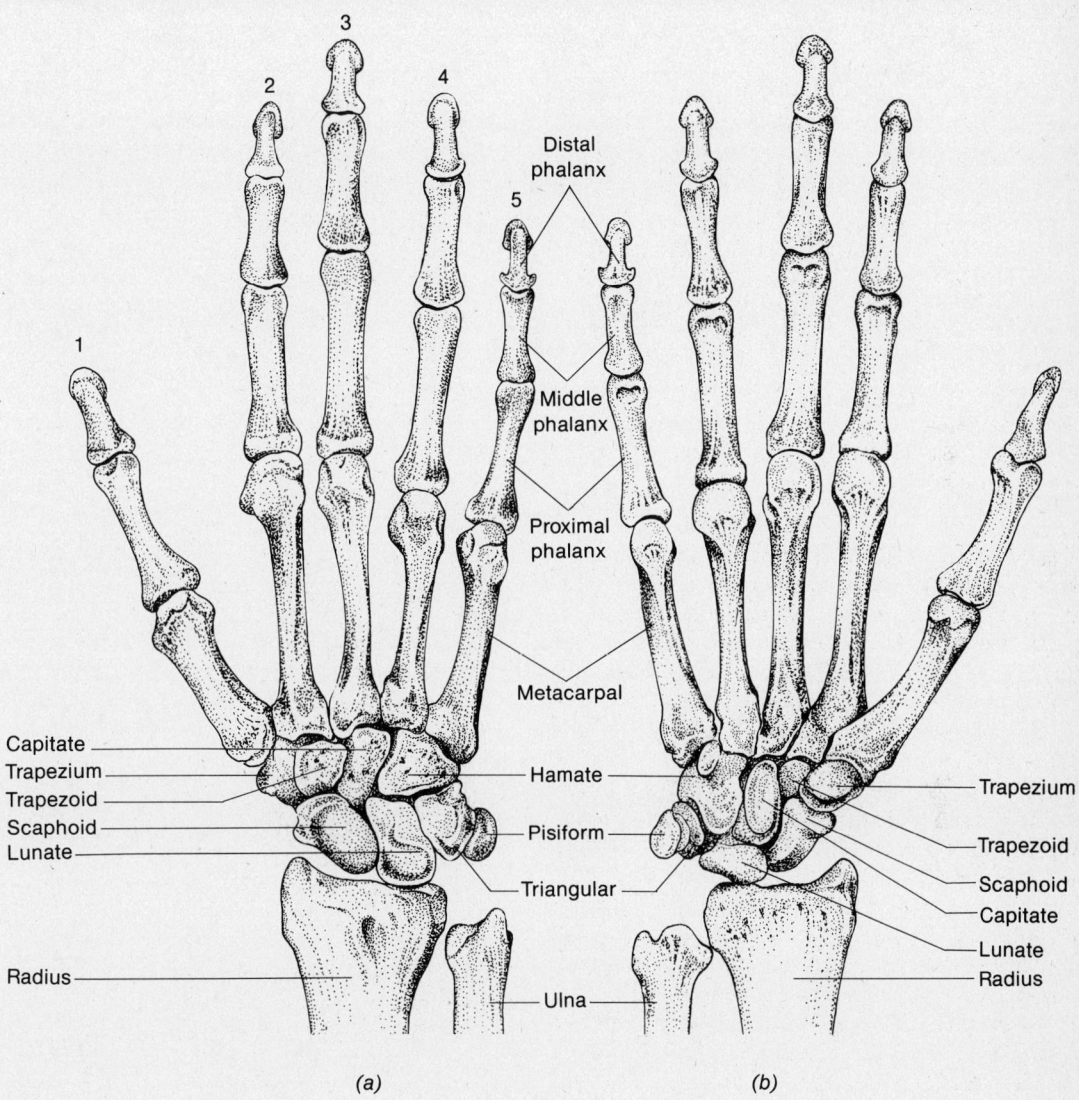

Fig. 5.15 Bones of the wrist and hand. (a) Right hand palm down. (b) Right hand palm up. The arrangement of these bones permits great flexibility of movement.

THE SKELETAL SYSTEM 61

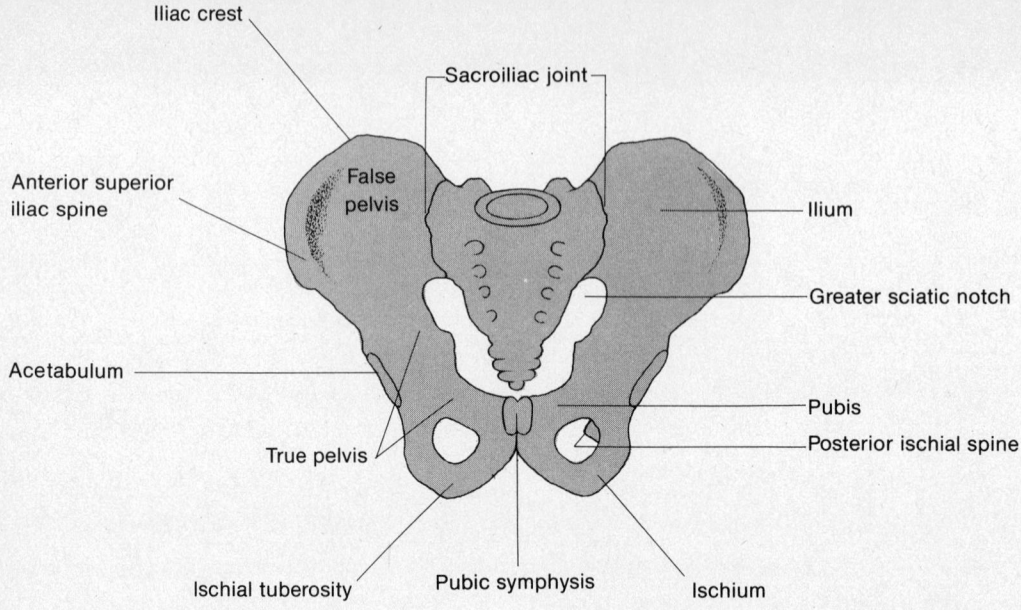

Fig. 5.17 The pelvic girdle. This series of bones forms the structural link between the lower axial skeleton and the legs. The sacroiliac joint is a slightly movable joint between the pelvis and spine.

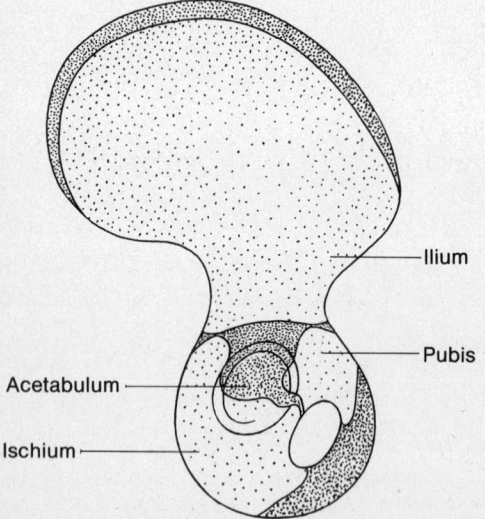

Fig. 5.18 Part of a human pelvic girdle at birth, showing the three embryonic bones that make up the girdle. The shaded area is cartilage that will later ossify.

The winglike flare of the ilium is called the *false pelvis,* while the lower ilium, upper ischium, and the pubis constitute the *true pelvis.* The two pubic bones are joined anteriorly at the *pubic symphysis.* This is a fibrocartilage joint of considerable importance in the female skeleton during childbirth.

A major difference in male and female skeletons is the size and structure of the pelvic bones, as compared in Figure 5.19. The female pelvis is broader and more flattened, with a larger and more circular bony ring *(pelvic canal).* During birth the infant's body must pass through this canal. The pregnant woman's pelvis is measured and in some instances X rays are used to determine if the canal is large enough to permit the passage of the infant's skull.

The *femur,* shown in Figure 5.20, is the longest and strongest bone in the body. It is the upper leg bone articulating at its proximal end with the pelvic acetabulum and at its distal end with the

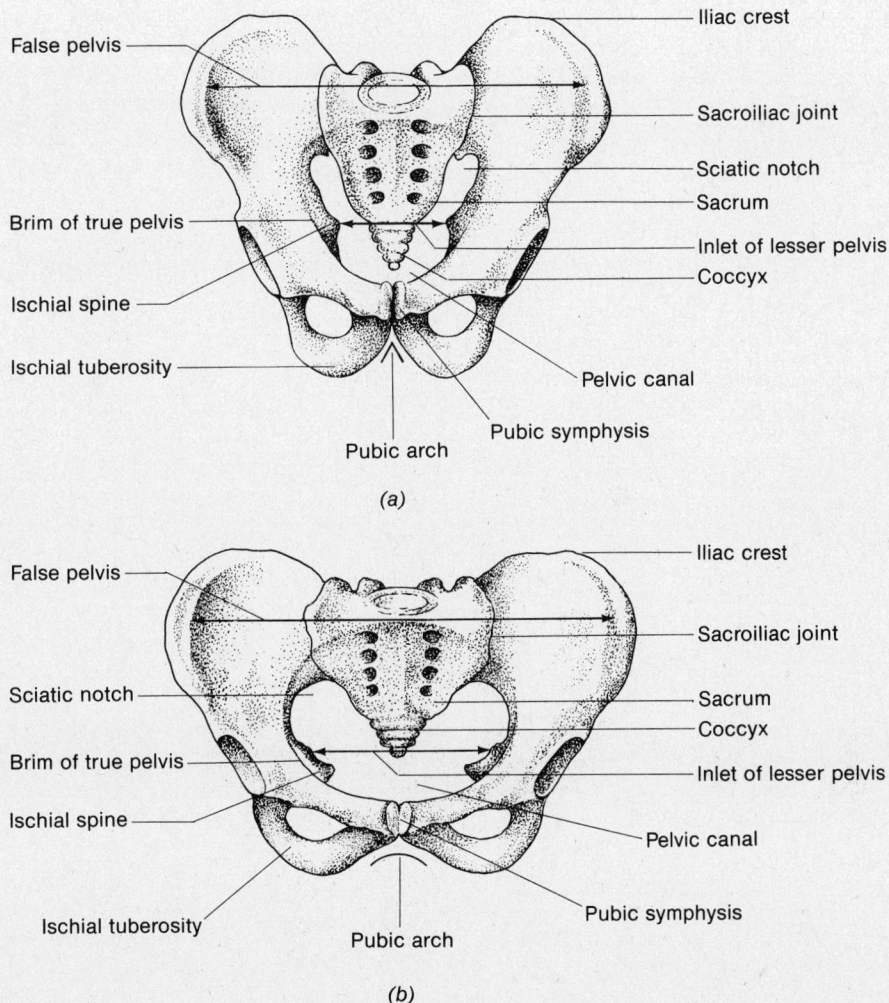

Fig. 5.19 Comparison of (a) male and (b) female pelvic girdles. The female pelvis is broader and more flattened with a more circular pelvic canal.

lower leg bones. The femur has two distinct projections at either end. At the proximal end, beneath the head are located two projections, the *greater* and *lesser trochanters.* Both serve as attachments for muscles. The two projections at the distal end of the femur are the *lateral* and *medial condyles,* which articulate with the tibia.

The *patella* (kneecap) is a small, triangular-shaped bone located just anterior to the knee joint. The patella does not articulate with another bone. It is buried in tendons of leg muscles and serves to protect the knee joint. The patella is called a *sesamoid bone,* from the name given to bony or cartilagenous nodules developing in tendons.

The *tibia* or shinbone, shown in Figure 5.21, is the larger of the two lower leg bones. It is also more superficially located, with only connective tissue and skin covering its anterior edge for most of its entire length. Its proximal articulation

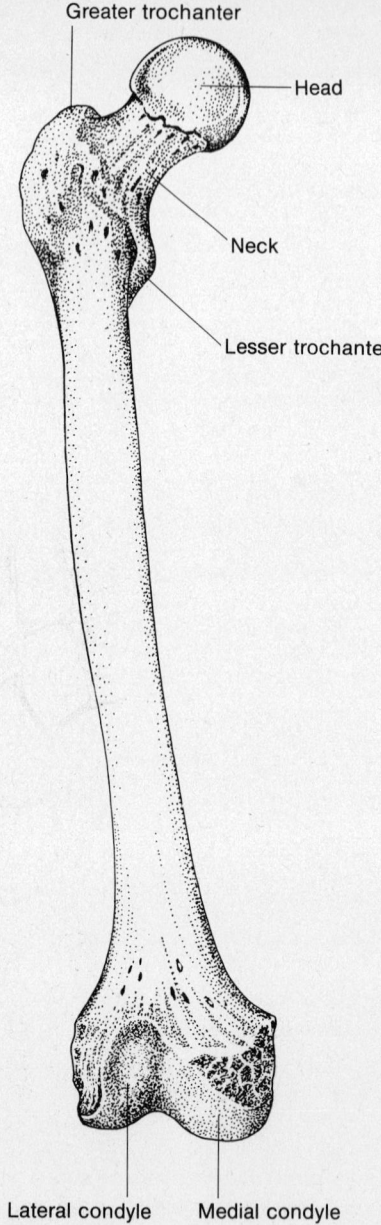

Fig. 5.20 Right femur, anterior view. This is the longest bone in the body. Its upper end articulates with the pelvis (acetabulum) and its lower end articulates with the tibia and fibula of the lower leg.

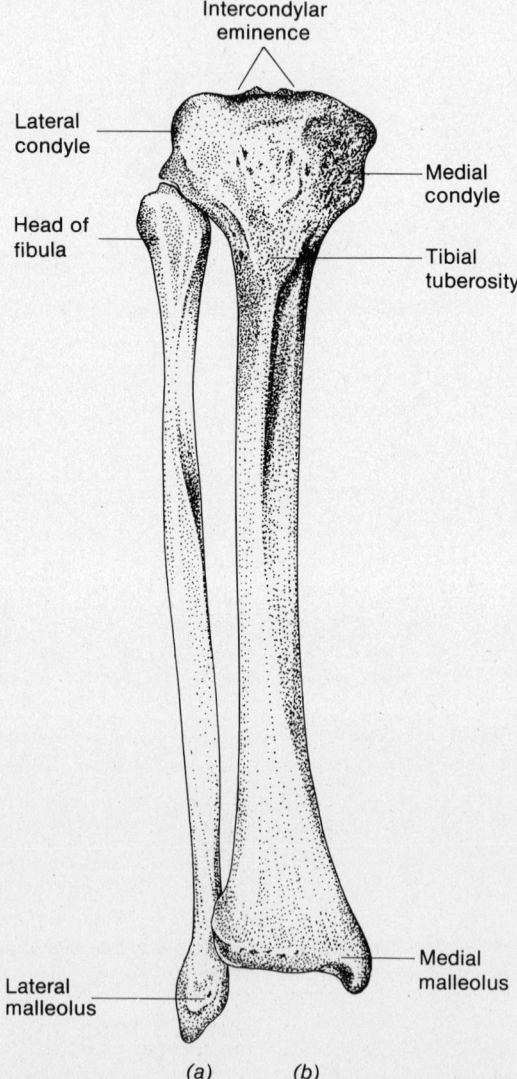

Fig. 5.21 Right *(a)* fibula and *(b)* tibia, anterior view. The tibia or shin bone is the larger of the two lower leg bones. It is also more superficially located with only connective tissue and skin covering its anterior edge for most of its length.

is with the condyles of the femur; distally it joins with one of the anklebones, the *talus*.

The *fibula*, shown in Figure 5.21, is the smaller of the two lower leg bones. Its proximal end articulates with the tibia. The entire bone, with the exception of a proximal prominence (the *lateral malleolus*), is encased in muscle. The X rays in Figure 5.22 compare the knee joint of a youth with that of an older individual.

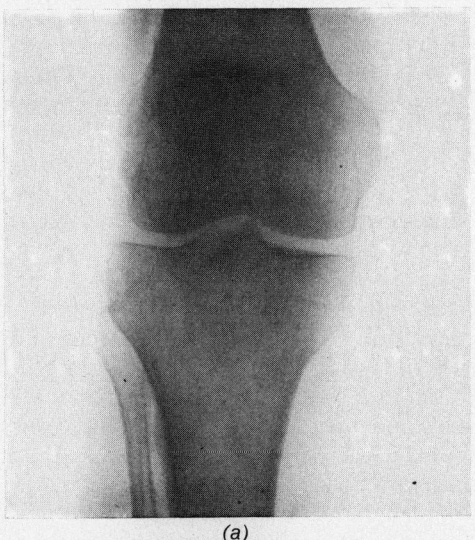

(a)

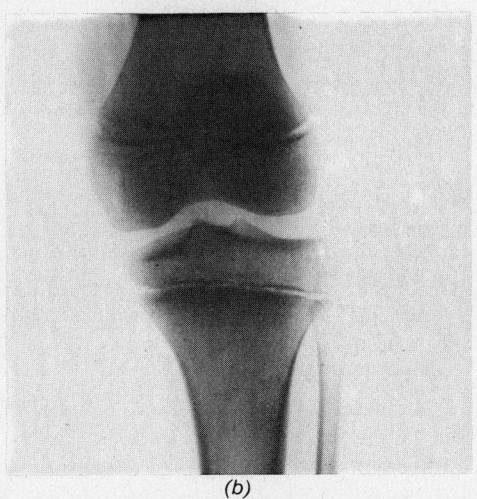

(b)

Fig. 5.22 The knee joint of a 65-year-old man (a) and (b) that of a 16-year-old male. Note the prominent epiphyseal lines (growth lines) in the younger person and the absence of these lines in the older one. (*Author's collection.*)

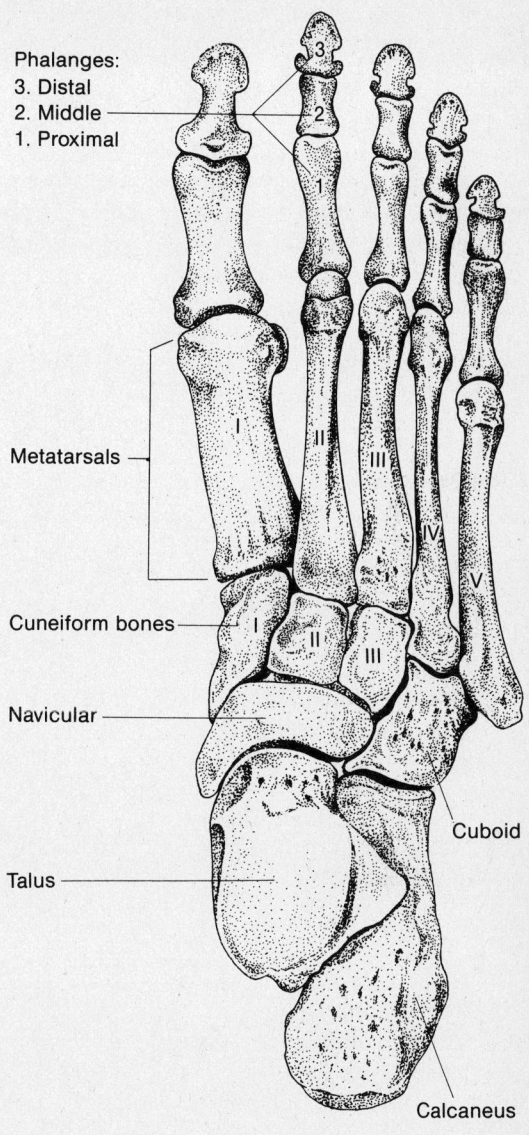

Fig. 5.23 Foot and ankle bones, top view. The ankle bones appear to be an ungainly, random arrangement of irregularly shaped bones. Functionally, these seven bones are precisely arranged to support the full weight of the body. In addition, they maintain adequate flexibility for walking, running, and jumping.

THE SKELETAL SYSTEM

Tarsals, or anklebones, as shown in Figure 5.23, appear to be an ungainly arrangement of irregularly shaped bones, but functionally the seven anklebones are well arranged to support the full body weight and maintain adequate flexibility for walking, running, and so on. The *talus* is the only anklebone that articulates with the tibia. The *calcaneus* shapes the heel of the foot. The *cuboid* articulates with the calcaneus and with the fifth and fourth metatarsals. The first, second, and third *cuneiform bones* join with the first, second, and third *metatarsals* and with the *navicular.* Each digit (toe) is made up of three phalanges except for the great (first) toe, making a total of 14 phalanges (the same number as is found in the hand).

Axial Skeleton
The axial skeleton, which contains all other bones in the body, may be divided into the *skull, vertebral column, sternum,* and *ribs.*

The Skull The cranium and face bones make up the skull. The cranium bones include:

1. Frontal (forehead bone)
2. Parietal (sides of the cranium)
3. Occipital (lower back portion of the cranium)
4. Temporal (beneath the parietal bones)
5. Sphenoid (beneath the junction of the parietal and frontal bones and anterior to the temporal bones)

The *frontal* bone makes up the forehead of the skull (the brow) and forms the upper part of the orbital (eye) cavity, as shown in Figure 5.24. The frontal bone is joined posteriorly with the two *parietal* bones and laterally with the *sphenoid* and *zygomatic* bones. The parietal bones form the top and sides of the skull and articulate with the *sphenoid* and *temporal* bones on the sides of the skull and with the *occipital* bone at the back of the skull.

The *zygomatic process* of the temporal bone articulates with the zygomatic bone. The ring of bone formed by this articulation forms the *zygomatic arch.*

Beneath the zygomatic process of the temporal bone is the *external auditory meatus,* a bony canal leading into the skull toward the middle ear. Immediately below the external auditory meatus is the *mastoid process* of the temporal bone. This process houses a large sinus. The mucous lining of this sinus may become inflamed causing a condition called *mastoiditis.*

Fontanels are soft spots in an infant's skull at the junction of skull bones, as shown in Figure 5.25. These open areas gradually fill in with bone, normally not entirely closing, however, until the brain has reached its maximum size. Occasionally the fontanels close prematurely, before the brain has reached its maximum size. This condition is called *microcephaly.* Excessive pressure is placed upon brain tissues, usually producing brain damage and mental retardation.

Hydrocephaly results when excessive fluid is secreted into the brain ventricles and other brain tissues. This keeps the fontanels open longer. However, pressure from the growing brain, along with the excessive fluids, may bring about brain damage and mental retardation. The excess fluid can be drained periodically to alleviate the condition.

Skull bones are joined by *synarthrotic* (immovable) *joints* called *sutures,* shown in Figure 5.24. The *coronal suture* joins the frontal bone with the parietal, sphenoid, and zygomatic bones, and the maxilla of the face. The *sagittal suture* joins the two parietal bones along the top of the skull. The *lambdoidal suture* joins the two parietal bones with the occipital bone at the back of the skull. The *squamous suture* joins the parietal bones on their inferiors, and their lateral edges with the temporal and sphenoid bones.

The Face Bones These bones give shape to the face and help form the oral, nasal, and orbital cavities, as shown in Figure 5.24. Bones of the face include the following:

1. Nasal (forms the ridge of the nose)
2. Maxilla (upper jaw, helps form the mouth, nose, and eye cavities)
3. Mandible (lower jaw)
4. Zygomatic arch (malar or cheek bone)

Nasal bones, besides forming the ridge of the nose, are also attached to cartilage, which forms the major portion of the nose. (Cartilage, being

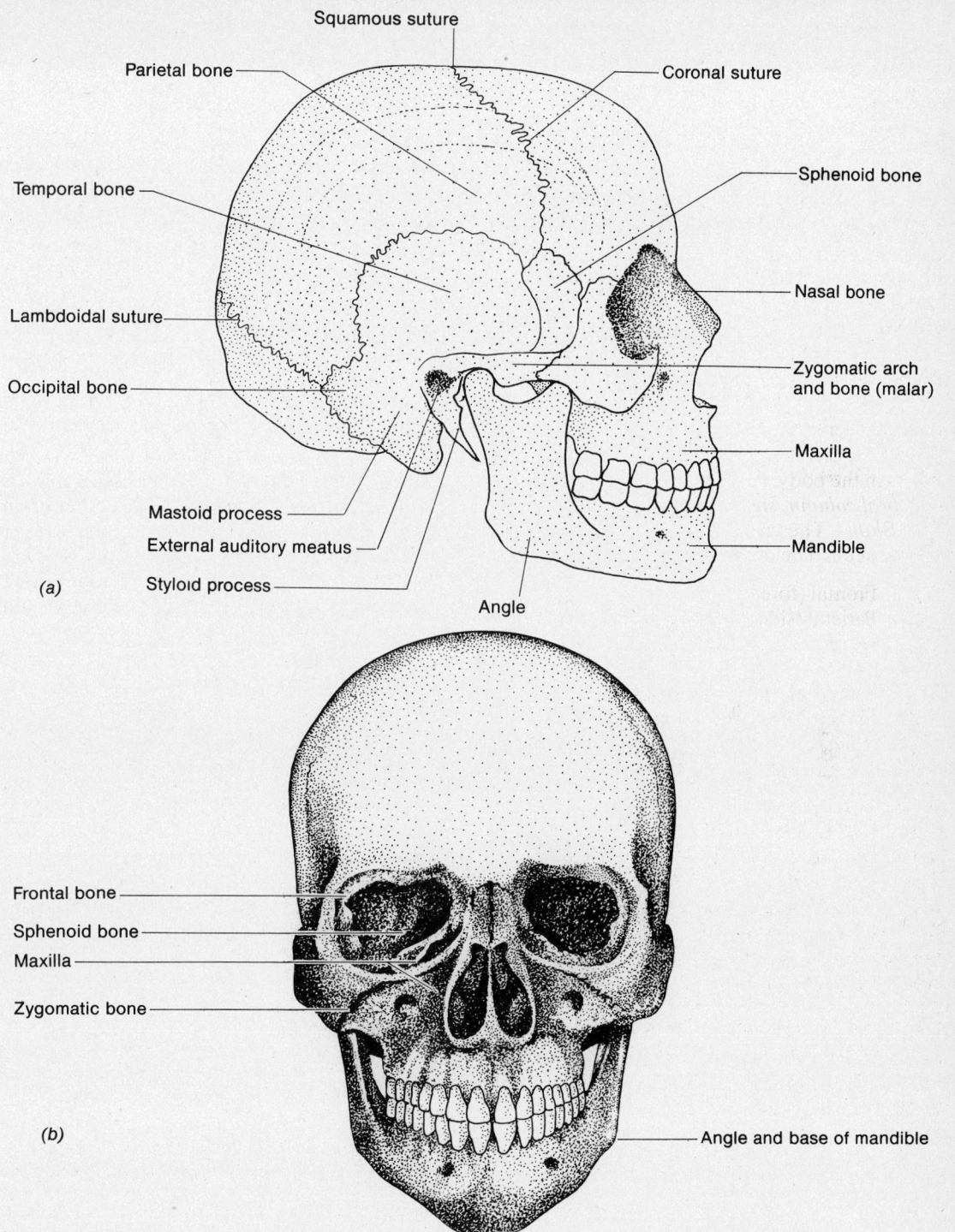

Fig. 5.24 Bones of the skull. (a) Lateral view. (b) Anterior view.

THE SKELETAL SYSTEM 67

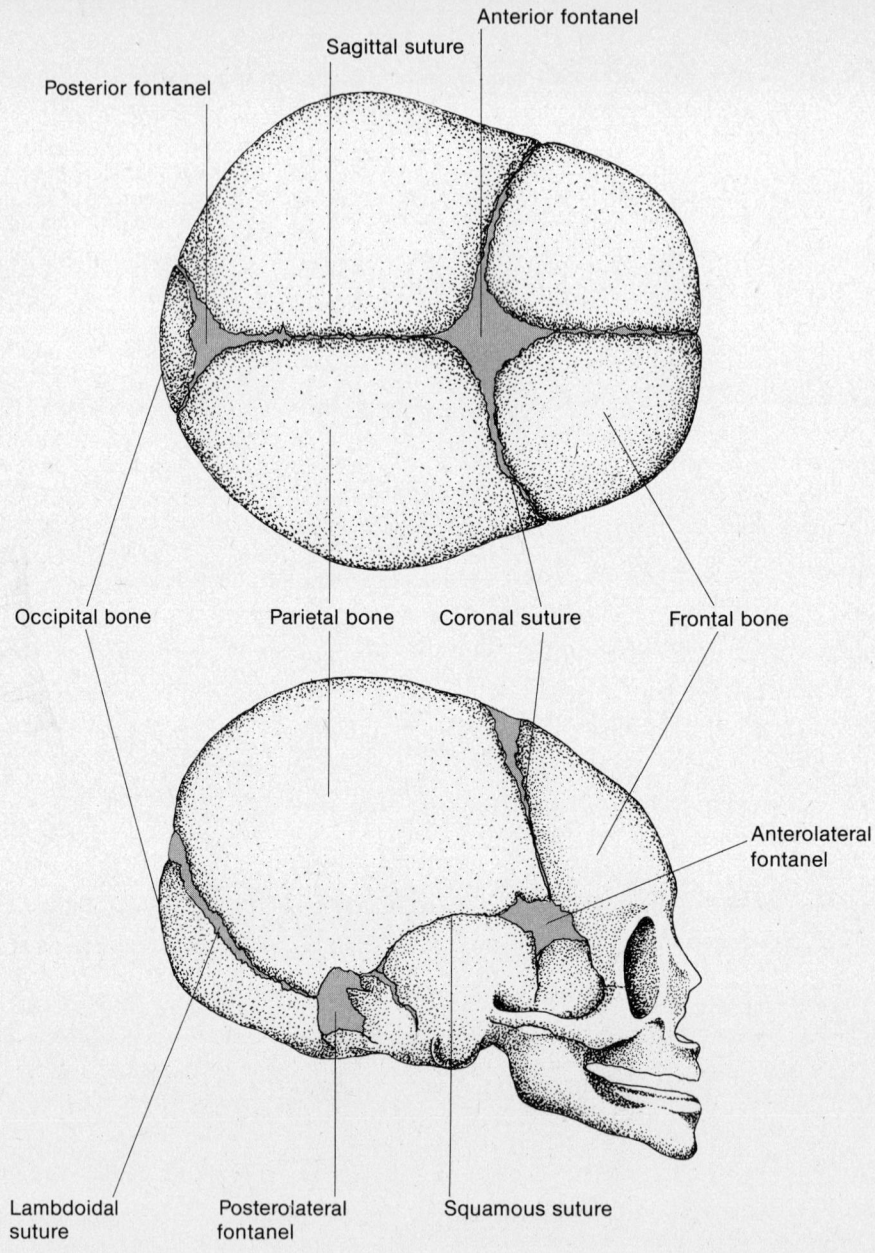

Fig. 5.25 Infant skull, top and lateral views, showing fontanels. These open areas gradually fill in with bone, however, normally not entirely closing until the brain has reached its maximum growth.

68 HUMAN FORM AND FUNCTION

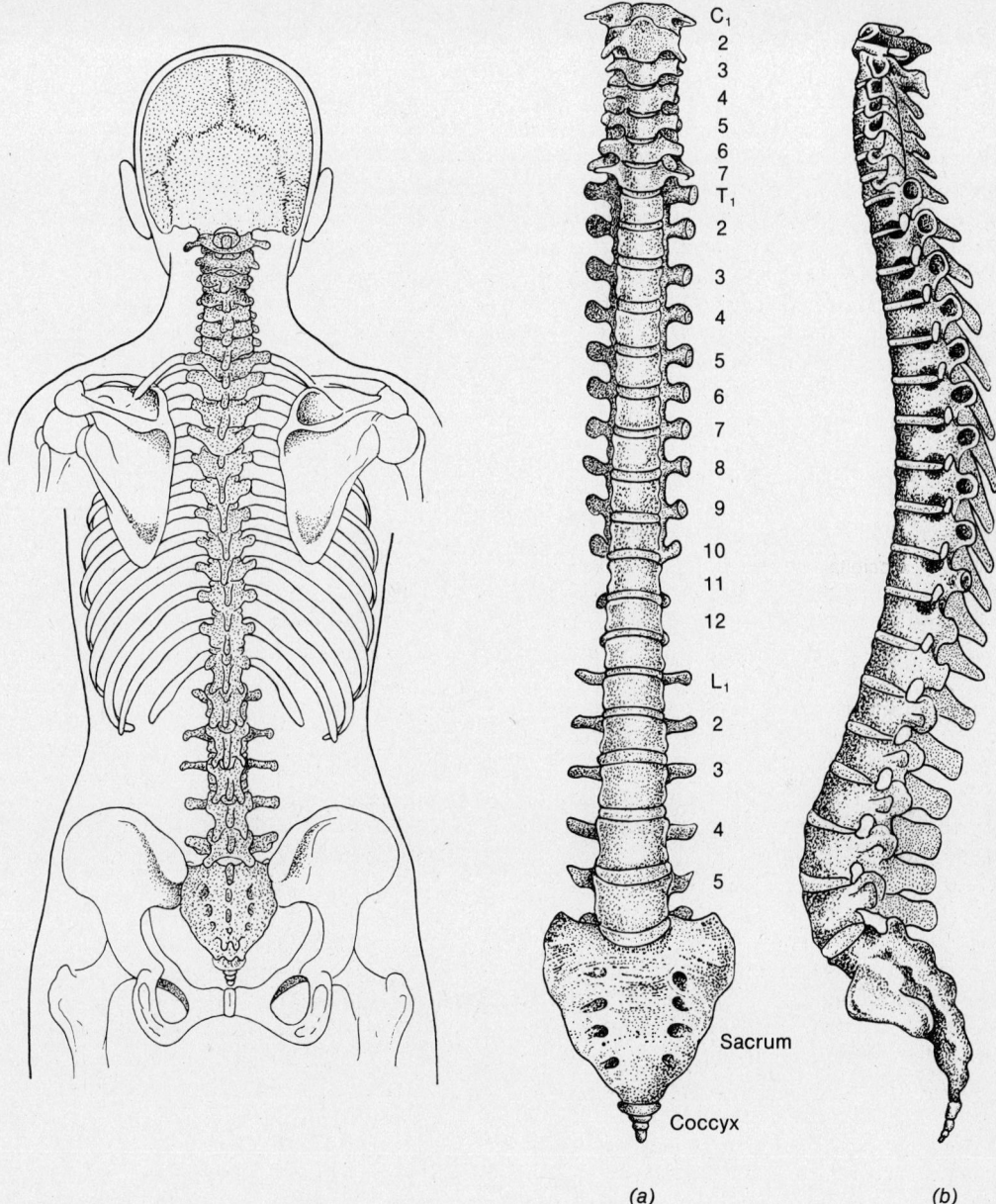

Fig. 5.26 (a) The vertebral column. (b) Posterior view. (c) Lateral view. The vertebrae numbered 1 through 7 are the cervicle vertebrae. Those numbered 8 through 12 are the thoracic vertebrae. Those numbered L_1 through L_5 are the lumbar vertebrae. The last vertebral bone, the sacrum, is a single bone composed of several fused bones. The tip end of the sacrum is called the coccyx.

THE SKELETAL SYSTEM

softer tissue that cannot be preserved, is missing on classroom skeletons.)

The *maxilla* is the upper jaw bone containing the upper teeth. It articulates with the sphenoid, zygomatic, temporal, nasal, and frontal bones.

The *mandible* is the lower jaw containing the lower set of teeth. Its *systoid process* articulates with and forms a movable joint with the temporal zygomatic process just anterior to the external auditory meatus. A large portion of the mandible is composed of flat bone surface, providing an attachment surface for the strong chewing muscles.

The *zygomatic bone,* or *malar,* forms the cheek bone. It articulates posteriorly with the temporal bone and anteriorly with the maxilla and frontal bones.

The Vertebral Column Twenty-six bones make up the vertebral column, illustrated in Figure 5.26. The arrangement of these 26 bones provides a support for the body, encloses and protects the spinal cord, serves as an articulation for ribs and as an attachment for muscles.

Rather than giving specific names to each particular vertebra, they are numbered, beginning with number one just beneath the skull. The first seven vertebra are called the *neck* or *cervical vertebra.* The first cervical vertebra is called the *atlas.* The second vertebra is called the *axis.* These two bones form a pivotal joint allowing the head to turn.

The next 12 vertebrae are *thoracic vertebrae.* Twelve pairs of ribs are attached to these vertebrae.

The next five vertebrae are called the *lumbar vertebrae.* Immediately below the fifth lumbar vertebra is the *sacrum.* This bone is composed of several other bones that in the adult are fused together. The last bone, also a series of fused bones in the adult, is called a *coccyx.*

The spinal column at birth has only a slight concave curvature anteriorly. A *cervical curvature* develops as the child learns to raise his head. Later, as the child learns to stand and walk, a *lumbar curvature* develops.

The Sternum and Ribs The *sternum* is the breastbone, to which ribs are attached, as shown in Figure 5.27. The sternum is made up of the *manubrium,* to which the first rib is attached, the body to which all the other true ribs are attached, and the *xiphoid process* is the posterior tip of the sternum. Ribs are attached to the sternum by *costal cartilage.* The *intercostal space* is where *intercostal muscles* are located that aid in breathing movement. True ribs are the upper seven pairs of ribs fastened to the sternum by cartilage. False ribs are the lower five pairs of ribs, which do not attach directly to the sternum. In fact, the last two pairs do not attach to the sternum at all, and therefore are called *floating ribs.*

Other bones not normally included in the previous groupings include the ear bones and the hyoid bone.

Ear bones include the *malleus* (hammer), the *incus* (anvil), and the *stapes* (stirrup). One other bone that does not lend itself to classification as either appendicular or axial is the *hyoid bone.* This U-shaped bone is located between the mandible and the upper part of the larynx. It does not form a joint with any other bone, being held in place solely by ligaments. Several muscles, including the tongue, are attached to the hyoid bone.

Effects of Hormones and Vitamins on Bone Growth

Bone physiology is influenced directly by the hormones of the *parathyroid* and *thyroid glands. Growth hormones* from the anterior pituitary have an indirect influence.

Vitamin D helps to increase the absorption of calcium from the small intestine into the blood.

Parathyroid Hormone In forming new bone, osteoblasts become osteocytes (mature bone cells). It is believed that parathyroid hormone acts upon osteoblasts and osteocytes, causing them to convert to *osteoclasts.* Osteoclasts, located in the endosteum, function in absorbing bone from the medullary cavity, increasing its diameter. In some manner, osteoclasts are capable of converting insoluble calcium and phosphorus of normal bone into a soluble form that may then be reabsorbed into the bloodstream and extracellular fluids. As greater numbers of osteoclasts are produced, the rate of bone ab-

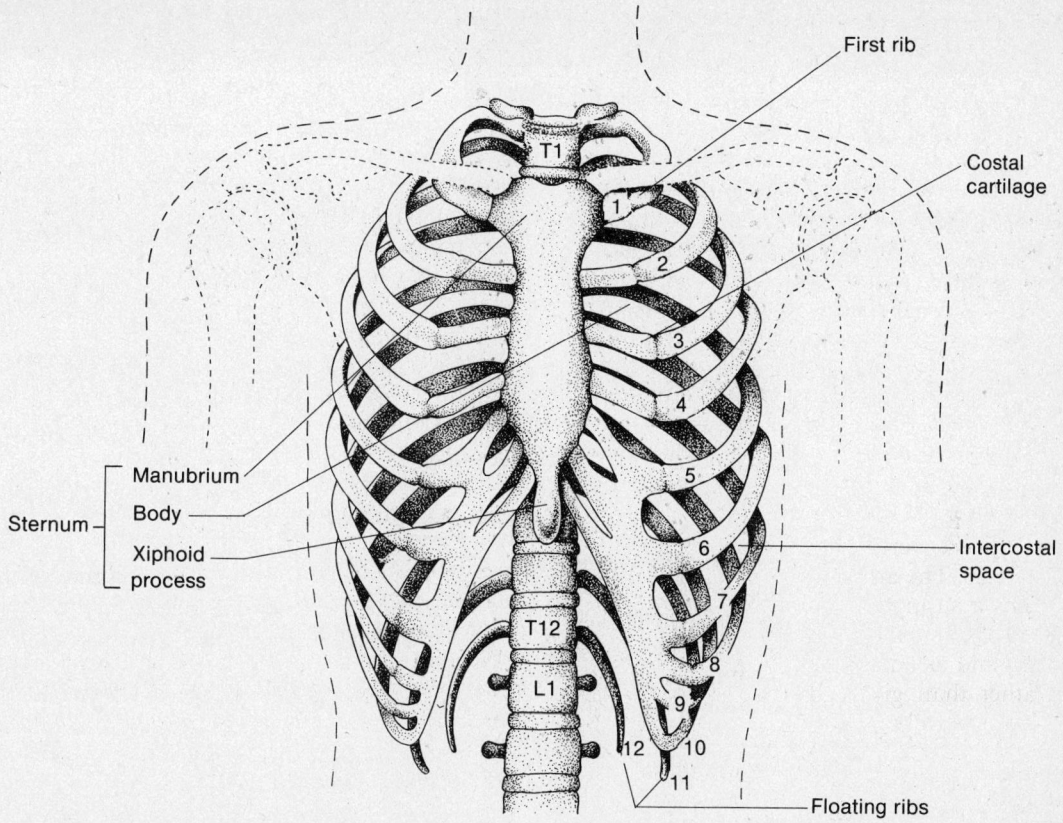

Fig. 5.27 The sternum and ribs. The seven pairs of upper ribs articulate anteriorly with the sternum.

sorption is increased. Higher than normal concentrations of calcium have dramatic effects upon nerves and muscles, as we will see later.

A low level of parathyroid hormone causes the opposite effect—calcium and phosphorus are deposited in bone at an accelerated rate. The depletion of these minerals from the extracellular fluid and blood also affects nerve and muscle physiology.

The delicate balance of calcium and phosphorus maintained between that deposited in bones and that in body fluids dramatically illustrates a homeostatic process.

Growth Hormone A deficiency or oversecretion of *growth hormone* secreted by the *anterior pituitary* affects bone growth. A deficiency of growth hormone in the young child inhibits bone growth as well as growth of other tissues. The result is *dwarfism*. The *pituitary dwarf* usually has a well-proportioned body without mental retardation.

An overabundance of growth hormone produces accelerated growth of bones and other tissue resulting in *giantism*.

If there is an increase in growth hormone production, possibly due to a tumor of the anterior pituitary after the epiphyseal cartilage has fused with the bone shaft, long bones do not increase in length. However, bones of the nose, chin, brow ridge, feet, and hands continue to

THE SKELETAL SYSTEM 71

increase in size. This condition is called *acromegaly.*

Other factors influencing both the rate and degree of growth include hormones from the thyroid gland, pancreas (insulin), and gonads (sex hormones). These hormones may play a decisive role even though an adequate quantity of growth hormone is available.

Vitamin D Vitamin D is indirectly involved in bone physiology, as it is essential for the absorption of calcium from the gastrointestinal tract into the bloodstream. A deficiency of vitamin D, particularly in children, produces rickets, due to the lack of proper absorption of calcium into the blood.

Calcium deficiency may also be a result of low levels of calcium in the diet.

FRACTURES

A *fracture* is a broken or cracked bone. Bones normally repair themselves. Children's bones heal much more rapidly than do those of adults, due to the fact that the calcium-depositing mechanism in children is more active. Children's bones proportionally have more organic material and less mineral content than adult bones. Also, the blood supply to children's bones is more extensive.

A common fracture of children's bones is called a *greenstick fracture* because the fractured bone resembles a broken green twig. As minerals become more heavily deposited in the adult, and even more so in old age, cracks across the bone or complete breaks are more common. A *simple fracture* is a cracked bone. The bone does not separate. A *compound fracture* is one in which the broken bone protrudes through the skin.

When a bone fractures, the periosteum is torn. Minute blood vessels are ruptured and tissue surrounding the fracture may be damaged. The space immediately around the fracture fills and swells with blood, producing a specialized cells called *fibroblasts*. These cells form stringlike fibers that penetrate and crosslace the hematoma, particularly in the area between the fractured ends. This process produces a raised knot of tissue around the fracture called a *callus.* In time the callus is invaded by osteoblasts, which will produce new bone in young children. The callus gradually shrinks as bone replaces it, eventually to such a degree that even X ray photographs in the adult may not detect the original break. Fractures occurring in adult bones may persist in maintaining some thickening at the fracture site.

Figure 5.28 shows X ray photographs of different types of fractures as well as a method of repairing a severe fracture.

BONE ARTICULATIONS

The function of bone joints is to provide flexibility of movement. Some joints allow little or no movement, but others allow a wide range of movement. Joints are broadly divided into three categories: *synarthroses, amphiarthroses,* and *diarthroses.*

Synarthroses

These are joints that allow no movement. The best examples of synarthrodial joints are those of the cranium, as shown in Figure 5.24.

During embryonic development and childhood the skull bones are not united, allowing for growth of the cranial cavity. By adulthood the skull bones have united and have formed rigid joints called *sutures.* Sutures are synarthrodial joints.

Amphiarthroses

Joints that permit a limited degree of movement are called amphiarthrodial joints. These are found between vertebrae and at the union of pubic bones.

Diarthroses

Joints that permit movement in one or more planes are called diathrodial joints. These include most of the familiar joints of the skeleton, such as those in the wrist, elbow, knee, and fingers. Diathrodial joint surfaces are covered with hyaline cartilage and are separated by an articular cavity.

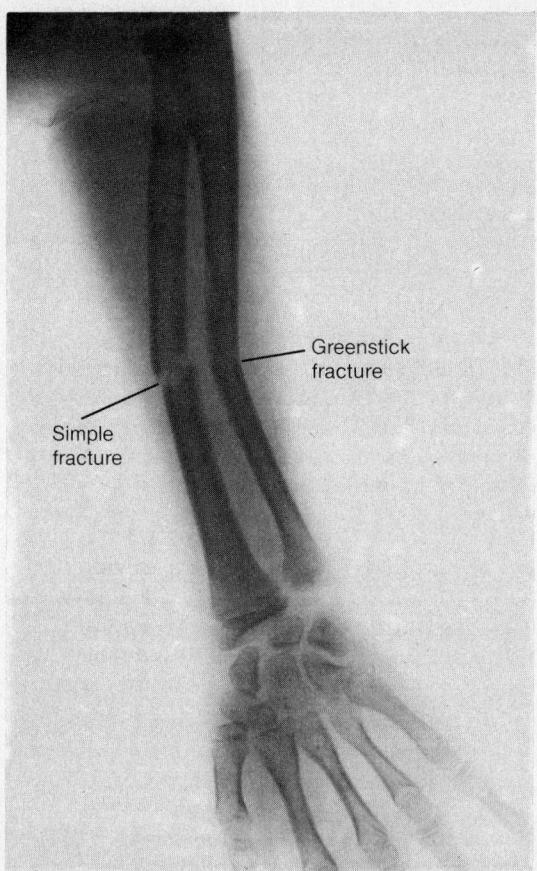
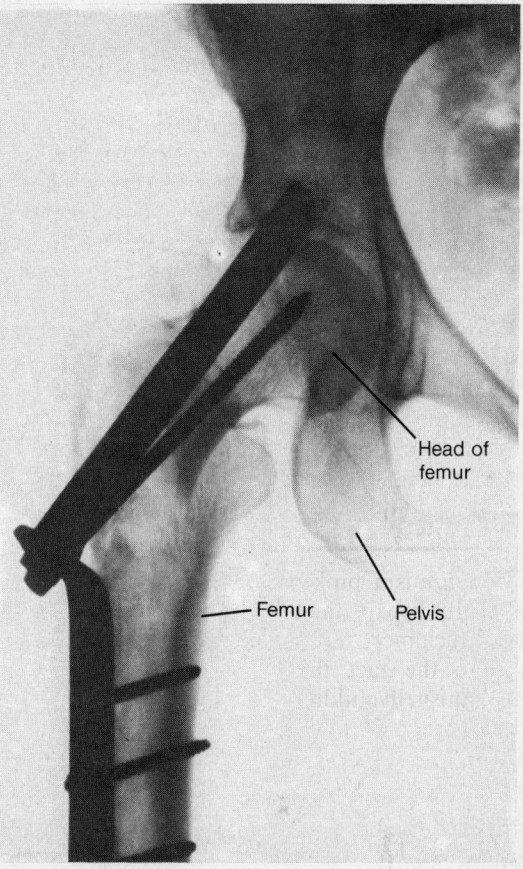

Fig. 5.28 X rays showing two different kinds of fractures and illustrating a device used to repair a shattered femur head. *(Courtesy Radiology Department, Normandy Osteopathic Hospital, St. Louis, Mo.)*

DISEASES OF THE SKELETAL SYSTEM

Some of the most common bone diseases are those related to some malfunction of calcium and phosphorus reabsorption and deposition. These malfunctions are associated with abnormal parathyroid activity or vitamin D deficiency.

Nonmicrobial Diseases

Osteitis fibrosa is a condition brought about by a hyperactive parathyroid and is characterized by severe decalcification of bone, which leaves it porous and weakened. Spontaneous fractures are common. In advanced states only slight stress will fracture the bone. The stress may be as minimal as that of normal movement of limbs.

Rickets is a disease normally associated with childhood vitamin D deficiency. Vitamin D is essential for the absorption of ingested calcium through the intestinal lining into the bloodstream. The normal diet of children may contain sufficient calcium, but without vitamin D, this calcium is excreted in the feces.

The bones of very young children have a high organic content and a relatively low mineral (calcium and phosphorus) content. The bones are flexible and may be easily deformed, particularly the bones which bear weight, such as the arms and legs. The ribs and sternum are also susceptible to deformation.

Rickets is alleviated with the administration of vitamin D. Direct sunlight stimulates the synthesis of vitamin D in the body. The ultraviolet fraction is the effective part of sunlight in vitamin D synthesis. Direct sunlight is important, since glass or other transparent window covering screens out the ultraviolet rays.

Osteomalacia is a condition that results from a vitamin D deficiency after normal bone growth is attained. Osteomalacia is often referred to as *adult rickets*. Bone calcium is reabsorbed into the blood, leaving the adult bone porous. This commonly occurs during pregnancy when there is a drain on the calcium of the woman's bones.

The overall effect of osteomalacia on bones is similar to that of hyperparathyroidism.

Microbial Diseases

Prior to the advent of aseptic technique, antimicrobial drugs, and other advances in infection prevention, the occurrence of infections of the bone was relatively common and greatly feared. Despite the scientific advances today, infections of the bone, bone marrow, and joints are conditions still feared because the chance of complete recovery from the infection and full use of the affected bone or joint may be limited. Bone often is more susceptible to infection and recovers from microbial invasion more slowly than most other body tissues.

Osteomyelitis is an infection of a bone that can result from a systemic infection that forms an abscess in a bone. It can also result from direct injury and subsequent infection of a bone. Any bone can be affected with osteomyelitis, but the most frequent sites are the femur, tibia, knee, and humerus. The most frequent cause of osteomyelitis is *Staphylococcus aureus*, but other causative organisms are the *Streptococcus, Pseudomonas, Proteus, Escherichia coli, Salmonella,* and *Neisseria gonorrhoeae*.

Although *arthritis* is more frequently associated with joint inflammation caused by the aging process, arthritis can also be caused by numerous microbial agents. Frequently microbe-caused arthritis develops after a localized infection in tissue other than bones and joints develops into a systemic infection or *septicemia,* a microbial infection and multiplication of microbes in the bloodstream. The pathogen or its toxic products then migrate to joints and arthritis develops.

The most common site for *tuberculosis* is the lungs, but tubercular lesions may occur in bones and joints such as the hips, knees, elbows, fingers, toes, and vertebrae. The vertebral lesions are known as *Pott's disease*. Although many believe that bone and joint tuberculosis is a secondary infection to tuberculosis of the lungs, in many persons affected with the former, no sign of pulmonary tuberculosis exists.

CALCIUM LEVEL INFLUENCES IN THE AGED

The progressive loss of calcium from aging bones presents the individual with a major health hazard. Not only do bones fracture more easily, they also heal more slowly.

Cartilages of the vertebral discs and ligaments atrophy and become progressively calcified. This results in joint changes such as spinal curvature and limits movement due to stiffness of joints.

Bones, particularly the vertebral segments and the femur neck, become *rarefied* (loss of structural bone). The loss of bone may begin as early as the early 30s and continue throughout a person's life.

SUMMARY AND REVIEW

The skeletal system is composed of bone, cartilage, and connective tissue. The skeletal system provides protection for other tissues, gives the body shape, permits movement, and is a reservoir for calcium and phosphorus.
A. Embryonic forerunners of bone include:
   1. hyaline cartilage
   2. fibroelastic cartilage
B. Structure of bone
   1. Bone is composed of microscopic cells called osteocytes arranged in a pattern called the Haversian system.

  2. The inorganic portion of bone is largely calcium and phosphorus.
  3. Bone marrow is of two types:
   a. Yellow marrow serves as a storage for fat.
   b. Red marrow produces red blood cells and some white blood cells.
 C. Bones are classified as:
  1. long bones
  2. short bones
  3. flat bones
  4. irregular bones
 D. Bone markings are identifiable ridges, openings, depressions, or cavities found on bones.
 E. The morphology of bones refers to those aspects of structure used to identify parts of bone. These include:
  1. epiphysis
  2. diaphysis
  3. medullary cavity
  4. endosteum
  5. periosteum
  6. articular cartilage
 F. Bones grow by the process of ossification. The embryonic pattern of cartilage is replaced by bone cells (osteocytes).
 G. The appendicular skeleton is made up of the arms, hands, legs, and feet.
 H. The axial skeleton is made up of the head, vertebrae, rib cage, and pelvic bones.
 I. Hormone and vitamin influences on bone: Parathyroid hormone aids in the conversion of osteoblasts and osteocytes to osteoclasts.
  1. Osteoclasts convert insoluble calcium to a soluble form.
  2. Soluble calcium is returned to the blood to aid in other body functions, such as nerve impulse transmission and muscle contraction.
 J. Bone articulations are classified as:
  1. Synarthrodial (immovable)
  2. Amphiarthrodial (slightly movable)
  3. Diarthrodial (freely movable)
 K. Diseases of bones include:
  1. osteitis fibrosa
  2. rickets
  3. osteomalacia
  4. osteomyelitis
  5. arthritis
  6. tuberculosis
 L. The major influence of aging on bones is the progressive loss of calcium.

6
The Muscular System

MAJOR CONCEPTS
- TYPES OF MUSCLE TISSUE
  - Skeletal Muscle
  - Smooth Muscle
  - Cardiac Muscle
- MICROSCOPIC STRUCTURE OF MUSCLE TISSUE
- NEURAL CONTROL OF MUSCLE CONTRACTION
- CHEMISTRY OF MUSCULAR CONTRACTION
  - Muscular Attachment
  - Muscles of the Head, Face, and Neck
  - Muscles of the Shoulder
  - Muscles of the Arm and Hand
  - Muscles of the Legs
  - The All-or-Nothing Law
  - Muscle Contraction and Heat Production
- CLASSIFICATION OF MUSCLES
- MAJOR MUSCLES OF THE BODY
- DISEASES OF MUSCLES
  - Nonmicrobial Diseases
  - Microbial Diseases
- MUSCLE REDUCTION WITH AGING

Muscle is the largest mass of tissue in the body. The three types—skeletal, smooth, and cardiac—all contract when stimulated. In this chapter skeletal muscle location, structure, physiology, and disease are emphasized.

Muscle is composed of *contractile tissue*. When properly stimulated, contractile tissue shortens. This shortening process accommodates all normal body functions associated with movement. We usually associate movement with locomotion; this, however, is but one of the many types of movement associated with body activities.

Muscles pump blood throughout the body, help pull air into the lungs, propel nutrients through the gastrointestinal tract, control the size of body openings, and help maintain posture. Muscles are also the major source of heat for the body. Other specific functions of muscles will be discussed as the need arises relative to specific organs.

TYPES OF MUSCLE TISSUE

Three kinds of muscle tissue are recognized and can be differentiated largely on the basis of cellular structure and arrangement. The three types of muscles, as shown in Figure 6.1, are *skeletal*, *smooth*, and *cardiac*.

Skeletal Muscle
Skeletal muscle represents the largest mass of tissue in the body. It is also called *striated* or *voluntary muscle*. Skeletal muscles, over which we have conscious (voluntary) control, permit the movement of skeletal parts in a coordinated pattern for such activities as walking or running—as well as guiding a surgeon's scalpel.

Smooth Muscle
Smooth muscle lines many internal organs, including the digestive tract, urinary tract, reproductive tract, and some glands. Smooth muscle is also referred to as *involuntary muscle*, since the brain exercises no voluntary control over these muscles—in other words, we cannot willfully contract these muscles. A third term, *visceral muscle*, is also used for smooth muscle, due to the fact that it lines the viscera.

Smooth muscle fibers vary in length from 20 to 200 microns and have a diameter of 3 to 9 microns. These fibers may be lengthened to as much as 500 microns in the lining of the uterus of a pregnant woman.

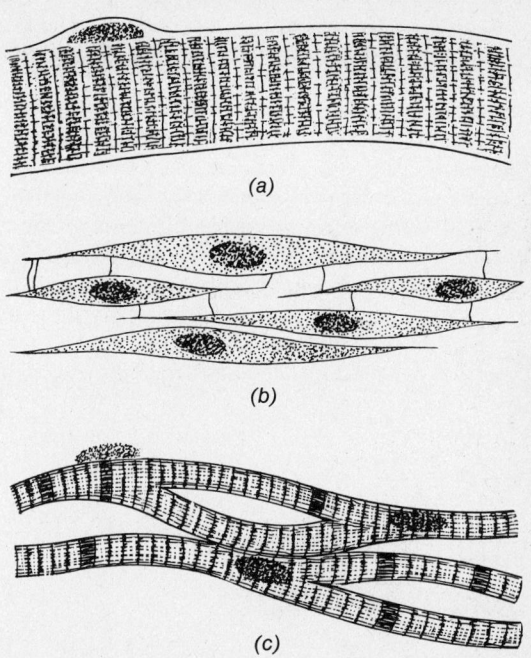

Fig. 6.1 Three types of muscle tissue. (a) Striated muscle (also called skeletal or voluntary), permits the capability of moving skeletal parts, contracts to pump blood, and maintains posture. (b) Smooth muscle lines many internal organs such as digestive, reproductive, and urinary tracts. This muscle is involuntary; we have no conscious control over its contraction or relaxation. (c) Cardiac muscle is a specialized muscular tissue found only in the walls of heart chambers. This muscle is also involuntary.

Cardiac Muscle
Cardiac muscle is specialized muscular tissue found only in the walls of the heart chambers. Although it is structurally similar to skeletal muscle, it is not under conscious control.

MICROSCOPIC STRUCTURE OF MUSCLE TISSUE

A single skeletal muscle fiber is composed of many smaller units called *myofibrils*. A single myofibril can be thought of as a single cell, as shown in Figure 6.2. The myofibrils are bundled together like a packet of straws. Several nuclei are located on the periphery of the bundles. A cell membrane, the *sarcolemma*, covers the nuclei and bundles.

Each myofibril is composed of even smaller fibers called *myofilaments*. Myofilaments are arranged in a pattern not unlike that produced if the teeth of two pocket combs are pushed together. Some of the darker and thicker myofibrils are composed of molecules of the protein *myosin*, while lighter, thinner ones are composed of another protein, *actin*.

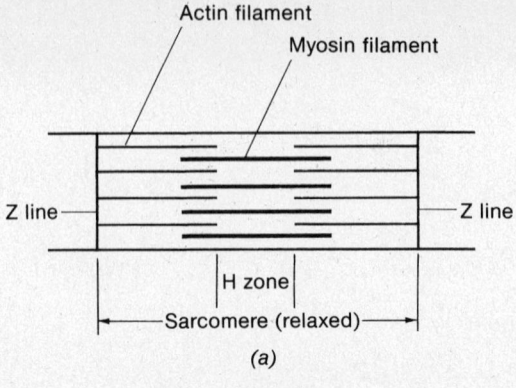

(a)

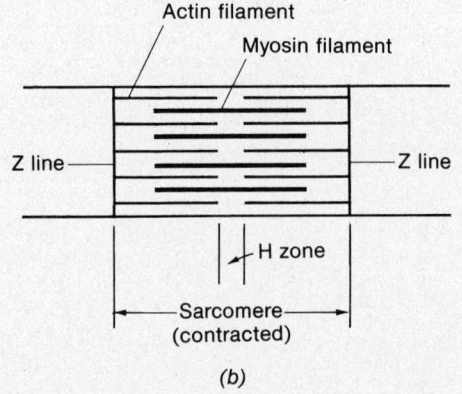

(b)

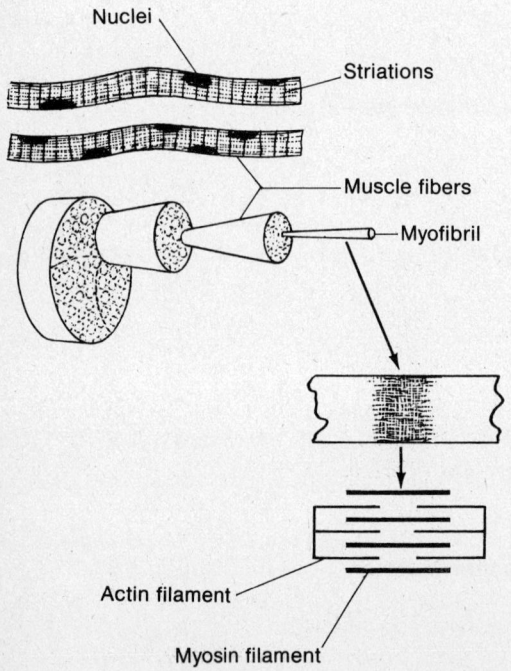

Fig. 6.2 Relationship between whole muscle, muscle fibers, and their components. Note the multinucleated nature of single muscle cells in the upper illustration. These cells collectively are arranged into myofibrils that in turn make up the individual muscle fibers. The myofibril is composed of simpler substances called actin and myosin.

Fig. 6.3 (a) Sarcomere in relaxed state. (b) Sarcomere in contracted state. A sarcomere is the functional unit of muscle. Each sarcomere is arranged end to end with other sarcomeres. The myofibril formed from this arrangement thus resembles a series of blocks laid end to end.

Applying the pocket comb analogy, it can be said that the teeth of one comb are made up of myosin, while the teeth of the other are made up of actin. During muscular contraction, the myofilaments slide past one another, thus shortening the fiber.

Myofibrils are made up of units called *sarcomeres*. A sarcomere is the functional unit of muscle. Each sarcomere is arranged end to end with other sarcomeres, so that the myofibril resembles a series of blocks laid end to end, as shown in Figure 6.3

Regions of each sarcomere are identifiable with electron microscope magnification, as shown in Figure 6.4. Each end of a sarcomere is called a *Z line*. Actin filaments are attached at

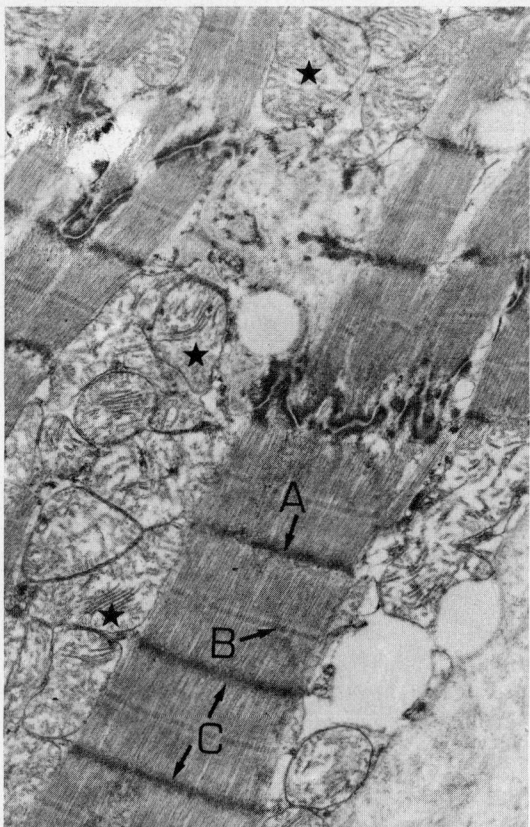

Fig. 6.4 In this electron micrograph of skeletal muscle, the Z line is shown at *(A)* and the H zone at *(B)*. At *(C)* the arrows define the limits of a single sarcomere. Note the large number of mitochondria, shown starred. Muscle contraction requires tremendous amounts of energy. Mitochondria are the sites of cellular respiration, which releases energy from nutrients. *(Courtesy Santiago Plurad.)*

the Z lines on one end and are free at the other end. When actin filaments slide inward, the Z lines are drawn closer together, thus shortening the length of each sarcomere. The cumulative effect of sarcomere shortening causes muscle shortening.

As sarcomeres shorten, actin and myosin filaments slide past each other. This produces a change in the structural appearance in each sarcomere. Note for instance, in Figure 6.3, what happens to the H zone when muscle is contracted. The H zone is the space between the ends of two actin filaments. Only the myosin filaments are microscopically visible. Thus the H zone appears to be a light line within the A band.

NEURAL CONTROL OF MUSCLE CONTRACTION

The stimulus initiating muscle contraction is usually supplied by the nervous system. The stimulus, called a *nerve impulse* arrives at the surface of a muscle by way of a *nerve pathway*. (More will be said about nerve impulses in Chapter 7.)

No direct connection exists between the nerve ending *(end plate)* and the sarcolemma of the myofibril, as can be seen in see Figures 6.5 and

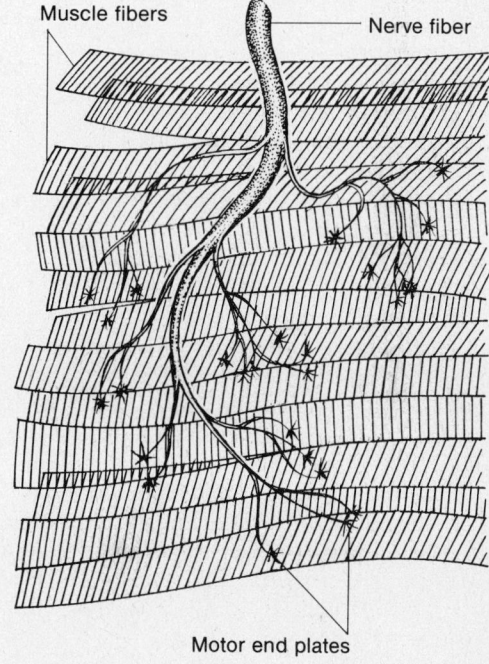

Fig. 6.5 Physical relationship between the nerve end plate and the sarcolemma of muscle fibers. The nerve end plates do not come into direct physical contact with the muscle fiber. When a nerve impulse arrives at the muscle a hormone is secreted that stimulates the muscle fiber to contract.

6.6. The nerve (neuron) end plate lies in a shallow concave region of the myofibril. This near connection between nerve and muscle is called the *myoneural junction*. A neuron together with that portion of a muscle myofibril it activates is called a *motor unit*.

When a nerve impulse arrives at the nerve end plate, it causes the release of a transmitter substance called *acetylcholine* from storage vesicles in the nerve end plate. Acetylcholine thus emptied into the space between the nerve ending and the muscle cell initiates the contraction process. The following sequence of events occurs:

1. Acetylcholine alters the permeability of the muscle cell membrane, allowing Na ions to enter. Normally, there is a greater concentration of Na ions on the outside of the cell membrane than the inside, and in this state the cell membrane is said to be *polarized*, with more positive ions on the outside. The arrival of acetylcholine and the rush of Na ions through the cell membrane into the cell causes the relative electrical charges on either side of the membrane to be reversed. In other words, there are now more positive electrical charges inside the membrane than outside. The membrane is now said to be *depolarized*.
2. The depolarization of the membrane, through some mechanism not yet understood, brings about the release of calcium ions that are stored in tiny sacs in the *sarcoplasmic reticulum*. The sarcoplasmic reticulum resembles the endoplasmic reticulum of other cells.
3. Calcium from these sacs unites in loose chemical combination with the myosin filaments, which brings about the sliding action of the actin and myosin filaments.

As long as acetylcholine remains in the gap of the myoneural junction, the muscle fiber remains contracted. However, it normally is not allowed to remain long—less than 2/1000 second. A second chemical, *cholinesterase* (an enzyme), is secreted by segments of the sarcolemma. Cholinesterase immediately destroys

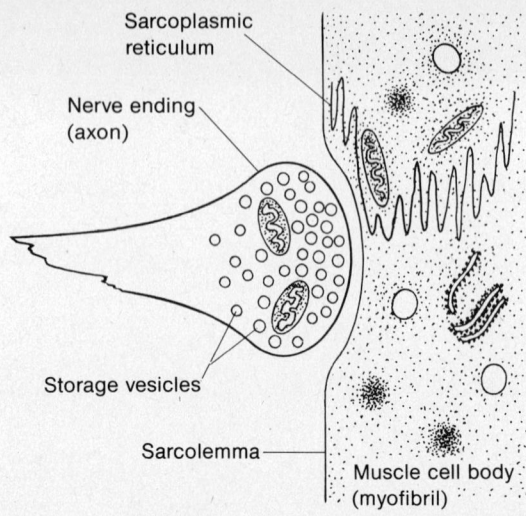

Fig. 6.6 An enlarged view of the myoneural junction where the nerve end plate lies near the sarcolemma.

acetylcholine, and the muscle filaments (actin and myosin) relax to their former positions. The events associated with muscle fiber relaxation include the following:

1. Cholinesterase destroys acetylcholine.
2. Calcium attached to actin fibers is released, returning to the sarcoplasmic reticulum sacs.
3. The cell membrane becomes repolarized, with a greater positive charge on the outside of the membrane.

The muscle cell is now ready to contract again if an impulse arrives releasing more acetylcholine into the myoneural gap. We will continue this discussion in more detail later in the chapter relative to muscular diseases.

CHEMISTRY OF MUSCULAR CONTRACTION

The previous discussion focused on general activities associated with contraction. We will now take a closer look at what happens chemically at the subcellular level.

When a muscle contracts, the myosin fibers remain stationary. Minute projections called *cross bridges* project outward from the myosin filament, as shown in Figure 6.7. When calcium is released from the sarcoplasmic reticulum, it

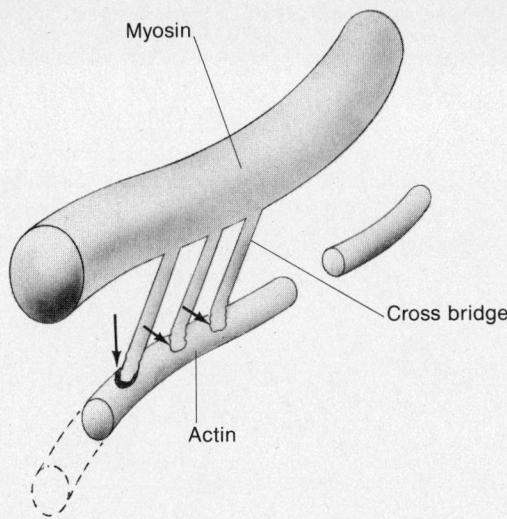

Fig. 6.7 Relationship of actin and myosin fibers and the action of the cross bridges.

forms a loose chemical union with myosin causing the cross bridges to connect momentarily with the actin filaments. The cross bridges then move, pulling the actin filaments inward, thereby shortening the sarcomere. Since sarcomeres connected end to end make up a muscle fiber, the fiber is shortened.

When the muscle fiber becomes polarized again, calcium returns to sacs in the sarcoplasmic reticulum and the cross-bridge activity ceases. The actin filaments now return to their former position (muscle relaxation).

We still do not know what processes are involved in causing the actin fibers to slide back to their former position.

The movement of cross bridges requires energy. This energy is supplied by the decomposition of ATP (adenosine triphosphate) molecules stored throughout muscle tissue. The union of calcium and myosin causes myosin to act as an enzyme *(ATP-ase)*, which aids in the release of energy from ATP. This process is illustrated as follows:

$$ATP \xrightarrow{Ca\ +\ myosin\ (ATP\text{-}ase)} ADP + P + energy$$

A large reservoir of stored ATP is available in muscle tissue. The energy chemically bound up in the ATP molecule originally comes from cellular respiration of ingested nutrients. As energy is released from foods, much of it is "captured" and stored as ATP molecules for some future use such as muscular contraction.

The energy released during cellular respiration is accomplished by two processes—aerobic and anaerobic respiration.

Aerobic respiration is the chemical process by which cells are capable of oxidizing glucose and other nutrients. Oxidation is a complicated process involving many intermediate substances before the final products are made available. The most important product is energy derived from chemical bonds. This is the energy used to do the body's work. Excess energy is stored in ATP molecules. Carbon dioxide and water molecules are by-products of the oxidation process. When the glucose molecule has been completely degraded to CO_2 and H_2O, all the former available energy of glucose has been released.

Sufficient oxygen must be available to cells to carry on aerobic respiration. Low level physical activities require less oxygen. The respiratory system is capable of supplying adequate oxygen under these circumstances.

Anaerobic respiration is the process muscle cells utilize if less than adequate oxygen is available. During strenuous exercise, muscle cells demand tremendous quantities of oxygen and ATP. Although the breathing rate is increased in an attempt to supply more oxygen, the muscles demand even more. With an insufficient quantity of oxygen, cells cannot complete the process of breaking glucose molecules apart.

Muscle cells, faced with a potential shortage of available energy, set in motion an alternate plan. This plan requires the synthesis of a new molecule, *lactic acid*. The process of changing glucose molecules to lactic acid involves a limited release of energy. This process, anaerobic respiration, can continue without oxygen.

The process of anaerobic respiration is not very efficient in terms of supplying adequate energy. Aerobic respiration releases vastly larger amounts of energy. Anaerobic respiration is therefore only a temporary method employed until oxygen is adequately supplied. In addition to relatively small amounts of energy made available through anaerobic respiration, other complications arise.

Excess lactic acid accumulates in muscle tissue, and is then, for the most part, absorbed into the bloodstream. The liver is capable of converting about one-fifth of the lactic acid into CO_2 and H_2O with the subsequent release of energy. The remaining four-fifths of the lactic acid is converted by the liver to glucose, which is returned to muscle tissue.

When anaerobic respiration continues over an extended period of time, so much lactic acid accumulates that the liver cannot convert it to other products. It is this accumulation that is thought to cause fatigue and to stimulate nerves that cause the sensation of muscle soreness.

Since the accumulation of lactic acid is due to a deficiency of oxygen, we say that the body has gone into *oxygen debt*. When we rest after vigorous exercise we continue to breathe deeply and more rapidly than normal. In doing so, oxygen is made available so that excess lactic acid can be oxidized. The process will then follow the normal aerobic respiration pattern.

Aerobic respiration yields a much greater quantity of ATP than the anaerobic process; however, if oxygen becomes deficient, the cells have no other option than to release energy in some other manner, although at a less efficient level.

Muscular Attachment

The basic unit of skeletal muscle is called a *fiber*, which varies from 1 to 300 mm in length.

All the connective tissue sheaths merge at the ends of the muscle forming a *tendon*, which attaches to bone or other muscle tissue.

All muscles do not attach directly to bones. However, as the most common function of muscles is to move bones, they therefore must be attached in such a manner as to produce some specific motion of one bone relative to another. The merging of all the connective tissue, together with other collagenous substances, form a

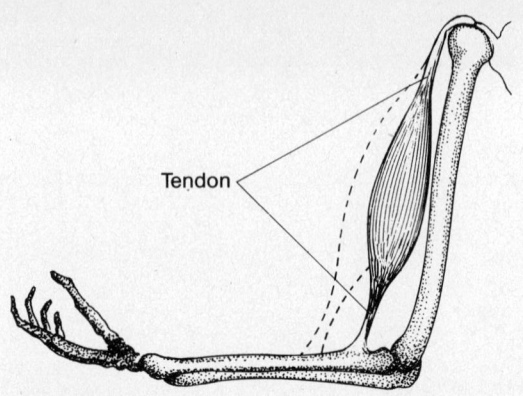

Fig. 6.8 Location of tendons in the upper arm. The merging of all connective tissue attached to muscle fibers together with other collagenous substances, form a tendon. A tendon thus attaches muscle to bone periosteum.

tendon, which in turn attaches to bone periosteum, as shown in Figure 6.8.

The fibrous connective tissue wrapping of muscle, rather than form a tendon, may form a sheetlike connection with similar tissues of other muscles. This type of connective tissue is called *aponeurosis*, and is illustrated in Figure 6.9. An aponeurosis is limited to flat, sheetlike muscles, while tendons form the attachment for bulky, massive muscles.

A thin connective tissue called *fascia* surrounds groups of muscles that have a common function. Fascia essentially holds muscles in place.

In order for a muscle to produce its action, one end must be attached to a relatively fixed, nonmovable bone. This attachment is called the *muscular origin*. The other end of the muscle is attached to the bone it moves. This attachment point is called the *muscular insertion*.

Muscles can only exert a pulling force, never a pushing force. Note in Figure 6.10 that one set of upper arm muscles pulls the arm upward, while another set pulls the arm down. Muscles that move the same bone in opposite planes are called *antagonistic muscles*. Obviously, when one antagonist contracts, the other must relax if movement is to be produced. There are instances when both sets of antagonistic muscles contract. This causes *muscular tetany*, which will be discussed in more detail later in the chapter.

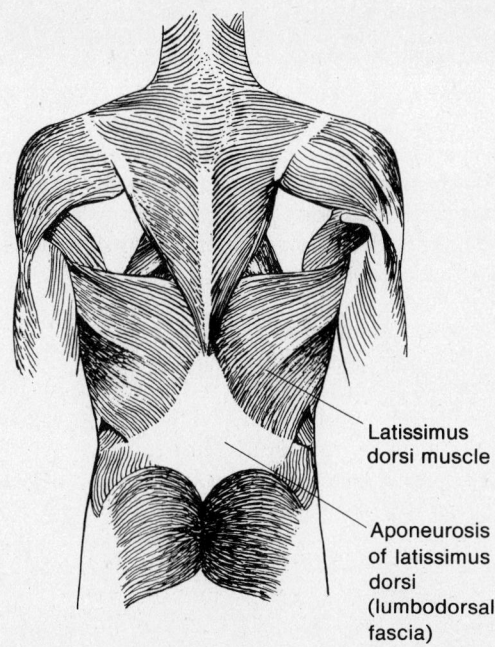

Fig. 6.9 An aponeurosis is a sheetlike tendon usually limited to large flat muscles. The term tendon is normally reserved for the attachment of large bulky muscles.

The All-or-Nothing Law

In discussing muscular contraction, we must be careful to differentiate between the contraction of a single fiber and the contraction of a whole muscle. Some stimuli reaching a muscle fiber are not strong enough to elicit a response from the fiber. Muscle fibers vary in their capability to respond to different intensities of stimuli.

The lowest level of stimulus which will elicit a response (contraction) from a particular muscle fiber is called the *liminal threshold* or *minimal stimulus*. A stimulus too weak to bring about a muscle fiber response is said to be *subliminal* or *subminimal*. A stimulus that is subliminal for one particular fiber may be liminal for others.

There is no need for a term to designate a stimulus greater than the liminal threshold level. Once this level is reached, the muscle fiber will completely contract. Stimulation of the fiber beyond the liminal level cannot produce greater contraction.

The *all-or-nothing law* of muscular contraction may be stated as follows: Under constant environmental conditions a single muscle fiber responds to a stimulus either by contracting to its limit or not at all.

How, then, are muscles capable of responding differently to variable work loads? The answer lies in the fact that muscles are an anatomical collection of muscle fibers, more specifically, a collection of motor units. When a stimulus reaches the fibers, all the fibers for which that stimulus is liminal will respond completely. If the stimulus is increased, other fibers will be recruited to help as their specific threshold stimulus is reached.

A kind of muscle contraction that appears at first to contradict the all-or-nothing law is called

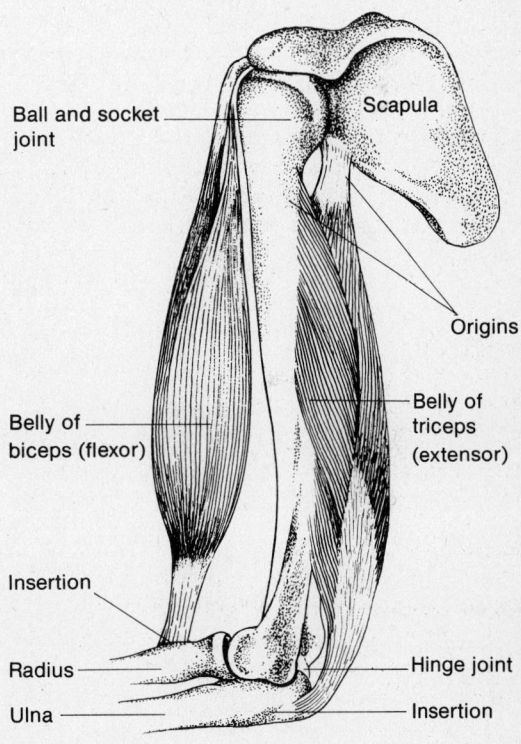

Fig. 6.10 A pair of antagonistic muscles, the biceps and triceps muscles of the upper arm. These are always pairs of muscles that move a bone in opposite directions. When one antagonist contracts the other one must relax, otherwise tetany results.

treppe. A muscle contracting due to a series of stimuli will increase its force of contractions momentarily until a maximum is reached. After the maximum force is reached, all contractions thereafter reach the same magnitude. This effect is due to ionic and possibly heat changes that occur in a muscle "warming up." The all-or-nothing law is not violated, since the environment is changing.

Tetany results when stimuli arrive at the muscle at a high rate, not giving the muscle time to relax between contractions. Figure 6.11 illustrates this effect. With gradual increase in the frequency of the stimulus, a point is eventually reached where no relaxation is possible. When this occurs, the muscle is *tetanized*. Muscular tetany may be induced in several ways, as we shall see in the discussion of muscular diseases at the end of this chapter.

Tonus is the constant partial contraction of skeletal muscle. This partial contraction keeps muscles in the "at ready" condition; in addition, the slight contraction provides body heat and helps support and hold other body tissues in place.

Muscle fatigue is a term that describes muscular inactivity due to insufficient energy. If an excised muscle is tetanized, eventually all contraction will cease since additional energy cannot be supplied. A similar but slower effect is produced in the body when the energy supply is less than adequate, as, for example, during vigorous exercise.

Muscle Contraction and Heat Production

Much of the energy released from ATP during muscular contraction does no useful work but rather is converted into heat. This is the heat that maintains a normal body temperature of 37°C (98.6°F) and is also the heat that the body must rid itself of if excessive. The control mechanisms that help maintain a constant body temperature are largely those of the nervous system, which will be discussed later. It may be pointed out here briefly that when the blood is only slightly chilled, muscular contraction exhibited as shivering may produce sufficient heat to restore the temperature to the proper level.

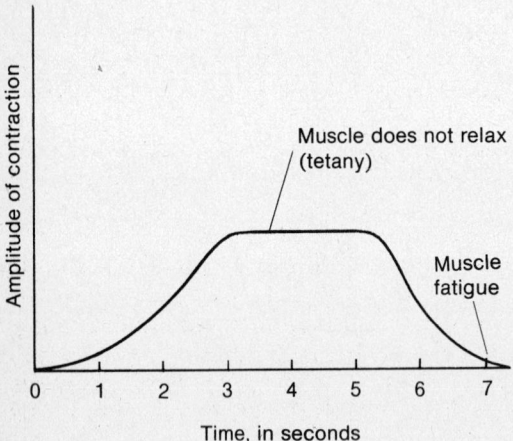

Fig. 6.11 Muscle tetany is produced by rapidly repeated stimulation of an excised muscle. An excised muscle is one that has been surgically removed from a living animal. The muscle will eventually fail to respond due to total fatigue.

Malignant Hyperthermia

The combination of certain physiological processes and some medications has long been known to cause serious medical problems and even death. One such combination is the often fatal reaction that occurs between susceptible persons and general anesthesia. The condition is known as malignant hyperthermia.

There appears to be little doubt that the condition is genetically linked. People with no knowledge of their susceptibility to the disease may enter a hospital for minor surgery. When the general anesthesia is administered, their muscles stiffen, the skin becomes cyanotic, and body temperature rapidly increases to 43°C or higher. The pulse rapidly accelerates, convulsions occur, and patients may die in as few as 10 minutes.

There are tests available (not at all medical facilities) to predetermine patients' susceptibility to malignant hyperthermia. Blood serum levels of creatine phosphokinase in these individuals are usually highly elevated from the normal 0 to 110 I.U. Conclusive tests are muscle biopsy and electromyography.

CLASSIFICATION OF MUSCLES

The name of a muscle is usually quite descriptive. The name may indicate the shape, location, origin, insertion, function, or it may indicate how many units make up a muscle. For example, the deltoid muscle of the shoulder is shaped like a delta. The trapezius muscle of the upper back is shaped like a trapezoid. The sacrospinalis muscle indicates its location, in the area of the sacrum and spinal column. The sternocleidomastoid not only indicates its general location, but also its origin, (the sternum) and its insertion (the mastoid process of the temporal bone). A muscle whose name indicates its function is the levator scapulae, which elevates or pulls up the scapula. This muscle contracts when you shrug your shoulders. The name quadraceps indicates that this muscle is made up of four (quad) units.

Muscles that function in a particular manner may be grouped into separate categories. For example, the muscles that elevate a body part are called *levator* muscles.

The following is a list of muscle categories commonly used.

Flexors: Bend joints or decrease the joint angle.
Extensors: Return joints from the flexed position.
Levators: Raise body parts.
Depressors: Lower body parts.
Adductors: Move body parts inward toward the body midline.
Abductors: Move body parts away from the body midline.
Pronators: Turn the palm of the hand downward.
Supinators: Turn the palm of the hand upward.
Sphincters: Rings of muscle that circumscribe and control body openings.
Rotators: Facilitate the rotation of one part relative to another.
Tensors: Make parts more rigid.

MAJOR MUSCLES OF THE BODY

There are more than 300 pairs of muscles in the body. We will pay particular attention to those illustrated in the diagrams that follow.

Only a small amount of attention will be spent on origins and insertions of muscles. This is not to discount the importance of such information, but rather to keep the text material within some bounds.

Muscles of the Head, Face, and Neck

Except for the muscles that help masticate foods and move the head, the primary function of head and face muscles is facial expression. These muscles differ from other skeletal muscles in that they do not support heavy loads. In fact, some facial muscles have the simple task of wrinkling the skin. Head and face muscles help exhibit the emotions of joy, hate, grief, anxiety, anger, and so on. Figure 6.12 shows the major muscles of the head, face, and neck. (Muscles that control eye movements are discussed in Chapter 8.)

Table 6.1 summarizes the location and function of muscles of facial expression, mastication, and muscles that move the head.

Muscles of the Shoulder

Except for an articulation between the humerus and scapula and the attending ligaments, the entire shoulder is held in position by muscles, as shown in Figure 6.13. These muscles move the shoulder forward or backward. They also elevate and depress the shoulder's other movements, including adduction and abduction. Table 6.2 summarizes the muscles of the shoulder.

Muscles of the Arm and Hand

These muscles, shown in Figures 6.14 and 6.15, permit a wide range of movement, including adduction, abduction, rotation, flexion, and extension. Keep in mind, however, that a muscle rarely brings about a specified motion alone; even the simplest movements may require the action of many muscles.

Table 6.3 summarizes the muscles that move the arm and hand.

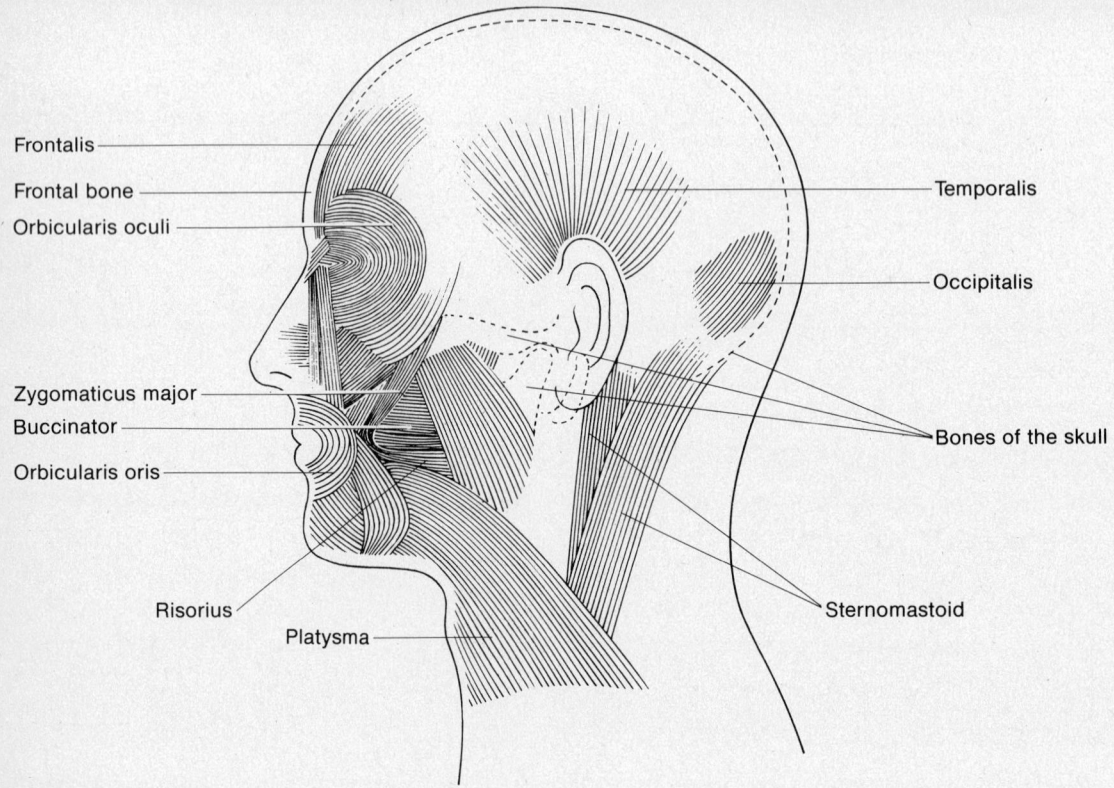

Fig. 6.12 Major muscles of the head, face, and neck. Except for the muscles that help masticate foods and move the head, the primary function of head and face muscles is facial expression. These muscles differ from most other skeletal muscles in that they do not support heavy loads.

Table 6.1 Muscles of the Head, Face, and Neck

MUSCLE	LOCATION	FUNCTION	MUSCLE	LOCATION	FUNCTION
Muscles of facial expression			Muscles of mastication		
Occipitalis	Covers occipital portion of skull	Draws the scalp backward	Buccinator	Main cheek muscle	Compresses the cheek, helps hold food against the teeth
Frontalis	Forehead	Draws scalp forward	Temporalis	Arises from the temporal fossa posterior to the zygomatic arch, inserts on the mandible	Closes the jaw
Orbicularis oculi	Encircles the eyelids	Closes the eye			
Orbicularis oris	Encircles the mouth	Closes the lips			
Zygomaticus major	Extends from zygomatic bone to mouth	Pulls the angle of the mouth upward and backward, as in laughing	Masseter	Exterior to temporalis	Assists the temporalis in closing the jaws
			Muscles that move the head		
Platysma	Large, sheetlike muscle covering most of the anterior portions of the neck	Pulls lower lip downward, tenses the neck	Sternomastoid	In the side of the neck, originates at the sternum, inserts on the mastoid process of the temporal bond	Flexes and rotates the head
			Splenis capitis	Upper neck	Extends the head

THE MUSCULAR SYSTEM

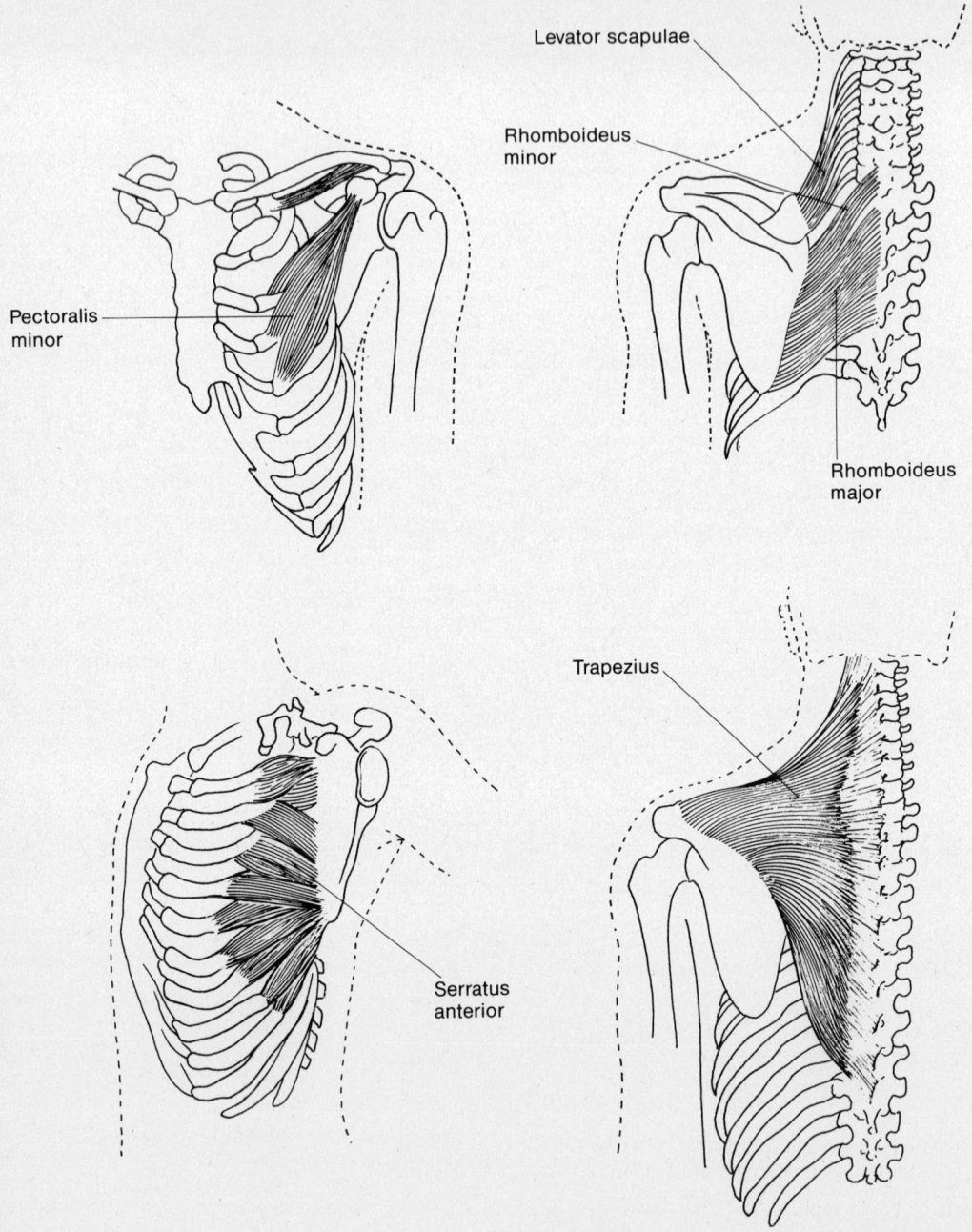

Fig. 6.13 Muscles of the shoulder include the pectoralis major and minor, levator scapulae, rhomboideus major and minor, serratus anterior and trapezius. Except for an articulation between the humerus and scapula and the attending ligaments, the entire shoulder is held in position by muscles.

88 HUMAN FORM AND FUNCTION

Table 6.2 Muscles of the Shoulder

MUSCLE	LOCATION	FUNCTION
Levator scapulae	Upper scapula and neck	Elevates the shoulder
Trapezius	Upper back and shoulder	Lowers and elevates shoulder
Serratus anterior	Lateral dorsal rib area	Moves shoulder forward
Pectoralis minor	Ventral, lateral thorax	Pulls shoulder downward
Rhomboideus major	Medial surface of scapula	Moves shoulder backward and upward

Table 6.3 Muscles that Move the Arm and Hand

MUSCLE	LOCATION	FUNCTION	MUSCLE	LOCATION	FUNCTION
Biceps brachii	Anterior surface of upper arm	Flexes forearm	Palmaris longus	Anterior surface of forearm	Flexes hand
Brachialis	Under the biceps	Flexes forearm	Extensor carpi ulnaris	Dorsal surface of ulna	Extends hand
Triceps brachii	Posterior surface of upper arm	Extends forearm	Extensor carpi radialis longus	Lateral surface of forearm	Extends hand
Brachioradialis	Lateral surface of radius	Flexes, pronates, and supinates forearm	Extensor digitorum	Posterior surface of forearm	Extends fingers
			Extensor carpi ulnaris	Posterior, medial aspect of forearm	Adducts wrist
Pronator teres	Proximal, lateral surface of radius	Pronates forearm	Pronator teres	Upper anterior forearm	Pronates forearm and hand
Supinator	Upper forearm	Supinates hand			
Flexor carpi radialis	Anterior surface of forearm	Flexes hand at wrist			

THE MUSCULAR SYSTEM

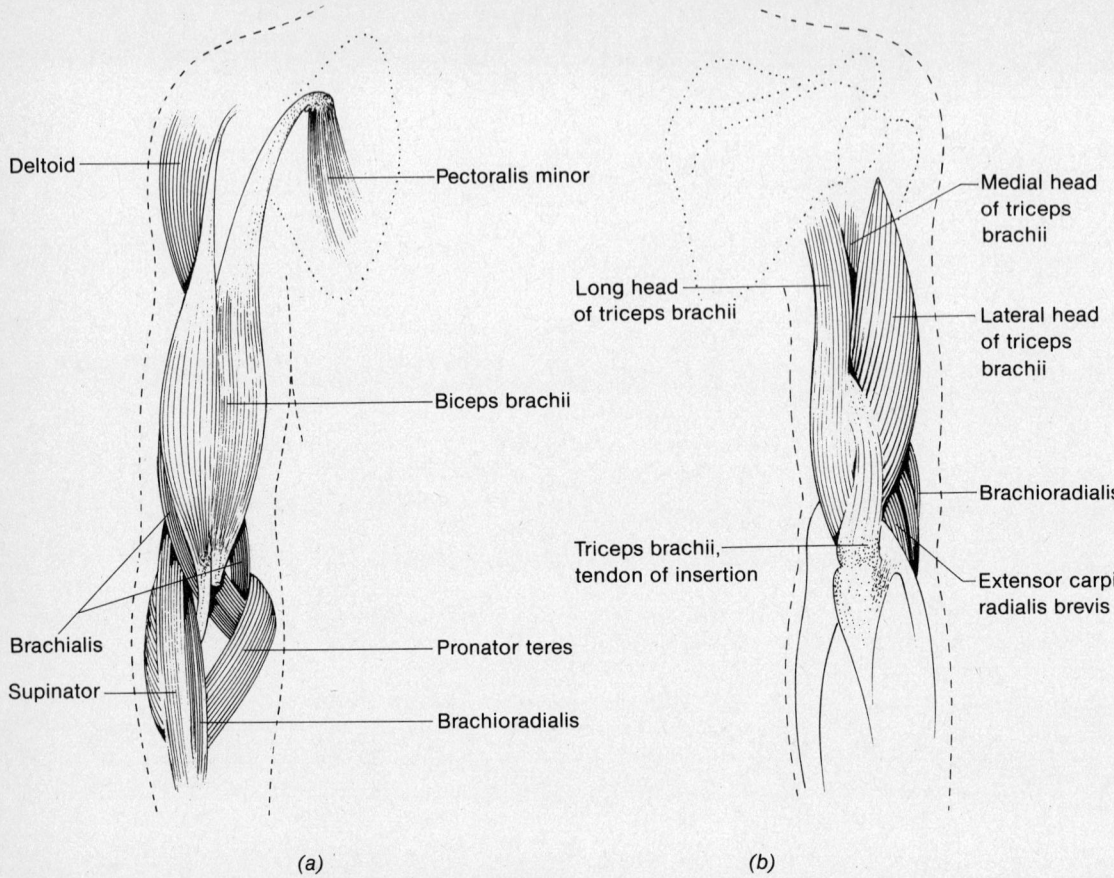

Fig. 6.14 The muscles that move the arm. (a) Anterior view. (b) Posterior view. These muscles provide abduction—movement away from the medial planes; adduction—movement toward the medial plane; rotation, flexion, and extension of the forearm.

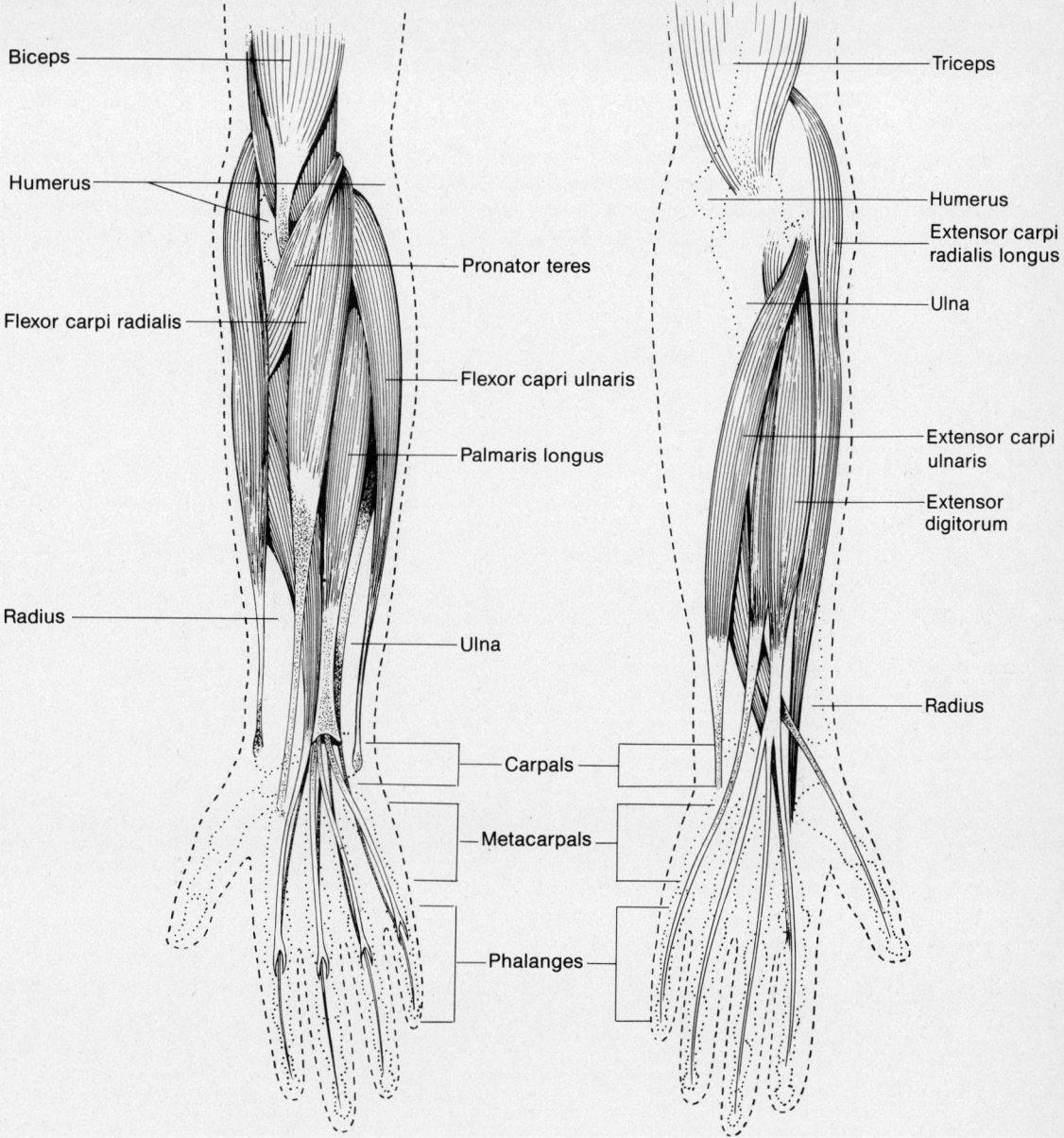

Fig. 6.15 Muscles that move the lower arm and hand. These muscles permit a wide range of movement including adduction, abduction, rotation, flexion, and extension.

THE MUSCULAR SYSTEM 91

Muscles of the Legs

Muscles of the legs are most commonly associated with locomotion and work. Some of the largest and strongest muscles in the body belong to this group. Figure 6.16 illustrates the upper leg muscles. Lower leg muscles basically provide for movement of the foot and toes. Figure 6.17 illustrates the lower leg muscles.

Table 6.4 summarizes muscles of the legs.

Table 6.4 Muscles of the Upper and Lower Leg

MUSCLE	LOCATION	FUNCTION
Gluteus maximus	Buttocks	Extends thigh
Tensor faciae latae	Lateral upper surface of thigh	Abducts thigh
Sartorius	Lies diagonally across thigh medially	Flexes thigh, helps the legs to be crossed
Rectus femoris	Anterior thigh	Extends knee
Vastus lateralis	Lateral thigh	Extends knee
Vastus medialis	Medial thigh	Extends knee
Vastus intermedius	Anterior thigh (deep)	Extends knee
Gracilis	Medial thigh	Adducts thigh
Biceps femoris	Posterior thigh	Extends thigh, flexes knee
Semimembranosus	Posterior thigh	Flexes knee
Semitendinosus	Posterior thigh	Flexes knee and extends thigh
Gastrocnemius	Calf of leg	Flexes knee and foot
Soleus	Calf of leg (deep)	Flexes foot
Peroneus longus	Lateral leg	Flexes and everts foot
Peroneus brevis	Lateral leg	Flexes and everts foot
Tibialis posterior	Posterior leg	Flexes and inverts foot
Flexor digitorum longus	Posterior tibia	Flexes toes
Flexor hallucis longus	Posterior fibula	Flexes big toe

DISEASES OF MUSCLES

Although muscle represents the largest tissue in the body, we know less about muscular diseases than we do about many other tissues. The study of muscular diseases is complicated by the fact that so many different factors influence muscular activity. These factors include the blood supply, nerves, fluid balance, ionic balance, and hormones.

Nonmicrobial Diseases

The following discussion pertains to only those muscular diseases which are not transmitted from one individual to another. They may be initiated by injury, nerve defect, or hormone imbalance. Some diseases of muscle are also congenital, while others may be passed on genetically from parents to offspring.

Traumatic myositis ossificans may result from a single serious injury or a series of injuries to a muscle. The injury possibly displaces osteoblasts from bone into muscle tissue. The displaced osteoblasts (bone-forming cells) carry out their preordained function even though they are in foreign tissue. The small bony accumulations among muscles creates a painful, though not normally disabling, disease.

Progressive myositis ossificans commences in early childhood with a soreness and swelling of the neck and back muscles. These muscles are gradually replaced by bony plates, and the body eventually becomes enclosed in a bony sheath, restricting respiratory movements and locomotion. An individual with this disease usually dies at an early age of complications resulting from dysfunctions of organs other than muscles.

Anterior tibial syndrome is a disease of young persons who have a common history of participating in vigorous athletic contests without previous athletic training. The activities may include long marches, football, jumping, and so on.

The anterior tibial muscles become painful and swollen with different degrees of tissue damage. Shinsplints common to athletes may be a minor manifestation if the disease is halted before extensive tissue damage results.

Congenital torticollis (wryneck) is noticeable as a cartilagenous lump on one side of the neck of an infant within 10 days after birth and is

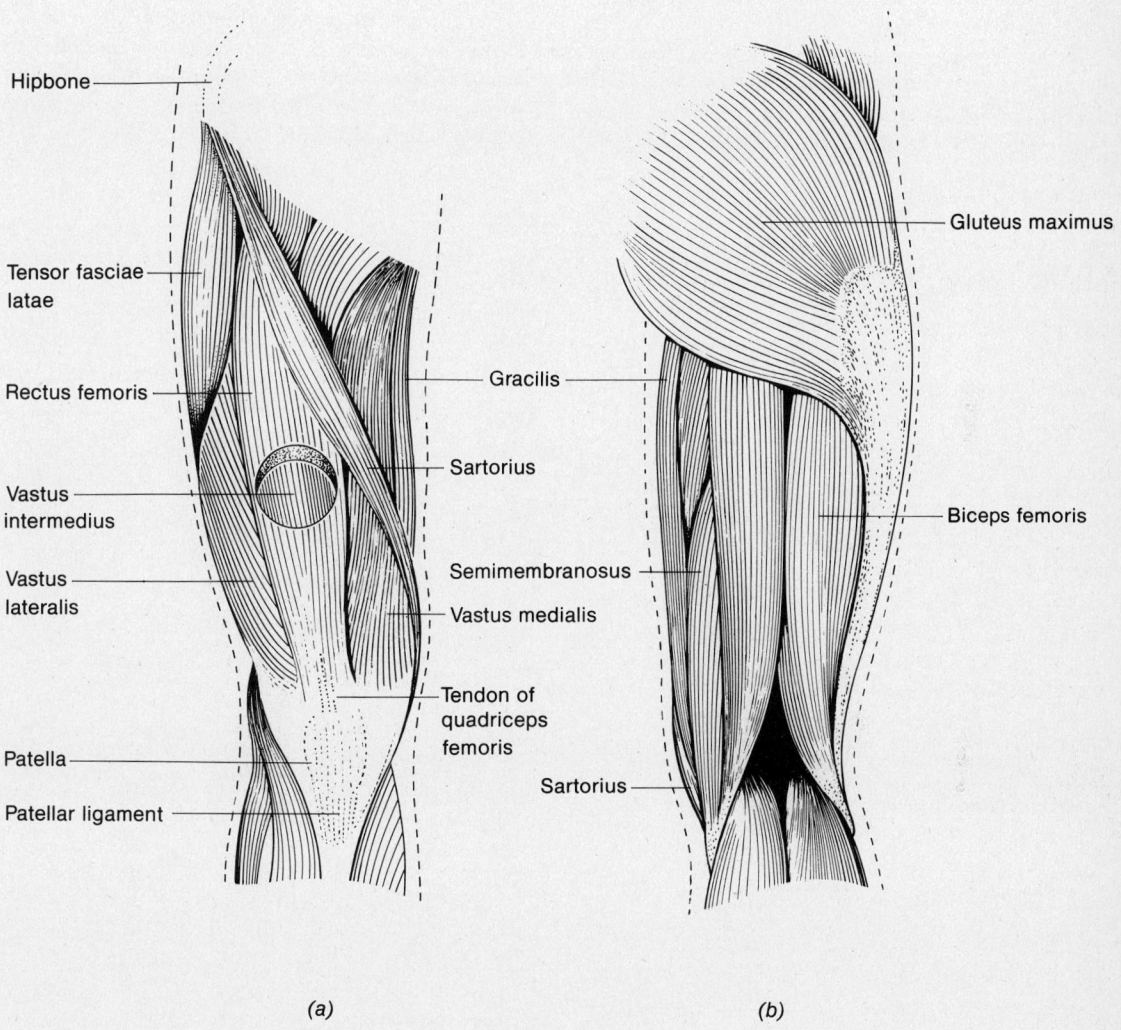

Fig. 6.16 Muscles of the upper leg. (a) Anterior view. (b) Posterior view. These muscles are usually associated with walking, running, and physical work. Belonging to this group are some of the longest, largest, and strongest muscles in the body.

THE MUSCULAR SYSTEM 93

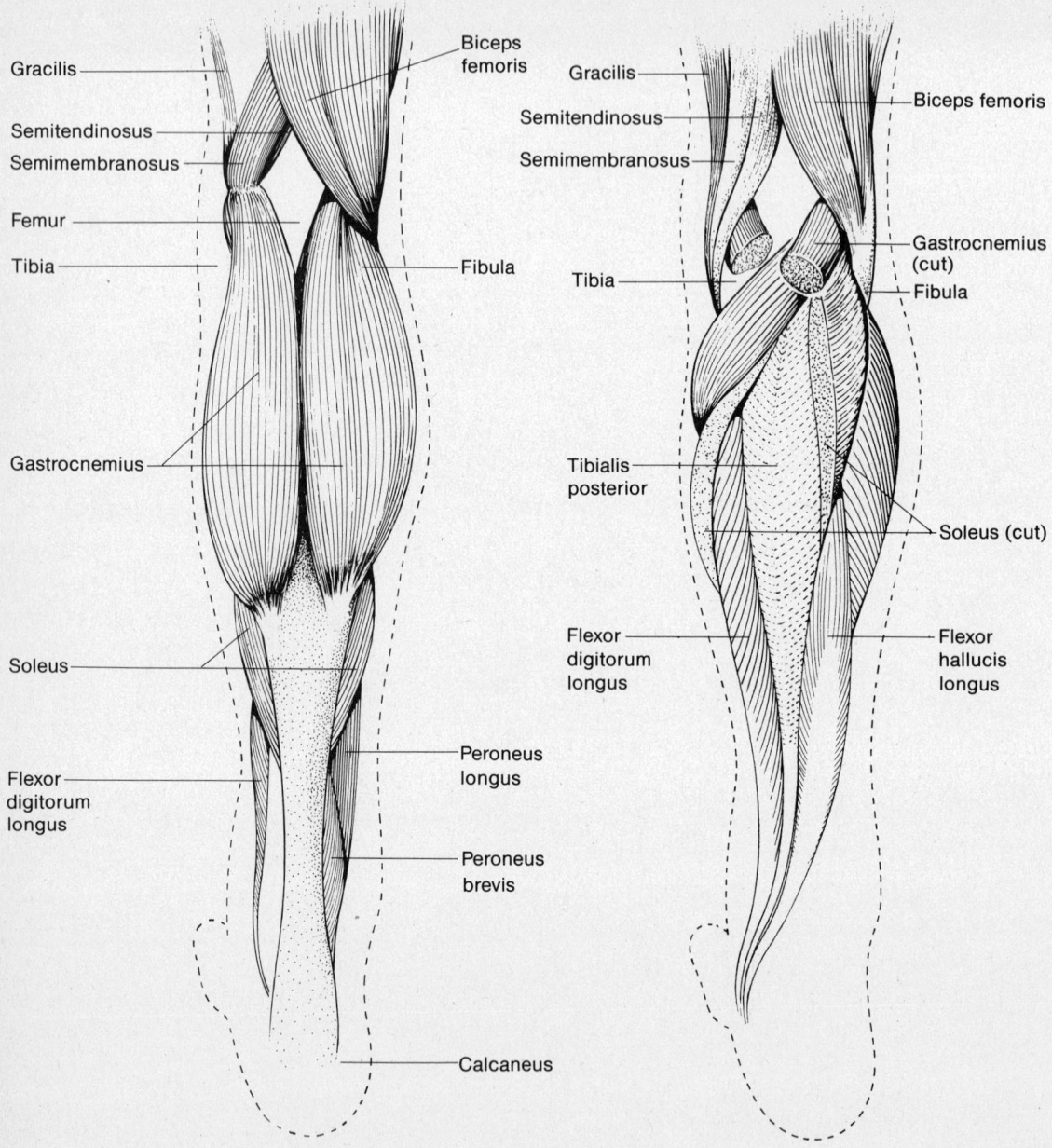

Fig. 6.17 Muscles of the lower leg. These muscles essentially provide for movement of the foot and toes.

apparently a result of injury to the sternomastoid muscle during birth. This injury causes the muscle to undergo necrosis (muscle deteriorates and dies), the muscle tissue being later replaced by fibrous connective tissue. As the individual grows, the sternocleidomastoid on the injured side cannot lengthen, resulting in the head being pulled downward and obliquely. This disability can be corrected by surgery.

Myasthenia Gravis: An Autoimmune Disease

Normal functions that occur at the neuromuscular junction require that a transmitter substance, acetylcholine, be secreted by neuron ends and received by muscle membranes. In myasthenia gravis, the muscle membrane receptors have been attacked by antibodies and are thus incapable of reacting to the influence of acetylcholine. Muscle paralysis is the consequence.

Research to control the disease is currently being directed in several different areas. Medication that increases the number of receptor molecules at the neuromuscular junction is effective in some patients. Medication that destroys antibody-producing cells at the muscle membrane appears to be helpful to some sufferers. A recent development is a process of cleansing the blood of the disease-causing antibodies. This process requires channeling the patient's blood into a centrifuge. Blood cells and platelets are returned to the body along with antibody-free plasma. Blood plasma containing the disease-causing antibody is discarded.

These procedures, used in conjunction with medication that suppresses antibody production, provide a hopeful future for those who suffer the debilitating effects of myasthenia gravis.

Myasthenia gravis is a myoneural junction disorder which affects skeletal muscle, but not cardiac or smooth muscle. The disease, a type of muscular dystrophy, is due to either a less-than-adequate secretion of acetylcholine into the synaptic gap of the myoneural junction or due to an excessive secretion of cholinesterase. In either instance, the quantity of acetylcholine reaching the muscular sarcolemma is not adequate to change the membrane polarity, and therefore the muscle cannot contract. The overall effect is great weakness of the major skeletal muscles.

Stress May Cause Protein Deficiency

Trauma, anxiety, fear, and other causes of stress may lead to decreased absorption of protein from the gastrointestinal tract into the bloodstream. Stress accelerates the chemical breakdown of muscle protein in excess of protein synthesis. Protein and amino acids are thus drained away from muscle. Amino acids are transported to the liver, where they are converted to glucose, which is used as an energy source. Prolonged stress thus aids in depleting muscle protein.

To further complicate the problem, stress often leads to a loss of appetite. Therefore, less protein as well as other nutrients is taken in. More stress leads to less muscle protein, which in turn leads to loss of appetite, which in turn leads to less protein being taken in. The loss of body protein thus continues downhill until the stress is removed.

Microbial Diseases

Most diseases of the muscular system are nonmicrobial in origin. This section will discuss only two diseases caused by microorganisms, *gas gangrene,* and *tetanus.*

Gas gangrene is caused by the bacterium *Clostridium perfringens* and certain other species of the genus *Clostridia.* Clostridia are anaerobic spore-forming bacilli and are natural habitants of the soil. *Clostridium perfringens* is a pathogen capable of producing a significant infection in the human body only when the muscle tissue contaminated by the organism has already been greatly traumatized and become *necrotic* (dead) as a result of an insufficient blood supply.

The action of *C. perfringes* on dead tissue produces toxic substances which may affect healthy tissue. The toxin has a hemolytic (red blood cell–destroying) agent that causes an anemia. (The term anemia refers to any condition that reduces the normal oxygen-carrying capacity of the blood.) The disease is characterized by the development of gas in the tissues surrounding the wound.

Tetanus is a muscular-nerve disease caused by the bacterium *Clostridium tetani.* These bacteria invade deep wounds where environmental conditions are favorable for their growth. Under these conditions, which include warmth, nutrients, and an anaerobic state, the bacteria produce toxins that are carried throughout the body by the bloodstream.

The disease is characterized by tetany, particularly of the neck and jaw (hence the common name lockjaw). Later stages of the disease produce paralysis of the thoracic muscles, usually causing death.

C. tetani is a common soil bacteria, and thus the disease is transmitted directly by contact with contaminated objects or by soil entering into deep wounds.

Tetanus: An Ancient Killer Still Present

Tetanus is an often fatal disease caused by toxin produced by Clostridium tetani, *a common soil bacterium.* Clostridium tetani *invades the body through puncture or laceration. Bacterial growth produces a local infection, and a lethal toxin, produced in dead tissue at the infection site, is absorbed into the bloodstream. This toxin eventually reaches the brain and spinal cord.*

Within two days after entrance of clostridium tetani the patient may experience pain in the jaw and back muscles and painful swallowing. During the next few days these conditions worsen. The jaw muscles tighten so that the mouth cannot open. Spasms of the upper respiratory structure inhibit normal respiratory functions. Spasms of the neck and back muscles pull the head backward and the spine into a painful curvature.

The death rate from tetanus is 30 to 70 percent. If the symptoms occur within two or three days after injury, the mortality rate approaches 100 percent.

Modern medical technology is of little help to one suffering from tetanus. Muscle relaxants to reduce muscle spasm intensity are helpful. Precautions to reduce the chances of pneumonia, atelectasis, and cardiovascular complications are standard clinical procedures. Other than these often ineffective measures, the patient's own defense mechanisms must try to win the battle.

MUSCLE REDUCTION WITH AGING

After the age of 20, muscle is progressively replaced with fat, at a rate of 0.44 percent per year. By age 70, one-quarter of the original muscle mass at age 20 is replaced by fat. Although the process of muscle wasting appears to be inevitable, it can be slowed to some degree by exercise. As physical activity normally decreases with age, muscular wasting increases.

SUMMARY AND REVIEW

A. Muscle is composed of contractile tissue that carries on functions such as:
 1. Respiration.
 2. Propelling nutrients through gastrointestinal tract.
 3. Pumping blood to tissues.
 4. Controlling the size of body openings.
 5. Production of heat.
B. Types of muscle include:
 1. Skeletal (striated or voluntary) muscle is the largest mass of tissue in the body.
 2. Smooth (visceral or involuntary) muscle lines many internal organs.
 3. Cardiac (heart) muscle is specialized tissue found only in the heart.
C. Microscopic structure of muscle tissue reveals that the muscle is made up of small units called myofibrils covered with a membrane called the sarcolemma. Each myofibril is composed of myofilaments, actin, and myosin.
D. Neural control of muscle contraction
 1. Nerve impulses arrive at the myofibril and secrete a transmitter substance, acetylcholine.
 2. The myoneural junction is the site of transmitter substance activity upon the sarcolemma of the myofibril.
 3. The overall effect of acetylcholine is to depolarize the sarcolemma. This initiates the chemical activity within each sarcomere.
E. The chemistry of muscular contraction
 1. Depolarization of sarcolemma causes calcium to be released from sarcoplasmic reticulum.

2. Calcium unites with myosin causing the cross bridges to momentarily attach to actin.
3. The cross bridges move the actin filaments, causing contraction.
4. ATP is degraded to ADP + P + energy. Muscle contraction is an energy-using process.
5. Aerobic respiration is an efficient oxygen-utilizing process that releases energy from nutrients.
6. Anaerobic respiration is the process by which a small quantity of energy is released in the absence of oxygen.
    a. Anaerobic respiration in muscle produces lactic acid accumulation in the muscle.
    b. Lactic acid accumulation leads to fatigue and muscle soreness.
    c. Oxygen debt results from the necessity of muscles, during vigorous and extended exercise, having to carry on anaerobic respiration.
F. Diseases of muscles
    1. traumatic myositis ossificans
    2. progressive myositis ossificans
    3. anterior tibial syndrome
    4. congenital torticollis
    5. myasthenia gravis
    6. gas gangrene
    7. tetanus
G. Muscle reduction with aging
    1. Muscle size is reduced.
    2. Muscle is replaced by fat.

7
The Nervous System

MAJOR CONCEPTS

THE NEURON
    Schwann Cells and Myelin Formation
    Nerve Fiber Regeneration
    Other Nerve Tissue
THE NERVE IMPULSE
COMMUNICATING PATHWAYS
    Reflex Arc
    Reflexes
    The Synapse
DIVISIONS OF THE NERVOUS SYSTEM
    The Spinal Cord
    The Brain
CRANIAL NERVES
THE AUTONOMIC NERVOUS SYSTEM
    The Sympathetic System
    The Parasympathetic System
DISEASES AND DISORDERS OF THE NERVOUS SYSTEM
THE AGING NERVOUS SYSTEM

The nervous system is the major control system for all other body systems. Working in conjunction with the endocrine system, the nervous system monitors and regulates specific functions such as muscle contraction, glandular secretion, and the reception of sensations. It is also responsible for processes involved in memory, learning, and behavior. Organization of the nervous system is discussed here, along with basic anatomical and physiological detail.

Every action the human body performs is initiated, coordinated, and completed by a complex cellular network called the nervous system. For convenience of study the nervous system is customarily divided into two parts. The *central nervous system* is composed of the brain and spinal cord. The rest of the human nervous system is called the *peripheral nervous system*.

The entire system is responsible for the initiation and integration of most body functions. These functions include muscular contraction, and the perception of light, sound, pain, and touch. In addition, the nervous system maintains homeostatic mechanisms responsible for coordination of all body activities. As external and internal changes occur, the nervous system senses these changes and initiates appropriate responses. These responses vary from subtle alterations on a minute-to-minute basis to major actions resulting in vigorous activity necessary to preserve a homeostatic balance, and indeed life itself.

The role played by the nervous system in integrating body functions cannot be over-emphasized. Just the simple act of swallowing a mouthful of food requires detailed and precise coordination and integration of body parts.

THE NEURON

The basic unit of the nervous system is the *neuron*. (See Figure 7.1.) This cell is capable of accepting a stimulus and transmitting information to distant parts of the body, eliciting an appropriate response.

Neurons differ structurally from other cell types. A neuron is made up of a cell body with threadlike extensions radiating from it. One of these extensions, called the *axon*, emerges from the cell body and may extend for a distance varying from a few millimeters to a meter or more. *Dendrites* are shorter processes radiating from the neuron. The cell also contains numerous dark staining granules called *Nissl bodies*, whose function is not known.

Three types of neurons are identified according to the arrangement of the attached processes. *Unipolar neurons* have a single process arising from the cell body, as shown in Figures 7.2 and

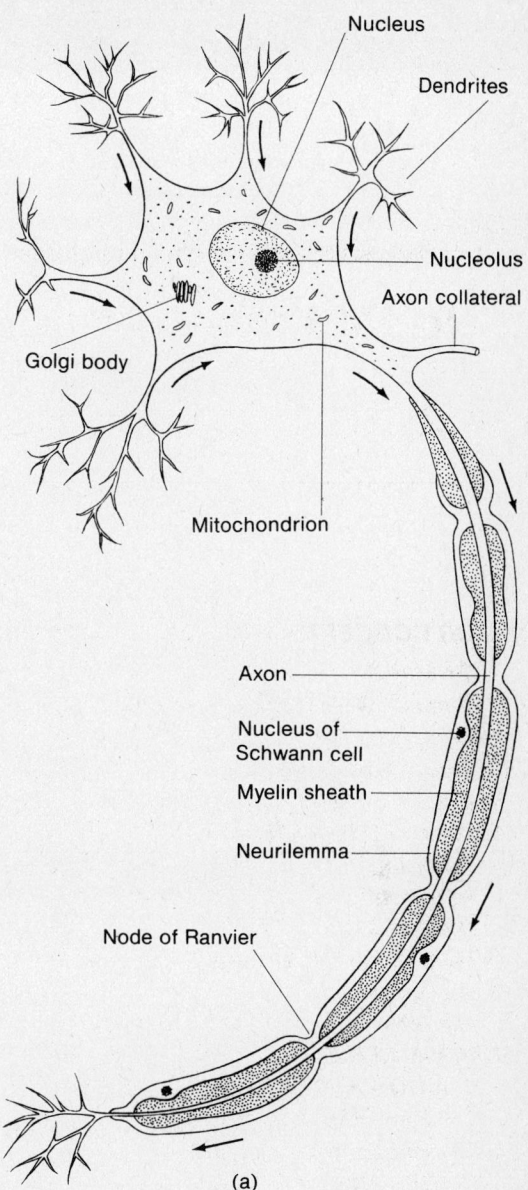

Fig. 7.1 A neuron, the basic unit of the nervous system. It is similar to the body cell in most respects in that it contains a nucleus, nucleons, mitochondria, and Golgi bodies. It differs from most cells in having cellular appendages called dendrites and axons.

THE NERVOUS SYSTEM 99

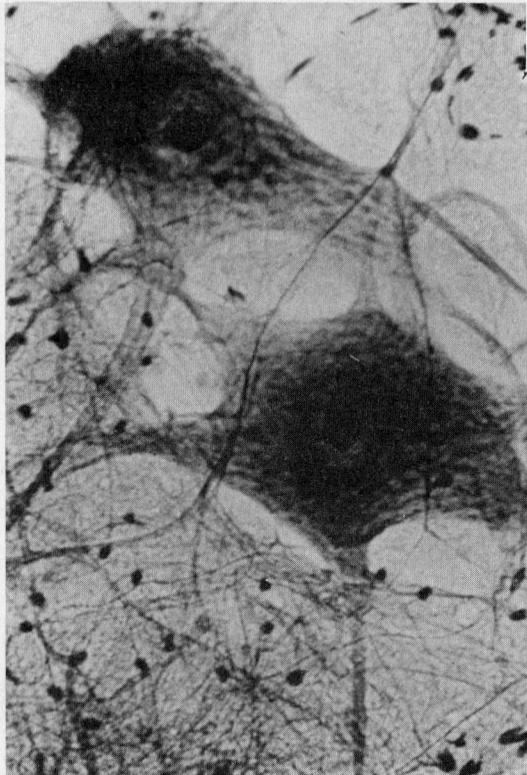

Fig. 7.2 A photomicrograph of neurons. *(Author's collection.)*

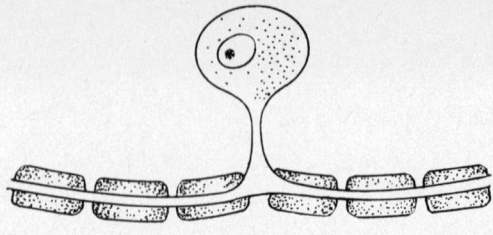

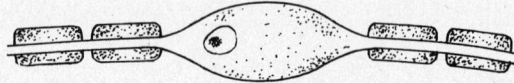

Fig. 7.3 Difference between unipolar neuron and bipolar neuron.

7.3. Neurons of this type are most often associated with the senses and are located near the spinal cord.

Another type of neuron, the *bipolar neuron,* has two extensions radiating from the cell body. These neurons are located in certain sense organs, including the eye, ear, and nose.

The most abundant type of neuron is the *multipolar neuron.* These neurons have a single, usually quite long, process and numerous short processes, as shown in Figure 7.1. Most multipolar neurons are located in the brain and spinal cord. The processes of these neurons carry nerve impulses away from the central nervous system to the viscera, skeletal muscle, smooth muscle, heart muscle, and glands.

Schwann Cells and Myelin Formation

Some nerve cells have specialized coverings over their axons and dendrites. These coverings are made up of a variety of structures.

A *myelinated nerve fiber* has these coverings while a *nonmyelinated fiber* does not. The myelinated fiber is usually larger and conducts a nerve impulse at a faster rate than the nonmyelinated variety.

Figure 7.4 shows a part of an axon. Notice the structure labeled *Schwann cell.* This cell has no direct nerve transmission function, but plays an important role in insulating the fiber and is instrumental in nerve regeneration.

Schwann cells wrap around segments of an axon, producing a laminated sheath. Schwann cells secrete an insulating substance called *myelin* between the laminations. Periodic gaps in the sheath are called *nodes of Ranvier.*

A myelinated and an unmyelinated nerve are shown in Figure 7.5.

Nerve Fiber Regeneration

Nerve cells are not capable of regeneration, nor can they increase in number once the nervous system is completely developed. We are born with most of the neurons which we will ever have; if neurons are destroyed, they are not

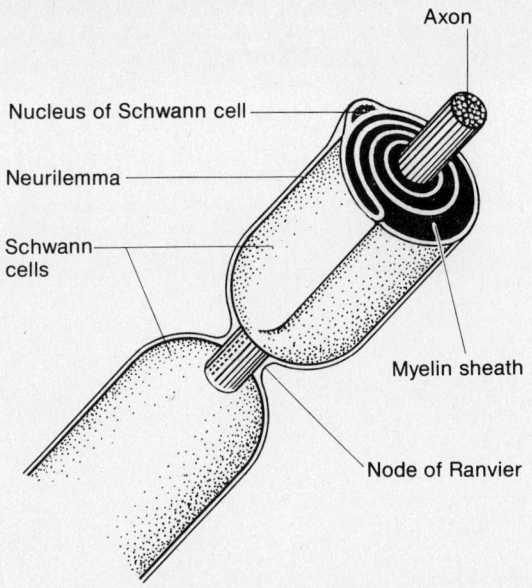

Fig. 7.4 A sectional view of an axon showing the relationship of the Schwann cell and axon. The Schwann cell has no function in nerve cell transmission; however, it plays an important role in insulating the nerve fiber and is also instrumental in the regeneration of nerve appendages.

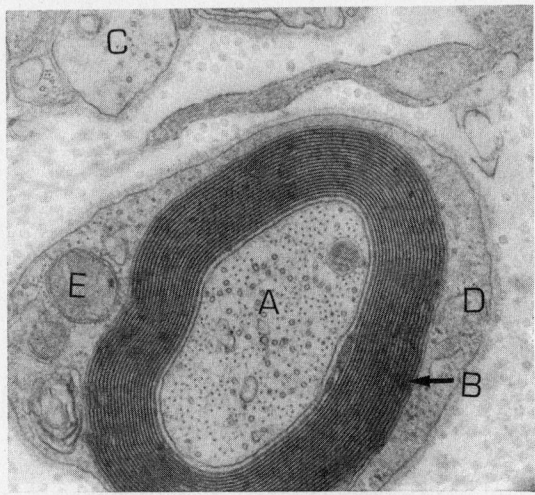

Fig. 7.5 Electron micrograph of (A) myelinated and (C) nonmyelinated nerves. The myelin sheath is shown at (B). Magnification is 12,300×. Also shown is the Schwann cell at (D) and the Schwann cell nucleus at (E). (Courtesy Santiago Plurad.)

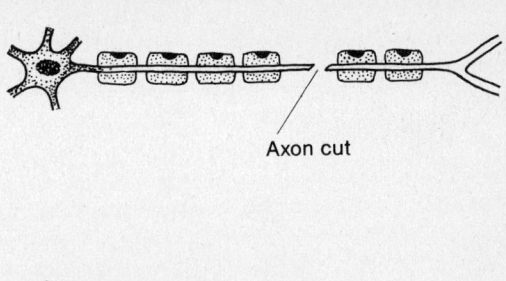
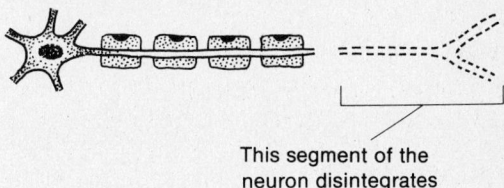
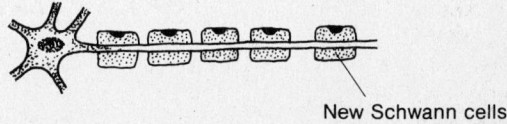

Fig. 7.6 Nerve fiber regeneration. After new Schwann cells are formed, the undamaged portion of the fiber grows into the newly recreated tube formed by the Schwann cells. At the injury site the neurilemma begins to regenerate, forming a new tube that grows in the direction of the previous nerve pathway. Once the tube is completely formed, the neuron fiber into this new tube reestablishes the connection of the neuron cell to its original site.

replaced by new cells. Parts of the neuron axons and dendrites, however, can regenerate if they are damaged.

If an axon is injured, the part separated from the cell body degenerates along with a portion above the injury site, as shown in Figure 7.6. The axon is covered by a layered tissue made up of Schwann cells. The outer surface of this layered tissue is called the *neurilemma.*

At the injury site the neurilemma starts to regenerate, forming a tube that grows in the direction of the previous nerve pathway. Once the tube is completely formed the neuron fiber grows into it, reestablishing the connection of the neuron to its previous site. In instances where the tissue has been severely damaged, the neurilemma and its nerve fiber cannot return precisely to the original site.

The process of nerve fiber regeneration is a rather slow procedure, and the time required is highly variable, often taking many months for completion.

Other Nerve Tissue
Associated with neurons are other specialized cells called *neuroglia*. The precise function of some types of glial cells is not understood. Some neuroglia serve as connective tissue that supports and protects neurons. Other types are phagocytic, thereby protecting neurons from microbial invasion. Table 7.1 summarizes the types of neuroglia and their functions.

Table 7.1 Neuroglia Types and Function

TYPE OF CELL	FUNCTION
Astrocyte	Provides a connective tissue that holds neurons together in the brain and spinal cord. Also aids in holding neurons closely attached to capillaries that nourish them.
Oligodendrocyte	Like astrocytes, these cells provide support and connective function for neurons. Oligodendrocytes also produce the axon covering called myelin.
Microglia	These cells have phagocytic action. They are capable of engulfing bacteria or other extraneous matter around neurons.

THE NERVE IMPULSE

When a dendrite is stimulated, an electrochemical signal (the message) is flashed in the direction of the cell body and continues uninterrupted to the distal end of the axon. This electrochemical signal is called a *nerve impulse*. When the impulse arrives at the axon end, a series of chemical reactions may stimulate new impulses in other nearby neurons located in the spinal cord or brain. The function of the brain and spinal cord is to monitor all impulses. This monitoring permits the body to make the appropriate response to environmental stimuli. For example, if someone touches your thumb, you may respond with a slight movement away from the stimulus. However, if this same thumb is struck with a hammer, a quite different response would be elicited.

A nerve impulse may be initiated by a variety of stimuli, including heat, cold, pressure, light, chemicals, and electricity. The impulse carried along the dendrite has the same intensity and travels at the same rate of speed regardless of the source of stimulation. The rate of speed is about 120 meters per second. Compared to the speed of electricity in a conductor (298,000 km per second), this is quite slow. Since, however, the distances over which impulses travel in the body are short, the time interval is measured in fractions of a second and seems instantaneous.

Figure 7.7, a generalized diagram of a portion of the axon of a neuron, will help explain how the electrical nature of a nerve impulse originates and travels along the nerve fiber. Before a nerve impulse is initiated, a nerve fiber has different electrical charges on its outer surface than on its inner surface. There are more positively charged sodium (Na^+) ions on the outer surface of the membrane. There are also more positively charged potassium (K^+) ions inside the cell membrane. Potassium ions can move out of and into the cell with relative ease. Sodium ions, on the other hand, cannot easily move across the membrane. This makes the outer surface more positively charged than the inner surface. An axon with these differential charges is said to be *polarized*.

Stimuli reaching the fiber causes its membrane to suddenly become highly permeable to sodium ions, which rush from the surface inward. The loss of positive sodium ions from the outer membrane makes it negative with respect to the inside. The reversal of charges on the membrane is called *depolarization*.

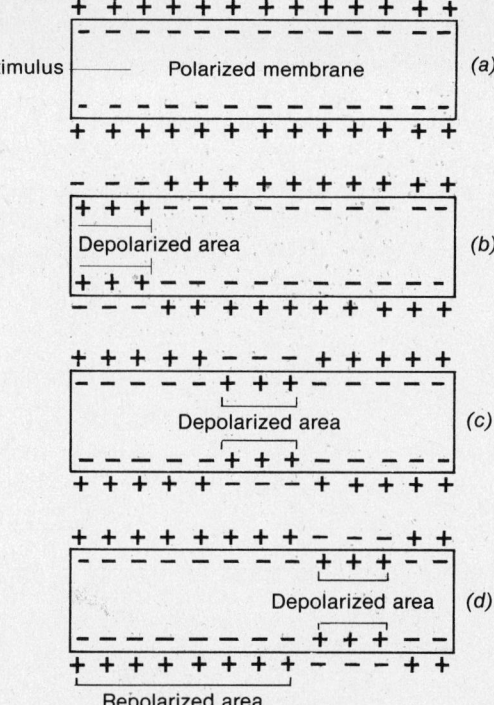

Fig. 7.7 Nerve impulse transmission. In (a) no impulse is moving along the fiber. In (b), (c), and (d), note the progression of the opposite charges along the fiber and the repolarization that occurs after the wave of depolarization has passed.

A nerve impulse is a wave of depolarization that travels along a nerve fiber away from the stimulus. As soon as the wave passes a segment of the nerve fiber, sodium ions are pumped back out of the fiber restoring its polarity with more positive ions on the surface of the membrane.

We do not know with certainty what a stimulus does to initiate the nerve impulse. Not every stimulus reaching the nerve fiber causes a nerve impulse to be started. The nerve fiber may be only partially depolarized, in which case an impulse is not initiated. The minimum strength of stimulus necessary to initiate a nerve impulse is called the *threshold of stimulation*. Different neuron fibers have different minimum threshold levels. Once this threshold is reached an impulse is always initiated. Once begun, a nerve impulse travels at the same rate of speed along the fiber regardless of the strength of the stimulus.

A stimulus greater than the minimum threshold initiates the same kind, speed, and strength of impulse. A very low level of stimulation (less than threshold) may partially depolarize the nerve fiber. A second less-than-threshold stimulus (subliminal stimulus) acting on the partially depolarized fiber may cause an impulse to be produced. In other words, subliminal stimuli may have a cumulative effect. Separately they would not initiate an impulse, but collectively they can depolarize the fiber, bringing about a nerve impulse, as shown in Figure 7.8.

COMMUNICATING PATHWAYS

The nervous system of the body is an accumulation of millions of neurons. Some receive stimuli (*sensory neurons*), some carry impulses to and activate organs (*motor neurons*), while others (*internuncial*) connect sensory and motor neurons. Internuncial neurons are generally confined to the central nervous system, where they

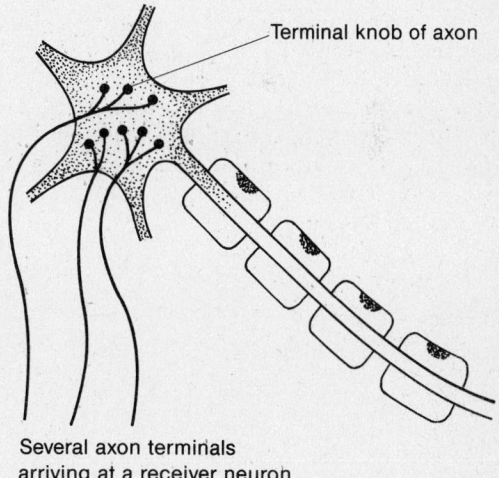

Fig. 7.8 Illustrates the cumulative effect of separate subthreshold stimuli. The stimuli from several axons may have to arrive simultaneously at the receiving neuron to provide a threshold stimulus.

act as pathways connecting sensory and motor neurons, as well as integrating the functions of both. One common pathway is called the *reflex arc*.

Reflex Arc

Reflexes are automatic, unlearned responses to specific stimuli which control many vital body functions. The pathway over which nerve impulses travel to produce reflex action is called a *reflex arc,* as shown in Figure 7.9.

The simplest reflex arc involves a sensory, an internuncial, and a motor neuron. An impulse originating in the sensory neuron dendrite is carried along the fiber entering the dorsal portion of the spinal cord. The sensory neuron cell body is located outside the spinal column. The sensory neuron axon enters the grey matter of the spinal column and communicates with the dendrites of an internuncial neuron. Notice that we did not say that the axon and dendrite are connected. A small space called a *synapse* separates the two nerve endings. A new impulse, initiated in the connecting neuron, is carried to another synaptic junction that communicates with a motor neuron. Another new impulse initiated in the motor neuron travels to a muscle, gland, or some other motor organ bringing about a response. The response made by motor organs to the impulse is called a *reflex act* or simply a *reflex*. The overall procedure is summarized in Figure 7.10.

Reflexes

Simple reflexes do not involve thoughtful activity, and the higher brain centers may or may not be stimulated. If you put your hand on a hot surface, sensory nerve endings in your hands are stimulated. This initiates an impulse that is carried along the reflex pathway to arm muscles that contract to pull the hand away from the hot object. This action is a sudden, automatic response that does not involve the brain or thoughtful activity. In reflex action the brain

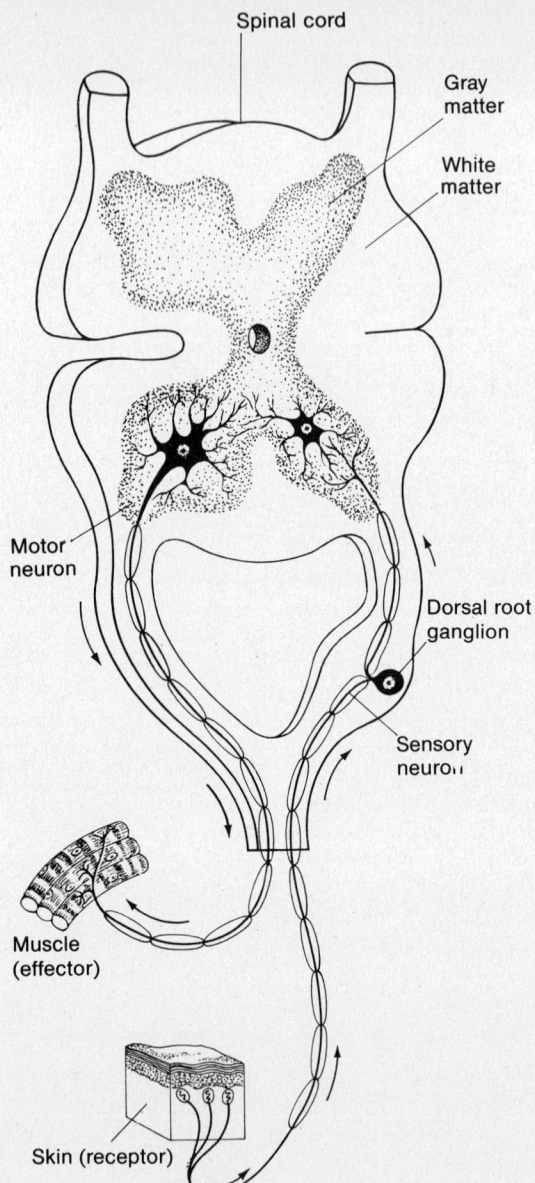

Fig. 7.9 The reflex arc is a pathway over which nerve impulses travel, thus initiating a reflex act. An impulse originating in the sensory neuron dendrite is carried along the nerve fiber entering the dorsal portion of the spinal cord. The sensory neuron axon communicates with an internuncial neuron, which in turn communicates with the dendrites of a motor neuron. The impulse thus communicated is transmitted to some tissue such as muscle or gland.

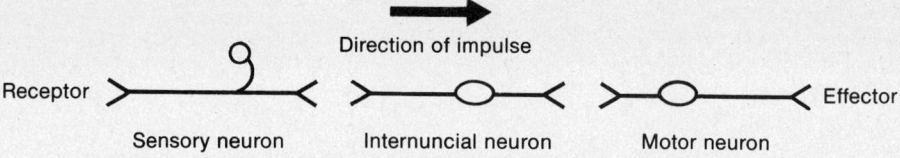

Fig. 7.10 Reflex arc pathway. This is a simplified representation of the conduction pathway involving the nervous system.

does not have to make a time-consuming decision. The action is automatic and prompt.

Reflexes are generally protective and may vary in intensity. For example, a weak stimulus such as an eyelash being touched results in eye blinking. The response to severe pain usually results in a vigorous contraction of many muscles that attempt to move the body away from the stimulus.

Reflexes have the added feature of being stereotyped responses to stimuli; their stereotyped behavior makes them useful in assessing nerve damage caused by disease or trauma.

Figure 7.11 illustrates how the brain may become involved in reflex activity. The impulse from the sensory neuron may make a synaptic junction with both a connecting neuron, which is a part of the reflex arc, and neurons leading toward the brain. Impulses such as hair on the arm being touched may be strong enough to elicit a simple reflex but too weak to initiate an impulse in neurons leading to the brain.

A strong impulse in a sensory neuron, such as one initiated by pain, causes impulses to be set up in neurons leading to the brain in addition to those traveling along the reflex arc. When these impulses reach the higher brain centers, an analysis of the stimulus is made. From this analysis we recognize what the stimulus was. This recognition may trigger other appropriate responses. For example, a loud noise may cause you to jump; however, when you recognize that the noise, say a dropped book, is not life threatening, no other response will be made.

The response to an explosive sound accompanied by a flash of light would be much more intense. Not only might you jump, but in addition you would no doubt take appropriate action to move away from the threatening situation. This latter type of activity is called a *conditioned response*, or a *learned response*. We have learned through experience that certain stimuli are harmful to the body.

Simple reflexes exhibited by infants and adults include sucking, swallowing, coughing, sneezing, and vomiting. Conditioned reflexes are too numerous to list since they are so intricately a part of normal activity. Not all conditioned

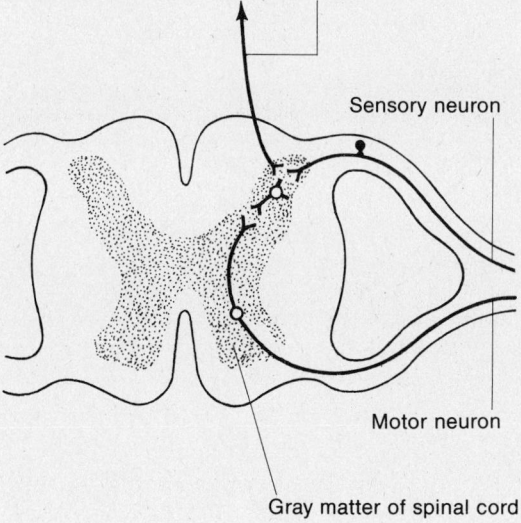

Fig. 7.11 How brain may become involved in reflex activity. Dendrites from brain cells may be stimulated by impulses arriving at the internuncial neuron. An impulse thus initiated arriving at the brain may be received and ignored or may stimulate the brain to counteract the reflex action.

THE NERVOUS SYSTEM 105

reflexes are necessarily in response to danger. For example, we have been conditioned to react, in a specified manner, to traffic signals and signs. Hospital personnel are conditioned to react "automatically" to specific situations.

Certain reflexes are important physiologically. These include postural reflexes, corneal reflexes, and others.

Postural reflexes involve nerve activity related to keeping the body in an upright position. Impulses for this action may arrive at the brain (cerebellum) from several sources, including the inner ear, pressure receptors in the skin, stretched muscles and tendons, and the retina of the eye. The brain coordinates all the arriving impulses and sends appropriate signals out to muscles, which contract sufficiently to keep the body in an upright position. This is an unconscious reflex activity; otherwise, we would constantly be making positional adjustments merely to sit on a chair.

It is interesting to observe the influence of stimuli received by the retina upon postural reflexes. A person standing on the edge of a cliff or building tends to lean backward away from the edge. The eyes relate to a normal walking surface extending in front of the body. When this surface is missing, impulses are sent to the brain and the spinal cord, which cause muscles to pull the body away from the edge. This reflex may also be influenced by conditioning; perhaps the memory of falling initiates other patterns of impulses traveling over nerve pathways which serve to protect the body from danger.

Clinically important reflexes are those that may be artificially induced to diagnose possible nerve pathway damage. These include such reflexes as the *corneal reflex*, which blinks the eye when the cornea is touched; the *pharyngeal* or *gag reflex*, which results when the upper pharynx is touched; the *abdominal reflex*, which results in the contraction of abdominal muscles when the skin of the abdomen is touched; and the *anal reflex*, which results in anal contraction upon the insertion of a rectal tube.

Deep reflexes involve the contraction of muscles when tendons are stimulated by physical stimuli.

The *biceps reflex* results from a sharp tap on the biceps tendon, causing the forearm to be flexed.

Tapping the triceps causes the forearm to be extended.

If the legs are crossed and the knee is tapped just below the kneecap (patella), the lower leg is extended. This is called the *patellar reflex*.

The *Achilles reflex* results from tapping the Achilles tendon (above the heel), causing the foot to be flexed.

Two other reflexes, called *visceral reflexes*, are the *pupillary reflex* and the *commensual light reflex*.

The pupillary reflex causes a constriction of the pupil when bright light is suddenly directed into the eye.

The commensual light reflex causes the pupils of both eyes to constrict when bright light is suddenly directed into one eye.

The Synapse

Thus far we have discussed impulse transmission along nerve pathways, such as the reflex arc, as if these pathways were continuous, connected filaments. We have made only the slightest reference to the communicating gap between neuron fibers—the *synapse*.

For years there was controversy as to how an impulse traveled from one neuron to another. Some believed that the transmission was electrical in nature, while others thought it was chemical. Experimentation by Otto Lowwi in 1921 showed it to be largely chemical. Subsequent experimentation indicates that some *transmitter substance* is released at the ends of axons into the synaptic gap. This substance, coming in contact with the dendrite end of another neuron, initiates another impulse having the same characteristics as the impulse originating in the axon, as shown in Figure 7.12.

One transmitter substance released from the ends of some axons is *acetylcholine*. This chemical rapidly diffuses across the synaptic space (about a millionth of an inch) and initiates a new impulse in the connecting neuron.

The amount of acetylcholine released into the synaptic space is determined by the rapidity of impulses arriving at the end of the axon. If only a few impulses arrive, the small quantity of acetylcholine released may be insufficient to initiate a response in the next neuron. In other words, a less-than-threshold stimulus is present. A threshold stimulus may be provided if other stimulated axon ends converge at the same synapse, each emptying acetylcholine into the synaptic space.

This method of perpetuating a nerve impulse along a nerve pathway permits the signal to travel in only one direction; in other words, only one end of a neuron fiber, the axon, can release acetylcholine, while the other end, the dendrite, can only receive the chemical stimulus and originate a new impulse.

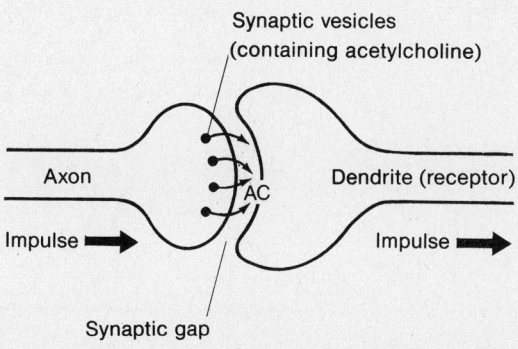

Fig. 7.12 The synapse. Acetylcholine (transmitter substance) is emptied from synaptic vesicles into the synaptic gap. The receptor dendrite is stimulated by acetylcholine and initiates a new impulse in the receptor.

If acetylcholine were permitted to remain in the synaptic space, it would repeatedly induce impulses in the next neuron. An enzyme, *cholinesterase*, secreted into the space effectively destroys the accumulated acetylcholine. The synaptic space is thus cleared of acetylcholine.

DIVISIONS OF THE NERVOUS SYSTEM

The nervous system is so diverse, both anatomically and functionally, that it will be necessary to divide the system into groups of activities, then summarize the interrelated functions of these groups later in the chapter.

In the reflex arc, sensory neurons carry impulses toward the central nervous system, the spinal cord, and the brain. Motor neurons carry impulses from the central nervous system outward to motor organs such as muscles. Motor neurons control the voluntary activities associated with the body, particularly those involving movement provided by skeletal muscles.

The Spinal Cord

The elongated, tubular mass of nerve tissue called the *spinal cord* lies in the dorsal region of the body and is surrounded by the bony vertebral column. The spinal cord in the adult body runs from the base of the skull to about the level posteriorly where the last rib is attached to the vertebral column. When the spinal cord is examined in cross section, a definite pattern of tissue arrangement can be seen, as shown in Figure 7.13.

The butterfly-shaped *grey matter* is largely composed of nerve cell bodies and nonmyelinated nerve fibers. The *white matter* is composed of myelinated nerve fibers.

The spinal cord is actually a collection of nerve cells and their fibers (axons and dendrites), functioning as traffic ways for impulses that produce reflexes, transmit information to the brain for coordination, and transmit information from the brain out to muscles and glands for appropriate action.

In a spinal cord cross section, one can see that the white matter is divided on each side into three *columns*—the *posterior, lateral,* and *anterior columns*. Within each column are found bundles of axons and dendrites which serve particular

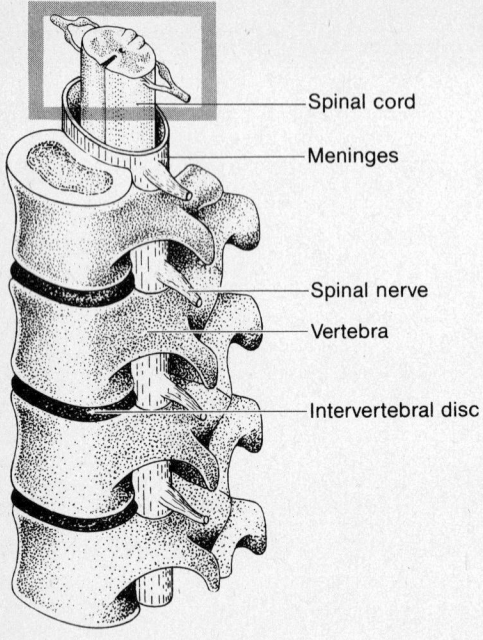

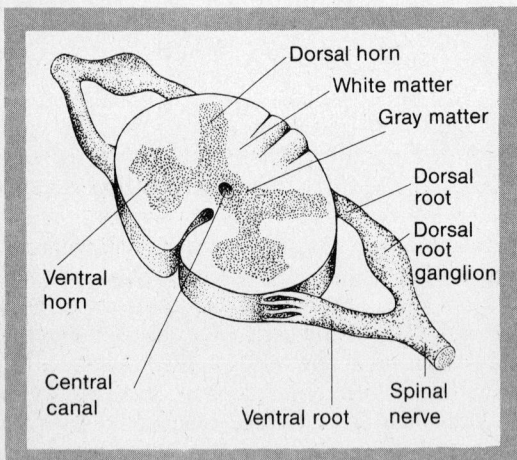

Fig. 7.13 Cross section of the spinal cord. The butterfly-shaped mass is largely composed of nerve cells and nonmyelinated nerve fibers. The white matter is composed of myelinated fibers.

body segments. These bundles are called *spinal tracts*. Some of these tracts carry impulses toward the brain, and hence are called *ascending tracts*, also known as *sensory pathways*. *Descending tracts*, also known as *motor pathways*, carry impulses away from the brain, down the spinal cord. Figure 7.14 illustrates spinal columns and tracts. Table 7.2 lists the major ascending and descending tracts and relative information about each.

Nerves and nerve fibers are not synonymous. A *nerve fiber* is a single extension of a single neuron. The extensions (axons and dendrites) may be very short or long as are some axons. For example, if you wish to move your big toe, a nerve impulse must travel over a single nerve fiber that originates at the neuron cell body located in the spinal cord and extends down the leg to muscles that move the toe. This nerve fiber may be 75 cm or more long.

A *nerve* is an accumulation of nerve fibers all originating in the same general area, wrapped together with connective tissue. Nerves may be nearly as long as the longest fibers, but they are much larger since they may be composed of hundreds of individual fibers. Nerves are easily visible to the unaided eye, while most nerve fibers are not. The individual nerve fibers function independently. Although fibers are bundled together, each one maintains its integrity and is

Table 7.2 Principle Spinal Tracts

DESCENDING TRACTS	FUNCTION
Corticospinal	Transmits impulses to skeletal muscles for voluntary action
Rubrospinal	Transmits impulses to skeletal muscles for involuntary action (posture, proprioception)
Reticulospinal	Skeletal muscle tone
ASCENDING TRACTS	
Spinothalamic	Pain, temperature, and crude touch recognition
Spinocerebellar	Muscle tone, posture
Gracile cuneate	Proprioception, pressure touch

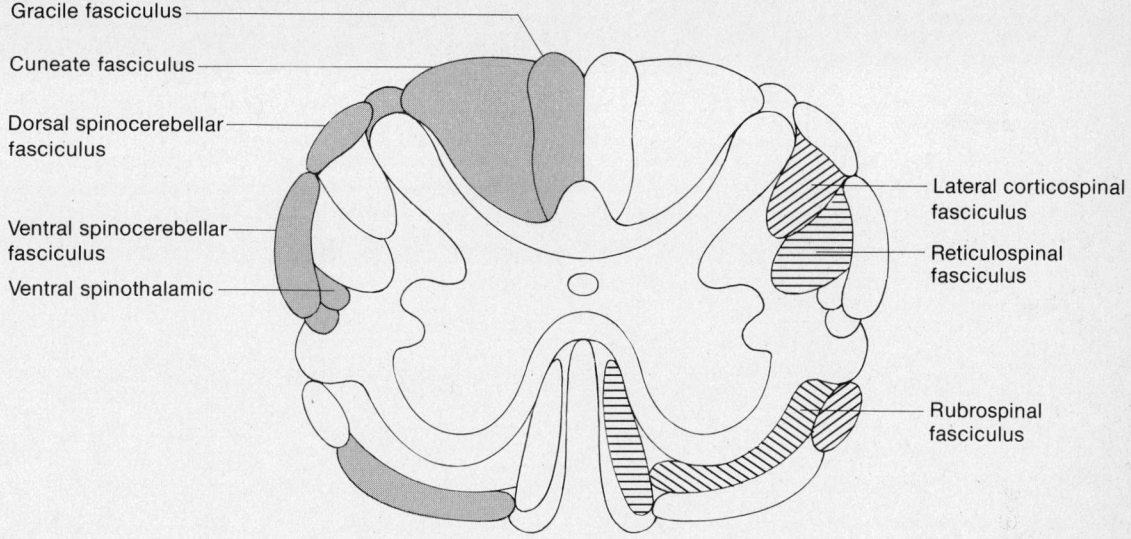

Fig. 7.14 Major ascending and descending nerve tracts. The ascending tracts, shown here as cross-hatched, carry impulses toward the brain. Descending tracts, shown shaded, are motor pathways carrying impulses away from the brain.

not influenced by the association with other fibers. Figure 7.15 illustrates the relationship between nerve fibers and nerves.

Spinal nerves, which are a part of the peripheral nervous system, are bundles of nerve fibers emerging from the spinal cord. There are 31 pairs of spinal nerves, named relative to spinal column levels. The first 8 are *cervicle*, the next 12 *thoracic*, then 5 *lumbar*, 5 *sacral*, and 1 *coccygeal*.

Meninges The spinal cord and brain are surrounded by a layered membrane called the *meninges*, which serves as a protective tissue for the delicate neurons making up the central nervous system.

The meninges, as shown in Figure 7.16 are composed of three layers. The outermost layer is the *dura mater;* the middle layer is the *arachnoid;* and the innermost layer is the *pia mater.* The pia mater adheres tightly to the surface of the brain and lines the vertebral canal, which houses the spinal cord.

The meninges are of special concern because they are susceptible to infection by certain bacteria causing the disease *meningitis*. This disease will be discussed at the end of this chapter.

Cerebrospinal Fluid The brain and spinal cord is bathed by a clear, colorless, watery fluid called *cerebrospinal fluid*. This fluid is secreted by special capillaries that line the brain ventricles.

Cerebrospinal fluid is constantly being secreted and then absorbed into the bloodstream, thereby maintaining a rather constant volume of about 200 ml in an adult. An excess of this fluid or a blockage of the normal fluid flow through the ventricles into the subarachnoid spaces creates a pressure upon brain tissues. An examination determining that the fluid pressure is excessive may indicate a brain tumor blocking the flow between brain ventricles.

If the fluid accumulates in the brain of an infant before the skull bones have fused, the head becomes enlarged, producing a condition known as *hydrocephalus*.

The primary function of cerebrospinal fluid is absorbing shock. Some of the force of a blow to

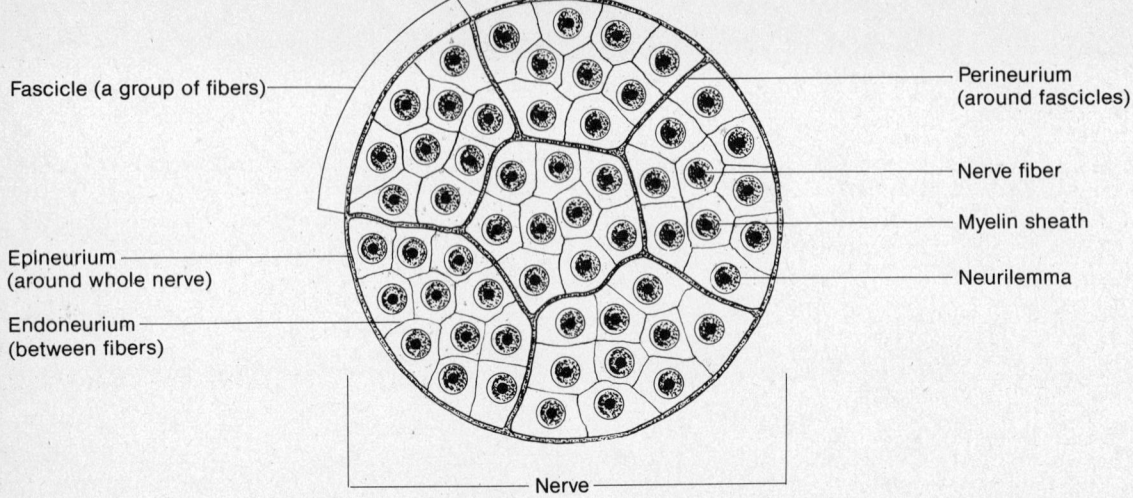

Fig. 7.15 A cross section of a large nerve. A nerve is an aggregation of nerve fibers wrapped together with connective tissue all originating from the same general area.

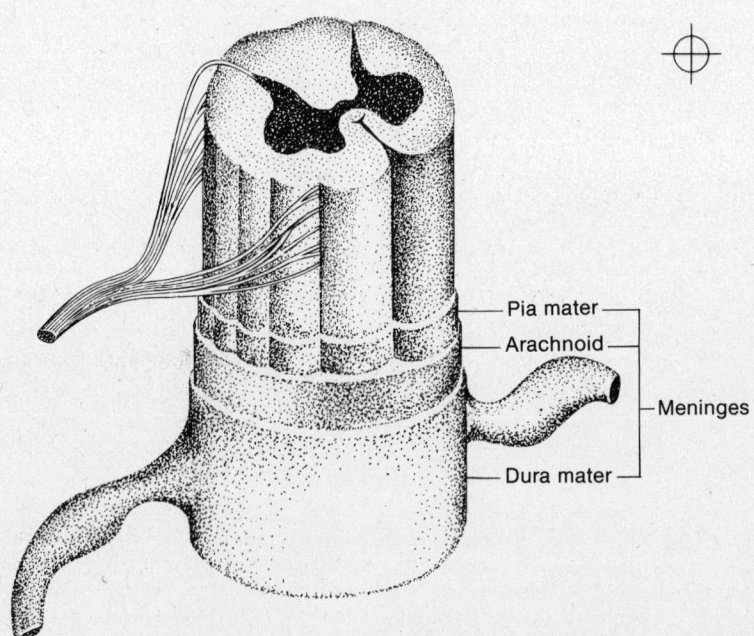

Fig. 7.16 The meninges is a membrane wrapped around the spinal cord. The membrane is composed of three layers. The outermost layer is the dura mater, the middle layer is the arachnoid, and the innermost layer is called the pia mater.

the head can be absorbed by the fluid, preventing injury to the brain tissues.

Because the fluid contains, in addition to water, some salts, glucose, protein, and a few white blood cells, it has been suggested that another function of the fluid is to serve as an intermediate source of nutrients and oxygen for brain cells and a temporary depository for metabolic wastes.

Examination of cerebrospinal fluid has become a routine clinical procedure. The fluid can be removed (or its pressure determined) by inserting a needle between the second and third lumbar vertebrae into the subarachnoid space (lumbar puncture) or into this same space through the foramen magnum (cisternal puncture).

Laboratory examination of cerebrospinal fluid is used to determine the chemical nature of the fluid, the presence of bacteria, and the number of white blood cells.

The Brain
The brain constitutes the largest mass of nerve tissue in the body. The average adult brain weighs between 1,300 and 1,400 g, depending on age, sex, and size of the individual. Men have a slightly larger brain than women.

The brain is an accumulation of neuron cell bodies, mostly located on its periphery, and neuron fibers, which make up the bulk of the brain. The peripheral cell bodies and their non-myelinated fibers make up the *cortex* of the brain. This area is also called grey matter. The rest of the brain, except for a few scattered islands of grey matter called the *basal ganglia,* is called *white matter.* This tissue is made up largely of myelinated nerve fibers that carry impulses to and from the cortex.

The brain is divided into five parts, each part having its own distinctive anatomy and function. The five divisions are: *cerebrum, cerebellum, midbrain, pons,* and *medulla.*

Cerebrum The cerebrum is the largest and most anterior part of the brain and is the center of all conscious activity. It is divided longitudinally (front to back) by a deep groove called the *longitudinal fissure,* as shown in Figure 7.17. The surface of each hemisphere is a mass of convolutions (folds). A furrow or depression between folds is called a *sulcus* and the ridge of each fold

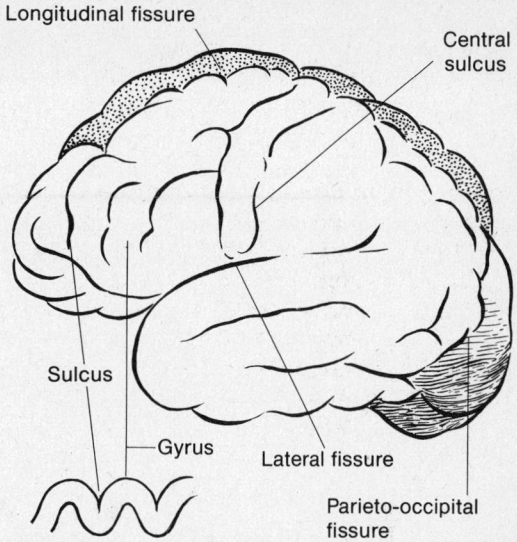

Fig. 7.17 Brain geography showing longitudinal fissure, sulci, and gyri. A furrow between folds is called a sulcus and the ridge of each fold is called a gyrus.

is called a *gyrus.* The folds in the cerebrum greatly increase the surface area.

Other deep fissures of the cerebrum include the *lateral fissure,* which separates the temporal lobe from each hemisphere; the *central sulcus* (fissure of Rolondo), which separates the frontal and parietal lobes; and the *parieto-occipital fissure,* which separates the parietal and occipital lobes. The lobes of the brain provide landmarks for a discussion of functions of the cerebrum.

The relative locations of the grey and white matter of the brain can be seen in a cross section of the cerebral hemispheres (see Fig. 7.18). Note the area labeled *grey matter* in the cortex and the more internal *white matter.*

A cross section of the brain through the longitudinal fissure, as in Figure 7.19, shows that the internal structure is composed of spaces called *ventricles.* These ventricles are normally filled with cerebrospinal fluid. This view also shows other internal structures such as the thalamus and hypothalamus.

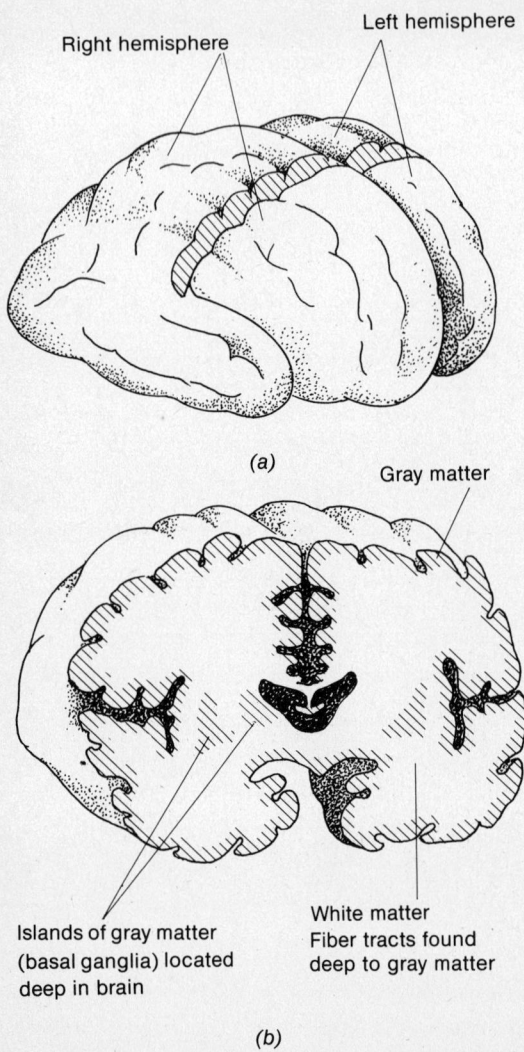

Fig. 7.18 Cross section of cerebral hemispheres. (a) View of brain partly cut. (b) View of cut frontal section.

The pituitary gland, just below the hypothalamus, is an endocrine gland and is not composed of nervous tissue. Its function, however, is closely regulated by the brain.

> ### A Brain Hormone from Bacteria
>
> *In 1977 researchers at the University of California in San Francisco succeeded in "tricking"* E. coli, *a common colon bacteria, into making the hormone* somatastatin. *This hormone, which is secreted normally by the hypothalamus, inhibits the pituitary glands in their release of hormones that regulate body growth and glucagon and insulin production.*
>
> *A gene was artificially constructed, then attached to a natural bacterial chain of genes.* E. coli *accepted the new code and produced somatastatin.*
>
> *The availability of relatively large amounts of somatastatin may be useful in the treatment of diabetes mellitus, abnormal bone growth, and pancreatitis. The researchers produced 5 mg of the hormone, equivalent to the quantity found in 500,000 sheep brains.*

Much of what we know about the cerebrum has been accumulated by experimenting with live animals and by observing patients with specific nervous disorders. If an animal such as a dog has its cerebrum removed, it appears to be quite normal. It can breathe and swallow food if food is placed directly in the mouth. However, it will not eat voluntarily even though food is placed in front of it.

Lesions or tumors in a particular area of the human brain may cause a malfunction in some specific activity, which in some instances results in severe mental problems.

Because the human cortex is devoid of pain receptors, a person undergoing brain surgery can withstand the operation with only local anesthesia. A specific part of the brain might then be stimulated and the person could be asked what sensation he felt. With all of these procedures the cerebral cortex has been mapped. Figure 7.20 illustrates a few of the areas of the brain and their specific functions.

The extreme posterior portion of the occipital lobe controls or has centers for controlling vision. Parts of the temporal lobe control the sensations of hearing, smell, and taste. In the frontal lobe are areas concerned with reasoning,

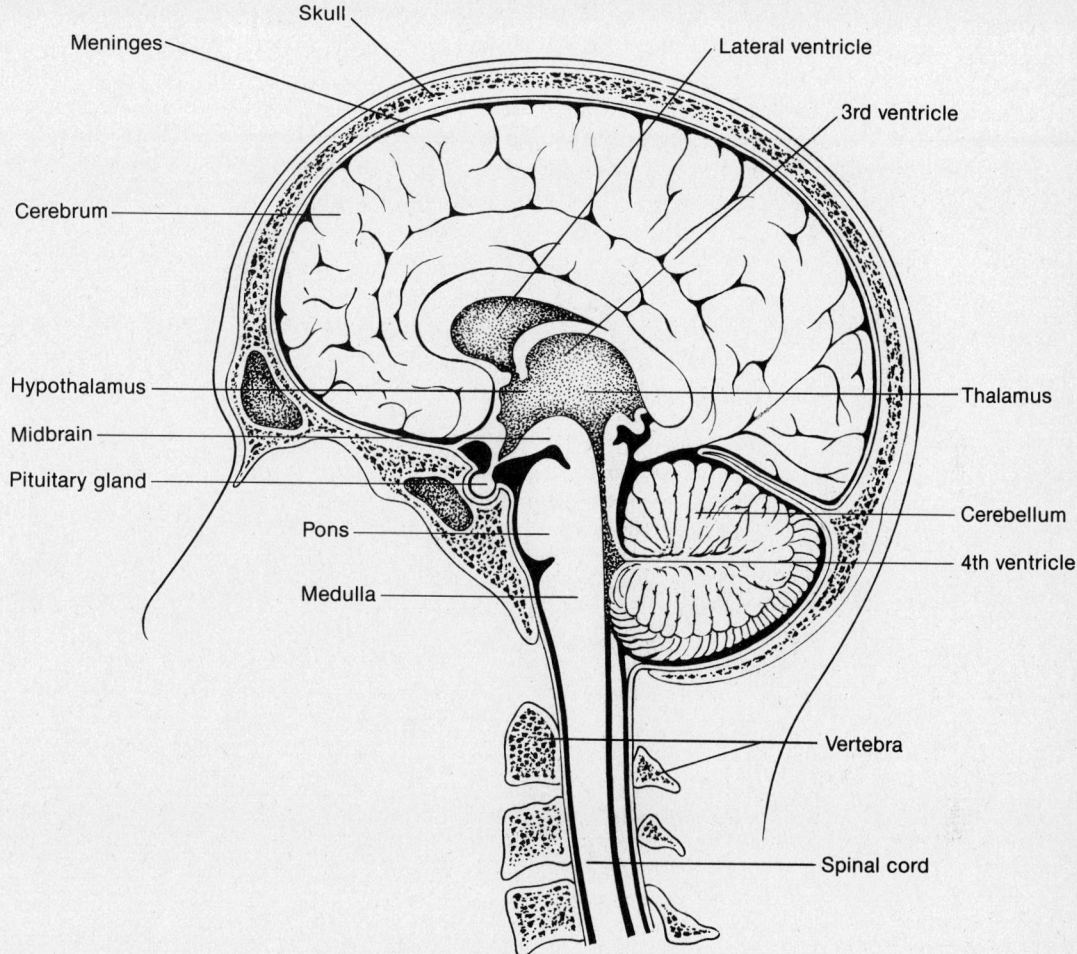

Fig. 7.19 A saggital section of the brain showing internal structures and ventricles. Ventricles are spaces filled with cerebrospinal fluid. The pituitary gland shown in this figure is not nerve tissue but rather is an endocrine gland which secretes hormones.

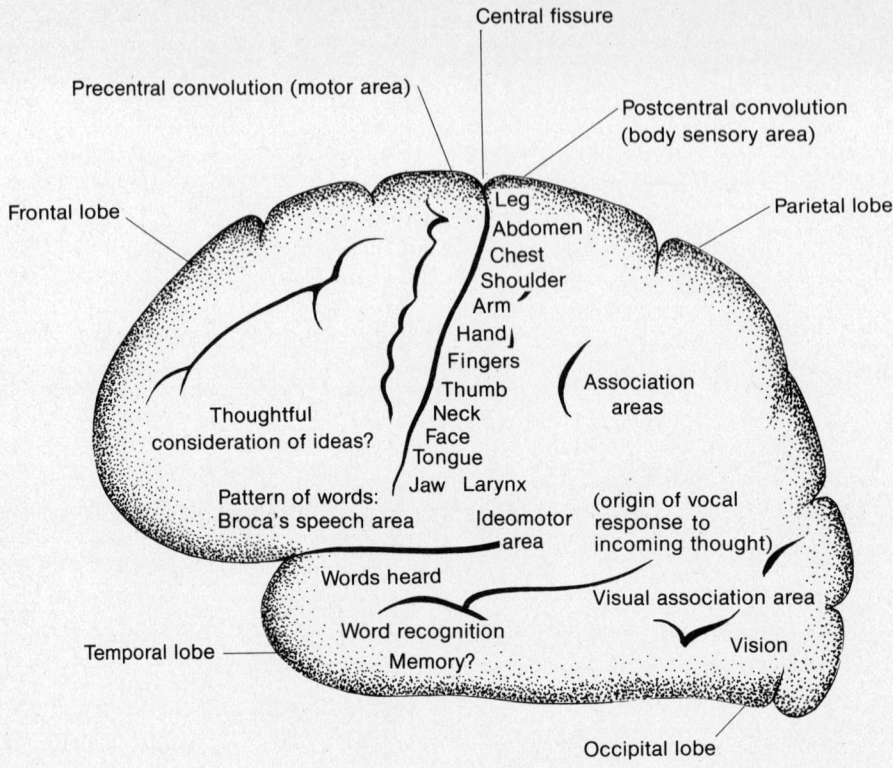

Fig. 7.20 The cerebrum and general locations of motor areas associated with specific body functions. The extreme posterior segment of the occipital lobe has nerve centers that control vision. Parts of the temporal lobe control the sensations of hearing, smell, and taste. The frontal lobe areas are concerned with reasoning, judgement, and abstract ideas. Just anterior to the central fissure are motor centers that control muscles of the leg, arm, neck, and head. Posterior to the central fissure are sensory centers for sensations received from the leg, arm, neck, and head.

judgment, and abstract ideas. Just anterior to the central fissure are motor areas that control muscles of the leg, arm, neck, and head, and just posterior to the central fissure in the parietal lobe are the areas where sensation is received from leg, arm, neck, and head.

Cerebellum The *cerebellum* lies immediately below the posterior part of the cerebrum. It is divided into two lobes and has a cortex of grey matter similar to that of the cerebrum. One difference between the structure of the cerebellum and the cerebrum is that the ridges of convolutions of the cerebellum appear as folds arranged in rows rather than the random arrangement of convolutions in the cerebrum.

By observing experimental animals it has been learned that the basic function of the cerebellum is one of coordination. An animal having had the cerebellum removed lacks coordinated movement such as walking, running, or, in the case of birds, flying. The importance of coordinated movement is obvious if we consider all the motions required for a simple task such as picking up a pencil and writing your name. Movements are required to reach out to the exact position to pick up the pencil, grasp it, arrange it between the fingers at a precise angle, and apply the sharpened end to the paper with just the right amount of pressure. The movement necessary to write your name requires precise control over the muscles of the shoulder, the arm, and the fingers.

> **Does the Right Side of the Brain Know What the Left Side Is Doing?**
>
> *The split-brain theory maintains that the left brain hemisphere controls the functions of speech, writing, mathematics, and logic. The right hemisphere is thought to deal with pattern recognition, abstract ideas, and so-called divergent thinking such as daydreams.*
>
> *There is evidence now that several functions once thought to be exclusive to the left brain can be carried out by the right brain. These functions include language skills, motor skills, and logic.*
>
> *In addition, it has been postulated that after the left hemisphere has analyzed a problem logically into its components, the right side may suddenly recognize a hidden pattern. This so called "Eureka!" phenomenon may challenge past beliefs that the left brain maintains so much control over right brain functions.*

These activities must be learned. Observe the uncoordinated action exhibited by children before they have become accustomed to writing, or catching a ball, or even walking.

As nerve pathways become established through the cerebellum, we attain the capability of precise muscular control to do intricate jobs.

One dysfunction of the cerebellum is exhibited in a person's difficulty in gauging the extent of muscular movement such as touching another person's fingertip on signal. There may be exaggerated tremor when there is a great intention on hitting a mark such as touching a fingertip or threading a needle. The individual may lose the capability of successive movement patterns such as pronation and supination of the hand in rapid successive movements.

Another indication of malfunction of the cerebellum is the so-called *rebound phenomena*. This is exhibited by having the individual push against another person's hand with slight pressure; when the opposing hand pressure is removed the normal individual's hand would move forward only a few inches before recovering balance. The individual with cerebellum malfunction could not check his or her forward movement as quickly.

To summarize, then, we can say that the cerebellum's function is to coordinate all muscular activity—all the movements we normally associate with everyday living. The input into this system includes sensory impulses from the ear, eye, and sense receptors. Impulses arriving from all these sources are filtered through the cerebellum. Some impulses continue on to the cerebral cortex for interpretation; impulses then are returned to the cerebellum, which coordinates this new information into some specific pattern.

Medulla   As the spinal cord progresses upward immediately through the foramen magnum, it expands into a bulbous triangular mass of tissue called the *medulla*. The medulla is more than just a continuation of the spinal cord since it contains nerve centers for critical body functions such as the cardiac rate, the vasomotor constriction area, and a regulator for the rate of respiration.

Impulses arriving at the medulla from areas of the spinal cord are transmitted to both the cerebellum and to higher brain centers for coordination of impulses.

Located within the medulla are nerve centers that regulate heartbeat, at least one phase of it, by inhibiting the rate. Nerve cells send out impulses along one of the cranial nerves, the *vagus*, which inhibits or slows down the rate of heartbeat. Other centers located elsewhere in the body will accelerate or speed up the heart rate. Therefore these centers are opposing or antagonistic to each other and so maintain the heartbeat rate required at the moment.

The medulla also contains nerve centers that control the size of blood vessels. As we will see in future chapters, there are many factors in the environment that require blood vessels to change their size. The smaller diameter attained by vasoconstriction results in higher blood pressure. The heart, pumping a constant volume of blood, has to push the blood through smaller diameter tubes, and therefore the pressure is increased.

The medulla contains centers that control the number of breaths taken in and exhaled per minute. This is influenced to a large degree by blood chemistry. If there is a deficiency of oxygen being breathed in or an excess of carbon dioxide in atmospheric air, the medulla is stimulated by the blood chemistry and sends impulses to the muscles that control breathing, such as the diaphragm and the intercostal rib muscles, to speed up their rate of contraction and relaxation.

As we exercise, we tend to breathe faster because muscular activity creates a higher concentration of waste products, including carbon dioxide, which must be removed at a faster rate. The increase in carbon dioxide in the blood stimulates the medulla, which in turn initiates impulses to the respiratory muscles, causing an increase in the breathing rate.

Thalamus The *thalamus* is located along the lateral sides of the third ventricle, as shown in Figure 7.19. It acts largely as a relay center for impulses that travel from the spinal cord to higher cortical centers. If centers in the cortex for perception of pain, touch, and pressure are destroyed, the thalamus is capable of making the body aware of these sensations, although in a very crude manner. When these impulses are passed on and registered in the cortex, finer and more discernable sensations are recognized.

Hypothalamus The *hypothalamus* is located beneath the thalamus and superior to the pituitary gland. In addition to acting as a relay center for impulses arriving from the spinal cord, it is directly responsible for controlling most of the body's so-called *vegetative responses*. These responses include temperature regulation, body water balance, and processes associated with autonomic function. The hypothalamus also secretes neurohormones that are emptied into the pituitary, where they are released in response to various stimuli.

The hypothalamus has centers that can both stimulate and inhibit specific body functions. For example, one center may increase heartbeat rate and constrict arteries, thus increasing blood pressure, while another area may inhibit these structures, decreasing heartbeat rate and dilating arteries, thus lowering blood pressure.

Some hypothalamic centers are responsible for wakefulness, while others are responsible for sleep.

Stimulatory and inhibitory signals may arrive at body tissues from the spinal cord or higher brain centers. If some higher cortical brain centers are destroyed, in experimental animals, the hypothalamus is freed of certain inhibitory signals and the animal will exhibit actions associated with rage. When other centers are inhibited the animal will exhibit an uncontrollable urge to eat; damage to other centers will cause an animal to ignore food to the point of starvation.

Table 7.3 illustrates a few of the body processes controlled by the hypothalamus.

Table 7.3 Hypothalamus Functions

1. Regulates and coordinates autonomic functions.
2. Aids in controlling and integrating responses made by visceral effectors.
3. Acts as major relay mechanism between cerebrum and autonomic centers.
4. Secretes hormones called releasing hormones.
5. Regulates amount of sleep.
6. Regulates appetite and the sensation of fullness.
7. Regulates temperature.

CRANIAL NERVES

Twelve pairs of nerves originate from the inferior surface of the brain. These nerves primarily maintain autonomic functions associated with muscles, sense organs, and glands of the head and neck regions.

The cranial nerves, shown in Figure 7.21, are numbered consecutively from front to back and are named relative to the structures they control.

The nerves are made up of nerve fibers, some of which are purely sensory, some purely motor, and some mixed. For example, the first cranial nerve, the olfactory, is made up exclusively of sensory fibers. Nerve endings stimulated in the

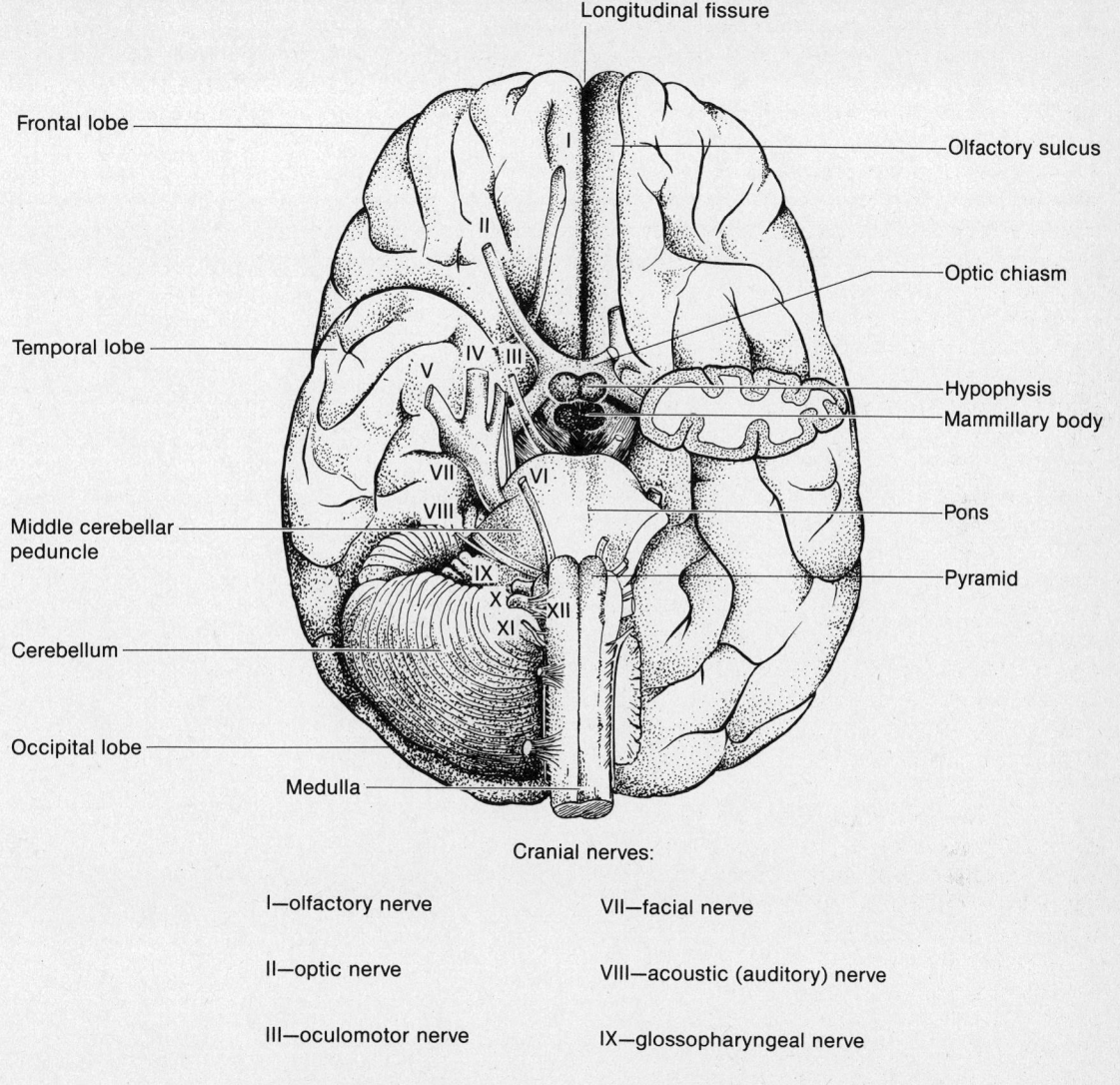

Fig. 7.21 A view of the ventral surface of the brain showing the route of cranial nerves. The nerves are numbered beginning anteriorly with the olfactory nerve and ending posteriorly with the hypoglossal nerve.

Cranial nerves:

I—olfactory nerve

II—optic nerve

III—oculomotor nerve

IV—trochlear nerve

V—trigeminal nerve

VI—abducens nerve

VII—facial nerve

VIII—acoustic (auditory) nerve

IX—glossopharyngeal nerve

X—vagus nerve

XI—spinal accessory nerve

XII—hypoglossal nerve

nasal membrane carry impulses to the brain, where the sensation is interpreted.

The third cranial nerve, the occulomotor, is composed of motor nerve fibers carrying impulses from the brain to muscles that move the eye.

Cranial nerve VII, the facial, has sensory nerve fibers carrying impulses initiated in the taste buds and also has motor nerve fibers that control some of the superficial muscles of the face and scalp. This is an example of a mixed cranial nerve.

Table 7.4 summarizes the functions of the 12 pairs of cranial nerves.

THE AUTONOMIC NERVOUS SYSTEM

The part of the nervous system that automatically controls such activities as smooth muscle contraction, cardiac muscle contraction, and glandular secretion is referred to as the *autonomic nervous system*. This part of the nervous system is further divided into two segments—the *sympathetic* and the *parasympathetic*.

The autonomic nervous system is not a separate system but rather a unique and different part of the overall system.

Figure 7.22 illustrates the structures controlled by the autonomic nervous system. Impulses are generated within the viscera and are transmitted to the spinal cord, then to the brain for coordination and integration. The higher brain centers then send impulses back to the viscera to correct or influence that particular organ.

The two divisions, the sympathetic and the parasympathetic, have opposite actions. In other words, when the sympathetic stimulates a particular organ, in most cases the parasympathetic has the opposite or inhibiting action on that same organ.

We previously discussed reflexes in terms of their protective nature. These reflexes are called *somatic reflexes* since they are reflexes essentially of muscles. The reflexes controlled by the autonomic nervous system are called *visceral reflexes*

Table 7.4 Functions of Cranial Nerves

NUMBER	NAME	FUNCTION
I	Olfactory	Odor detection
II	Optic	Vision
III	Oculomotor	Contracts four of the six muscles which move the eyeball, the eye lid muscles, and muscles that control the pupil and lens focus
IV	Trochlear	Contracts superior oblique muscle of the eye
V	Trigeminal	Influences muscles of mastication and sensations of the head and face
VI	Abducens	Abducts the eye
VII	Facial	Influences taste receptors, facial glands, facial expression, secretions of nose, tear glands, and mouth
VIII	Vestibular acoustic	Sound perception, body and joint movements
IX	Glosso-pharangeal	Taste, middle ear, and throat sensations
X	Vagus	Sensory stimuli to digestive, circulatory, and respiratory organs. Autonomic stimuli to all thoracic and abdominal viscera.
XI	Spinal accessory	Stimuli to neck and shoulder muscles and to the larynx
XII	Hypoglossal	Stimuli to tongue and neck muscles

since they are reflexes of internal organs. There are several differences between the nerve pathways of visceral reflexes and somatic reflexes. Table 7.5 illustrates these differences.

The Sympathetic System

Cell bodies in the sympathetic nervous system arise from the thoracic and lumbar segments of the spinal cord. The axon of an originating neuron leaves the spinal cord and terminates at an area near the spinal cord called the *ganglionic chain*. The ganglionic chain is an accumulation of cell bodies and synapses. As the axon ending enters the ganglionic chain it synapses with another neuron that in turn goes directly to an organ or a gland. The neuron leading toward the

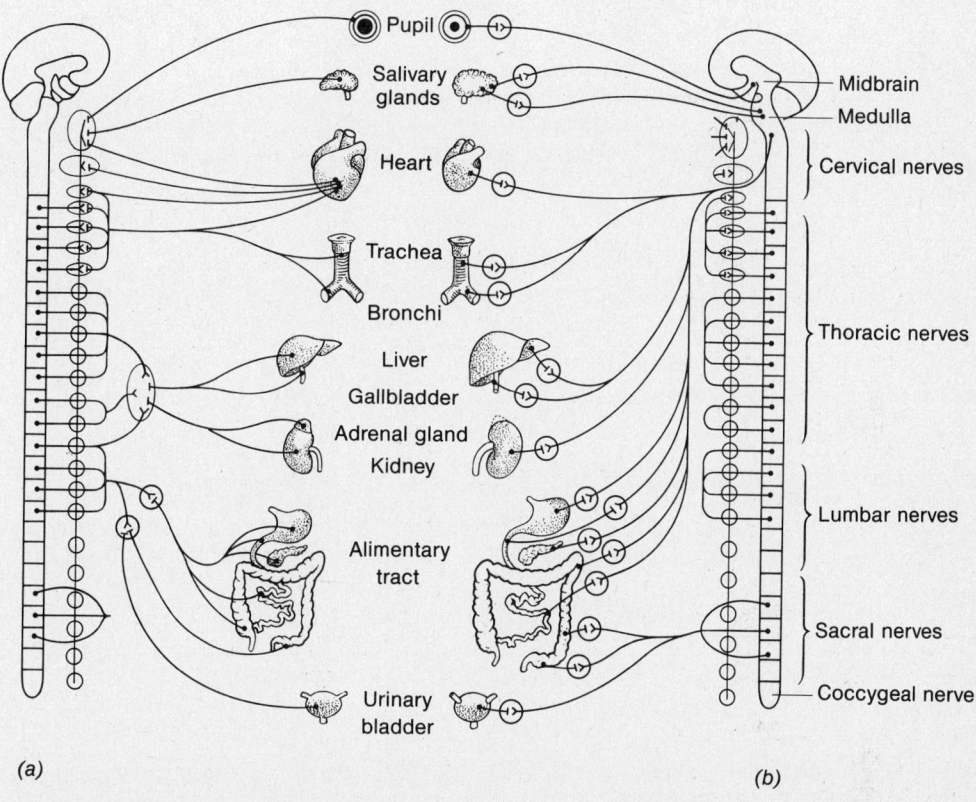

Fig. 7.22 General design of the autonomic nervous system. (a) The sympathetic system is responsible for preparing the body to meet stress or emergencies. Accelerated pulse, increased breathing rate, and adrenal gland stimulation are a few processes initiated by the sympathetic system. (b) The parasympathetic system generally provides actions opposite those induced by the sympathetic system.

ganglionic chain is called the *preganglionic neuron* or *presynaptic neuron*. The neuron leading away from the ganglionic chain is called the *postganglionic neuron* or *postsynaptic neuron*. These terms will have a greater significance with regard to their secretions, as we will see in a moment.

Note in Figure 7.22 that although one preganglionic neuron may arise from a segment of the thoracic or lumbar regions of the spinal cord, it may synapse with several postganglionic neurons that fan out to many parts of the body. This permits a more complete and diffuse network of organs that are indirectly innervated by single preganglionic neurons.

The sympathetic nervous system can be thought of as a part of the nervous system that produces protective reflex actions. This protective nature may prepare the body by providing it with energy so that it may meet some emergency. The preparedness function of the system is not its sole purpose, since there are many conditions that stimulate the sympathetic system which result in no apparent preparedness on the part of

Table 7.5 Differences Between Visceral and Somatic Reflexes

VISCERAL REFLEXES	SOMATIC REFLEXES
1. Neurons originate in the spinal cord and brain stem.	1. Neurons originate in or near the spinal cord.
2. Neurons form a union with a second neuron that innervates a muscle or gland.	2. Neurons arise in the spinal cord and go directly to the effector organ.
3. Neurons are divided into preganglionic and postganglionic neurons.	3. No preganglionic and postganglionic divisions.
4. The system is divided into the parasympathetic and sympathetic divisions.	4. No subdivisions.

the body for an emergency. The protective nature of the system is, however, of primary importance.

Table 7.6 shows that the parts of the body which are stimulated are those that appear to prepare the body for an emergency. For example, the pupil of the eye is dilated so that more light may be admitted; blood vessels constrict, producing a higher blood pressure that pushes the blood about faster to organs such as muscle; the heart rate is increased and the heart pumps with a greater force, pushing blood about more quickly to the tissues; and so on through the list.

The reactions that result from stimulation of the sympathetic nervous system are paralleled almost exactly by the effect of hormones from the adrenal glands (adrenal medulla). The hormone, *adrenalin (epinephrine)*, prepares the body to meet emergencies.

The difference between sympathetic nerve action and hormone action is actually only one of speed of reaction. The sympathetic nervous system is very quickly stimulated and the reactions to this stimulus are immediate. Stimulation of the sympathetic nervous system may in turn stimulate the adrenal glands to empty epinephrine into the blood stream.

This, however, is a slower process. Not only must the hormone first be emptied into the bloodstream, it must then be carried to a specific part of the body, where it then stimulates a particular structure to bring about a proper response.

The hormone, however, produces a longer-lasting effect. Once a hormone is dumped into the bloodstream it is not easily or quickly eliminated. It continues to show its influence for some time afterwards. In contrast, the preparedness nature of the body brought about by stimulation of the sympathetic nervous system may be terminated quickly with the effects not as long-lasting.

At this point we will examine the specific neurons making up neural pathways that connect the spinal cord with a particular organ and look

Table 7.6 Organ and Tissue Response to Stimulation from the Autonomic Nervous System

ORGAN OR TISSUE	SYMPATHETIC	PARASYMPATHETIC
Iris	Dilates pupil	Constricts pupil
Digestive tract	Inhibits peristalsis	Increases peristalsis
Urinary bladder	Relaxes bladder	Contracts bladder
Heart muscle	Increases heart rate	Decreases heart rate
Sweat glands	Increases perspiration	(No nerve fibers from parasympathetic)
Islet cells of pancreas	Increases secretion of insulin	Decreases secretion of insulin
Adrenal medulla	Stimulates secretion of epinephrine	(No nerve fibers from parasympathetic)
Saliva glands	Inhibits saliva secretion	Stimulates saliva secretion
Liver	Stimulates glucose release	(No nerve fibers from parasympathetic)

at what is occurring at the synapse at the ganglionic chain and at the organ that is innervated.

All preganglionic neurons of the sympathetic system secrete *acetylcholine* at the ends of the axon into the synaptic gap. This substance in turn stimulates the dendrites of the postganglionic neuron and sets up an impulse in that neuron which travels to an organ or gland. The axon end plates of the postganglionic neuron secrete another substance called *epinephrine* or *norepinephrine*. This substance influences the organ or gland in its own particular fashion. (The term "influences" is used here rather than "stimulates," since often the organ that receives the impulse may be inhibited rather than stimulated. For example, parts of the digestive system may be inhibited. The peristaltic waves of the intestine may be slower and the secretion of digestive enzymes may be either reduced or terminated, since these are not critical processes in emergency reactions.)

Acetylcholine is deactivated by the enzyme cholinesterase, as happens at the myoneural junction. Once acetylcholine has been eliminated, the synapse then is cleared and another impulse then may cause the axon endings to release more of this substance into the synaptic gap, initiating a new impulse in the postganglionic neuron.

The Parasympathetic System

The actions of the parasympathetic system are basically opposite to those of the sympathetic system, as shown in Table 7.6. One basic difference between the two systems is that, like the sympathetic system, all preganglionic neurons secrete acetylcholine, but in addition all postganglionic neurons of the parasympathetic system also liberate acetylcholine at the axon ending. Another difference between the two systems is the arrangement of the neuron associations.

In the parasympathetic system the neurons originate in the cervicle and sacral regions of the spinal cord. These neurons do not synapse in ganglia near the spinal cord but rather go directly to the organ where they synapse in ganglia, either near the organ which they innervate or within the organ itself. In other words, the preganglionic neuron might be very long, originating in the spinal cord and traveling its entire length to the organ to which it carries impulses.

Within that organ we would find a ganglion with a postganglionic neuron leading away from it. However, the postganglionic neuron would be very short, since it would be going or traveling only within the tissue.

By way of definition, neurons that secrete acetylcholine are called *cholinergic neurons;* those which secrete epinephrine are called *adrenergic neurons.*

DISEASES AND DISORDERS OF THE NERVOUS SYSTEM

The nervous system by design is so intimately involved in all body functions that it is often difficult to assign a disease to the nervous system.

As we proceed through the discussions of body systems we will have ample opportunity to relate many body malfunctions with nerve disorders. The following discussion is concerned with only a few major nerve disorders. Some are of microbial origin, while others are caused by physical injury.

Cardiovascular disease (stroke) is caused by the rupture or blockage of blood vessels carrying nutrition and oxygen to brain tissue. Unlike muscle tissue, the brain cannot carry on anaerobic respiration and go into oxygen debt to be later repaid. Brain tissue dies in a matter of minutes, never to be regenerated, if oxygen or nutrients are withheld or absent.

The extent of brain damage resulting from a stroke depends upon which arteries are affected and consequently which parts of the brain are affected. The condition may prove fatal in some instances. In milder forms, the damage may be limited to different degrees of paralysis, speech impediment, hearing loss, memory lapse, or loss of or diminished eyesight.

Cerebral palsy is the name given to brain injury involving the fetus. Specific causes are not known. However, maternal-fetal physical relationships, X rays and radiation, and maternal

German measles have all been implicated. The disease is not progressive, but the damage to brain tissue is irreversible, with most children exhibiting some level of mental retardation. Some studies indicate that since children with cerebral palsy also may suffer vision and hearing difficulties, they may only appear to be retarded.

Epilepsy is a convulsion or seizure caused by massive nerve discharges involving many parts of the body. The causes involve infectious diseases, drugs, physical damage to the brain, hypoglycemia, and a host of other conditions.

The seizures of epilepsy are classified as *grand mal* or *petit mal*. In grand mal, the seizure may produce unconsciousness and last for several minutes to an hour or more. Petit mal may last for only a few seconds, with no unconsciousness.

Parkinson's disease (syndrome) is a progressive central nervous system disease. The common name "shaking palsy" describes the tremor and uncoordinated movements exhibited by patients with the disease. Specific causes are unknown, but chemical, physical, and microbial factors may be implicated.

Syphilis, an infectious venereal disease, may seriously affect the nervous system. Toxins of the agent, *Treponema pallidum*, may attack the spinal cord, the meninges, and the brain tissue.

Meningitis (meningococcal) is caused by a bacterium, *Neisseria meningitis*, and is highly infectious. The disease is normally associated with inflammation and damage to the meninges; however, other tissues such as blood vessels may be affected. Meningitis is spread by direct contact or from discharges from the mouth or nose. The disease at first resembles a cold, which quickly (within 24 hours) progresses into high fever, neck and shoulder pain, and severe headache. Circulatory shock and death follow unless prompt medical treatment is provided.

THE AGING NERVOUS SYSTEM

The major changes in the nervous system that can be attributed to aging are those affecting the central nervous system. Loss of memory, slower speeds of response, and more time required to choose between alternative responses are all processes involving the brain and spinal cord.

Some mental activities, on the other hand, may show little or no decline. The speed of nerve impulses along nerve fibers is not significantly reduced. Comprehension and vocabulary in the absence of disease may remain unimpaired in the elderly.

Extreme mental deterioration is known as senility. All elderly persons, however, are not senile. Senility is a pathological condition, aging is not.

SUMMARY AND REVIEW

A. The basic unit of the nervous system is the neuron. Neurons are specialized cells that transmit instructions by way of impulses to body tissues. The major function of the nervous system is to integrate body functions. This integration is accomplished by three types of neurons called monopolar, dipolar, and multipolar neurons.

B. Neurons are structurally and physiologically similar to other cells except for the fact that they possess specialized processes called axons and dendrites.
  1. Neurons are capable of transmitting messages from one segment of the body to another.
  2. Neurons are incapable of reproduction; however, their processes can be regenerated.
  3. The nerve impulse is orginated due to a change in the permeability of the cell membrane toward sodium ions.
  4. A neuron process (axon) becomes depolarized. This depolarization progressively runs the full length of the neuron, resulting in the release of transmitter substances.
  5. Transmitter substances such as acetylcholine are capable of altering the permeability of cell membranes for certain ions such as sodium and potassium.

C. Reflex pathways are those over which impulses travel. They activate mechanisms that are generally protective to other body tissues.

D. The spinal cord is a mass of nerve tissue that extends down from the brain posteriorly to approximately the level at which the last rib attaches to the spinal column. Thirty-one pairs of spinal nerves emerge from the spinal cord.
   1. The first eight pairs of spinal nerves are called cervicle nerves.
   2. The next 12 pairs of spinal nerves are called thoracic nerves.
   3. Five lumbar nerves, five sacral nerves, and one coccygeal nerve are below the thoracic nerves.
E. The brain is the largest mass of nerve tissue in the body. It is composed of the cerebrum, cerebellum, and medulla.
   1. The cerebrum is the center of all conscious activity.
   2. The cerebellum controls most coordinated activities.
   3. The medulla contains centers for such body functions as heart rate, vasoconstriction, and respiratory rate.
F. Cranial nerves (12 pairs) originate from the inferior surface of the brain. Primarily, they maintain autonomic functions associated with muscle contraction, sense organs, and glands of the head and neck.
G. The autonomic nervous system controls the self-running portions of the body. It is divided into two segments.
   1. The sympathetic nervous system initiates reflexes that prepare the body to meet emergencies or stress.
   2. The parasympathetic nervous system has functions which oppose actions of the sympathetic system.
H. Diseases and disorders of the nervous system.
   1. cardiovascular disease
   2. cerebral palsy
   3. epilepsy
   4. Parkinson's disease
   5. syphilis
   6. meningitis

8
Coordination: The Senses

MAJOR CONCEPTS
   THE EYE
      Anatomy of the Eye
      Chemistry of Vision
      Muscles that Move the Eye
      Tear Secretion and Function
      Light and Image
      Diseases Affecting the Eye
   THE EAR
      The External Ear
      The Middle Ear
      The Inner Ear
      The Vestibular Apparatus
      Disorders and Diseases of the Ear
   SENSE OF SMELL
   SENSE OF TASTE
   AGING OF THE SENSES

Senses are those body structures which receive stimuli from the environment, both external and internal, and transmit information to the central nervous system. In this chapter, major emphasis is placed upon the reception of light and sound and how this information is transmitted to the brain for interpretation.

Sense organs are those parts of the body that receive stimuli from the external and internal environment. In addition, these organs transmit, by nerve impulses, information to the brain for interpretation.

The sense organs discussed in this chapter include the eye, the ear, and the smell and taste receptors. The skin as a receptor organ was discussed in Chapter 4.

THE EYE

Comparing the structure and function of the eye with a camera has perhaps been overdone; however, certain similarities are worth pointing out. Both the eye and the camera have openings that admit light in some regulated manner, an optical system that projects a light image upon a sensitive surface, and mechanisms that bring the image into sharp focus on this light-sensitive surface.

Anatomy of the Eye

The eyeball, part of the optic nerve, and muscles that move the eye are housed in the bony socket of the skull called the *orbit*. The orbit is lined with a connective tissue called the *bulbae fascia*, which along with fluids provides an almost frictionless surface for the intricate and varied eye movements. Small fat deposits in the orbit provide a cushion against shock. Figure 8.1 illustrates a section through the eye. The eye is approximately spherical in shape, filled with clear fluids.

Vitreous humor is a fluid with a jellylike consistency found in front of the retina and posterior to the lens. It is this fluid that gives the eyeball its shape.

Aqueous humor is a more watery fluid than vitreous humor and is found between the lens and the cornea. This fluid is constantly being replaced at the rate of about 10 to 12 ml per minute. The fluid is in constant motion because of its high production rate and must have an exit from the eye. This is provided by the *canals of Schlemm*. A serious eye disease, *glaucoma*, results when these canals become physically blocked so that fluid cannot leave. The increase of fluid produces a sudden and often damaging increase in fluid pressure in the eye. More about glaucoma is included later in this chapter.

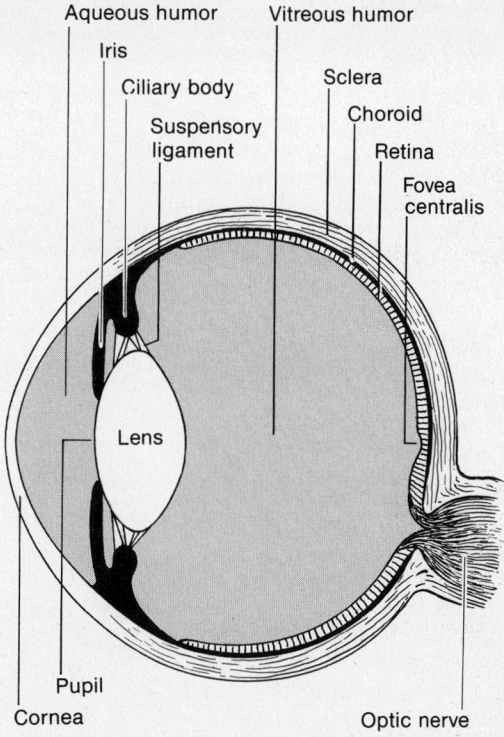

Fig. 8.1 Sectional view of the eye. The cornea and lens gather light and project it onto the retina. Special light-sensitive cells in the retina convert the projected image into nerve impulses. These impulses, carried by the optic nerve, are interpreted by the brain as vision.

The *sclera* is a tough, white, fibrous connective tissue that covers the eyeball, giving it a great deal of protection. In the anterior portion of the eye, the sclera becomes transparent, forming the light entry area of the eye called the *cornea*.

The layer interior to the sclera is called the *choroid layer*. The choroid layer consists of arteries, veins, and pigment. Interior to the choroid coat is the *retina*.

The retina consists of photo-receptor cells, nerve fibers, and connective tissue. Figure 8.2 illustrates the arrangement of nerve cells in the retina.

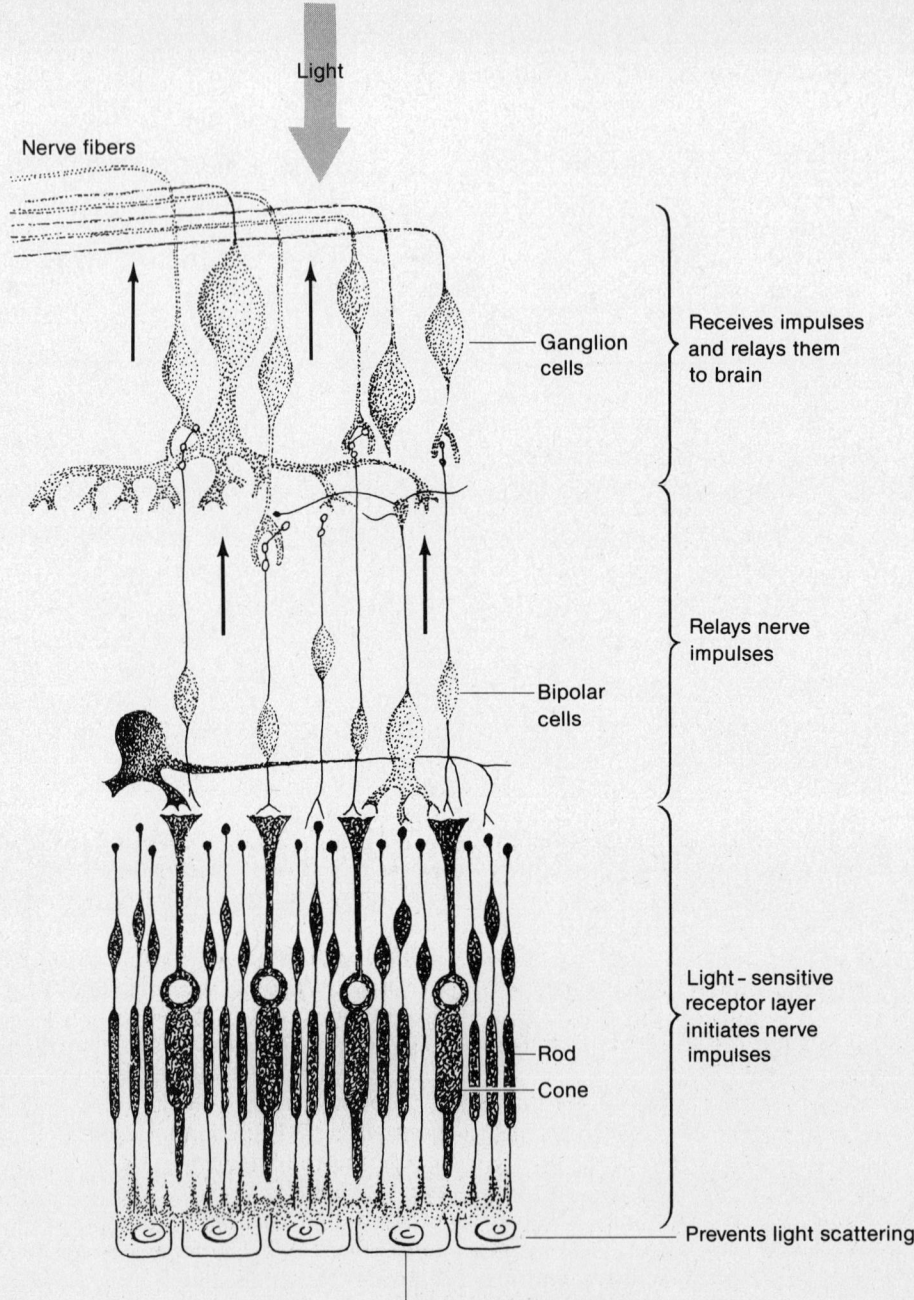

Fig. 8.2 The arrangement of light-sensitive cells in the retina and their physical relationship to the optic nerve. The retina, containing rods and cones, receives light after it is filtered throuth two layers of cells. The wisdom of the body might be questioned here. From an engineering standpoint, a better model might have been designed. But perhaps we should not complain—it works.

The receptor cells called *rods* and *cones* are so named because of their general shape. Cones are the receptor cells most directly related to daylight and color vision. The *fovea centralis* is an area where the cones are highly concentrated and is the area of sharpest vision. The eye in focusing upon an object directs the light entering the eye onto the fovea.

The *lens* is a biconcave-shaped, clear, transparent tissue lying between the two fluid layers. It is held in position by *suspensory ligaments* that are processes of the *ciliary body*. The major function of the lens is to help focus the image sharply upon the retina. As we view distant or nearby objects, the lens must change shape if the image is to be seen sharply. The lens shape change (more convex or less convex) is accomplished by the tension on the suspensory ligaments. The term *accommodation* describes the lens shape change.

With age, the lens loses its power of accommodation and the eyes must be fitted with lenses with two or more specialized areas for viewing far and near objects—bifocal eyeglasses.

The lens may change in other ways with age, including loss of its transparency. The lens becomes milky, shutting out much of the light that would normally enter the eye. This condition is called a *cataract*. In severe cases the lens has to be surgically removed.

The *iris* is also an extension of the ciliary muscle. It is the colored part of the eye and contains an opening called the *pupil*. Muscles attached to the iris can change the size of the pupil depending upon several factors including light intensity, drugs, or sensations from the sympathetic and parasympathetic nervous system.

The *cornea* is the window into the eye. It is a continuation of the sclera. It is this portion of the eye that is most critically involved with the refraction of light into the eye which helps focus the image on the retina. As light enters the eye and goes from the medium of air into the medium of the corneal tissues, the light ray is bent slightly. This bending effect, called *refraction*, is the process by which the light ray is projected onto the retina.

The surface of the cornea may lose its regular shape due to damage or disease. When this happens, the light rays are not uniformly carried to the retina. This condition is called *astigmatism*. If the astigmatism is minor it may be corrected by specially ground lenses. If it is too severe to be corrected by lenses, the cornea may have to be surgically removed and replaced by a corneal transplant.

The *conjunctiva* is a layer of tissue that covers the inside of the eyelid, progresses posteriorly, then reflects forward over the cornea, covering it completely. Inflammation of this protective membrane produces the condition called *conjunctivitis*.

Chemistry of Vision

There are many unanswered questions concerning the precise mechanism by which light image is converted into nerve impulses that are transmitted to the brain. We know more about this function in the rods than we do in the cones, though it is believed that similar mechanisms operate in both.

The rods contain a pigment known as *rhodopsin*, also called *visual purple*. Rhodopsin in the presence of light breaks down chemically into two substances, *scotopsin* and *retinene,* as shown in Figure 8.3. Retinene is a compound formed from vitamin A. In dim light scotopsin and retinene combine to form rhodopsin. This substance is essential for vision in dim light.

You have no doubt experienced the sensation of walking into a dark room from bright light and being unable to see for a few moments. Bright light bleaches out or destroys the rhodopsin, and it takes a few minutes for it to be resynthesized in the darkened room. Once it has resynthesized, you are able to see in very dim light.

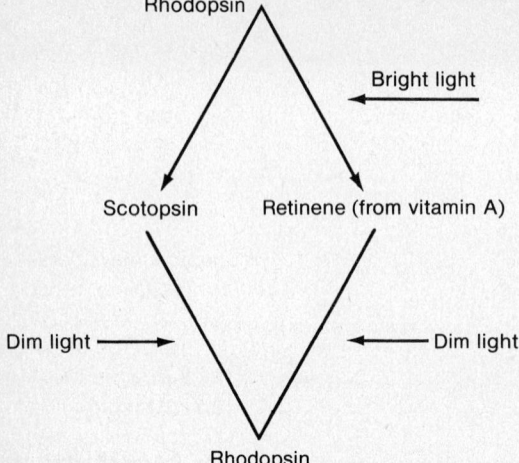

Fig. 8.3 Chemicals involved in low light vision and the effect produced by bright and dim light. Rhodopsin found in rods breaks down chemically in bright light to scotopsin and retinene. Since rhodopsin is essential for dim light vision, a person is temporarily blinded in dim light by a flash of bright light. In dim light scotopsin and retinene recombine to form rhodopsin.

The rods are associated with night vision or very dim light vision. Persons suffering diseases that may not permit the utilization of vitamin A in the formation of rhodopsin have the condition called *night blindness*. Night blindness may also occur simply as a result of vitamin A deficiency.

Muscles that Move the Eye

Three pair of voluntary muscles, as shown in Figure 8.4, permit the eyeball to be moved in any desired direction. The *inferior rectus muscle* pulls the eye downward and the *superior rectus muscle* pulls the eyeball upward. The *lateral rectus muscle* abducts the eyeball and the *medial rectus muscle* adducts the eyeball. Two other muscles,

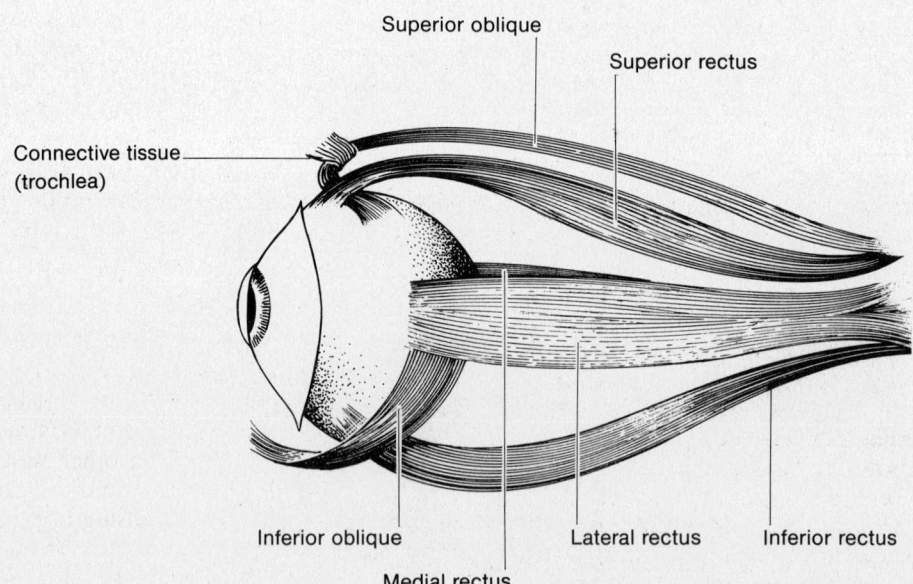

Fig. 8.4 Muscles that move the eye. The inferior rectus muscle pulls the eye downward, the superior rectus muscle pulls the eye upward. The lateral rectus abducts the eye and the medial rectus muscle adducts the eye. The inferior oblique muscle and the superior oblique rotate the eyeball.

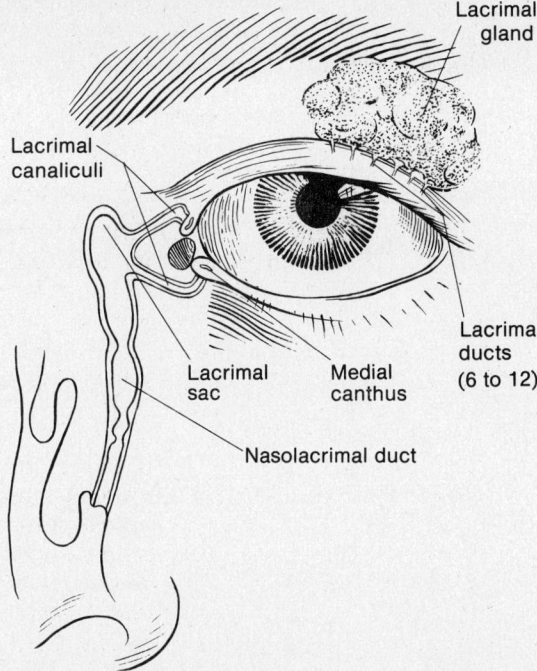

Fig. 8.5 Lacrimal glands and their physical relationship to the nasal ducts. Lacrimal (tear-secreting) glands are located in the anterior orbit. Six to twelve ducts carry tears from these glands to small openings along the rim of the eyelid, where they empty over the cornea.

which rotate the eye, are the *inferior oblique muscle* and the *superior oblique muscle*.

The unique attachment of the superior oblique muscle facilitates rotating the eye. Note in Figure 8.4 how the superior oblique muscle passes through a pulleylike arrangement of connective tissue (trochlea) so that the eye is rotated rather than pulled upward.

Tear Secretion and Function

The glands that secrete tears are called *lacrimal glands*. These glands are located in the anterior orbit, fitting into a small depression in this bony socket, as shown in Figure 8.5. Six to twelve ducts carry tears from the glands to fine openings along the rim of the eyelids. Tears wash down and across the eyeball toward the medial corner (medial canthus) of the eye, eventually emptying into lacrimal sacs located in the posterior nasal passages. A duct leading from the lacrimal sac toward the nasal openings is called the *nasolacrimal duct*.

The major function of tears is to provide lubrication for the eye. As the eye blinks it has a windshield wiper action, carrying fluid down over the eyeball that keeps the membranes from drying out. Tears also have a *lysozyme function,* having the capability of destroying some kinds of bacteria.

Light and Image

Figure 8.6 shows how light from an object passes through the cornea and lens to the retina, where it casts an image. The retinal cells are stimulated by the different qualities of light from the object. The stimulation converts the chemicals of the receptors into an electric stimuli that is carried to the brain.

In the normal eye, the light from an object passes through the refractive cornea and lens and forms an inverted image on the retina. The brain converts the inverted image so that the object is perceived as being upright.

If the eyeball for some reason is too long, the image is focused in front of the retina, and a fuzzy image is seen. This condition, called *myopia* or *nearsightedness,* may be due to defects in the cornea or the lens. It can be corrected with biconcave lenses, which spread the image rays wider apart so that they focus more exactly upon the retinal surface.

If the eyeball is too short, so that the image is focused behind the retina, the condition is called *hyperopia* or *farsightedness.* The correction for this defect is biconvex lenses.

Diseases Affecting the Eye

Many diseases associated with the eye are not caused by microorganisms. Included among these are those associated with diabetes. The small capillaries in the retina of a diabetic tend to form *microaneurysms.* In other words, they balloon outward due to weakened capillary walls. These may burst, releasing blood into the vitreous humor. When this occurs, the person for all general purposes goes blind, since the blood absorbs light coming through the pupil toward the retina. If the hemorrhage is halted, the blood and other materials are absorbed and the vi-

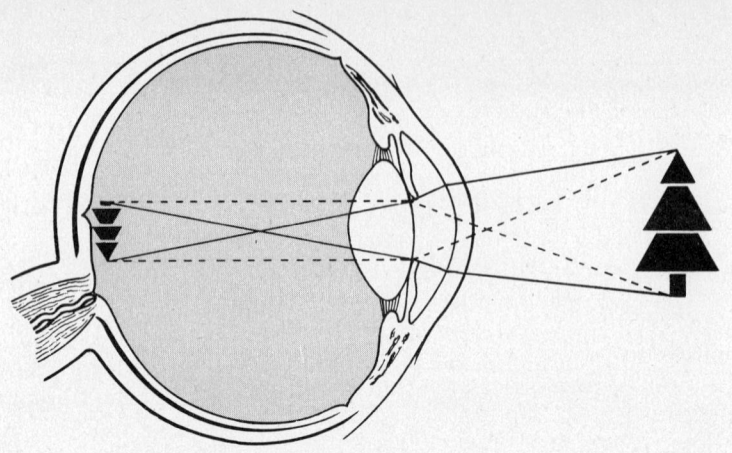

Normal eye

(a)

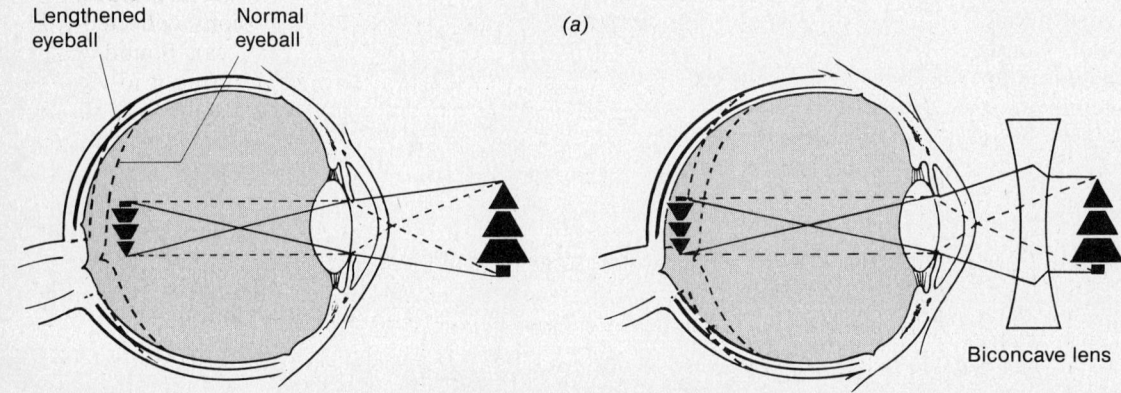

Nearsighted eye (myopia)

(b)

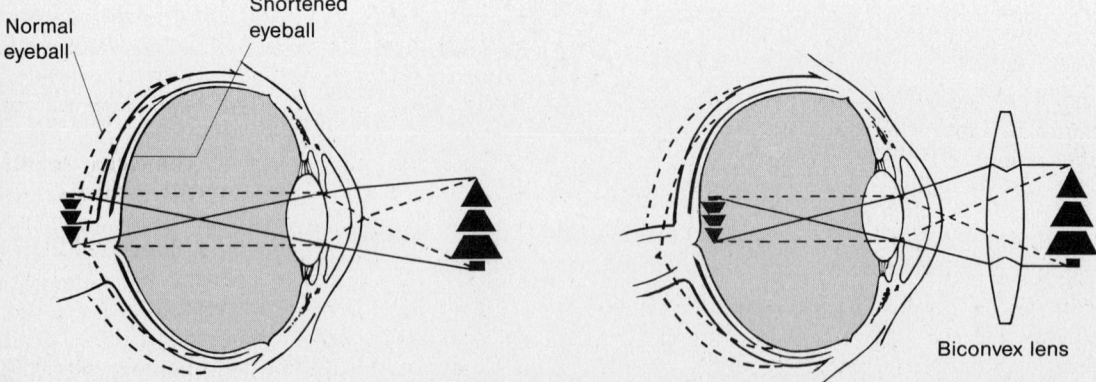

Farsighted eye (hyperopia)

(c)

Fig. 8.6 (a) Normal eye and the light pathway. (b) Nearsighted eye (myopia), showing how the defect can be corrected with a biconcave lens. (c) Farsighted eye (hyperopia), showing how the defect can be corrected with a biconvex lens.

treous humor becomes clear once again. The diabetic, susceptible to this type of eye condition, has to be watched closely to prevent the microaneurysms from rupturing.

Another condition occasionally associated with diabetes, but which can result from other causes, is *retinal detachment*. In this condition the retina simply blisters or peels away from the choroid layer. When this occurs the retina has to be surgically tacked back into position.

Conjunctivitis is an inflammation of the conjunctiva that turns the inner eyelids a bright red color. Conjunctivitis may be of several types, some due to an allergic reaction, others caused by microorganisms.

Since the conjunctiva comes into contact with microbes more than most other tissues it is more susceptible to infection. Table 8.1 indicates some types of bacteria that cause conjunctivitis.

In *glaucoma*, aqueous humor accumulates between the lens and the cornea, creating greater-than-normal pressure in these areas. This is often due to the *ducts of Schlemm*, shown in Figure 8.7, being physically blocked so that the fluid that normally escapes by this route is contained. Glaucoma can be diagnosed by determining the pressure placed upon the cornea by excess fluids. Pressure created in this fashion is transferred to nerves of the eye, causing extreme pain and discomfort. So much pressure may be placed on the eye that the small retinal capillaries may be closed. When this happens, the retinal cells receiving nourishment from these capillaries die, and blindness results.

THE EAR

The human ear has the dual functions of sound reception and equilibrium (sense of balance). These two unrelated physiological functions are accomplished with a high degree of efficiency by the ear. The reason for the anatomical proximity of the ear structures that carry out these different functions is yet to be determined.

We will begin the study of the ear with the more familiar function of sound reception.

The ear, shown in Figure 8.8, may be anatomically divided into three sections. The *external ear* includes the outer projection and the auditory canal. The *middle ear* contains an arrangement of three small bones. The *inner ear* contains the physical and neural structures essential for the conversion of sound waves into nerve impulses.

The External Ear

Structures of the external ear collect sound waves from the environment and direct them into the *auditory canal* toward the *eardrum (tympanum)*. The eardrum, a thin membrane that physically separates the external and middle ear, receives the physical vibrations collected and transmitted by the external ear. Sound waves beat against the eardrum, causing it to vibrate. The intensity and quality of sound is faithfully received and passed along to the middle ear structures.

Sudden sharp sounds may be of such intensity as to cause the eardrum to rupture. Sound waves of less than rupturing intensity can be conditioned to some degree by small muscles attached to the middle ear bones.

The Middle Ear

Structures of the middle ear include three tiny bones, the *malleus (hammer), incus (anvil)*, and *stapes (stirrup)*, that aid in the transmission of sound waves from the external ear to the inner ear. The three bones are precisely arranged to receive and amplify the vibrations from the eardrum and faithfully transmit them to the inner ear.

The middle ear cavity is connected to the upper throat by the *eustachian tube*, which serves to equalize pressures on either side of the eardrum, as shown in Figure 8.8. Increased atmospheric pressure on the eardrum causes the eardrum to bulge inward, increasing the pressure in the middle ear. Swallowing or yawning opens the pharynx end of the eustachian tube, allowing

Table 8.1 Common Causes of Conjunctivitis

DISEASES	INFECTION AGENT	HOW TRANSMITTED	DISEASES	INFECTION AGENT	HOW TRANSMITTED
Conjunctivitis, acute bacterial	Haemophilus aegyptius	Direct contact with infected person or insects	Leptospirosis	Numerous species of Leptospira	Contact with water contaminated by animal excrement and urine
Gonococcal opthalmia neonatorium	Neisseria gonorrhoeae	Infected mother to child at birth	Diphtheria	Corynebacterium diptheriae	Contact with infected person or nasopharyngeal discharges
Trachoma	Chlamydia	Contact with infected person			
Rubeola (red measles)	Rubella virus	Discharge of nose and mouth of infected person	Staphylococcal disease	Staphylococcus aureus	Contact with carrier nasal secretions

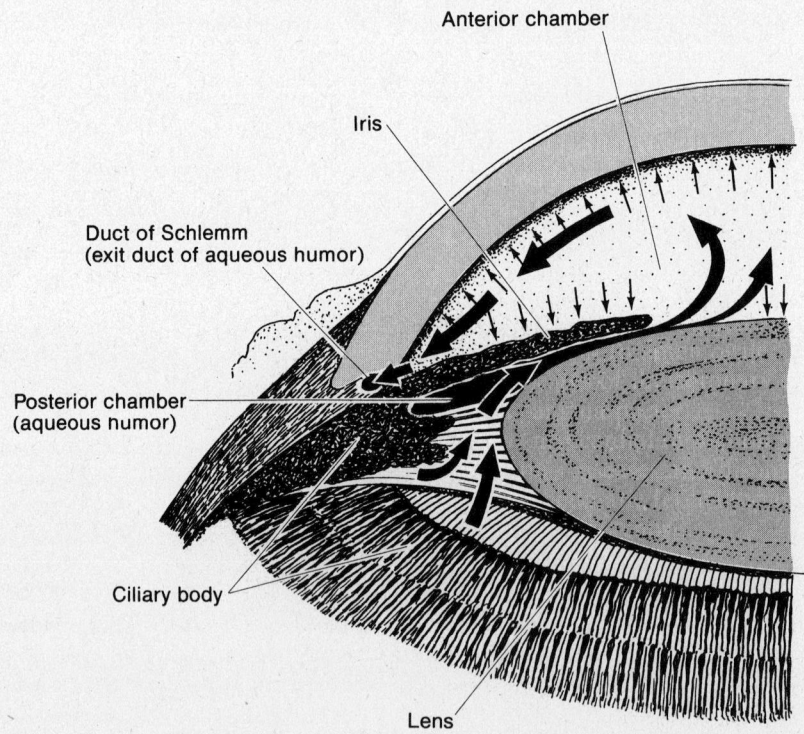

Fig. 8.7 Ducts of Schlemm.

the pressure to equalize. Unfortunately, the eustachian tube also serves as a portal of entry into the middle ear for infective microorganisms.

Sinuses located in the mastoid bone of the skull are connected to the middle ear. Microorganisms thus have a route from the pharynx through the eustachian tube to the middle ear, then to the mastoid sinuses. Infection and inflammation of the mastoid sinuses is called *mastoiditis*.

The Inner Ear

The inner ear is a fluid-filled space shaped like a snail shell. The inner ear, or *cochlea*, contains specialized cells capable of being stimulated by

---

Noise: The Controllable Stress

Noise is defined as any loud, unmusical, or disagreeable sound. Certain sounds may be agreeable to some individuals and disagreeable to others. The fact remains that disagreeable sounds that constitute noise can create stress. The irritating bang of firecrackers, the penetrating noise of food choppers, dishwashers, vacuum cleaners, and so on, can and do cause physiological changes in the body. Among the physiological changes that can be documented are increased pulse rate, hypertension, sympathetic nervous system stimulation, and psychological disorientation.

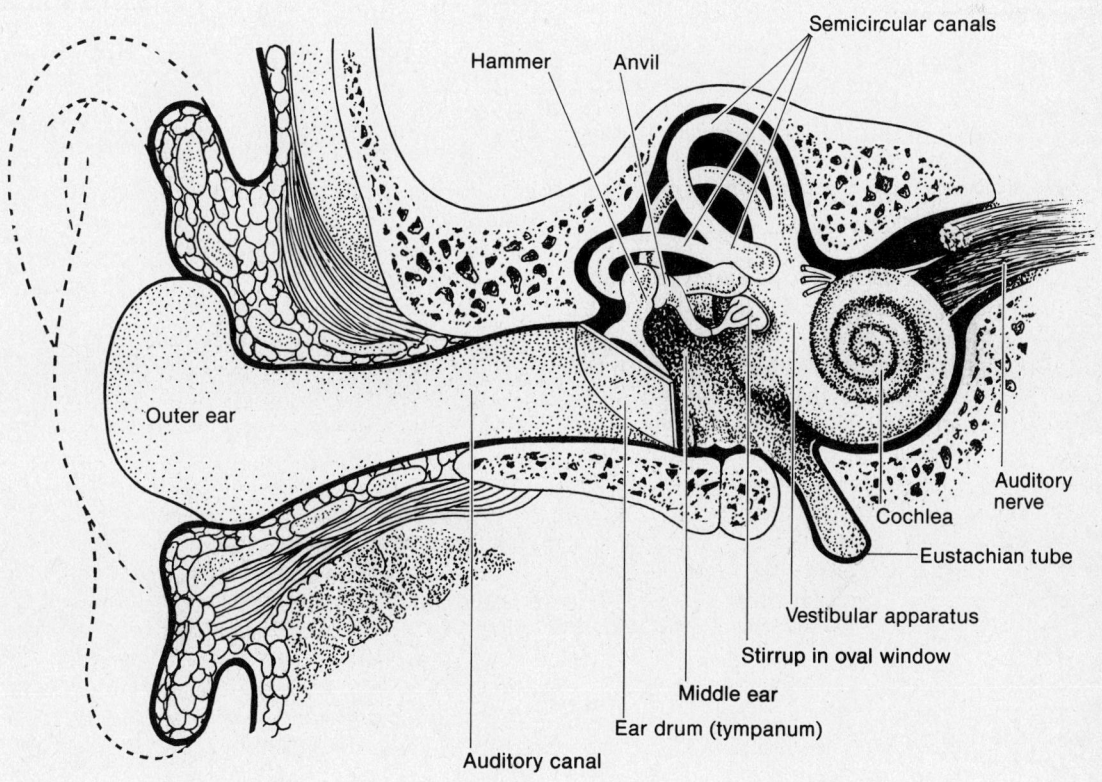

Fig. 8.8 A section through ear structures showing the external, middle, and inner hearing apparatus. Sound waves are physically transmitted across all these structures, finally being converted into nerve impulses in the cochlea.

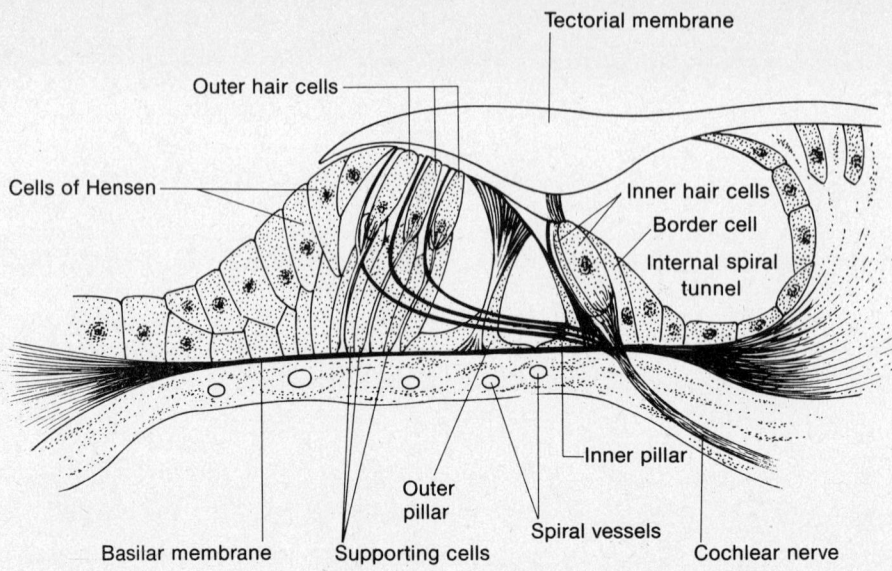

Fig. 8.9 The organ of Corti. This structure located within the cochlea contains minute hair cells. When properly stimulated (correct sound frequency), a nerve impulse is initiated.

sound vibrations. The vibrations are converted into nerve impulses that are relayed to the brain by way of the auditory nerve. Figure 8.8 shows the location of the cochlea. Sound vibrations enter the cochlea through the *oval window,* which is in contact with the stapes. The oval window membrane vibrates in harmony with sounds entering the ear, and these vibrations are transmitted throughout the cochlea by the fluid within it.

Within the cochlea is a specialized structure called the *organ of Corti* that contains a shelf or partition (the *basilar membrane*) made up of specialized cells called *hair cells.* The structure is shown in Figure 8.9. Each cell has a fine hairlike structure protruding from it. These hairs are particularly sensitive to specific sound frequencies. For example, middle C on the piano when struck emits 256 vibrations per second. Certain hair cells are stimulated by 256 vibrations per second, while others are stimulated by other frequencies.

When hair cells are stimulated, the physical sound vibration is converted into nerve impulses. These impulses travel from the hair cells by way of the auditory nerve to the brain, where the sensation of sound is produced.

The human ear can detect sound frequencies as low as 50 cycles (vibrations) per second and as high as 18,000 cycles per second. The detection of the higher frequencies is normally reduced with age.

The Vestibular Apparatus

The inner ear contains structures that are responsive to head position and movement and help us maintain an upright position. These structures, grouped into what is called the *vestibular apparatus,* include the *semicircular canals,* the *ampulla, sensory cells,* and fluids.

The *semicircular canals,* shown in Figure 8.10, are membraneous, tubular canals bent into a semicircle. A set of three canals is found in each ear. Since all three canals are arranged at right angles to each other, head movement in any direction would involve at least one plane.

The semicircular canals are filled with a fluid called *endolymph.* The movement of endolymph aids in the stimulation of receptive cells (also hair cells). When the fluid is set in motion, which

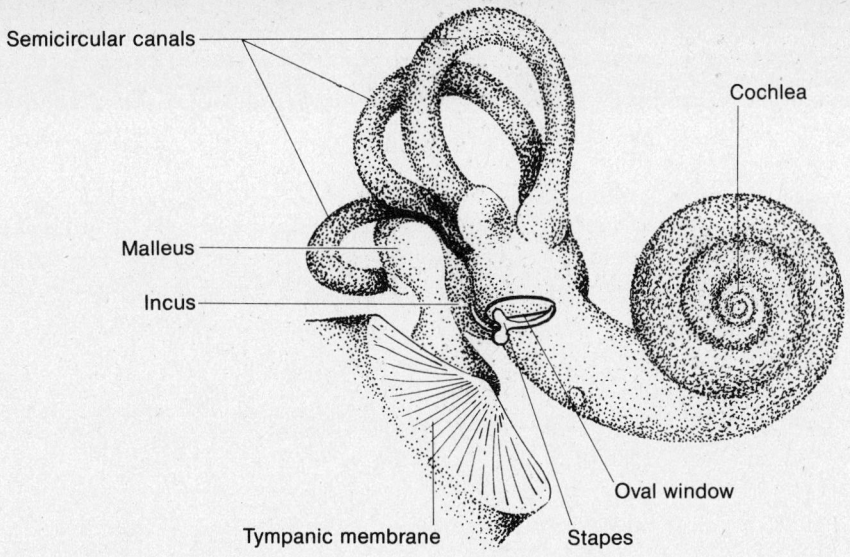

Fig. 8.10 Vestibular apparatus.

happens when the head is moved in any direction, the disrupted hair cells send an impulse to the brain informing it of the new head position.

When a person spins around rapidly, many of the hair cells are stimulated and so many impulses arrive at the brain simultaneously that the brain cannot sort out the separate motions. A feeling of dizziness is produced by the spinning motion.

Disorders and Diseases of the Ear

Conduction deafness may result from disease or accidents. The hearing apparatus that conducts sound waves toward the inner ear is damaged in some manner. Ear channels may be blocked by obstructions such as ear wax or objects that find their way into the ear accidentally. The middle ear bones may become fused through disease, thereby losing their amplifying function. Often in conduction deafness there is only diminished hearing capability rather than complete deafness. A perforated eardrum will result in partial loss of hearing.

Nerve deafness results from *acoustic nerve* damage, preventing impulses from reaching the brain. Nerve deafness may be caused by disease or some drugs.

Microbial diseases caused by bacteria, viruses, yeasts, or fungi may occur in all parts of the ear. The following is a brief description of a few of them.

Mastoiditis is an inflammation of cells that line the mastoid sinuses of the skull. The mastoid processes are located just behind and slightly below the ears. This disease is potentially dangerous due to the close proximity of the mastoid sinuses to the brain and the meninges. Infection from mastoiditis may spread to these other tissues. Bacteria may gain entry into the mastoid process by two routes—the eustachian tube-middle ear connection, and the blood.

Otitis media is an inflammation of the middle ear that may be caused by a variety of microorganisms, many of which gain entry by way of the eustachian tube during respiratory infection. Seventy-five or more viruses are known to be involved in respiratory infections that may ultimately spread to the middle ear. Bacteria that may infect both the external and internal ear include species of *Staphylococcus* and *Streptococcus*.

Pseudomonas aeruginosa is a common soil and water bacteria especially abundant in water contaminated by human or animal sewage. It is directly or indirectly responsible for some types of intestine, skin, urogenital, respiratory, eye, and ear infections. The bacteria may cause otitis media (respiratory route) or *otitis externa,* external ear infection, if a person swims in contaminated water.

SENSE OF SMELL

Less is known about the sense of smell (olfaction) than any of the other senses. It is presumed that the detection of odors is a much less critical process in man than is hearing or seeing. The sense of smell in some animals is highly developed. We depend upon the sense of smell to only a limited degree for protection, such as detecting potentially dangerous gases including ammonia and chlorine. One of the most dangerous gases, carbon monoxide, is odorless.

The lining of the nasal cavities contains small patches, 3 to 5 sq cm, made up of ciliated cells that are stimulated by odors. Nerves stimulated by these cells carry impulses to the brain by way of the first cranial nerve.

The smelling apparatus loses its effectiveness if a specific odor persists for a time. The nerves apparently become fatigued and cannot be stimulated. Other odors that may be introduced, however, may be detected. Cells fatigued due to one odor remain receptive to other odors.

SENSE OF TASTE

This is another sense about which we know very little. As with the sense of smell, taste sensation has a rather low priority among the senses in the human body. Only substances in solution can be tasted. The substances are limited to four groups—sour, salty, bitter, and sweet. *Taste buds* located on the tip, sides, and back of the tongue contain sensitive cells. When stimulated, these cells send impulses by way of several cranial nerves to taste centers in the cerebrum.

Deodorants: The Big Cover-up

The use of deodorants by Americans during the past decade has reached an amazing level. Perfumes, colognes, soaps, powders, mouthwash, hair spray, and so on all contain deodorants. Ingredients in these products stimulate nasal receptors so efficiently that some body odors are masked. Sweat glands secrete copious amounts of water along with some body wastes. When sweat first appears on the skin surface, it is relatively odorless. Normal skin bacteria convert parts of the secretion to malodorous compounds that "polite" societies find objectionable.

Body odors may vary depending upon what we eat, what we wear, how active we are, and our state of health. Even anxiety about emitting body odor may increase the rate of perspiration.

As long as we find normal body odor objectionable, companies that produce deodorants will remain pleased.

AGING OF THE SENSES

Many changes occur in the eyes of aging persons. The conjunctiva becomes more sensitive to irritation. The eye lens may lose its power of accomodation. The iris may react more slowly, thus making bright light almost unbearable. Cataracts also commonly occur in aging persons.

Hearing impairment is a typical and common factor affecting a majority of the elderly. The eardrum becomes less flexible and free movement of the middle ear bones is reduced.

SUMMARY AND REVIEW

The senses include the eye, ear, the vestibular apparatus, sense of smell, and sense of taste.

A. The eye functions to collect light and convert it into nerve impulses interpreted by the brain as vision. Parts of the eye include:
 1. cornea
 2. aqueous humor
 3. vitreous humor
 4. retina
 5. optic nerve
B. The chemistry of vision involves rhodopsin, a pigment that breaks down in the presence of light to:
 1. scotopsin,
 2. retinene (formed from vitamin A)
C. Muscles that move the eye:
 1. inferior rectus (pulls the eye downward)
 2. superior rectus (pulls the eye upward)
 3. lateral rectus (abducts the eye)
 4. medial rectus (adducts the eye)
 5. inferior oblique (rotates the eye)
 6. superior oblique (rotates the eye)
D. Tears are secreted by lacrimal glands. Ducts carry tears into the nasolacrimal duct, then into the nose.
E. The shape of the eyeball determines where the image is focused. In a normal eye the image falls on the retina. If the eyeball is distorted, two conditions result:
 1. Myopia. The eyeball is too long, and the image falls in front of the retina.
 2. Hyperopia. The eyeball is too short, and the image falls behind the retina.
F. Diseases of the eye include:
 1. conjunctivitis
 2. glaucoma
G. The ear receives sound waves and converts them to nerve impulses. These impulses are carried to the brain by the auditory nerve, where they are interpreted as sound. The ear is divided into three parts:
 1. external ear
 2. middle ear
 3. inner ear
H. The vestibular apparatus located above the inner ear aids in the perception of balance.
I. Disorders and diseases of the ear:
 1. conduction deafness
 2. nerve deafness
 3. mastoiditis
 4. otitis media
 5. otitis externa
J. The sense of smell (olfaction) detects atmospheric odors. Stimulated nasal receptors carry impulses to the brain via olfactory nerves.
K. The sense of taste involves receptors located on the tongue. The sensations of sourness, saltiness, bitterness, and sweetness can be tasted.

9

The Heart and Blood Vessels

MAJOR CONCEPTS
  THE HEART
    Work Done by the Heart
    Forces Affecting the Heart's Work Load
    Blood Pressure Measurement
    The Physiology of the Heart
    Heart Sounds
    The Electrocardiogram (ECG)
  ARTERIES, VEINS, AND CAPILLARIES
    Arteries
    Veins
    Capillaries
  CIRCULATION PATTERNS
    Pulmonary Circulation
    Systemic Circulation
    Fetal Versus Adult Circulation
  DISEASES OF THE CIRCULATORY SYSTEM
    Congenital Diseases
    Diseases of Coronary Arteries
    Other Diseases and Disorders
  HYPERTENSION IN THE ELDERLY

The average-sized person has approximately 5 liters of blood flowing in a continuous circuit. This circuit includes a muscular pump—the heart—and the vessels that lead from and to it. Arteries carry blood from the heart to cells, and veins return blood to the heart. The distribution of blood to cells from the arteries is provided by capillaries. Blood flows into veins from capillaries.

This chapter discusses the structure, location, and names of blood-circuit parts and some of the regulating influences imposed by the nervous and endocrine systems.

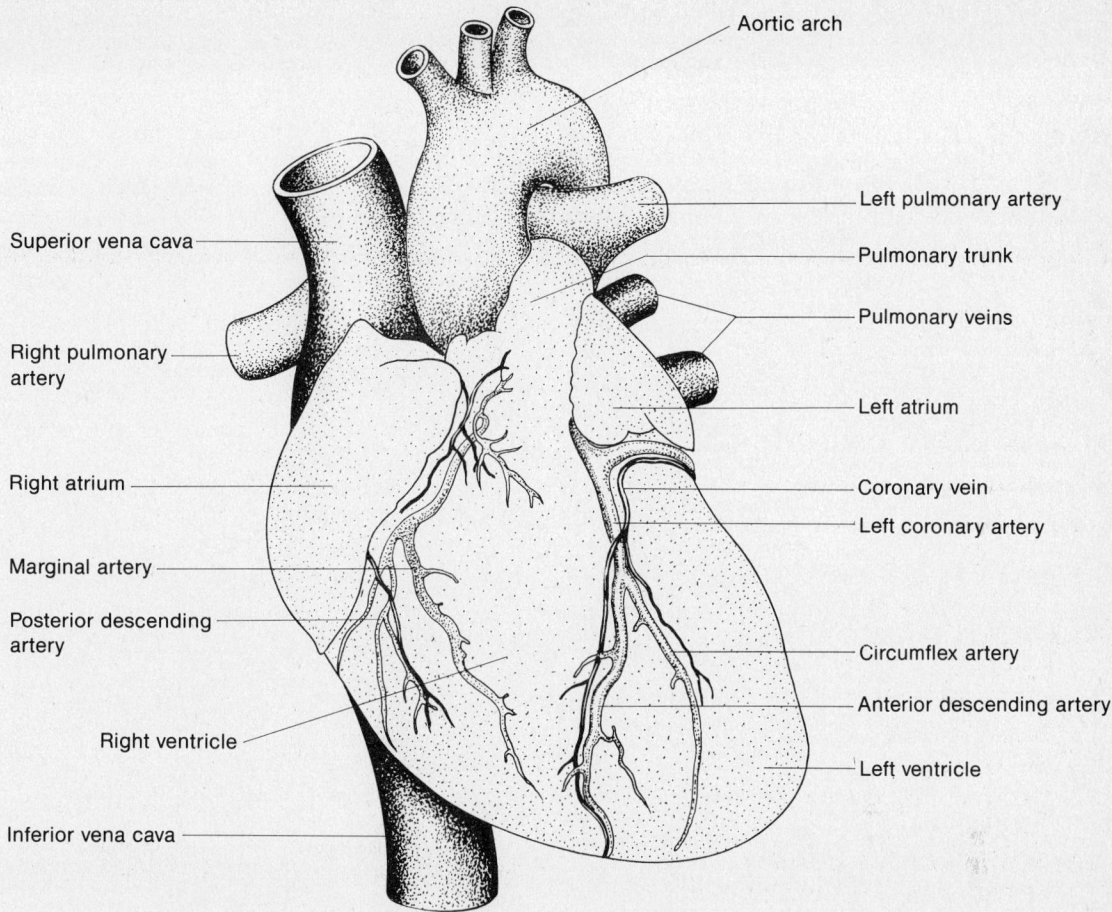

Fig. 9.1 External view of the heart. The inferior vena cava and superior vena cava are large veins returning blood to the heart. Blood leaves the heart through the pulmonary artery to the lungs and through the aorta to body tissues. The coronary artery carries nutrients and oxygen to the heart tissue itself.

The cardiovascular system is made up of several specialized tissues and organs, including the heart, arteries, veins, and capillaries.

The function of this system is to pump and transport blood from the heart to all body tissues and return blood to the heart.

THE HEART

The heart is a muscular organ housing four cavities—the right and left *atria*, and the right and left *ventricles*. Cardiac muscle makes up the greatest mass of the heart. A thin layer of epithelial tissue, the *pericardium*, covers the heart and secretes a lubricating fluid that permits frictionless movement of the heart in the mediastinum. The *mediastinum* is the space between the two lung cavities. The heart lies in this cavity. The inner lining of the atria and ventricles is made up of another epithelial tissue called *endocardium*.

The right and left cavities are separated by a muscular partition, the *septum*. The right side of the heart receives blood returning from body tissues and pumps blood to the lungs for oxygenation and the release of carbon dioxide. This blood then returns to the left side of the heart to be pumped to the tissues.

The circulatory pathway from the right side of the heart to the lungs and back to the left side of the heart is called the *pulmonary circulation.* The pathway from the left side of the heart to the tissues and back to the right side of the heart is called the *systemic circulation.*

The valves located between the atria and ventricles, as shown in Figures 9.1 and 9.3, include the *tricuspid* on the right side and the *mitral (bicuspid)* on the left side. Both valves function to prevent the backward flow of blood, permitting the flow of blood in only one direction.

Figure 9.2 shows two artificial heart valves implanted in the heart.

Blood pressure on the atrial side forces the valves open, but blood pressure on the ventricular side forces them shut. The valves are prevented from turning inside out, that is, into the atrial space, by stringlike fibers called *chordae tendineae.* These fibers extend from papillary muscles that are continuous with the wall of each ventricle at its base, as shown in Figure 9.3. Another set of valves is located where the blood leaves the ventricles. These are the *semilunar valves.* The semilunar valve on the left side, the *aortic semilunar valve,* prevents the back flow of blood into the left ventricle during relaxation. The semilunar valve on the right side, the *pulmonary semilunar valve,* prevents blood from reentering the right ventricle when it relaxes.

The *right ventricle* pumps the blood past the right semilunar valve into the *pulmonary artery,* then to the lung capillaries, where carbon dioxide is given off and oxygen accepted. The oxygenated blood then leaves the lung capillaries and enters the *pulmonary veins,* which empty into the *left atrium.* The left atrium contracts, forcing the blood past the mitral valve into the left ventricle. Contraction of the left ventricle squeezes the blood past the aortic semilunar valve into the *aorta,* which carries the oxyge-

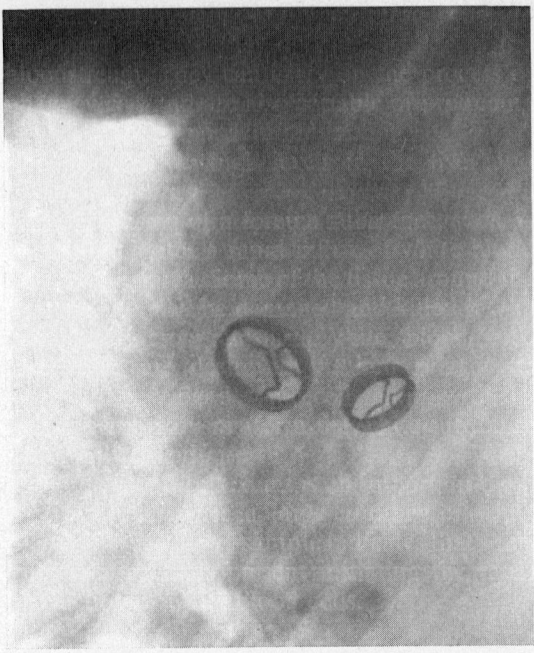

Fig. 9.2 X ray of artificial heart valves. The valve on the left is the aortic semilunar. The valve on the right is the mitral valve. *(Courtesy Radiology Department, Normandy Osteopathic Hospital, St. Louis, Mo.)*

nated blood away from the heart toward the tissues. The left ventricle is much more muscular than the right ventricle. Because body tissues offer greater resistance to blood flow than do the lung capillaries, the left ventricle develops accordingly.

Work Done by the Heart

The function of the heart is to force blood into the arterial system. If the capacity of each ventricle is approximately 60 ml and the heart rate is 72 beats per minute, then an amount equal to 4,320 ml, or 4.32 liters, of blood is pumped per minute. At this rate, the heart would pump 259.2 liters (about 85 gallons) of blood per hour.

The human heart begins beating long before birth and continues to beat for an average of 72 years. It has been estimated that a normal heart beats 36,000,000 times per year. A little arithmetic will dramatically show that the heart is a pumping mechanism capable of an astounding amount of work. A healthy existence depends on the precision with which it works.

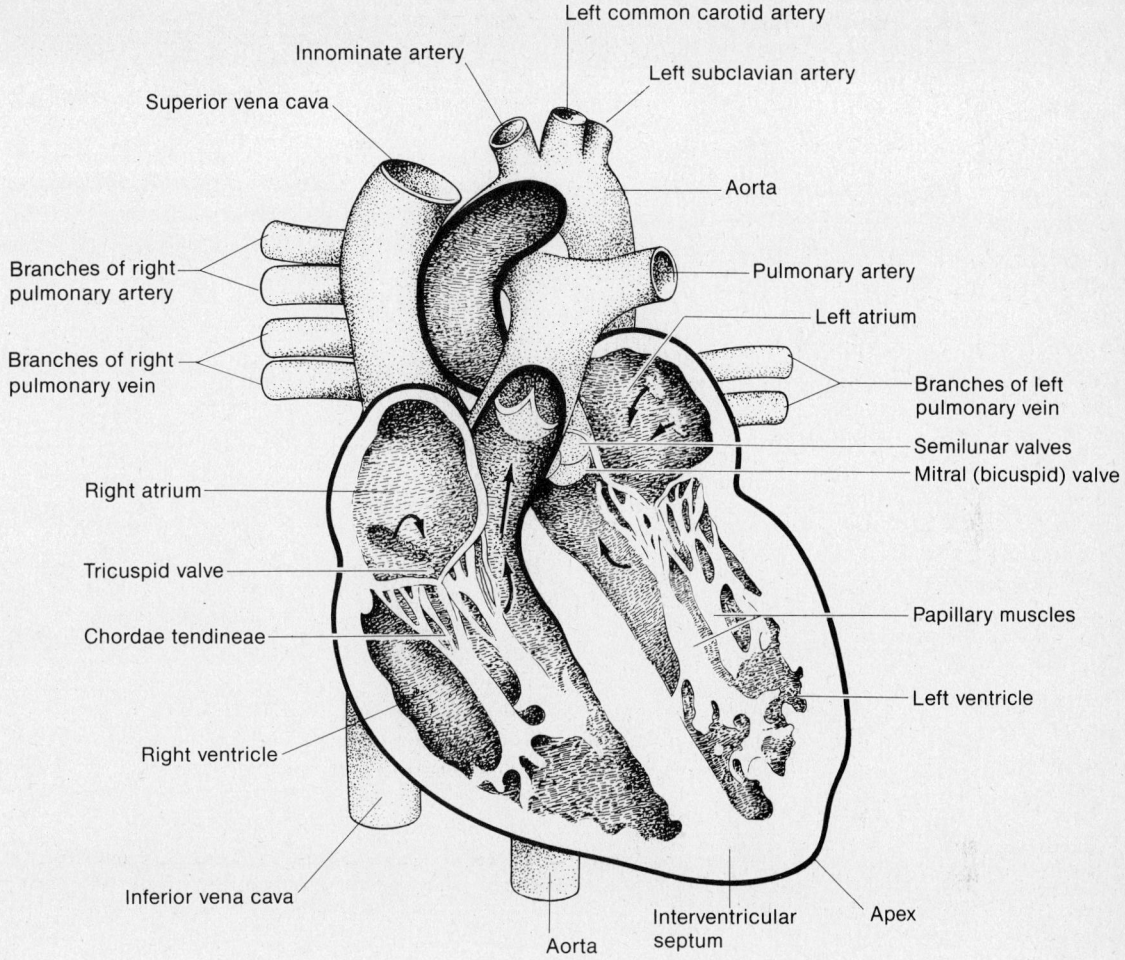

Fig. 9.3 Longitudinal section through the heart showing the four chambers, vessels entering and leaving the heart, and heart valves. The tricuspid valve on the right side of the heart aids in keeping blood moving toward the lungs. The mitral valve on the left side keeps blood moving toward the aorta.

Forces Affecting the Heart's Work Load

The circulatory system, of which the heart is a part, is a closed pressure system. High pressure in one part of the system causes blood to flow toward a lower pressure area. The highest pressure is found in the heart ventricles. The one-way valves of the heart prevent a back flow of blood under high pressure. Pressure is created in the system by several factors, such as the volume of blood being pumped, the viscosity of the blood, the effectiveness of the heart's pumping action, and the resistance to the flow of blood throughout the circulatory system.

Cardiac Output The volume of blood being pumped out of the left ventricle per unit of time is commonly called *cardiac output*. Blood flow indicates how much blood flows past a specific point during a specified time. In experimental animals an incision can be made in the aorta and a flow meter attached. In humans, however, other methods are employed, avoiding these surgical procedures.

A greater flow of blood requires the heart to work harder. Two factors working either independently or in combination will produce a greater flow of blood: (1) the ventricles may be pumping at a faster rate, or (2) the ventricles may be pumping at a normal rate with greater contracting force.

Viscosity More blood cells per ml of whole blood *(higher hematocrit)* cause greater friction as the blood cells flow past each other in a vessel. It is believed that blood cells flow in layers. A layer of cells near the vessel lining will move slower, due to greater friction, than will a layer near the center of the vessel. These layers, or laminations, sliding past each other produce friction. This friction in fluids is called *viscosity*. A higher viscosity of the blood requires a greater force to pump it throughout the system; therefore, the viscosity of blood is a factor partially determining the contractile effort that the heart must assume in maintaining a specific pressure.

Effectiveness of the Heart as a Pump In order to maintain pressure throughout the circulatory system, the heart has to be an effective pump, which simply means it has to do its normal work. Factors that interfere with normal function decrease its effectiveness. These factors include slow or irregular contraction patterns, incomplete closure or opening of the atrioventricular or semilunar valves, and the inability of the heart to adapt to changing situations.

Resistance to Blood Flow Factors other than blood viscosity also provide resistance to the flow of blood through the circulatory system. The force of gravity pulling downward on the entire body, including the blood, serves as a resisting force when the heart has to pump blood upward against gravitational attraction. This occurs in pumping blood to levels anatomically above the heart, such as the shoulders, lungs, and head region. It also is a factor involved in the return of blood to the heart from lower body levels.

Gravitational force is less when we lie down, greater when we stand up, and is the major factor producing a dizzy feeling when we suddenly jump to our feet after lying down. The blood is suddenly pulled downward away from the brain. The heart must compensate for this force by beating faster or with more force or a combination of both.

Another resistance to blood flow is the length of the flow path. When weight is gained, more capillaries are formed, thus increasing the flow path and the resistance, making more work for the heart.

The flow of blood, as we have seen, depends upon several contributing factors. Essentially the rate of flow depends upon the total pressure divided by the total resistance in the structure (heart, vessels, or capillaries) being studied.

Blood Pressure Measurement

There are many types of blood-pressure indicating devices. Two types commonly used include a direct method of determination and an indirect method.

The Strain Gauge (Direct Method) The strain gauge is an electronic device for determining pressure. A needle is inserted into an artery, vein, or heart cavity. Blood flows into the needle through an attached tube to the indicating instrument, where it creates a pressure equal to the pressure in the structure under study. This pressure alters electrical signals within the instrument, which causes deflection of a pointer on a scale calibrated in units of pressure. This method of blood-flow measurement is commonly used during surgery.

Sphygmomanometer (Indirect Method) For indirect blood pressure measurement, a *sphygmomanometer,* as shown in Figure 9.4, is used. This is the *blood pressure cuff* that most of us at some time have had contact with. The cuff is actually a flat, rubberized fabric balloon, one end of which is attached by a tube to a column of mercury or other pressure-sensing device. The other end is attached to a small hand air pump. The cuff is wrapped around the arm just above the elbow and inflated. The increased pressure in the cuff pushes against artery walls with sufficient presssure to cause them to collapse. The *brachial artery* is normally utilized in this method. If the air is gently released from the equal to the *systolic blood pressure* (pressure of ventricular contraction). This pressure is normally sufficient to raise the column of mercury to a level of 120 mm. As the air continues to be released from the cuff, more blood is allowed to pass, since the constriction is less. The rushing of the blood past the constriction produces a characteristic tapping sound. This tapping sound is due to the force of the blood pushing against the partially constricted arterial walls with each heart pulse.

Continued reduction of the cuff pressure will produce a point where there ceases to be a constriction in the arterial wall. When this occurs the tapping sound ceases. The cuff pressure is now approximately equal to the pressure normally exerted by the arterial wall during ventricular relaxation, which is called *diastolic pressure*. Diastolic pressure is normally about 80 mm of mercury. The difference between systolic and diastolic pressure is called the *pulse pressure,* which is normally about 40 mm of mercury. In other words, 120 systolic minus 80 diastolic equals 40 pulse pressure.

Pulse pressure is not constant. It is highly variable, and many factors may significantly influence it. Systolic and diastolic pressures can be greatly increased or decreased, depending upon what factors are affecting the heart or the circulatory system.

The Physiology of the Heart

Heartbeat As we saw in a previous chapter, skeletal muscle cells contract when stimulated. Heart muscle cells contract similarly, but with a few exceptional characteristics:

1. Skeletal muscle cells contract independent of one another, heart cells contract as a unit.
2. Heart muscle has its own built-in contractibility and has evolved a rhythm for this contraction.
3. During an action potential (discussed in Chapter 6), the depolarization phase is spread out over a longer period of time.

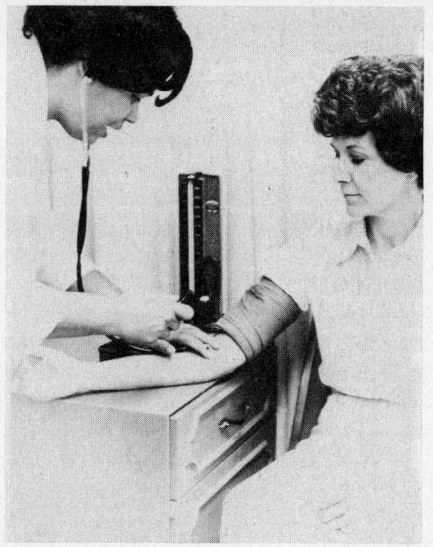

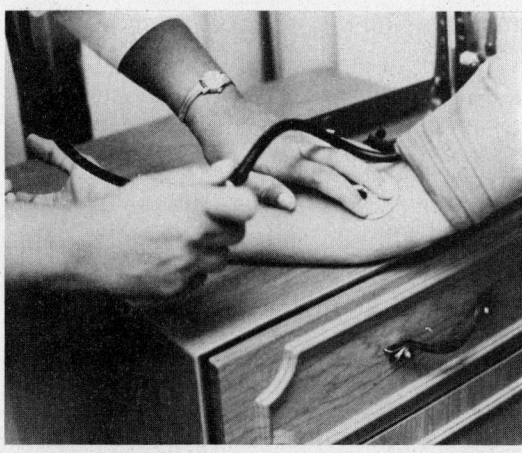

Fig. 9.4 *(a)* Blood pressure measurement with the blood pressure cuff (sphygmomanometer). *(b)* A stethoscope placed over the brachial artery picks up the sound of blood rushing past the constricted artery. This reading in millimeters of mercury is called the systolic pressure.

cuff, a point will be reached when the blood pressure is just sufficient to push blood past the constriction. The rushing of blood through this constriction produces a low, muffled sound.

A stethoscope placed just below the cuff on the brachial artery will pick up this sound. At this point, the pressure in the cuff is very nearly

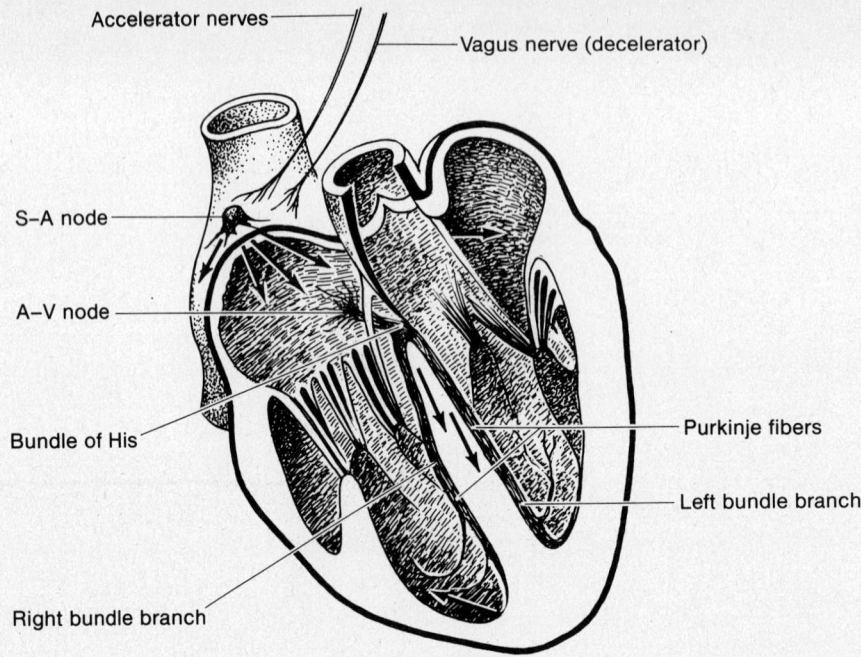

Fig. 9.5 Heart structures involved in the rhythmic contraction of heart muscles. When a signal is discharged from the S-A node, a nerve impulse spreads downward throughout cardiac muscle of both atria causing contraction, which forces blood past both heart valves. The signal arrives at the A-V node, stimulating it to initiate a new signal, which travels down the ventricular septum (via bundle branches), then enters the ventricular cardiac muscle, causing contraction blood upward into the major arteries.

Unit Contraction of Heart Fibers Within certain limits, all heart muscle fibers respond to a single stimulus. If one fiber is stimulated, a wave of contraction spreads over the entire heart, as shown in Figure 9.5. The limits imposed involve the septum, which separates the atria from the ventricles. The fibers of both atria contract as a unit; the stimulating signal, however, apparently cannot be conducted across the septum and therefore stops at this barrier. This does not mean that the stimulating signal is not transmitted to the ventricles, but rather that it is transmitted by a different route. When the signal reaches the ventricles, a wavelike contraction of all the fibers of both ventricles occurs.

Inherent Contractibility of Heart Tissue A bit of heart tissue, either from atria or ventricles, if removed from a living animal, will continue to rhythmically contract for some time without external stimulation. This is not true of skeletal muscle. Excised skeletal muscle will contract only if stimulated by electrical shock or strong chemicals. This built-in contractability is a specific heart muscle phenomenon. We now know that the stimulating signal for each heart muscle contraction originates in a small bit of specialized tissue called the *S-A node (sinoatrial node)*.

The S-A Node The S-A node is located along the inner surface of the right atrium near the entry of the superior vena cava, as shown in Figure 9.5. Precisely how cells in the S-A node generate stimulating signals is still not known. We do know that action potentials are created, but the mechanism that causes the self-firing of these potentials is still unknown.

When a signal is discharged from the S-A node, the impulse spreads out in a concentric pattern from the upper regions of both atria toward the ventricular septa. This results in atrial muscle contraction that begins at the upper levels and, through a wavelike motion, progresses downward. This squeezing motion forces the blood out of the atria into the ventricles.

A defective S-A node may fail to send stimulating signals in an orderly manner. In such instances an artificial pacemaker may be implanted.

The Artificial Pacemaker

An artificial pacemaker, shown in Figure 9.6, is an electronic device capable of establishing and maintaining a heartbeat pattern compatible with and essential for life itself. Pacemakers are implanted under the skin on the anterior chest area, with electrodes leading to the heart. Minute electrical impulses stimulate heart tissue to contract in a precise, regulated manner. Persons suffering from heart attacks, heart block, slow ventricular rates, or congenital heart disease can all be aided by artificial pacemakers.

The electronic signal emitted by a pacemaker may be disrupted or altered by the nearness of other electronic devices, including microwave ovens and electric motors. A patient with an implanted pacemaker therefore must exercise caution where such devices are in use.

The A-V Node After the ventricles become filled due to the action of venous blood pressure and atrial contractions, they contract, forcing the venous blood from the right ventricle toward the lungs and from the left ventricle toward body tissues. Here again we find a bit of specialized tissue that generates a stimulating signal for ventricular contraction. This tissue is called the *A-V node* (atrioventricular node) and is located between the right atrium and the right ventricle along the septum and near the tricuspid valve.

The A-V node differs from the S-A node in that it has a conducting pathway over which the impulse travels, as shown in Figure 9.5. From the

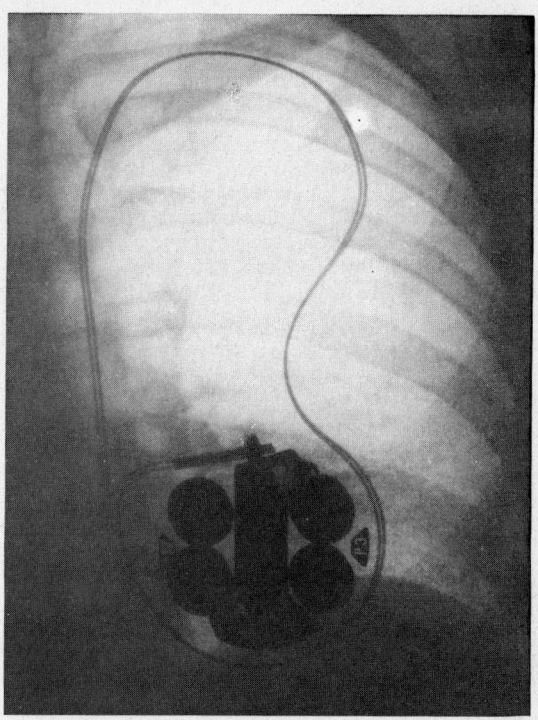

Fig. 9.6 X ray of an artificial heart pacemaker in position. Note the electric lead (wire) leaving the pacemaker and entering ventricular tissue. (Courtesy Radiology Department, Normandy Osteopathic Hospital, St. Louis, Mo.)

A-V node, specialized conducting tissues, the bundle of His, carries the impulse to the septum. Here the bundle of His splits into two branches (right and left), each respectively carrying the impulse to the right and left ventricles. The branches are made up of specialized bundles of fibers called *Purkinje fibers*. Purkinje fibers carry the signal along the pathway and into the inner heart muscle cells. The contraction wave that results from this pattern of signal transmission again provides a squeezing effect that forces the blood out of the ventricles.

S-A Node and A-V Node Coordinate Function Since the S-A node generates the initial signal and maintains a fairly constant rhythm, it is referred to as the "pacemaker." The adult human heart has an average beating rate of 72 beats per minute. This rhythm is passed along to the A-V node, then to the specialized pathways to activate ventricular contraction in a rhythmic fashion. When the S-A node sends out an impulse, the atria contract; there is not, however, an immediate A-V node response. A slight delay occurs before the A-V node sends a signal to the ventricle muscle fibers. This allows time for both atria to empty their contents into the ventricles before ventricular contraction occurs. This delay keeps the atria and ventricles from contracting at the same time.

The A-V node has an inherent, rhythmic, signal-producing mechanism of its own. It is slower (60 to 65 beats per minute) and is therefore overridden by the more rapid impulses arriving from the S-A node. If the S-A node is diseased or damaged so that it cannot originate the rhythmic signals, the A-V node takes over the job and produces the lower rate pattern. In this condition, the atrial fibers no longer receive signals and therefore do not contract. The heart may continue to function without atrial contraction, however, since venous blood, entering the superior and inferior vena cava, is under sufficient pressure to fill the relaxed ventricular cavities.

Atrial and ventricular muscle each are capable of rhythmic contraction without the influence of either the S-A node or the A-V node. If both nodes cease functioning, a rhythmic pattern of 20 to 40 beats per minute is initiated by atrial muscle. If this is eliminated, ventricular muscle establishes a rhythmic pattern of 10 to 30 beats per minute. Of course, unless corrected, a human would not live long with such a slow heartbeat, but it has been established that these rhythmic contraction patterns do exist. Each lower rate is overridden by the higher rate. In a normally functioning heart, the highest rate is provided by the S-A node (70 to 80 beats per minute), and it overrides all lower-rate-producing tissue.

Neural Control of Heartbeat Although the S-A node establishes a pattern and rhythm for heart contraction, the heartbeat rate can be altered by nerve stimulation. Both the sympathetic and parasympathetic divisions of the autonomic nervous system may influence heart rate.

Sympathetic neurons originating in the *cardiac acceleratory center* of the medulla send impulses to a network of nerve fibers located near the aortic arch. Impulses are then sent over other neurons to the S-A node, A-V node, and other heart tissue, causing an increase in heart rate. Parasympathetic neurons originating in the *cardiac inhibitory center* of the medulla also send impulses to the cardiac plexus by way of the vagus nerve. These impulses inhibit S-A node impulses, causing a decrease in heart rate.

Pressoreceptors are neurons that are stimulated by pressure or stretch. A mass of these receptor neurons is located in the *carotid sinus*, an enlarged portion of the internal carotid artery. The carotid arteries carry blood to the brain. When blood pressure in the internal carotid is increased to a certain level, the arterial wall expands and stretches. Pressoreceptors thus stimulated send impulses to the cardiac inhibitory center in the medulla. The heart rate is decreased, which decreases the cardiac output, thus decreasing the pressure in the carotid sinus.

Chemical Receptors that Regulate Heart Rate Excess carbon dioxide in the blood inhibits the inhibitory center of the medulla. The acceleratory center is thus released to some degree to send impulses to the heart, speeding up the rate. The breathing rate is also increased, and therefore a greater volume of blood arrives at the lungs enabling the blood to lose its CO_2 at a faster rate. Some blood vessels contain receptors that are stimulated by low oxygen levels. These receptors send impulses to the cardiac acceleratory center of the medulla. The heart rate increase causes a larger volume of blood per minute to flow into and out of the lungs, thus increasing the rate of oxygen taken up by the blood.

Some chemicals produced in the body may accelerate or inhibit heart rate. Norepinephrine secreted by postganglionic fibers of the sympathetic nervous system accelerates heart rate. These fibers may be stimulated by stress conditions such as fear or anxiety. The postganglionic neurons of the parasympathetic nervous system release acetylcholine, which inhibits the S-A node, thus reducing the heart rate.

Heart Sounds

The sound of a throbbing heart romantically portrayed in song turns out to be the rather unromantic sounds associated with closing heart valves, shown in Figure 9.3.

The tricuspid valve, located between the right atrium and right ventricle, is a one-way valve that permits blood to flow from the atrium into the ventricle. In other words, when the filled right atrium is contracted, blood under pressure forces the tricuspid valve open and the blood gushes into the ventricle. The valve is so designed, however, that when the ventricle contracts the valve is snapped shut. This prevents blood from reentering the atrium.

The blood under pressure of ventricular contraction forces open another one-way valve, the right semilunar, which permits the blood to enter the pulmonary artery leading to the lungs. The same sequence of events occurs on the left side of the heart. Blood entering the left atrium from the pulmonary vein forces the mitral valve open permitting blood to enter the left ventricle. As the left ventricle forcibly contracts, the mitral valve is slammed shut, prohibiting a return flow of blood into the atrium.

The first heart sound one detects—the *lub* of *lub-dub*—is the sound of the synchronous closing of the mitral and tricuspid valves, along with sounds set up by the vibration of contracting ventricles and opening semilunar valves. As the ventricles relax after forcing the semilunar valves open, the blood under high pressure attempts to gush back into the relaxed ventricular cavities. The sudden relaxation of the ventricles and the high pressure of blood attempting to return to the ventricles closes the semilunar valves. This creates the second heart sound—the *dub* of *lub-dub*.

The exact cause of heart sounds has been debated for years. The earliest suggestion was that the sounds were caused by the slapping together of the closing valves. The nature of the closing valves is such that they probably do not slap together, since the blood attempting to rush back into the previously emptied cavities cushions the closing motion of the valves. There is, therefore, probably little if any noise from this source.

The current explanation is as follows: (1) The tricuspid and mitral valves close and at about the same time the semilunar valves of the aorta and the pulmonary artery open. (2) The blood gains momentum as it rushes from the atria into the ventricles. This rushing action sets up turbulence within the ventricle. This turbulence causes the walls of the ventricles to vibrate. (3) The ventricular contraction forces the tricuspid and mitral valves to close and blood surges back against the valves, attempting to reenter the atria. This causes the valves to bulge back into the atria. The force of the bulging attempts to push the blood back in the ventricle. This push-and-pull effect on the blood in the ventricle produces further turbulence that causes more vibration of the ventricle walls.

These vibrations set up in the valves themselves and in the ventricle walls eventually reach the chest wall, where it is in contact with the heart. The sounds produced are those related to the first heart sound, the *lub* of *lub-dub*. The second heart sound is related to the closing of the semilunar valves.

When the ventricles contract, they force blood into the aorta and the pulmonary artery under considerable pressure. When the ventricles relax, the blood from the pulmonary artery and aorta first attempts to rush back into the relaxed ventricular cavity. This backward rush of blood closes the semilunar valves and causes them to bulge back into the ventricular spaces. This bulging stretches the semilunar valves, producing a counterforce, pushing the blood back into the aorta and pulmonary artery against the backwards rush of the pressures of these two vessels. These forces pushing in opposite directions set up turbulences that cause the walls of the vessels and portions of the heart muscle to vibrate. These vibrations are transmitted into the chest wall and produce the second heart sound, the *dub* of *lub-dub*.

These two most prominent heart sounds can be detected without a stethoscope by placing your ear on a person's chest. A stethoscope, however, is essential for listening for other less obvious sounds produced by defective heart valves.

Another instrument, called a *phonocardiograph,* picks up heart sounds and converts them into electrical currents, which in turn activate an inked stylus marking on a moving paper belt. This provides a permanent record of heart sound and also is useful in detecting levels of sound lower than the stethoscope is capable of picking up.

A third heart sound is associated with the turbulent action of blood gushing into the ventricles. A fourth, much less obvious sound is produced by the contraction of the atria and the rush of blood through the open valves.

An electrocardiogram (discussed later in the chapter) cannot indicate whether heart valves are functioning properly. Its only function is to indicate heart electrical rhythm patterns.

Heart Murmurs If any of the heart valves fail to open completely, blood is forced through a smaller than normal opening. This produces an abnormal sound. If a valve does not completely close, blood is permitted to flow back into the atria or the ventricles, also producing a characteristic abnormal sound. These sounds are called *heart murmurs.* Proper diagnosis can generally locate the defective valve and, where necessary, surgery may be used to correct the defect.

Other Structural Abnormalities Structural abnormalities of the heart cause extensive circulatory problems throughout the body. The heart may be incapable of pumping sufficient blood to essential organs, causing such abnormalities as *interatrial defect, interventricular defect,* and *patent ductus arteriosus.* These conditions are often related to some congenital defect.

The Electrocardiogram (ECG)

An *electrocardiograph* is an instrument capable of detecting electrical changes that occur during rhythmic contraction of the heart. The electrical charges are normally picked up from the surface of the body at specific places called *leads*. The charges are then amplified and are used to activate a sensitive electromagnetic inked stylus that marks on a moving paper belt. The paper is calibrated to show the magnitude of voltage change along the vertical axis and the time sequence in seconds along the horizontal axis. The paper belt is normally set to travel at a rate of 25 mm per second. Figure 9.7 illustrates one of 12 different leads used in the recording of an ECG.

The electrical impulses originating in the heart are not only conducted to all the heart cells but are also transmitted throughout the body. The leads are attached to conduction electrodes that, when held tightly on the skin surface, are capable of picking up the electrical charges initiated by the heart tissue. A firm connection must be made between the electrode and the skin. Most often, hair is shaved from the area on which the electrode will be placed. The electrode is held in place by a belt webbing or a strong rubber band.

An ECG can diagnose a defective heart ailment only to the extent that a specific ailment in some way influences normal rhythmic electrical potentials initiated in the heart. An ECG cannot diagnose a defective valve nor can it help anticipate heart attacks before they happen. After a heart attack, such as a thrombosis, some heart tissue dies; this in turn changes the electrical pattern of the normal heart. An ECG may then be used to locate the damaged area. Figure 9.7 shows a normal ECG and two abnormal ones.

ARTERIES, VEINS, AND CAPILLARIES

The circulatory system includes a network of vessels that carry blood to all parts of the body. It is just as critical that the most distal cell in the big toe receives nutrients and oxygen as it is for other cells in the body closer to the heart. The nutrients and oxygen that blood carries must be exchanged at the cell membrane for carbon dioxide and other metabolic wastes. These sub-

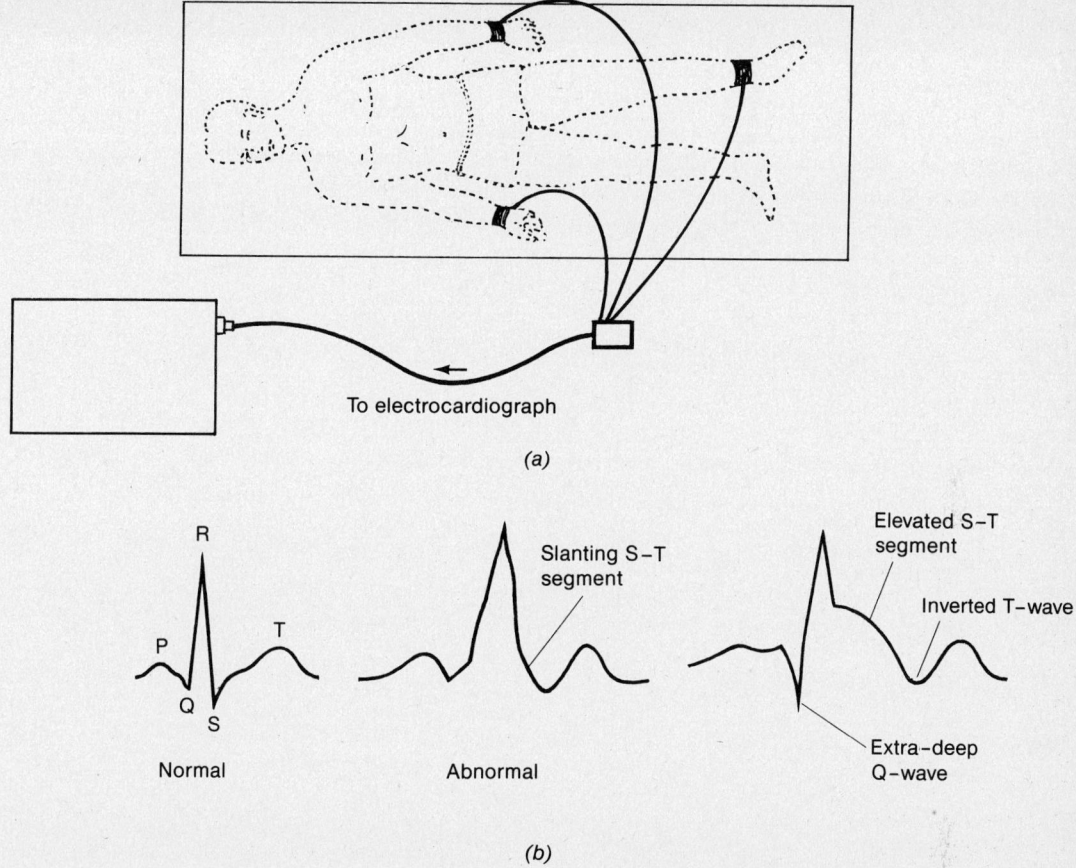

Fig. 9.7 (a) Electrocardiograph showing one arrangement of leads. (b) A normal and two abnormal ECGs.

stances are returned by a series of vessels to the heart.

Arteries, veins, and capillaries are the conducting vessels. Each has different structural characteristics and different functions.

Arteries

An *artery* is defined as a blood vessel that carries blood away from the heart. Arteries are elastic muscular tubes through which the blood flows outward to all cells. Figure 9.8 illustrates a cross section of an artery.

An artery has an inner lining called *endothelium* surrounded by an elastic membrane. External to this membrane is a muscular coat composed of smooth muscle and an outer coat of connective tissue. The internal elastic membrane not only gives additional strength to the walls of the artery but also aids in pumping the blood throughout the system. As blood leaves the left ventricle and enters the aorta, the vessel walls are stretched and expanded by the pressurized blood. The force created by the stretching process helps push the blood through the arterial system. The aortic walls must be able to withstand at least the pressures created by the left ventricle of the heart. Occasionally, a segment of an artery becomes weakened and balloons out as a dilated sac, called an *aneurysm*.

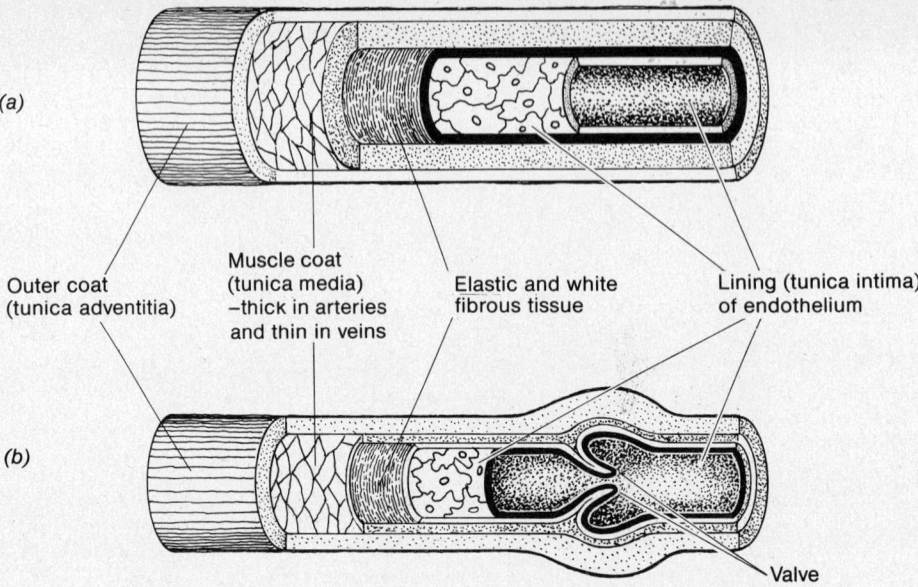

Fig. 9.8 Sectioned view of (a) an artery and (b) a vein. An artery is lined with endothelium, which is surrounded by elastic tissue. External to this tissue is a muscular coat of smooth muscle, then an outer covering of connective tissue. Veins are less muscular and contain little or no elastic tissue.

Aneurysms occur most frequently in the aorta, where the pressure is the greatest. If an aneurysm bursts, a fatal hemorrhage results.

Veins

A *vein* is defined as a vessel that carries blood toward the heart. Compare the structure of the vein shown in Figure 9.8 with that of the artery. Note that the vein has a less muscular wall and contains little or no elastic material. The reason for the structural difference is that veins have much less blood pressure to contend with.

The pressure in veins of a person lying in a horizontal position is practically zero. The pressure in veins is considerably higher in the leg of a person in a standing position. In order for blood to get from the lower extremities back to the heart, it must be pumped upward against gravity. The pressure in the lower leg veins is about 95 mm of mercury in the standing position as compared to almost zero in the horizontal position.

Blood returning from the head and shoulders toward the superior vena cava has a negative pressure of about −10 mm of mercury. The force of gravity helps pull the blood downward.

Since the pressure of blood in the veins is normally quite low, veins have specialized valves that tend to keep the blood moving in one direction, toward the heart. Figure 9.8 shows one of these valves. The flow of blood is dependent upon pressures coming from the blood system, and to a great degree upon muscular contraction. As muscles move and contract, blood is squeezed in the veins. It can only flow toward the heart because the valves will not permit blood to flow in the other direction.

Varicose veins result from a breakdown of the valves within veins, which permits the pressures that exist due to gravity or pregnancy to stretch the blood vessels and enlarge them, causing disfiguring bluish marks.

Capillaries

Between the arteries and veins at the cellular level are the *capillaries*. Capillaries are microscopic blood vessels, the lining of which is only one cell thick. Figure 9.9 is a diagram of a capillary bed.

Arteries leaving the heart divide into smaller vessels called *arterioles*, which empty their contents of oxygenated blood into the capillary beds surrounding cells. The capillary vessel is only large enough to accommodate red blood cells passing through in single file. Even single file some of the red cells have to squeeze through the tiny capillary openings by folding.

Oxygen diffuses from the red cell through the capillary membrane into interstitial cell space (fluid space around cells), then into body cells. Food nutrients follow the same general pathway.

Waste products and carbon dioxide reverse this procedure. These waste substances exit by diffusion through the cell membrane, enter the interstitial cell space, then diffuse into the capillary, where they are carried by the blood back to small divisions of the veins called *venules,* and then back to the heart by the venous system.

The capillaries are the most important part of the circulatory system, since this is where oxygen and nutrients are taken into the cells and waste products and carbon dioxide are removed.

CIRCULATION PATTERNS

Circulation throughout the body is normally divided into two categories. The circulation of blood from the heart to the lungs and back to the heart is called *pulmonary circulation*. Circulation from the heart through arteries to all parts of the body via the capillary beds and back by the venous system to the right side of the heart is called *systemic circulation*. Figure 9.10 illustrates these two circulatory patterns.

Pulmonary Circulation

The simplest circulatory pattern is the pulmonary. In this, deoxygenated blood leaves the heart and passes through the *pulmonary artery* to the lungs. The pulmonary artery divides as it leaves the heart, with one segment, the right pulmonary artery, leading to the right lung. The other, the left pulmonary artery, goes to the left lung. Each artery divides into smaller arteries (arterioles) that empty into capillaries that surround the lung cells. It is here that oxygen from the air we breathe diffuses into these capillaries.

Carbon dioxide leaves the capillaries and enters the lung spaces, later to be expelled. The oxygenated blood then collects into larger vessels, and these into still larger vessels called *pulmonary veins*, which return the blood to the left side of the heart.

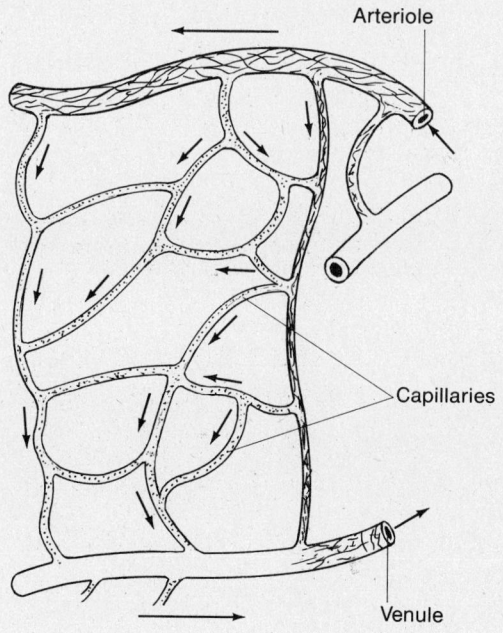

Fig. 9.9 A capillary bed, showing the direction of blood flow through it. Capillaries are microscopic blood vessels with walls only one cell thick, across which nutrients and oxygen pass to enter cells. Waste substances and carbon dioxide from cells enter capillaries to be carried away.

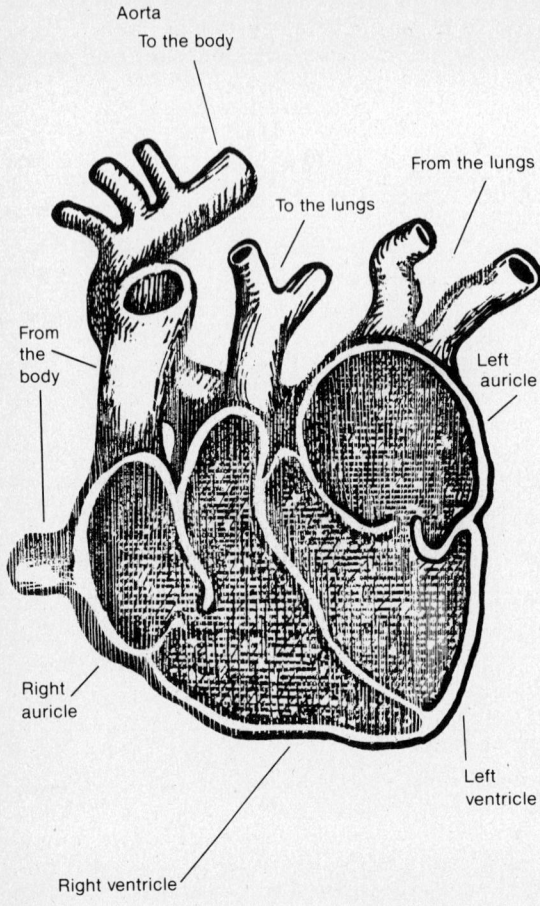

Fig. 9.10 A diagram of pulmonary and systemic blood circulation patterns. Pulmonary circulation flows from the right ventricle of the heart to the lungs, then to the left atrium. Systemic circulation flows from the left ventricle outward through the aorta to body tissues, capillaries, and veins, returning blood to the right atrium.

Systemic Circulation

This pattern of circulation is much more complicated than pulmonary circulation. The one large artery leaving the left ventricle of the heart is called the *aorta*. Immediately past the aortic semilunar valves two small branches of the aorta arise. These are the *right* and *left coronary*

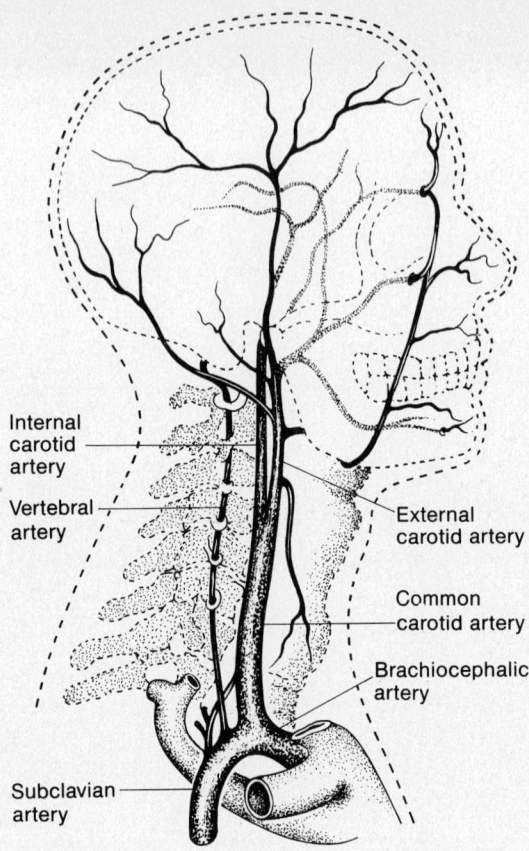

Fig. 9.11 Arteries of the head arising from the right subclavian artery. These arteries, the common carotid and vertebral, along with their counterparts on the left side, distribute blood to the neck and head.

arteries, which supply heart tissue with nutrients and oxygen. The right coronary artery divides into two smaller branches, the *posterior descending artery* and the *marginal artery*. The left coronary artery also divides into two smaller arteries, the *anterior descending artery* and the *circumflex artery*. The aorta continues upward for a short distance and then forms a U-shaped curve to the left and travels downward.

Arteries of the Head, Shoulder, and Arm As the aorta rises above the heart, making its U-turn, two arteries branch off. On the left, the *subclavian artery* travels down into the left arm becoming the *axillary artery* and further down becoming the *brachial artery*. At about the elbow level, the brachial artery divides into the *radial*

152 HUMAN FORM AND FUNCTION

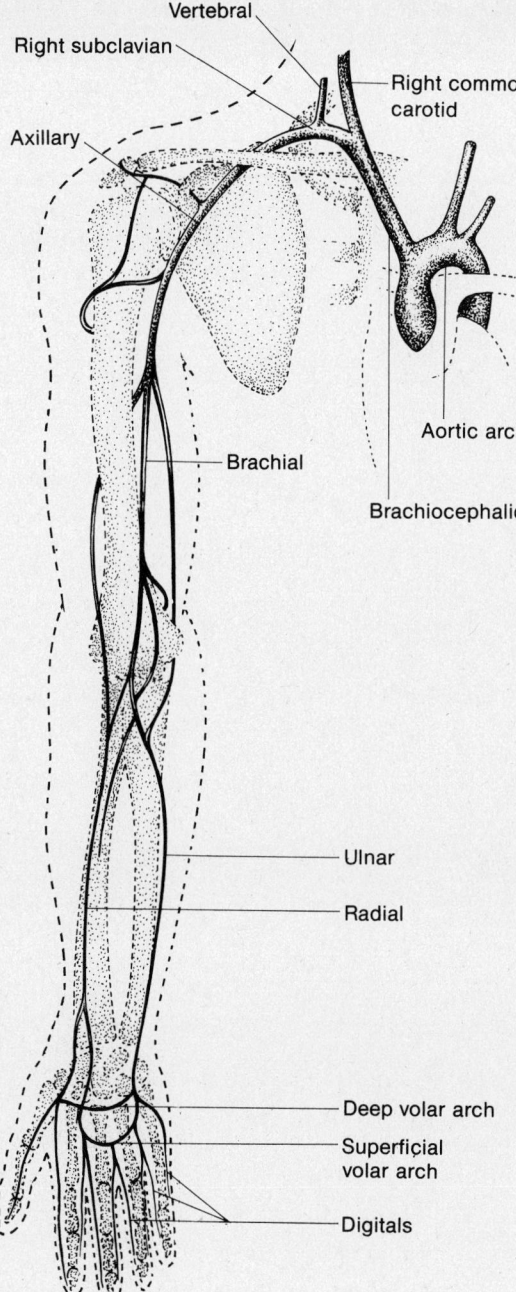

Fig. 9.12 Arteries of the right shoulder and arm. As the right subclavian artery enters the arm it is called the right axillary artery. Just above the elbow it becomes known as the right brachial, which subdivides into the right ulnar and right radial arteries. These arteries unite, forming the right palmar artery, which branches into the fingers.

and *ulnar arteries*. These two arteries later reunite in the palm of the hand to form the *palmar arteries* from which the *digital arteries* of the fingers arise.

Also arising from the aortic arch and extending upward toward the head is the *left common carotid artery*. The first branch of the left subclavian also extends into the head region. It is called the *left vertebral artery*. The common carotids and vertebrals supply blood to the brain, face, and muscles of the head and neck region, as shown in Figure 9.11.

The *common carotid artery* in the upper neck region divides into two segments, the *external carotid* and the *internal carotid*. The branches of the external carotid supply blood to the muscles and tissues of the face and neck. The internal carotids and the vertebral arteries continue upward, passing through the foramen magnum, and supply blood to the brain tissues. The two vertebral arteries, one from either side, continue to the base of the brain where they unite to form the *basilar artery*. The basilar artery is connected by a ring of arteries with the internal carotid arteries. This ring of arterial connections between the internal carotids and the basilar arteries is called the *circle of Willis*.

Returning to the aortic arch once again, we will now trace the vessels on the right side of the body, as shown in Figure 9.12.

At about the same location that the left common carotid arises from the aortic arch, another branch leaves the arch providing blood for the right side of the body. This artery is called the *innominate artery*. This is a very short artery almost immediately subdividing into the *right common carotid artery* going to the head and the *right subclavian* that has a pathway very closely resembling that of the left subclavian artery. As

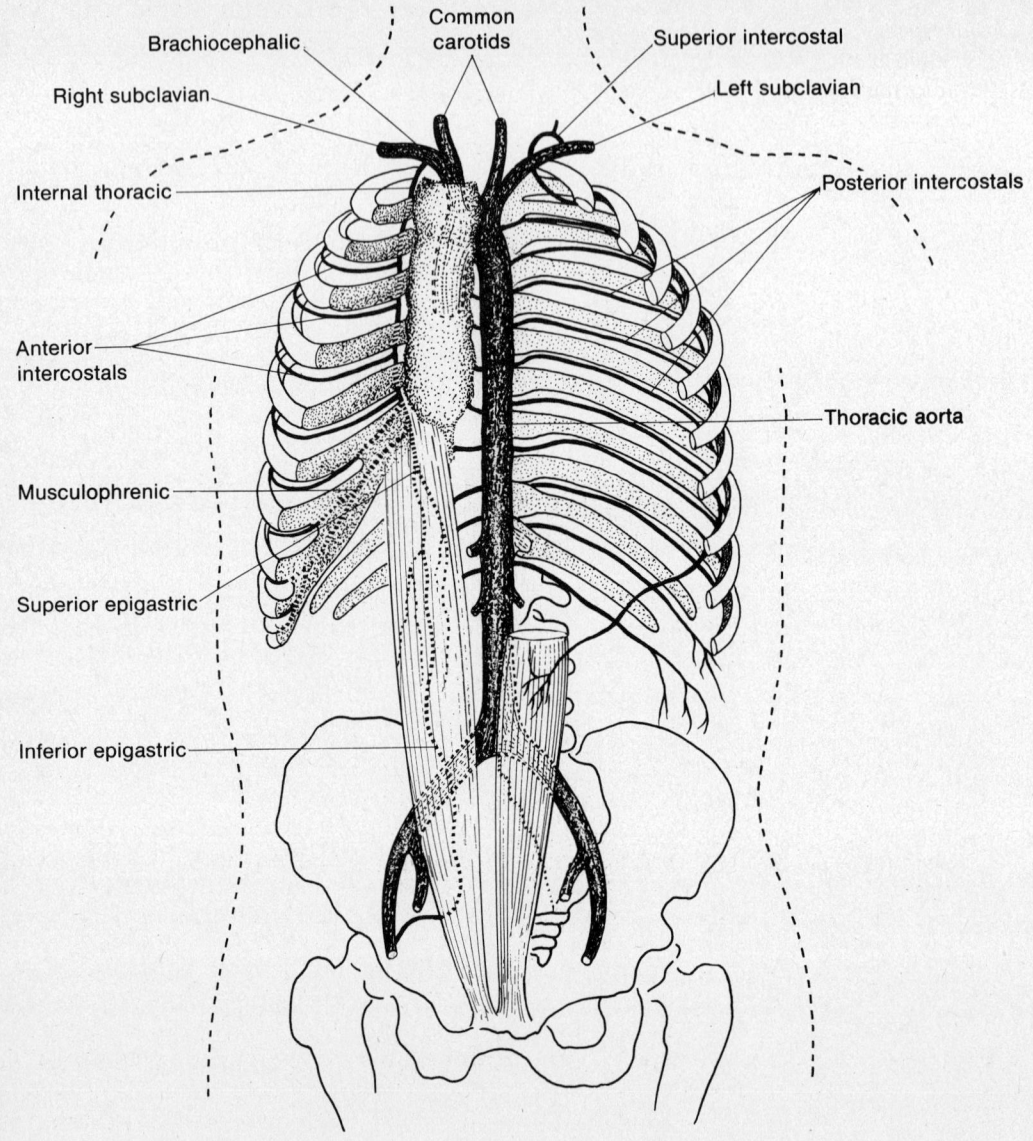

Fig. 9.13 Arteries of the chest. The thoracic aorta, as it descends, sends paired branches into the rib muscles. These are the intercostal arteries. Other branches of the aorta include arteries to the bronchial tubes, lungs, esophagus, and heart pericardium.

the right subclavian enters the arm it is called the *right axillary artery*; lower in the arm above the elbow it is called the *right brachial artery*, which subdivides into the *right ulnar* and *right radial*, these arteries uniting to form the *right palmar artery*, which has branches extending into the fingers. The names of these arteries are the same as those on the left side, so in referring to them one must state whether they are on the right or the left side of the body.

Also arising from the right subclavian is the *right vertebral artery*, which runs upward parallel with the *left vertebral artery*. These arteries eventually fuse in the lower brain as the *basilar artery*. The left and right common carotids both give off branches in the lower head and upper neck region called the *external* and *internal carotids*.

Arteries of the Chest and Abdomen The descending aorta sends off many small branches to the rib muscles. These paired vessels are called the *intercostal arteries*, as shown in Figure 9.13. Other branches of the aorta include arteries to the bronchial tubes and lungs, the esophagus, and heart pericardium.

Many small arteries branch off the aorta as it descends into the abdominal region. Some of these arteries supply blood to the muscles of the abdominal wall. The larger branches supply the viscera.

The first large arterial branch of the descending aorta is the *celiac artery* (see Figure 9.14), which almost immediately branches into the *left gastric*, the *hepatic*, and *splenic arteries*. The left gastric supplies blood to the esophagus and the stomach muscles along its lesser curvature.

The hepatic artery heads in the direction of the liver, supplying it with blood, and sending off branches that include the *right gastric* and arteries to the duodenum, pancreas, and stomach.

The *splenic artery* supplies the spleen as well as the pancreas, stomach, and part of the anterior small intestine.

Immediately below the celiac-aortic junction is the *superior mesenteric artery*, which supplies the small intestines and the first part of the large intestine. These arteries are interlaced throughout the mesentery (thin connective tissue) that holds the intestines in position.

Arteries of the Trunk and Lower Extremities The *renal arteries* arise just below the origin of the superior mesenteric arteries. These paired arteries carry blood to the kidneys.

Gonadal arteries arise from the lower aorta beneath the renal arteries and supply the gonads (the testes of the male, and the ovaries of the female).

The *inferior mesenteric artery*, a large branch of the lower aorta, supplies blood to most of the colon (large intestine).

The aorta continues to descend to the level of about the fourth lumbar vertebra where it divides into the *right* and *left common iliac arteries*. The common iliac arteries, as they progress posteriorly toward the leg region, divide into two smaller arteries, the *external* and *internal iliac arteries*. The internal iliac artery, also called the *hypogastric artery*, gives rise to many small branches that supply the pelvic muscles and the lower viscera. The external iliac artery continues into the lower extremity, becoming known as the *femoral artery*.

As the femoral artery progresses downward behind the knee, it becomes known as the *popliteal artery*. Just below the knee, the popliteal artery divides into three branches, the *anterior* and *posterior tibial arteries*, and the *peroneal artery*. The posterior tibial artery and the peroneal artery descend into the sole of the foot, forming the *plantar artery*. The anterior tibial artery descends and supplies the anterior surface of the foot and becomes the *dorsalis pedis artery*, which branches over the top of the foot.

Venous Circulation In studying the vessels carrying blood from the various parts of the body back to the heart, it is helpful to remember that the names of veins many times are the same as those of the arteries that take blood to a particular tissue from the heart. Our study of veins will begin at the most extreme part of the body, the foot, and work upward toward the heart.

In the lower leg, as shown in Figure 9.15, we find the *anterior tibial, posterior tibial, peroneal veins*, and *small saphenous* uniting to form the *popliteal vein* just behind the knee. The upper extension of the popliteal is called the *femoral vein*, which continues upward and unites with

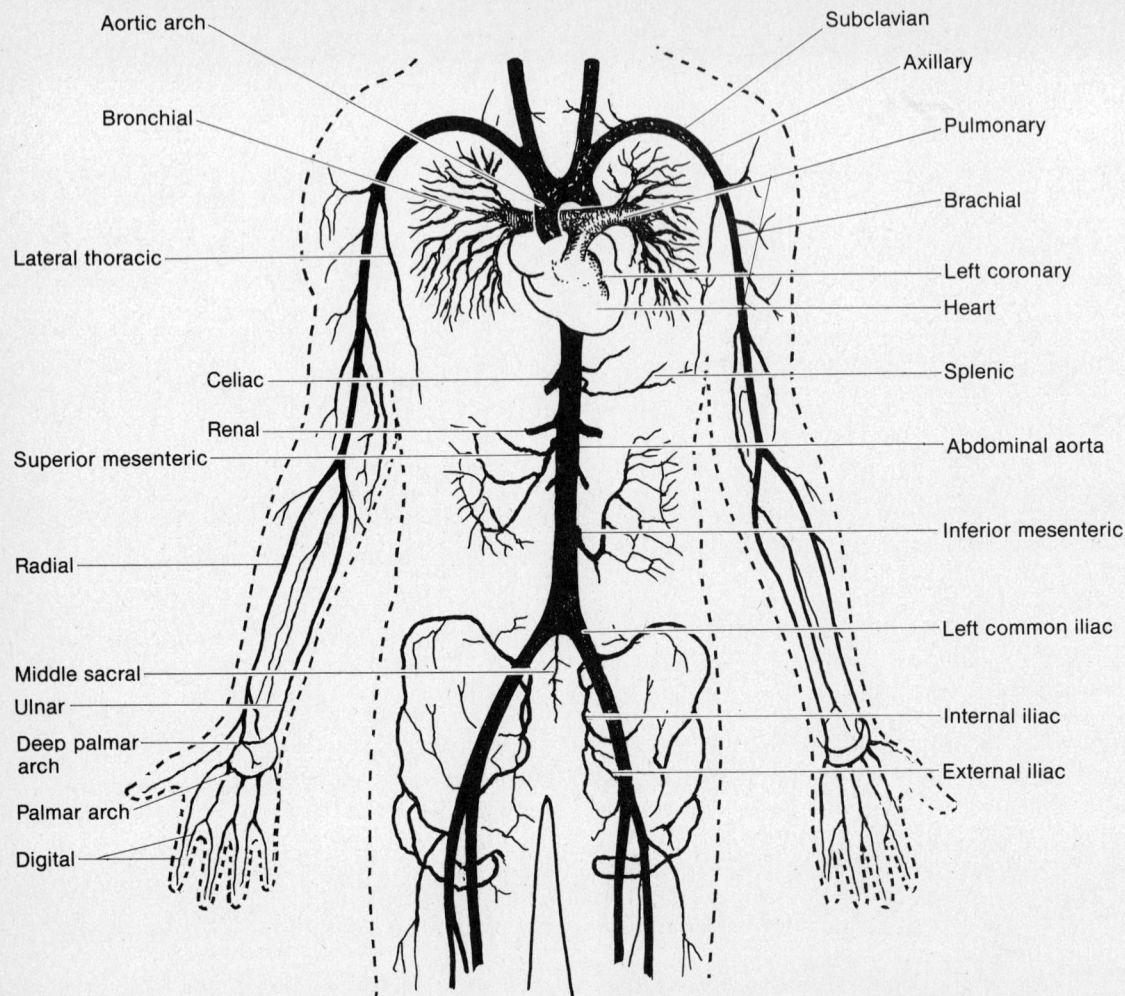

Fig. 9.14 Arteries of the abdomen and upper trunk. Many small arteries branch off the aorta as it descends into the abdominal region. Some of these arteries supply blood to muscles of the abdominal wall. Larger branches of the aorta supply the viscera.

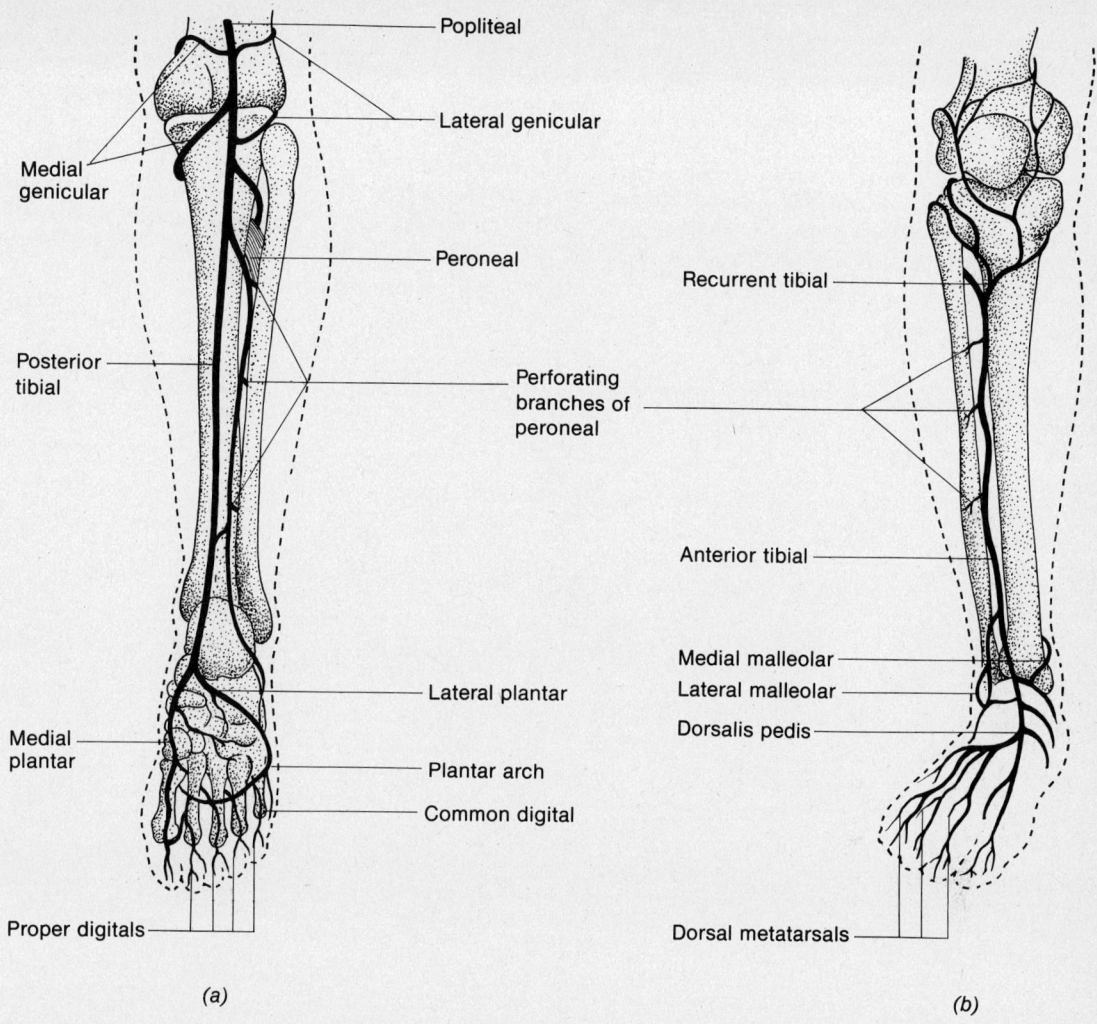

Fig. 9.15 Major arteries of the leg and foot. (a) Arteries of the posterior leg and foot. (b) Arteries of the anterior leg and foot.

THE HEART AND BLOOD VESSELS 157

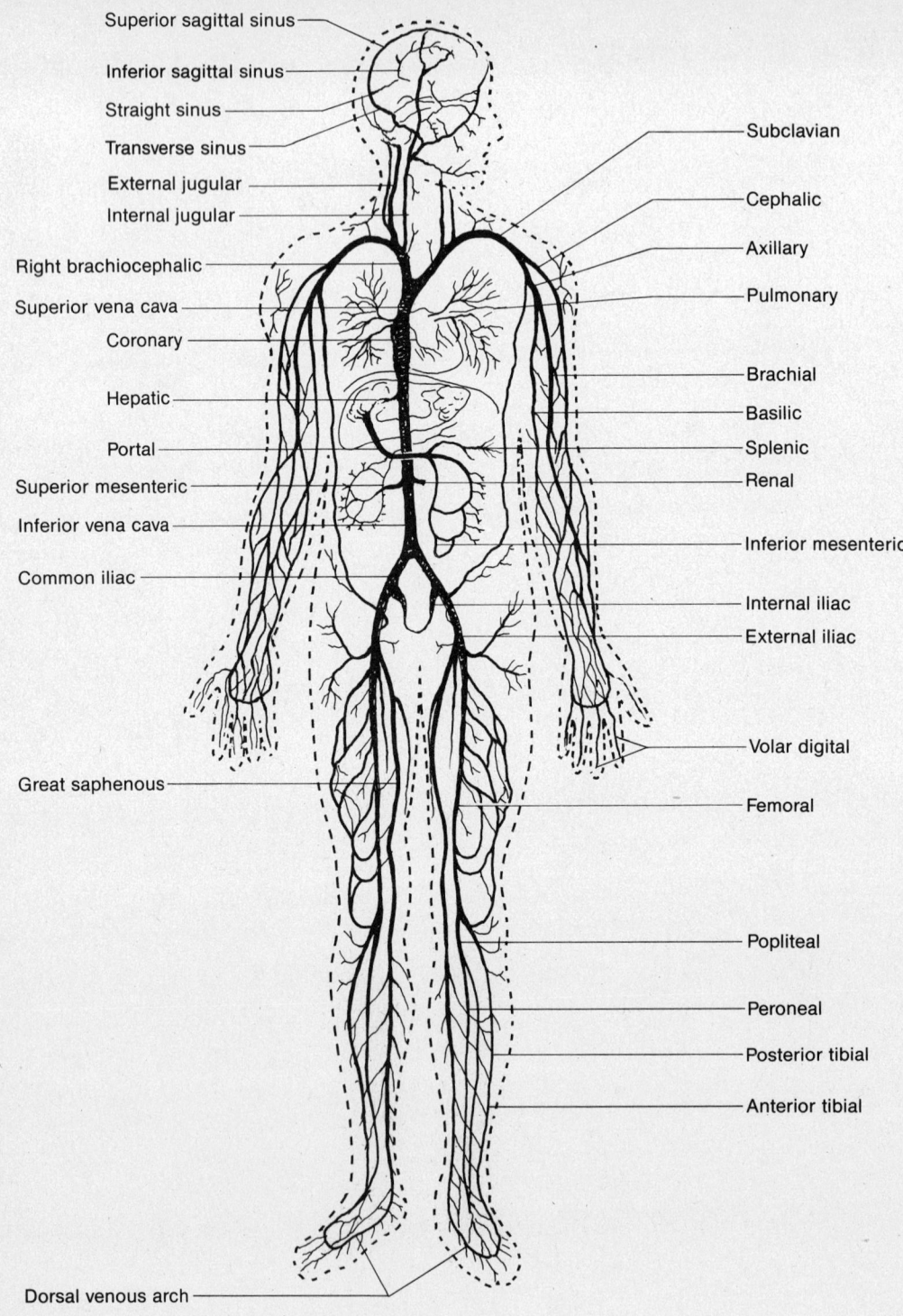

Fig. 9.16 Major veins of the body. In studying veins it may be helpful to remember that the name of a vein is often the same as that of the artery that supplies blood to a particular tissue.

the *great saphenous*. The union of these two veins forms the *external iliac,* which combines with the *internal iliac* in forming the *common iliac vein*. The right and left common iliac veins unite to form the *inferior vena cava,* which carries blood back to the right side of the heart from the lower extremities.

The veins of the upper extremities are classified as superficial or deep. The deep veins are the *radial, ulnar, brachial, axial,* and *subclavian*. The superficial veins include the *basilic* on the medial aspect of the arm and the *cephalic* on the lateral aspect. The cephalic and basilic veins of the lower arm are joined by the *median cubital vein*. The *axillary vein* is formed in the shoulder by the union of the cephalic and basilic veins. The axillary vein becomes the subclavian. The subclavian veins (right and left) unite with the internal jugular veins to form the *brachiocephalic veins*, which drain into the superior vena cava. Figure 9.16 illustrates the major veins of the body.

The Hepatic Portal System   A *portal system* is one that connects two different beds of capillaries. The blood capillaries, which carry nutrient materials from the intestine, drain into a vessel called the *hepatic portal vein*. This vein carries all the food nutrients absorbed from the intestine into liver capillaries. The blood, heavily laden with nutrients, enters the liver, which is able to control the concentration of these nutrients in the bloodstream, storing some of them and allowing others to continue on their way by way of the *hepatic vein* to the heart.

Fetal Versus Adult Circulation
The fetus depends upon maternal tissues to provide nutrients and oxygen. This means that the digestive processes including the secretion of digestive enzymes are not functioning in the fetus; neither is the oxygen-providing apparatus, the lungs, in operation. Oxygen supplied to fetal blood diffuses across the mother's placental membranes and is carried to the fetus by blood vessels within the umbilical cord. Figure 9.17 offers a comparison of fetal and adult circulation.

The major difference in the circulation of the blood in the fetus and the adult involves the heart and adjacent blood vessels. The umbilical vessels carrying oxygen from the placenta pass through the liver by way of the portal vein. As the blood enters the liver, certain nutrients are removed or stored as the fetal metabolism requires, then the blood passes into the inferior vena cava. Blood entering the right atrium from the inferior vena cava would in the normal adult pass through the tricuspid valve into the right ventricle. A small amount of blood does pass along this route. However, most of the blood passes through an interatrial opening called the *foramen ovale,* This opening between the two atria permits the blood to flow from the right side of the heart to the left side of the heart.

Since the right side of the heart is involved only in pumping blood to the lungs and back to the heart, and since the lungs are not operating in the fetus, there is no reason for blood to flow to the lungs. Although a small amount of blood does enter the right ventricle, flowing out the pulmonary artery on its way to the lungs, this blood is diverted back into the aorta by the *ductus arteriosus*. Most of the blood passing across the interatrial connection into the left atrium is squeezed through the mitral valve into the left ventricle, then pumped into the aorta and throughout the body.

Shortly after birth and within the first months after birth, circulatory changes occur in the infant. The foramen ovale closes because there are now equal pressures on either side of this flaplike opening, thus closing it.

Some conditions do not permit the proper closure of this opening, allowing it to persist into adult life. If the opening does persist, the blood is not properly oxygenated, since the opening serves as a bypass to the lungs.

Another structural change occurring shortly after birth is the closure of the ductus arteriosus. Pressure from the right ventricle tends to keep the ductus open. Once the lungs become operable, however, the blood can flow more easily into

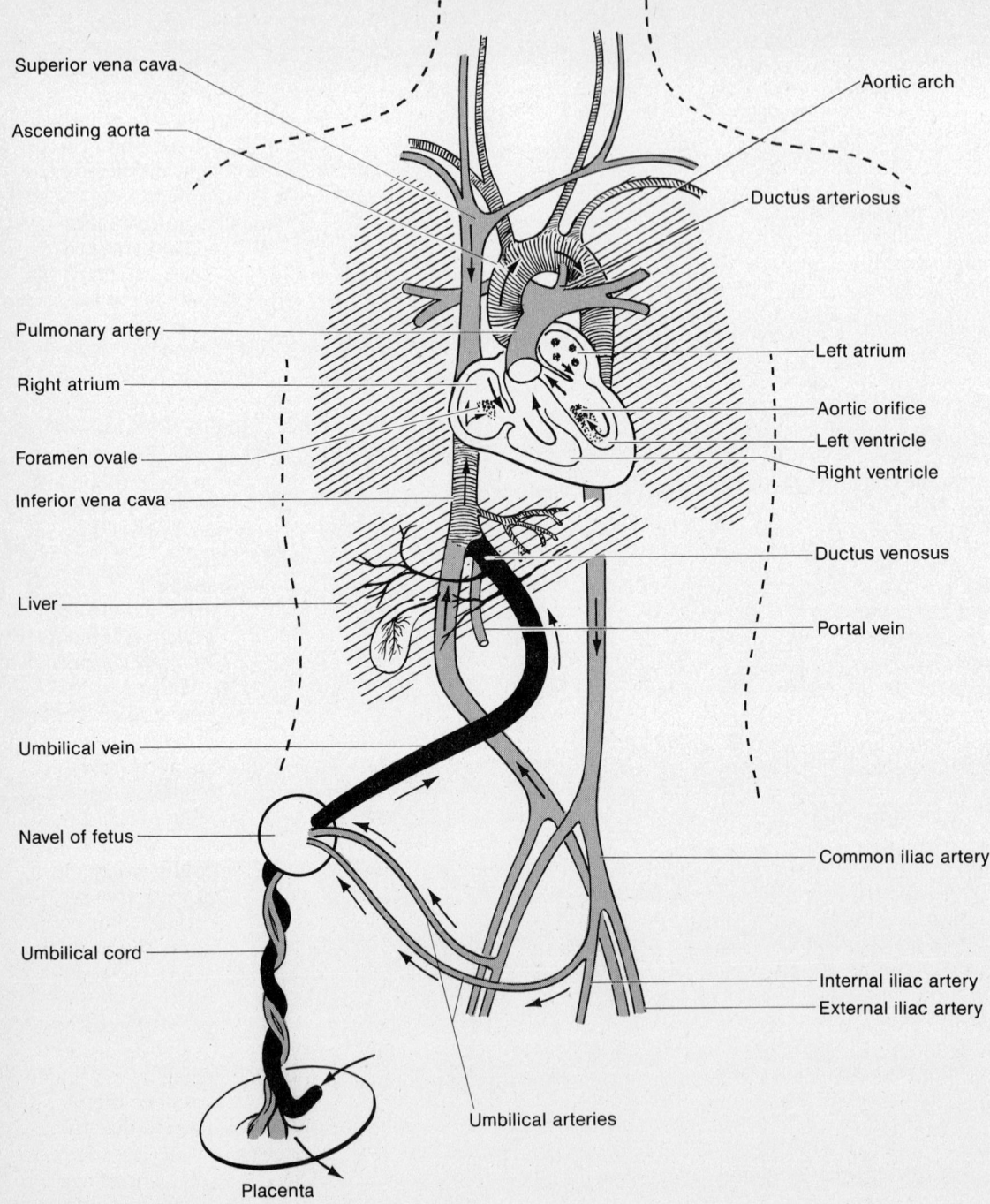

Fig. 9.17 Fetal circulation. Three major differences can be noted between fetal and adult circulation: (1) the presence of the foramen ovale between the atria; (2) the presence of the ductus arteriosus, which shunts blood into the aorta from the pulmonary artery; and (3) the presence of the umbilical cord transporting blood to and from the placenta.

the pulmonary arteries. Therefore the ductus, held open by pressure, closes since the pressure into it is reduced. This eventually closes so completely that in the adult it appears as a ligament.

DISEASES OF THE CIRCULATORY SYSTEM

Congenital Diseases

Congenital diseases are due to defects that occur during fetal development.

In the congenital condition, known as *Tetralogy of Fallot*, the aorta originates from the right ventricle or it overrides the ventricular septum. In either case, blood from the right ventricle then is pumped into the aorta, bypassing the pulmonary arteries to the lungs; this results in a very low oxygen-accumulating system. The pulmonary artery is usually *stenosed* (blocked). This adds to the problem of low oxygen availability since blood from the pulmonary artery flows to the lungs to pick up oxygen. If this artery is partially closed or stenosed, then less blood can get to the lungs to pick up oxygen.

In Tetralogy of Fallot, there is also usually a ventricular septum defect. This simply means there is an opening between the right and left ventricles. There is also an occasional interatrial defect. As discussed previously, in a normal infant the foramen ovale usually closes. In an infant with this condition, however, the opening remains. The right ventricle enlarges due to the extra workload placed upon it. Obviously, the Tetralogy of Fallot condition is quite serious, and can only be corrected by surgery. The surgery involves so many sites within the heart that a series of procedures are required usually involving several years. The child with this condition exhibits cyanosis, a stubby growth of the fingers, and has difficulty breathing—even having to squat to take a deep breath. If this condition goes unattended, the child will die.

Transposition of the pulmonary artery and aorta results in cyanosis. In this condition, the aorta arises from the right atrium, thus blood arriving from the tissues is pumped back to the tissues without going to the lungs. Oxygenated blood arriving at the left atrium from the lungs is returned to the lungs by the pulmonary artery.

Ventricular septal defect is an opening in the septum that separates the right and left ventricles. This permits oxygenated blood from the left ventricle to mix with unoxygenated blood from the right ventricle. The blood thus mixed will arrive at tissues with less than adequate oxygen. A similar mixing of blood occurs in *atrial septal defect*.

Patent ductus arteriosus results from failure of the ductus arteriosus to close after birth. Blood, which normally should go to the lungs to pick up oxygen, is shunted into the aorta, thus bypassing the lungs. Similar consequences result from the failure of the foramen ovale to close. Blood from the right atrium continues to directly enter the left atrium. This also creates a lung bypass.

Diseases of the Coronary Arteries

Diseases of the coronary arteries include *angina pectoris* and *myocardial infarction*. Angina pectoris causes severe pain in the upper left portion of the chest and shoulder. It is caused by insufficient quantities of blood carried by the coronary arteries. It may be brought on by even mild exercise, cold temperatures, anxiety, and other emotional disturbances. Nitroglycerine provides temporary relief due to its effect of dilating the coronary arteries.

Myocardial infarction results when heart tissue fails to receive sufficient oxygen to remain alive. Where in the heart the infarction occurs and how much heart tissue dies are the factors that determine whether or not a person lives after a myocardial infarction.

Circulatory shock results from reduced blood flow to the tissues. It may be caused by decreased cardiac output or reduced venous blood return. A myocardial infarction may reduce the pumping action of the heart so severely that tissues, including heart tissue, fail to receive adequate nutrition and oxygen. Severe hemorrhage reduces blood volume *(hypovolemia)*, and thus results in less blood pressure. Less pressure means that nutrients and oxygen are not adequately supplied to tissues.

Other Diseases and Disorders

Rheumatic Fever This disease is normally initiated by streptococcal infection. This may begin as a sore throat, scarlet fever, or middle ear infection. Streptococcal bacteria release toxins against which the body forms antibodies. These antibodies react not only against the streptococci, but also against many normal body proteins, particularly those found in the heart valves.

The valves become infected and may hemorrhage, and large lesions form along the inflamed edges of the heart valves. These edges are very thin and are instrumental in making the proper closure between the atria and ventricles and between the ventricles and the arteries leaving the heart. The mitral and aortic valves are the ones most commonly affected. The edges of these valves become hemorrhaged and stuck together; later, scar tissue develops causing a fusion of the valve edges. This condition is called *valvular stenosis*. A stenosed valve will not permit the normal flow of blood from one space to another because of the constricted nature of the valve. If the valve edges become so scarred and thickened that they cannot close adequately, blood will be permitted to return from one space to another. This particular condition is called *valvular regurgitation*.

Murmur of Aortic Stenosis If the semilunar valve in the aorta is stenosed, up to 450 mm of mercury pressure may be produced by the ventricle in pushing the blood into the aorta. Due to the stenosed condition of the aortic valve, the blood is inhibited in progressing normally into the aorta. The aortic pressure may be normal, but the volume of blood is reduced. This extreme pressure by the ventricle produces a nozzle effect, spraying blood into the aorta as if from a high-pressure hose. Turbulence due to the spraying effect causes intense vibrations to be set up in the aortic wall. These vibrations are so great that they can be felt with the hand if it is placed over the heart or in the upper neck region and can actually be heard without a stethoscope.

Murmur of Aortic Regurgitation In this condition, the semilunar valves have become so thickened and scarred that they cannot close adequately. When the ventricles relax (diastole), blood is permitted to reenter the ventricle with a blowing sound. This sound is due to the turbulence caused by the back rush of blood into the relaxed ventricle. These two sounds, the murmur of aortic stenosis and the murmur of aortic regurgitation, are decidedly different and can be easily detected.

Abnormal Circulatory Patterns in Valvular Heart Disease Muscles increase in size due to the degree of work load placed upon them. Heart muscle is not an exception to this rule. If the ventricles are required to contract with greater force due to some valvular disfunction, they increase in size. *Hypertrophy* (enlargement) of the left ventricle is a common condition resulting from valvular heart disease.

In an attempt by the body to balance the deficiency of fluids that the rest of the body receives, due to valvular disfunction, there is an increase in total blood volume. The increase in blood volume creates an even greater work load upon the heart, which in turn has the tendency to accelerate hypertrophy of the left ventricle.

If the mitral valve is defective so that it permits the blood to return from the left ventricle into the left atrium, pressures are created that resist the incoming blood through the pulmonary veins from the lungs. This back pressure is distributed along the lung capillaries, forcing some of the fluids back into the lungs, resulting in *pulmonary edema*. The left ventricle simply cannot keep up with the work demand, and blood dams up in the left atrium and in the lungs.

Arteriosclerosis Blood vessels in normal young adults have flexible walls with elastic tissue that can repeatedly expand and contract with each pulse. Age and other factors tend to reduce this elasticity. Arteriosclerosis is the condition commonly called "hardening of the arteries." Some arteries come to resemble bonelike tubes incapable of flexing in response to ventricular contraction. This stretching absorbs some of the pressure of ventricular contraction, helping to maintain a pulse pressure of about 40

> ### Arteriosclerosis: The Time Bomb
>
> *Almost half of all human beings die of arteriosclerosis. Deaths may result from thrombosis of coronary arteries or hemorrhage of blood vessels in other organs of the body. Although the precise cause of arteriosclerosis is not known, the following list summarizes some of what is known about the disease.*
>
> *1. More men under the age of 50 die of arteriosclerosis heart disease than do women of the same age. This may indicate that the male sex hormones accelerate the disease or perhaps that female sex hormones inhibit the disease.*
> *2. Severe diabetes or severe hypothyroidism accelerates the disease.*
> *3. Atherosclerosis, the forerunner of arteriosclerosis, is definitely related to hypertension. Almost twice as many hypertensives will die of the disease as compared to individuals with normal blood pressures.*
> *4. High-fat diets, containing large quantities of cholesterol and saturated fats, increase the tendency for arteriosclerosis.*
> *5. Being overweight increases the chance for the disease.*
> *6. Genetics may prove to play a major role in arteriosclerosis since it is observed to run in families.*
>
> *Other coronary risk factors include heavy smoking, obesity, and stress. As is often the case, when two or more of these factors are found to be operating the effect is greater than the sum. Smoking, hypertension, and elevated serum cholesterol constitute the major risk factors. Potentially fatal coronary problems may be reduced as much as 50 percent with a 10 percent reduction in the three major risk factors. Obesity and stress are often related. A person may eat compulsively when stress-producing situations arise. Overeating, eating large quantities of carbohydrates and fats, and lack of proper exercise all may lead to obesity.*
>
> *Stress may also be involved with hypertension. Individuals who feel hurried, are chronic worriers, feel pressed for time, or who feel pressures owing to job, family, or cultural influences are prime candidates for hypertension and high coronary risk.*

mm of mercury (120 systolic minus 80 diastolic). Blood in arteries affected by arteriosclerosis may be under a pulse pressure of 100 mm of mercury or more.

Atherosclerosis Factors such as improper diet, smoking, and lack of proper exercise have been suggested as causes of the thickening of the lining of arteries. This condition is called *atherosclerosis*. Fatty deposits containing cholesterol appear on the inner walls of arteries. These become gradually interlaced or covered with fibrous connective tissue that may also contain deposited calcium.

An *occlusion* (block) in an artery such as the coronary artery may form due to a piece of the deposit breaking away. This material may then block a smaller artery, resulting in the death of heart tissue distal to the block. This condition is commonly called a *heart attack*. The severity of the attack depends upon how much heart tissue is deprived of oxygen and nutrients. A blocked large artery is more serious (often fatal) than an occlusion of a small artery.

Endocarditis (Bacterial) The lining of the heart chambers and valves is called *endocardium*. Certain bacteria, such as *Streptococcus viridans* and *Staphylococcus aureus*, may be released into the bloodstream during surgery or from a local infection. These bacteria may collect and reproduce on previously damaged heart valves, causing more damage to the valves. The valves may become so severely damaged that they fail to function, initiating congestive heart failure.

Aneurysms An *aneurysm* is a dilated segment of a weakened arterial wall. Blood flowing into the dilated artery loses some of its pressure, which affects tissues distal to the aneurysm. The aneurysm may also burst and prove fatal if it occurs in a large artery or in brain tissues.

HYPERTENSION IN THE ELDERLY

With aging, certain vascular and organic changes are evident. Atherosclerosis of the major arteries increases. Arterial walls become less elastic and therefore less capable of distension to accommodate pulse pressure. The aorta and major arteries thus become high-pressure tubes.

The coronary arteries are not excluded from this process. Thus heart tissue itself is subjected to higher pressure but less actual blood volume. The influence of arterial resistance causes the left ventricle to increase in size. Increased size involves increased work by the heart and an increased need for oxygen availability through inadequate coronary arteries.

This vicious cycle may continue until the heart can no longer meet the increased demands. The consequence often is cardiac failure.

SUMMARY AND REVIEW

A. The cardiovascular system includes the heart, arteries, veins, and capillaries.
   1. The heart is a muscular organ housing four cavities—the right and left atria, and the right and left ventricles.
   2. Arteries carry pressurized blood away from the heart to the tissues.
   3. Veins carry low-pressure blood from the tissues back to the heart.
   4. Capillaries are minute vessels that surround cells, providing them with nutrients and oxygen.
B. Factors that influence heart action:
   1. blood volume
   2. neural control
   3. congenital defects and diseases
C. The heartbeat is initiated by impulses arising from the S-A node.
   1. Impulses from the S-A node cause atrial contraction and A-V node stimulation.
   2. A-V node impulses travel from the bundles of His to bundle branches to Purkinje fibers.
   3. Purkinje fibers penetrate heart ventricle fibers and bring about ventricular contraction.
D. Heart sounds are the result of heart valve closure. The first heart sound is caused by the closing of tricuspid and mitral valves. The second sound is caused by the closing of the semilunar valves.
E. The electrocardiogram (ECG) is used to detect changes in electrical patterns in heart tissue.
F. Circulation from the heart to arteries, arterioles, and capillaries is called arterial circulation. The blood flow patterns from the capillaries to venules, veins, and the heart is called venous circulation.
G. Fetal circulation differs from adult circulation.
   1. The fetal ductus arteriosus shunts blood into the aorta from the pulmonary artery. This bypasses the lungs.
   2. The fetal foramen ovale results in blood passing directly from the right atrium into the left atrium, thus bypassing the lungs.
   3. The placenta takes the place of lungs for supplying oxygen.
H. Cardiovascular diseases and disorders
   1. Congenital diseases
      a. Tetralogy of Fallot
      b. transposition of pulmonary artery and aorta
      c. ventricular septal defect
      d. patent ductus arteriosus
      e. patent foramen ovale
   2. Diseases of the coronary arteries
      a. angina pectoris
      b. myocardial infarction
   3. Other diseases and disorders
      a. rheumatic fever lesions
      b. aortic stenosis
      c. aortic regurgitation
      d. arteriosclerosis
      e. atherosclerosis
      f. endocarditis
      g. aneurysms
I. Hypertension in the elderly. Major causes include:
   1. atherosclerosis
   2. loss of elasticity in arterial walls
   3. ventricular muscle increase in size owing to higher pressures caused by 1 and 2.

10
Blood

MAJOR CONCEPTS
BLOOD FUNCTIONS
RED BLOOD CELLS (ERYTHROCYTES)
   Number of Red Blood Cells
   Red Blood Cell Formation
   Vitamins and Red Blood Cell Production
WHITE BLOOD CELLS (LEUCOCYTES)
PLATELETS AND THE CLOTTING MECHANISM
   The Blood Clot
   Hemophilia and Other Disorders
BLOOD GROUPS
   How Blood Is Typed
   Cross Matching
   The Rh Factor
   ABO Incompatibility
DISEASES AND ABNORMALITIES OF BLOOD COMPONENTS
   Anemia
   Sickle Cell Anemia
   Erythroblastosis Fetalis
   Iron Deficiency
   Polycythemia
   Leukemia
   Leukopenia
AGING OF BLOOD

Blood is the vital fluid that transports nutrients to cells and carries away cellular waste. In addition to nutrients, other essential substances are transported by blood. These include oxygen, hormones, minerals, vitamins, and water.

Although blood is defined as a tissue, this chapter will describe several different functions maintained by structural blood components.

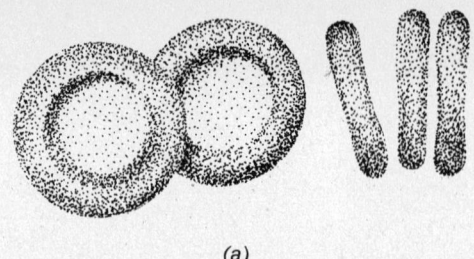

(a)

GRANULAR LEUKOCYTES

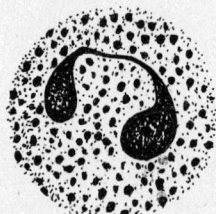
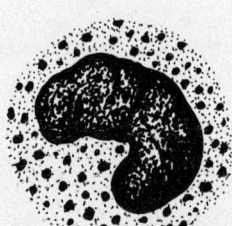
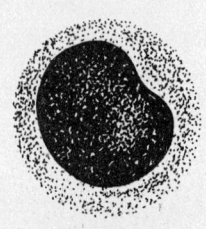

Neutrophil Eosinophil Basophil

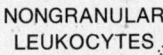

NONGRANULAR LEUKOCYTES

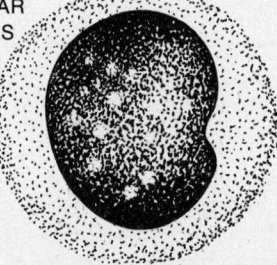

Lymphocyte Monocyte

(b)

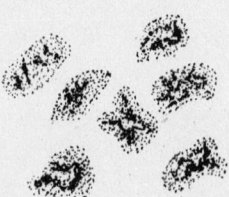

(c)

Fig. 10.1 Types of cells found in blood. (a) Only one type of red cell, the erythrocyte, exists. (b) Five kinds of white cells are found—neutrophiles, eosinophiles, basophiles, lymphocytes, and monocytes. (c) Platelets, particles that lack a nucleus, are fragments of red blood cells.

The overall function of the circulatory system is to provide nutrients, oxygen, minerals, ions, vitamins, and water for all body cells. In addition, the blood acts as a heat distributor, a carrier for hormones, a regulator of fluid volumes, and a buffer to maintain an equilibrium between acids and bases in tissue fluids. In this chapter we will deal exclusively with blood, its composition and function. Unless otherwise stated, the quantitative data given is related to normal adults.

BLOOD FUNCTIONS

Blood is a complex tissue having a high degree of diversity with respect to its function.

The average amount of blood in the body is about 5 liters, most of which is pumped from the heart to the tissues and back to the heart in approximately one minute.

The following is a survey of blood functions.

1. Nutrition Blood is a carrier of the products from the digestive processes. These include all the food nutrients that the body utilizes.

2. Respiration The oxygen-carbon dioxide exchange materials are carried throughout the body. Oxygen is picked up from the lungs by the blood, delivered to tissues, where it is exchanged for carbon dioxide, which in turn is delivered back to the lungs to be expelled from the body.

3. Fluid Balance The consistency of the quantity of fluids in and around tissues is a critical factor. Blood serves as a homeostatic tissue controlling tissue fluid levels.

4. Acid-Base Balance Hemoglobin is a good chemical buffer. Buffers are substances that equilibrate or neutralize excess acids or bases.

5. Excretion Substances that are thrown off by the body as waste materials are removed from tissues by the blood and carried to the kidneys where they are excreted.

6. Protection Blood contains specialized cells and secretions (antibodies) that can destroy or neutralize bacteria and their by-products.

7. Temperature Regulation Blood carries heat throughout the body to all tissues. This heat-carrying capacity of blood maintains normal body temperatures throughout the body.

8. Carrier of Hormones Endocrine glands produce hormones that are emptied directly into the bloodstream and are carried to all body tissues.

Blood is composed of erythrocytes (red cells), leukocytes (white cells), platelets, and plasma, as shown in Figure 10.1. Figure 10.2 shows one type of white blood cell.

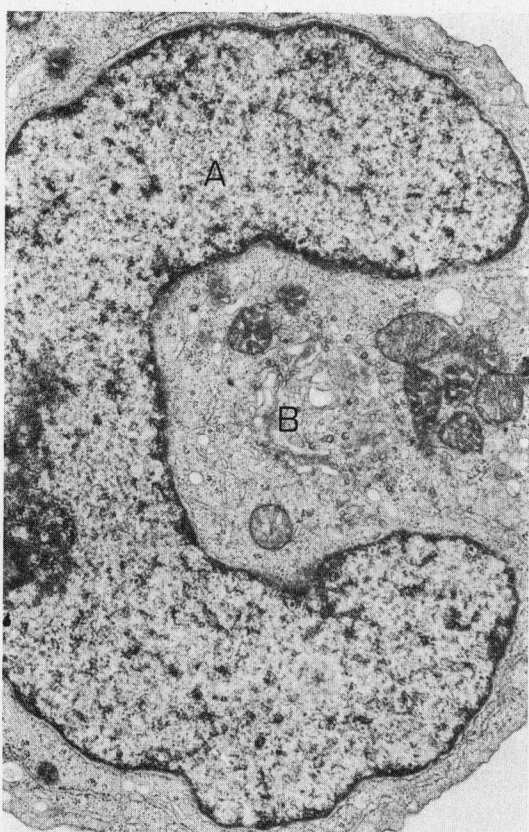

Fig. 10.2 Electron micrograph of a monocyte, a type of white blood cell. Monocytes are large cells which function as phagocytes. Monocytes make up about 6 to 8 percent of all white blood cells found in normal whole blood. The structure at *(A)* is the nucleus; *(B)* is part of the cytoplasm. *(Courtesy Santiago Plurad.)*

RED BLOOD CELLS (ERYTHROCYTES)

An *erythrocyte* (red blood cell) is a biconcave, disc-shaped cell with a diameter of about 8 microns. It has a thickness of about 2 microns nearest its outer edge and about 1 micron in the center. Erythrocytes originate from nucleated cells in red bone marrow, the liver, and spleen. As they normally mature, the nucleus is lost.

Erythrocytes have a relatively short life (average 120 days) and cannot reproduce themselves.

The characteristic shape of an erythrocyte permits the cell to be greatly deformed without destroying it. The cell has to squeeze through narrow passageways in the capillaries often smaller in diameter than the erythrocyte itself. The cell can be folded and its shape otherwise distorted to permit its passage through the capillaries. As the erythrocyte becomes older or diseased, its cell membrane becomes fragile. As fragile cells attempt to penetrate small capillaries, they are broken apart and destroyed.

Number of Red Blood Cells

The number of erythrocytes in the blood varies with many factors, including sex, level of sustained activity, age, and altitude. The average red blood cell count is about 5,000,000 cells per cubic millimeter (mm^3). Women have somewhat less than this (4,600,000 per mm^3) and men more (5,400,000 per mm^3).

A red blood cell count is obtained by diluting a precise quantity of blood, then spreading this diluted blood over a finely marked glass slide. With microscopic examination, one can count red blood cells within a given space and multiply this by the dilution factor and the divisions on the glass slide. This is an average count, since one is not counting every single cell. This average count is important, however, in the diagnosis of many body malfunctions, since the red blood cell count may be drastically increased or decreased by disease or other influencing factors.

Normal blood is composed of about 40 percent (by volume) cells and 60 percent plasma. The term *hematocrit* is used to indicate what percent (by volume) of whole blood is cells. A normal hematocrit is about 40 percent. A hematocrit is determined by filling a test tube with whole blood, then centrifuging it. Blood cells and other formed particles, being heavier than the plasma, will be pulled to the bottom of the tube. One then simply observes the tube afterwards, determining what percent of the tube's length is packed with cells, as shown in Figure 10.3.

A normal hematocrit for males is about 45 percent and for females about 42 percent. White blood cells (WBC) and other cellular components of the blood will also be separated from the plasma. However, their number is so small (about 9,000 WBC per mm^3) compared to the number of red blood cells that they do not produce a significant deviation in the reading.

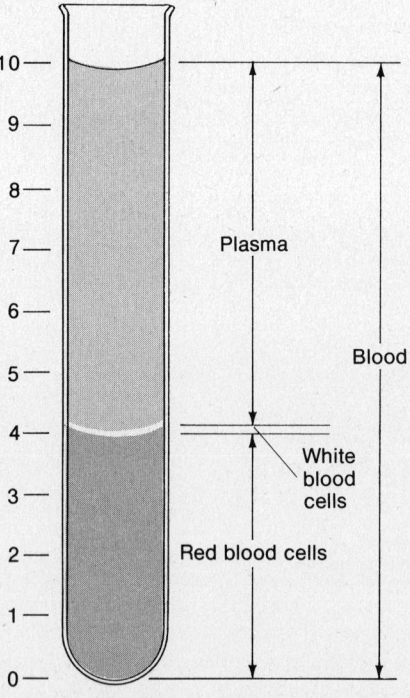

Fig. 10.3 Hematocrit determination. The tube shown here after centrifugation has its lower 40 percent packed with red cells. This is read as a hematocrit of 40.

Red Blood Cell Formation

During fetal development, red blood cells are formed mainly in the liver, spleen, lymph tissue, and bone marrow. In the adult, the spleen, liver, and lymph tissue cease to function in this activity and the major tissue for red blood cell production is *red bone marrow* located in flat bones, such as the ribs and sternum. The long, round bones of the extremities during growth from childhood to adult become filled with fatty substances and lose their ability to produce red blood cells. Under certain severe conditions that produce body stress, the spleen and liver may reactivate their role in red blood cell production.

Several factors can stimulate increased production of red blood cells. Among these are anemia, the destruction of red bone marrow (by X ray or nuclear particles), prolonged cardiac malfunction, living at high altitudes, and prolonged increased physical activity. In most of these instances, increased red blood cell production is stimulated by low oxygen levels in the blood. This may be due to insufficient hemoglobin content of red blood cells as found in certain types of anemia, a low hematocrit, or the body's need for more oxygen during exercise. At high altitudes there is less oxygen available than at sea level.

The body's control of blood oxygen levels is an excellent example of homeostasis. The kidney, in response to a low blood oxygen concentration *(hypoxia)*, secretes a substance called *erythropoietin* into the bloodstream. Red bone marrow is stimulated by erythropoietin to produce more red blood cells. More red blood cells carry more oxygen, which in turn decreases the stimulation of the kidneys to produce erythropoietin. This is also called a *feedback homeostatic mechanism.* In this case, it is a negative feedback since an increase in the oxygen level, due to increased red blood cell production, brings about a decrease in the stimulating substance. This homeostatic process is diagrammed in Figure 10.4.

Vitamins and Red Blood Cell Production

Vitamin B_{12} is essential for the normal production and maturation of red blood cells. Vitamin B_{12} deficiency apparently affects both the structure of cells and their ability to mature. This vitamin deficiency over an extended period causes red blood cells to be oval- or round-shaped, lacking the biconcave structure due to the retention of a nucleus. Also, the cell membrane is thinner and more fragile than normal. Both conditions reduce the cell's durability in squeezing through capillaries, where it is easily ruptured or broken. If large numbers of the cells are destroyed in this manner, anemia is the result.

Maturation failure also causes anemia *(pernicious anemia).* If development of a red blood cell is halted before the cell has reached maturity, it cannot function effectively in its oxygen

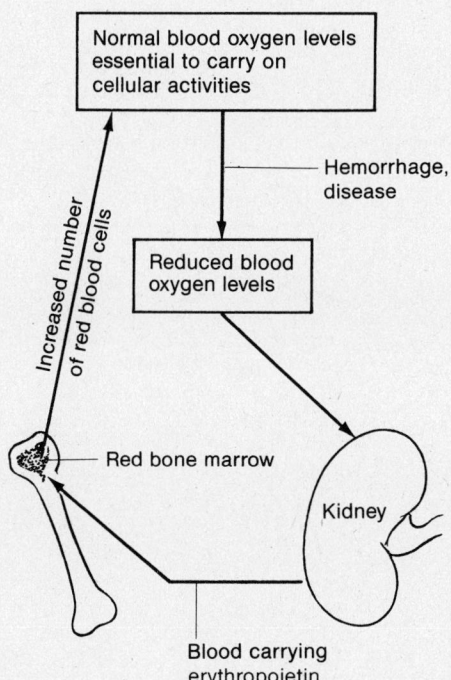

Fig. 10.4 A summary of the homeostatic process resulting in the restoration of adequate oxygen supplies to tissues. The kidney in response to hypoxia secretes the hormone erythropoietin. Red bone marrow is stimulated to produce more red blood cells. More red blood cells can carry more oxygen, thus restoring the proper level of oxygen to the tissues.

carrying role, and anemia is the result. Maturation failure is also caused by vitamin B_{12} deficiency.

Deficiency of vitamin B_{12} in many instances is due not to the lack of this vitamin in foods eaten but rather to its lack of absorption in the intestine. The stomach wall normally secretes a substance called *intrinsic factor*, which in some manner combines with vitamin B_{12} and protects it from digestion along with the other foods. The vitamin B_{12} intrinsic factor complex is then absorbed by the capillaries of the intestinal wall. If intrinsic factor is absent, vitamin B_{12} is digested, and therefore destroyed before it has time to be absorbed.

Hemoglobin

The oxygen carrying capability of red blood cells is the specific responsibility of a complex blood protein called *hemoglobin*. A molecule of hemoglobin is made up of two molecules—one called *heme*, which contains iron (Fe^{++}), and another called *globin*, which is a blood protein. These two molecules (heme and globin) combine to form the hemoglobin molecule. It is the iron segment that forms a loose attachment with oxygen, thus giving hemoglobin the ability to pick up oxygen in the lung capillaries and then release it in the tissue capillaries. Figure 10.5 shows a schematic arrangement of the hemoglobin molecule.

Iron does not chemically bond with oxygen; it only forms a loose attachment. Because oxygen is carried in the molecular (O_2) state, it is ready to be used when it reaches the cells. Each hemoglobin molecule is capable of attaching to four oxygen molecules. This peculiar attachment between hemoglobin and oxygen varies in its bonding power depending upon the oxygen concentration in the tissue in which hemoglobin is located.

In the lung capillaries, where the oxygen concentration is very high, hemoglobin has a great affinity for oxygen. In other tissue, such as leg muscle, where the oxygen level is quite low, hemoglobin readily releases oxygen, which is absorbed into the muscle cells for metabolic utilization. The hemoglobin molecule then is carried back to the lung capillary by the blood where it picks up another "load" of oxygen.

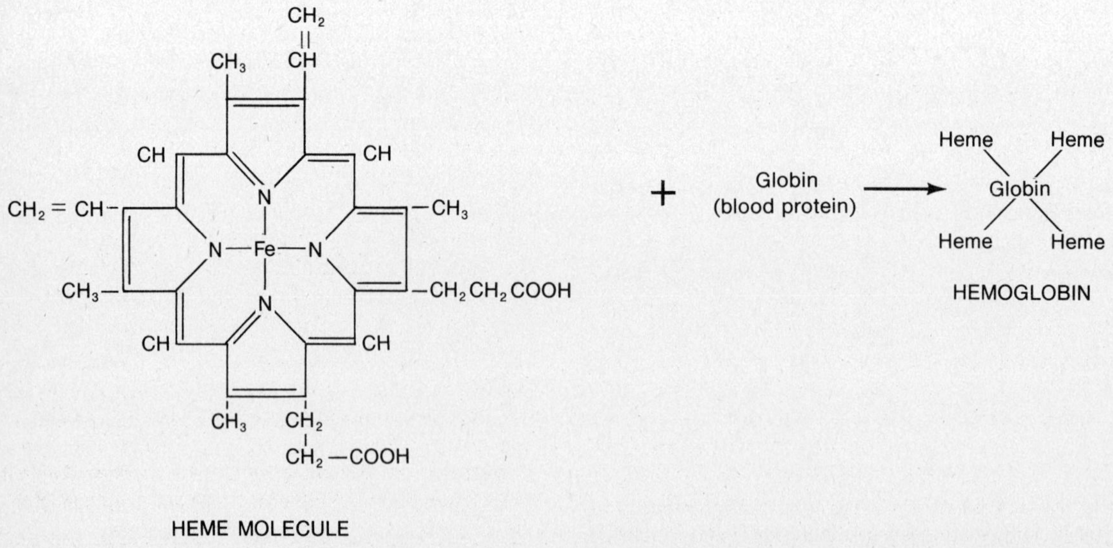

HEME MOLECULE

Fig. 10.5 Hemoglobin is a complex molecule composed of two smaller molecules, heme and globin. Heme binds with oxygen. The hemeoxygen complex is carried by globin to cells.

Iron is important not only for the synthesis of the hemoglobin molecule but also for the synthesis of some enzymes and muscle proteins. Approximately 65 percent of all the iron absorbed by the body is utilized in the synthesis of hemoglobin. About 4 g of iron are normally found in the body at any one time.

The absorption and metabolism of iron is a long and involved story, and it will suffice to simply say that there are substances secreted by the intestine wall that combine with iron and carry it to the blood-forming sites. Excess iron is combined with a blood protein. These molecules (iron and protein) are then stored in cells throughout the body, with a particularly large amount stored in the liver. These storage reservoirs then can release iron as needed for the synthesis of hemoglobin and other substances.

As red blood cells become old and fragile, they rupture easily when pushing through narrow capillaries or the pulp of the spleen. White blood cells also phagocytize old red blood cells. When a red blood cell ruptures, its hemoglobin is released into the blood plasma. The iron portion becomes disassociated from the molecule and is picked up by the specific carriers and circulated for synthesis of more red blood cells or stored in the liver and other cells. The heme portion of the hemoglobin molecule combines with bile produced by the liver to form a substance called *bilirubin*. The purpose of this combination apparently is to rid the body of a waste product. Bilirubin gives bile its characteristic greenish color. It enters the intestinal tract along with bile and is excreted with other body wastes. A concentration of this pigment colors the feces and urine.

WHITE BLOOD CELLS (LEUKOCYTES)

White blood cells are called *leukocytes*. Their basic function is that of protecting the body against the invasion of foreign substances, particularly proteins. Bacteria or other foreign substances gaining access to body tissues are rapidly neutralized or destroyed by leukocyte activity. The different kinds of leukocytes are named according to their shape, nuclear morphology, staining qualities, or occasionally, where they are formed.

Three types—the *eosinophiles, basophiles,* and the *neutrophiles*—are grouped into a major classification called *polymorpholnuclear granulocytes,* as shown in Figure 10.6. This term simply means "cells having various shaped nuclei and containing granules within the cytoplasm." These cells are commonly called *polys.*

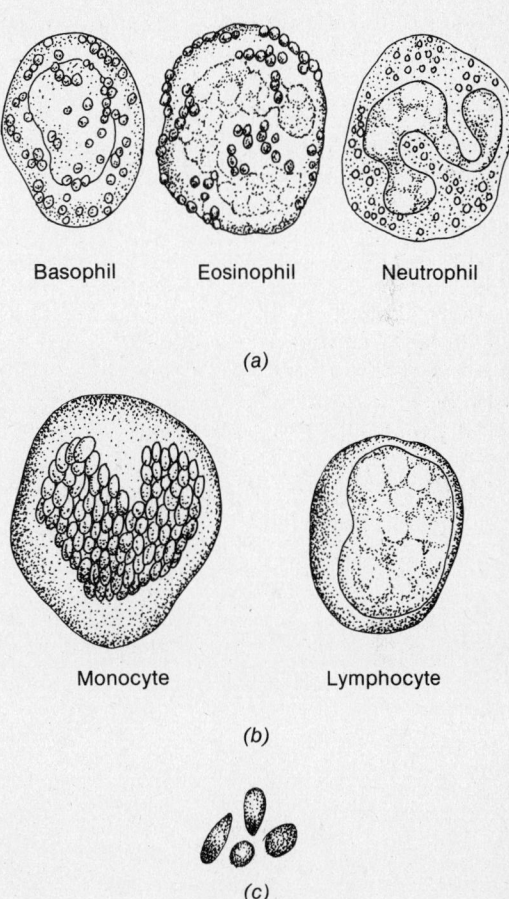

Fig. 10.6 Types of leukocytes. *(a)* Granulocytes. *(b)* Agranulocytes. *(c)* Platelets. The different kinds of leukocytes are named according to their shape, nuclear morphology, staining qualities, or occasionally where they are formed. The three types in *(a)* are often referred to as "polys," from the word polymorphonuclear granulocyte.

Other types of leukocytes are *lymphocytes* and *monocytes*.

The number of white blood cells in the body is much less than that of the normal red blood cell count. A normal white blood cell count is about 9,000 cells per ml of whole blood, much less than the 5,000,000 or so red blood cells per ml in a normal red cell count.

White blood cells are formed in red bone marrow and in specialized tissue called *lymph nodes*.

The life span of white blood cells is very short compared to that of red blood cells. White blood cells live for an average of 12 hours; however, during infection this may be reduced.

White blood cells have a variety of functions. Many of them possess ameboid action; these cells actively engulf and digest foreign substances, such as bacteria. Certain chemical substances are released into the bloodstream by damaged tissue. These substances attract white blood cells. The white blood cells follow the concentration of this chemical substance to the site of damaged tissue. Upon reaching the traumatized tissue, white blood cells digest and neutralize the damaged cells. This process by which cells are attracted along a chemical gradient is called *chemotaxis*.

The process by which a white blood cell ingests foreign substances is called *phagocytosis*. White blood cells are induced to attack certain kinds of foreign substances that have a rough surface, a positive electric charge, or which may be covered by a chemical attractant.

White cells may either neutralize the foreign substance by covering it so that it no longer is capable of further activity, or may secrete enzymes that physically disintegrate the substance. A single neutrophil (a type of leukocyte), for example, can digest 5 to 25 bacteria before it is itself deactivated and destroyed. This destroyed cell is itself then phagocytized by other white blood cells, thereby ridding the body of the components of the destroyed cell.

Lymphocytes are produced in the lymph vessels, lymph channels, and some of the linings of the vascular system. Many of these cells are stationary; that is, they do not circulate in the bloodstream. They can carry on the process of phagocytosis and are also capable of producing antibodies. Lymphocyte production is also carried on at fairly high levels in the liver and spleen.

The invasion of a foreign substance, such as bacteria, initiates a rapid influx of white blood cells into the area. Some of the tissue around the invaded site may be damaged and killed by the invading microorganisms. White blood cells themselves are destroyed in their phagocytic activity. These dead white cells along with the dead microorganisms form the white substance commonly called *pus*.

White blood cell count is an important diagnostic process since a large number of white blood cells in a blood sample indicates that some infectious agent is present initiating the rapid production of white blood cells.

PLATELETS AND THE CLOTTING MECHANISM

In addition to red blood cells and white blood cells, another type of particle found in whole blood is the *platelet*. Platelets are particles that lack a nucleus. They are now considered to be fragments of red blood cells. Their function is critical, since they contain a substance necessary for blood clotting. We do not completely understand the process by which this substance is released from platelets when tissue is traumatized; however, it is evident that something is released that initiates the clotting process. In addition to supplying the initiating clot substance, many platelets become trapped in the clot meshwork, thereby giving it added strength and bulk.

Platelets, in addition to their role in clotting, also prevent blood from leaking out of ruptured blood vessels. The process utilized to stop the flow of blood from a ruptured vessel depends upon, to a large degree, how large the rupture is. Very minute ruptures may occur within small vessels without our knowledge. These may be plugged by the platelets themselves, without the formation of a clot. As platelets circulate throughout the body and come in contact with roughened surfaces such as a ruptured vessel, the rough nature of the ruptured area causes the

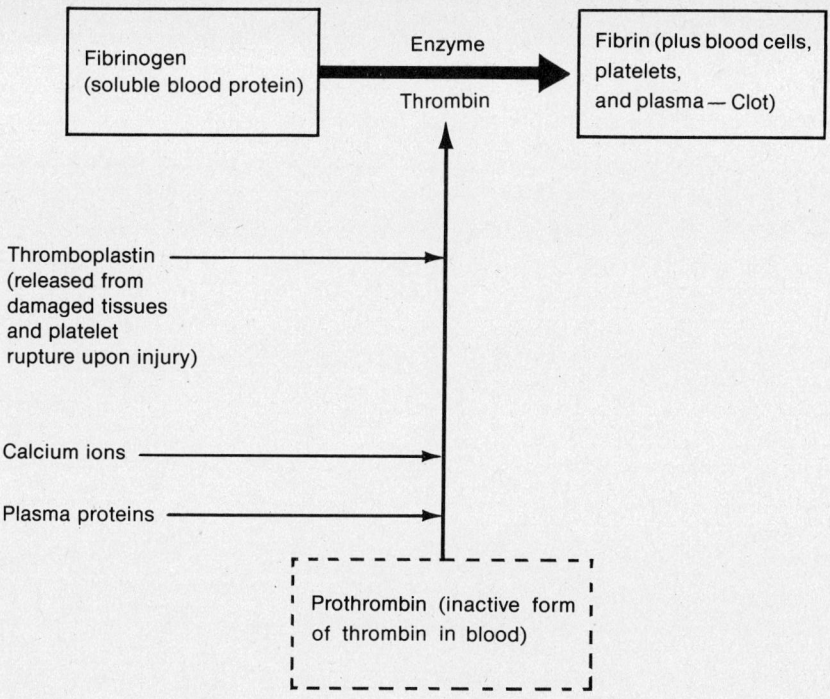

Fig. 10.7 In the blood-clotting process, prothrombin is acted upon by other plasma proteins, calcium ions, and thromboplastin. The product of this chemical action is an enzyme called thrombin. Fibrinogen in the presence of thrombin is converted to fibrin. Fibrin, along with trapped blood cells and platelets, forms the blood clot.

fragile platelets to undergo some drastic changes.

They send out amoeboid arms and become sticky, which allows them to stick to each other and to the walls of the ruptured vessel. This piling-up effect literally forms a plug to prevent the loss of blood from the vessel. This is not a clot—many other processes are required to initiate the clotting process.

This platelet plug process alone is not sufficient to stop the flow of blood from large ruptures since the blood pressure would tend to flush the plug away. The process is an on-going one since these minute ruptures probably occur quite regularly. Persons who for some reason have a very low blood platelet count are susceptible to minute hemorrhages that occur just beneath the skin or throughout the entire body with the related loss of blood from the vessels.

The Blood Clot

The blood-clotting mechanism is much more involved than the platelet plug. The process is quite complicated, and its chemistry has not yet been completely determined. Some of the more obvious facts about blood clotting, however, are known and these are summarized in Figure 10.7.

When a rupture or a cut occurs in a blood vessel that is too large for the platelet plug to control, platelets rupture by coming in contact with the roughened part of the wound. The broken platelets release a substance called *thromboplastin*. This substance apparently is necessary to initiate the formation of the clot threads. A blood protein synthesized in the liver, *prothrombin*, is a normal protein constituent of blood plasma and constantly circulates with the rest of the whole blood.

Thromboplastin from the ruptured platelets acts upon prothrombin, converting it to an enzyme called *thrombin*. This enzyme initiates another chemical reaction that converts still another blood protein, called *fibrinogen*, to the threadlike material *fibrin*.

Fibrin creates the stringy network that forms the clot. As more fibrin is produced, it forms a net of threadlike fibers in the area of the ruptured blood vessel. This net entraps both red and white blood cells and blood platelets, forming a barricade against the further loss of blood through the rupture. After the clot has formed of sufficient size to stop the flow of blood, it undergoes a physical change called *clot retraction*.

Clot retraction occurs as the fibrin threads shorten. This pulls the clot tighter together, making it denser and stronger. The sides of the clot, attached to the ends of the ruptured vessel, pull the ends tighter together. In the process, the liquid portion of plasma, which is called *serum*, is squeezed out.

The time required for a clot to form depends on several factors, one of which is how much prothrombin is available in the system. Vitamin K is an essential ingredient for the synthesis of prothrombin in the liver. Liver disease also has an effect on the level of prothrombin production. In addition, it is known that calcium ions are required for certain phases of blood-clot formation. The clotting time is variable depending upon these factors. The normal blood clot should occur in 2 to 5 minutes.

Hemophilia and Other Disorders

Persons suffering from the hereditary blood disease *hemophilia* either are deficient in the quantity of prothrombin required to form the blood clot or are deficient in some of the factors that initiate the conversion of prothrombin to thrombin.

Hemophiliacs can be protected for short periods by blood transfusions from persons having normal blood. However, this is a very short-range protection because once the substance required for clotting in the normal blood of the transfusion is used up or disintegrated, the only alternative is more transfusion.

There are many unanswered questions concerning blood clotting. For example, we do not know precisely what causes the blood clot process to be initiated. It is apparently due to the breakage of blood platelets coming in contact with roughened vessel surfaces. The lining of the vessels of the circulatory system is quite smooth and apparently this prevents clots from forming within blood vessels where no rupture has occurred.

There are also other protective substances that prevent the formation of blood clots where they are not needed. One of these substances, which is secreted by many tissue cells in the body, is called *heparin*. This substance is called an *anticoagulant*. In some manner, it prevents the formation of some of the substances that produce a blood clot.

Mosquitoes, other blood-sucking insects, and leeches inject a similar type substance into the bloodstream which acts as an anticoagulant; this prevents the blood from clotting and allows the insect to suck up the blood without the interference of a clot forming.

Occasionally blood clots form within vessels or part of a formed clot breaks loose and is carried away by the bloodstream. If this clot enters small vessels, such as the arterioles or capillaries, it may be too large to pass through, thereby blocking the passage of blood into that segment. If the blood supply is blocked, the tissue beyond this blocked area cannot receive oxygen or nutrients and the tissue dies. It depends upon how large an area is affected as to whether this will cause the death of the individual. For example, if a segment of the heart is blocked from receiving blood, this may or may not prove fatal depending upon how much of the heart tissue and what part of the heart is affected.

This is true also for the brain, where it may be even more critical, since certain arteries blocked in this area may cause the death of nerve cells, which cannot reproduce themselves. Once they are destroyed the person has lost the use of them for life. These destroyed nerves may be those which innervate muscles that move the arms or legs, or those in other special sense areas, such as speech, hearing, or vision.

BLOOD GROUPS

The transfusion of whole blood from one person to another is a commonplace procedure. However, it is essential for the donor and recipient to have compatible blood types. Compatibility of blood types means that blood from the donor will mix with that of the recipient, producing no ill effects. When incompatible types of blood are mixed, blood cells of the recipient are attacked by chemical substances of the donor's blood.

The nature of incompatible blood types is known to involve genetic factors that are transmitted from parent to offspring. Two such factors are designated *A* and *B*. If only A factors are carried, the person is said to have type A blood. If only B factors are carried, the blood type is B. If both A and B factors are carried, the blood type is AB, and if neither A nor B is carried the blood type is O. These factors behave like antigens if they, by transfusion, come in contact with a specific concentration. An *antigen* is a foreign protein, one which the body is incapable of producing. Antibodies are special proteins synthesized by the body that deactivate antigens.

To distinguish the antigens of blood groups from other kinds of antigens, they are referred to as *agglutinogens*. Group A blood has type A agglutinogens carried by the blood cells. In addition, blood serum of group A blood contains substances called *agglutinins* (antibodies) that will react with type B blood. These are called *anti-B agglutinins*. Group B blood has type B agglutinogens and *anti-A agglutinins*. Group AB has both A and B agglutinogens, however, no agglutinins are present. Group O has neither agglutinogen, but has both anti-A and anti-B agglutinins.

The basic concern in blood transfusion is whether or not the donor's blood will be *agglutinized* (the sticking together of red blood cells) by the recipient's plasma agglutinins.

Since type O blood has no antigens, it can be given to all other blood types. Persons with type O blood are therefore called *universal donors*.

Type AB has no agglutinins, and therefore can receive blood from all other types. Persons with AB blood are *universal recipients*. Table 10.1 shows which types can be transfused and which types cannot be.

Table 10.1 Which Blood Types Can Be Transfused

BLOOD TYPE	CAN RECEIVE FROM	CANNOT RECEIVE FROM
A	A, O	B, AB
B	B, O	A, AB
AB	A, B, AB, O	—
O	O	A, B, AB

It is always most desirable to transfuse an identical blood type. However, if the identical type is not available, other types within the limits just discussed may be used.

Since type O has both agglutinins, anti-A and anti-B, why does this not create a problem when type O blood is given to a person with type B? The answer lies in the fact that the agglutinins (dissolved in the plasma) in type O are diluted as they enter and mix with the recipient's blood. The agglutinins must be of specific concentration *(titer)* before they cause agglutinization. Why then does not the same principle follow when type A is given to type O? Type A, the donor, is bringing in antigens that are immediately agglutinized since they enter and mix with agglutinins that are at the proper titer level for agglutinization.

The basic principle is that the cellular reaction of the administered blood is the deciding factor in blood transfusion.

How Blood Is Typed

Type A blood has anti-B agglutinins in its plasma. When all the cells and protein fractions are removed from whole blood, the remaining portion containing the agglutinin is called *serum*. The serum from type A blood is called anti-B sera.

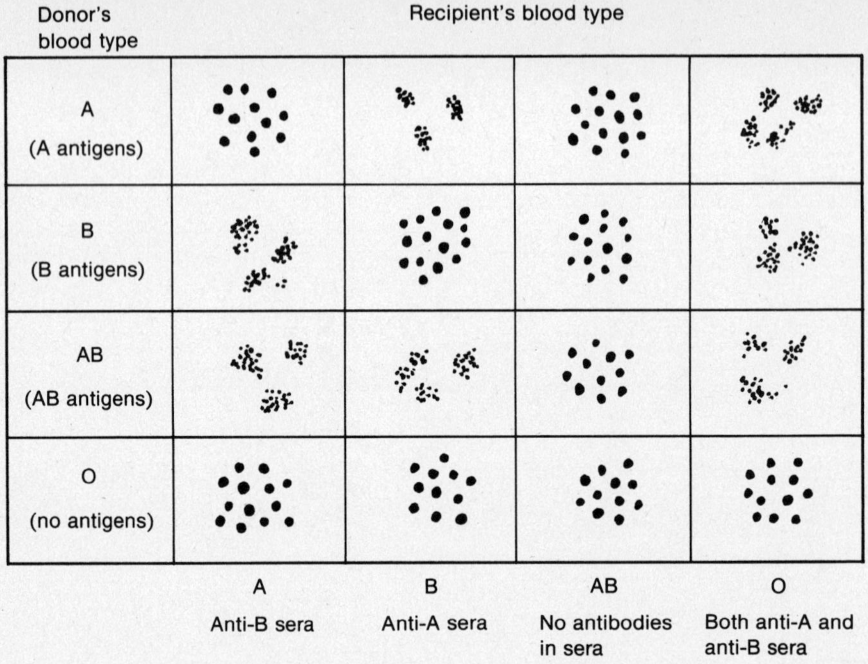

Donor's blood type	A — Anti-B sera	B — Anti-A sera	AB — No antibodies in sera	O — Both anti-A and anti-B sera
A (A antigens)	no clumping	clumping	no clumping	clumping
B (B antigens)	clumping	no clumping	no clumping	clumping
AB (AB antigens)	clumping	clumping	no clumping	clumping
O (no antigens)	no clumping	no clumping	no clumping	no clumping

Fig. 10.8 A summary of blood compatibility with respect to A and B factors. No clumping indicates compatibility of blood types. Clumping indicates incompatibility.

In blood typing, two small drops of the fresh blood to be tested are placed on a glass slide. A drop of anti-B sera is mixed with one drop of blood on the slide and a drop of anti-A sera is mixed with the other. If cell clumping is observed with anti-B sera, the blood type is B. If clumping is observed with anti-A sera, the blood type is A. If clumping occurs in both drops of blood, the type is AB. If clumping does not occur in either blood sample, the type is O. Figure 10.8 summarizes the blood-typing process.

Cross Matching

Blood-typing serum is usually obtained from a pharmaceutical supply house. If it is available, a specific blood type can be determined in a matter of minutes. If typing serum is not available, another procedure can be used to determine the compatibility of donor and recipient blood.

A sample of the potential donor's blood is obtained and cells and some proteins are removed. This provides a serum having unknown agglutinins. This serum is then added to a sample of the recipient's whole blood and the mixture is observed for clumping. If none occurs, there is reasonable assurance that donor and recipient blood are compatible. To double check, however, a cross match is made by obtaining a serum sample from the recipient and mixing it with donor whole blood and again observing for cell clumping. If none occurs, then compatibility of the two types of blood is even more assured.

The Rh Factor

In addition to the factors relative to the ABO series, there are many other factors that directly or indirectly influence processes involved with blood transfusions and particularly those involved with pregnancy. One such factor is called *Rh factor,* named Rh because it was first discovered in the rhesus monkey.

The Rh factor is different from the ABO factors in one major respect—antibodies against the factor do not originate naturally in the body. In the ABO series, type A blood has anti-B agglutinins and type B blood has anti-A agglutinins. There are no natural agglutinins for the Rh factor. The Rh factor is present if any one of eight different agglutinins are present. About 85 percent of the Caucasian population has the Rh factor, and about 15 percent lack the factor.

The problem with pregnancy occurs when the father of the child is Rh positive and the mother is Rh negative. The child may be either Rh positive or Rh negative. If the child is Rh positive, and if the fetal and maternal bloods become mixed, the mother may become sensitized by the Rh positive factor of the fetus's blood. During the normal course of pregnancy, the fetal and maternal blood are kept separated by the placental tissues. However, during birth, abortion, or miscarriage, the placental tissues are torn and hemorrhaged. A mixture of the infant's blood and maternal blood will occur at this time. When the Rh positive blood of the infant mixes with the maternal blood, the mother's blood will form antibodies against the Rh positive blood of the child. These antibodies are produced only if the antigen is present.

Once the sensitization has occurred, the mother's body will continue to build antibodies against the Rh positive factor. This normally has no effect upon the child that produced the sensitizing reaction. The child is Rh positive, and has suffered no ill consequences of the mother being Rh negative.

However, if a successive pregnancy results within a short time, and if the second child is also Rh positive, the mother's antibodies have a tendency to agglutinize the child's blood. The first child is not affected because there were no previous antibodies against the Rh factor, but the second Rh positive child will face the problem of antibodies previously formed against the first child. This situation produces a disease called *erythroblastosis fetalis.*

Until the mid-1960s, the procedure to correct this condition was to replace the child's blood. By replacing the blood, the antigens and antibodies could be removed and no serious aftereffects were observed.

The prevention of erythroblastosis in infants with Rh negative mothers is now a common hospital procedure. The Rh negative mother having an Rh positive child is immunized within 72 hours after delivery with serum from a woman who has previously been sensitized by the Rh factor. The serum in the innoculation contains antibodies that will neutralize the Rh positive factor. Antibodies given this early after birth will react with the antigens that would tend to sensitize the new mother and eventually eliminate them from her system. A second child born to her regardless of the Rh factor would have no difficulties. However, if the second child were Rh positive, the mother might again be sensitized. It is a normal hospital routine to immunize the mother who is Rh negative.

The child with erythroblastosis fetalis is usually anemic at birth, jaundiced, and has an enlarged liver and spleen due to the mechanisms that attempt to build more red blood cells to offset the anemic condition. The resulting anemia is usually not the cause of death of the child having this disease. Death is usually the result of conditions produced by kidney malfunction.

The Rh negative male normally has few problems with the Rh factor, but there are some possible difficulties. If an Rh negative male receives a transfusion of Rh positive blood, he may be sensitized and his body will build antibodies against the factor. This in itself causes no difficulty. However, if after the antibodies have been formed a second transfusion of Rh positive blood is given, the same problems that exist in erythroblastosis fetalis could be produced. It would be rare indeed today if a hospital made this serious error in blood transfusions twice in succession. Not only is blood typed according to the ABO series, it is also typed to the Rh factor. Some rare types of blood, however, may not be available and under certain circumstances, the Rh factor is considered less a problem for the male than is the ABO series. Since AB type blood is the least common of the blood types, and since Rh negative is more uncommon than Rh positive, AB negative is the rarest blood type.

ABO Incompatibility

Although an individual's blood may be donor-recipient compatible with respect to A, B, and Rh factors, other factors found in blood may cause transfusion problems. In the past, these have been overlooked or have gone unrecognized. These factors when present are of little medical significance since they rarely cause serious transfusion problems. These factors have no naturally occurring agglutinins; therefore they cause a problem only if a second transfusion of the incompatible type is transfused.

ABO incompatibility is a condition affecting a fetus having A or B type blood whose mother has type O blood. Type O blood has both anti-A and anti-B agglutinins. Agglutinins are transferrable across placental membranes, and therefore the fetal agglutinins may be agglutinized during pregnancy, creating different degrees of seriousness depending on how much transfer has taken place. ABO incompatibility generally causes fewer problems than Rh incompatibility.

DISEASES AND ABNORMALITIES OF BLOOD COMPONENTS

The study of blood diseases and blood abnormalities is called *hematology*. The science has grown over the past 50 years to such a degree that specialities have evolved. These specialities require experts in areas such as blood chemistry, red cell enzymes, and blood clotting.

Many blood diseases are the result of secondary disorders rather than a direct influence upon blood components. Those diseases having a direct effect upon the blood are rare, particularly in the Western Hemisphere.

Anemia

The term *anemia* can be defined as a lower than normal hemoglobin concentration of blood. Normal ranges have been established for populations of comparable age and sex. This definition cannot be taken as an absolute, owing to some influencing factors (see box). Basically, anemias can be classified into the two categories of either low numbers of red blood cells or impaired hemoglobin synthesis.

Anemia Caused by Factors Other than Reduced Hemoglobin Concentration

1. Changes in plasma volume. An individual with normal plasma volume and normal hematocrit may experience an increase in plasma volume. A hematocrit determination taken during the increased plasma volume would indicate a false anemia.
2. Reduced hemoglobin–oxygen affinity. Factors such as pH, temperature, and the partial pressure of oxygen in the tissues may in some manner reduce the attachment of the oxygen molecule to the hemoglobin molecule.
3. The presence of genetically substituted portions of the hemoglobin molecule. The improperly formed hemoglobin molecule may become an inefficient oxygen carrier.
4. The quantity of hemoglobin carried by red cells. Less hemoglobin means less oxygen-carrying capacity.

A low level of red blood cells may be due to many factors. One is the destruction of red bone marrow that normally produces red blood cells. This tissue may be destroyed by X rays, fissionable by-products of atomic radiation, or disease.

Some anemias may be a result of a combination of the low number of red blood cells and the inadequate synthesis of hemoglobin. In all types of anemias, the overall result is a reduced oxygen-carrying capacity for the blood.

The typically anemic person appears pale and weak and suffers from a loss of energy.

A low hematocrit may be induced by several factors. One may be just simple loss of blood from the body—for example, severe hemorrhage. This reduces the total number of red blood cells, of course, and it also reduces plasma content. However, the body has mechanisms that tend to pull fluids from the interstitial cell spaces back into the blood, thereby replacing lost plasma to a certain extent.

Red blood cell production is a longer range project for the body. Some conditions may bring about too rapid an increase in the number of red blood cells. When this occurs, the red blood cells are not given time to completely mature. These immature red blood cells are more fragile and of a different shape than normal red blood cells and are easily destroyed in the passage through the small capillaries or through the pulp of the spleen.

Instant Blood Analysis

The use of computers in medical diagnosis is not new. However, the ingenious utility of such devices is capable of saving patients time and money. The conventional procedure for blood analysis is for the physician to drain a few ml of blood and send the sample to a laboratory. Several days later, the laboratory analysis is returned to the physician, who in turn calls the patient to report the findings.

A computerized blood analysis machine is currently being tested that will provide almost instant information concerning blood chemistry. Using this machine, the doctor inserts a syringe-type needle into a patient's arm. Electrodes connect the syringe contents to a desktop computer. The doctor then has almost instant information available concerning blood gases, chemical warnings of diseases, and ion imbalances.

Future medical computers may be able to detect disease-producing organisms before they have time to induce the disease, identify chemicals involved in antigen-antibody reactions before a crisis is produced, and monitor transplant patients for signs of tissue rejection.

Sickle Cell Anemia

Sickle cell anemia is a particular type of anemia in which the red blood cells are deformed due to a genetic trait. These cells are fragile and lack the biconcave shape of normal cells. They are easily destroyed, thereby lowering the total red blood cell count and producing an anemic condition.

In sickle cell anemia the cells contain an abnormal type of hemoglobin. When this hemoglobin is exposed to relatively low levels of oxygen concentration the cells become sickle shaped and fragile. This produces a vicious cycle. As the malformed red blood cells are destroyed, the oxygen concentration in the tissues is even further lowered, which brings about the production of more sickle-shaped cells. The oxygen-carrying capacity of the blood may become so reduced that the person afflicted with this disease cannot survive.

Sickle cell anemia is an inherited disease predominantly found among the black population of the United States. The disease, however, is not exclusively limited to blacks. Other persons whose ancestors are of the so-called malaria belt of the world also may be affected. These areas include parts of Europe around the Mediterranean and the tropics of Africa and Asia.

The perpetuation of the disease, handed down through centuries, may be due to some resistance a carrier has for malaria.

Erythroblastosis Fetalis

One other kind of anemia is the disease of newborn children suffering what is commonly called the *Rh factor*. The disease, called *erythroblastosis fetalis*, is due to an Rh positive child being born to an Rh negative mother. Maternal tissues will produce antibodies against the Rh positive factor in the child. When this occurs, the typical antigen-antibody reaction is initiated and the child's red blood cells are agglutinized or destroyed.

Other kinds of anemia may be brought about by certain diseases, transfusion reaction, or certain kinds of drugs.

The diseases malaria, histoplasmosis, and syphilis may cause destruction of the red cells, thereby causing anemia.

Some bacteria, due to their excreted toxins, may cause *lysis* (cell destruction) of red cells. These bacteria include *Staphylococcus, Streptococcus, Pseudomonas,* and others.

Iron Deficiency

Normal red blood cells cannot be produced unless an adequate supply of iron is available at the blood-forming sites. Iron deficiency may result from inadequate intake in foods, defective absorption in the intestines, and defective or inadequate blood protein *(transferrin)* responsible for attaching to absorbed iron.

Polycythemia
Too many red cells (6,000,000 to 8,000,000 per ml) cause circulation problems. This condition, called *polycythemia*, may be mild, with the red cell count within the ranges given above, or severe, with the red blood cell count reaching 11,000,000 per ml or more. Blood with this high a count is sluggish in moving through capillaries, taking up to double the time to circulate from the heart tissues and back to the heart. (See box for other problems caused by polycythemia.) The higher red cell numbers (hematocrit of 70 to 80 percent) are generally caused by a tumerous condition of red bone marrow.

A mild polycythemia may be caused by living at higher altitudes. The reduced oxygen concentration at higher elevations induces the blood cell forming sites to be stimulated.

Effects of Polycythemia on Circulation

1. *Blood viscosity increased.*
2. *Venous return to heart slowed.*
3. *Blood volume increased.*
4. *Capillaries plugged.*
5. *Renal capillaries plugged, resulting in renal failures.*
6. *Cyanosis, a blue skin color in Caucasians owing to the sluggish movement of blood past the tissues. In this condition, hemoglobin loses more oxygen than normal. Cyanosis in dark skin is difficult to observe. Inspection of the lips, nail beds, and conjunction usually will reveal cyanosis if it exists.*

Leukemia
Leukemia is a serious white blood cell disease. In leukemia, white blood cells are produced at an alarmingly high rate. The term *cancerous cell* is applicable in this case since the rapid increase is similar to that of other cancerous tissue. The increase in white blood cells is so rapid that the red blood cells are literally starved of nutrient materials and therefore either killed or not produced in sufficiently high numbers, causing serious anemia. A leukemic cell produced at such a high rate does not mature properly, losing much of its ability to attack invading foreign substances. Leukemic cells, having ameboid action, travel throughout the body and concentrate in the liver, lymph tissues, spleen, and the lungs, where they impair the normal function of these systems. These activities, along with the resulting anemia and loss of energy, which is drained, maintaining this cancerous-type growth, often result in death.

Leukopenia
Leukopenia exists when the white cell count is below 4,000 per ml. The condition may result from red marrow damage (along with anemia), enlarged spleen *(splenomegaly)*, or bacterial infection such as typhoid fever, tuberculosis, and viral diseases (measles, infectious mononucleosis, and hepatitis).

AGING OF BLOOD

Most blood components constantly wear out and are routinely replaced. This process is maintained throughout life; therefore blood does not age like other tissues.

The whole process of replacement may, however, be slowed down due to the aging of other tissues. For example, the red bone marrow becomes slightly less capable of replacing blood cells as a person ages.

Factors other than aging that may also contribute to some alteration of blood components may include improper nutrition, inadequate respiratory function, and impaired kidney function.

SUMMARY AND REVIEW
A. Blood provides tissues with nutrients, oxygen, enzymes, and hormones. In addition, cellular waste is removed, certain substances are buffered, and body heat is distributed.
B. Erythrocytes (red blood cells)
 1. Structure is a biconcave disc about 8 microns in diameter. Mature cells lack a nucleus.
 2. Normal number is about 5,000,000 per mm^3.

3. Formed in red bone marrow. Other blood forming sites, particularly in the fetus, include the spleen, liver, and lymph tissue.
4. Vitamin B_{12} is essential for red blood cells to mature properly.
5. Hemoglobin carried on red blood cells is the oxygen carrier.

C. Leukocytes (white blood cells)
1. Cells are larger than red blood cells and are nucleated.
2. White cells vary in size and shape of nuclei. They also vary in staining qualities.
3. Different types include (the first three are polymorphonuclear granulocytes):
   a. eosinophiles
   b. basophiles
   c. neutrophiles
   d. lymphocytes
   e. monocytes

D. Blood clotting is initiated by platelet rupture and the release of thromboplastin. Thromboplastin acts upon the blood protein prothrombin converting it to thrombin. Thrombin aids in converting another blood protein, fibrinogen, to fibrin, the clot material.

E. Blood groups depend upon genetic factors and are classified as types A, B, AB, and O. The blood group is named for the factors carried on red blood cells.
1. Blood group A carries agglutinogen A and anti-B agglutinin.
2. Blood group B carries agglutinogen B and anti-A agglutinin.
3. Blood group AB carries both A and B agglutinogen and neither agglutinin.
4. Blood group O carries neither agglutinogen and both anti-A and anti-B agglutinin.

F. Rh factor is another genetic factor that may result in transfusion problems for the recipient. The factor is also important in fetal development. A mother who is Rh negative who has an Rh positive child may be sensitized, causing her immune system to produce anti-Rh positive antibodies. These antibodies may diffuse across placental membranes and cause fetal red blood cells to clump. This condition is called erythroblastosis fetalis.

G. Diseases and abnormalities of blood components
1. anemia
2. sickle cell anemia
3. iron deficiency
4. polycythemia
5. leukemia
6. leukopenias

H. Aging of blood. Blood as a tissue is not adversely influenced as one ages. However, blood components may undergo alteration owing to the aging of tissues that produce them.

11

The Lymphatic System

MAJOR CONCEPTS
   THE LYMPHATIC SYSTEM
   THE RETICULOENDOTHELIAL SYSTEM
      The Liver
      Red Bone Marrow
      Tonsils
      The Thymus Gland
   THE ANTIGEN-ANTIBODY REACTION
   KINDS OF IMMUNITY
      Natural Immunity
      Acquired Immunity
      Passive Immunity
      Autoimmunity
   DISEASES OF THE RETICULOENDOTHELIAL SYSTEM
      Hodgkin's Disease
      Leukopenia
      Leukemia
   AGING OF THE IMMUNE SYSTEM

Although identified here as a system, lymphatic and reticuloendothelial cells are scattered throughout the body. These cells, located in the liver, red bone marrow, tonsils, lymph nodes, and thymus gland, give the body its immunity to invading microorganisms. This chapter discusses the location of immunity-producing cells and the mechanisms of the immune response.

A small quantity of fluid normally exists between blood capillaries and cells. When excess fluid collects here, it must be removed and returned to the blood. This is the function of the lymphatic system.

THE LYMPHATIC SYSTEM

The lymphatic system is a comprehensive arrangement of vessels whose purpose is to return fluid to the bloodstream previously "lost" in the capillary beds. The circulatory system is almost, but not quite, a closed system. In the capillaries, fluid from the blood leaves and bathes cells to provide nutrition and oxygen. It is into this fluid that cells empty carbon dioxide and other metabolic wastes.

The fluid that leaves the capillaries on the arterial side and reenters the capillaries on the venous side is called *interstitial fluid.* It makes up about 15 percent of body weight. Fluid leaves capillaries at the cell level. However, some portion of the fluid is not returned immediately to the bloodstream. A system of blind tubules reclaims this "lost" fluid and returns it to circulation. This system is the *lymphatic system.* The fluid reclaimed is carried along *lymph channels* within which are located valves that permit the fluid to flow only toward the heart, as shown in Figure 11.1.

Lymph channels collect into larger vessels called *lymph ducts,* which carry lymph to the left subclavian vein and the right subclavian vein.

The movement of lymph is accomplished by muscular contraction. As muscles contract, they squeeze the lymph channels. The one-way valves permit lymph flow only in one direction. In the absence of muscular contraction, lymph fluid accumulates in the tissue, causing it to swell, a condition called *edema.*

Lymph channels are much more permeable than veins. Protein molecules cannot easily cross

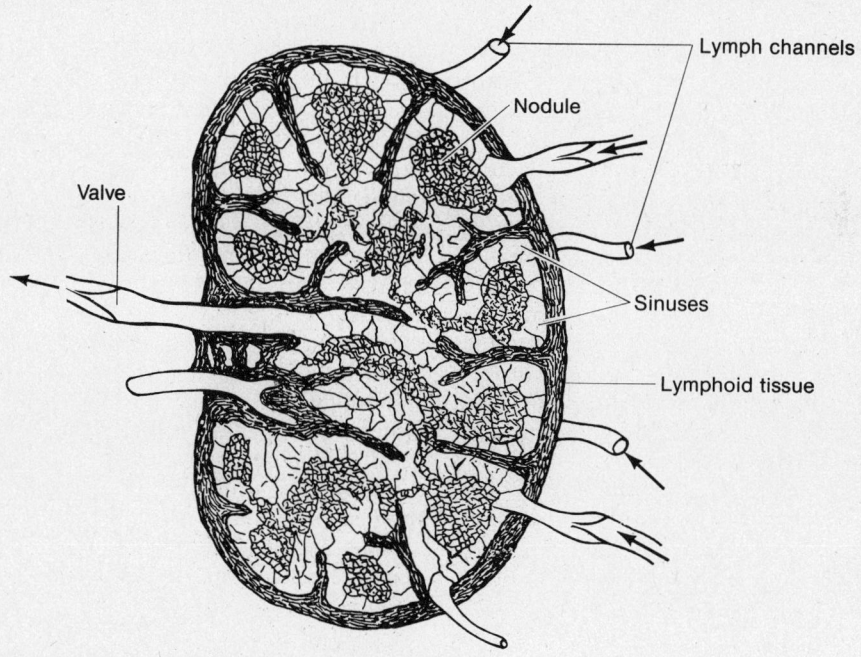

Fig. 11.1 A sectioned view of a lymph node and lymph channels. Note the arrangement of valves in the channels permitting a flow of lymph in only one direction. The sinuses are filled with lymphoid tissue and lined with phagocytic cells. The lymphoid tissue filters out bacteria and other particulate matter, which is then attacked and destroyed by the phagocytic cells.

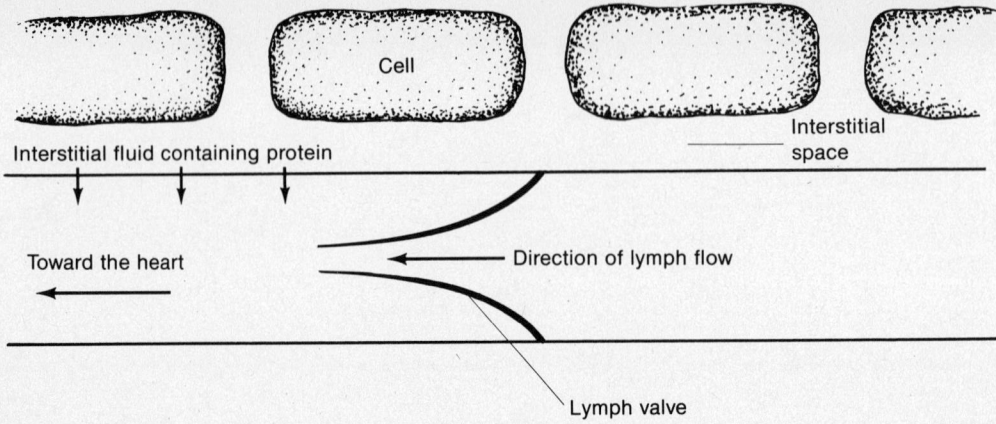

Fig. 11.2 Excess interstitial fluid, which contains protein molecules, is absorbed into lymph channels. These channels, containing one-way valves, conduct lymph into veins.

capillary or vein membranes, but a small number do and these become part of the interstitial fluid. These protein molecules cannot be reabsorbed by the capillary and therefore, if not removed, they would accumulate to high concentrations. Lymph channels are sufficiently permeable to absorb proteins, thereby removing them from the interstitial spaces, as shown in Figure 11.2.

If the protein molecules were left in the interstitial fluid, osmotic pressures within the tissues would reduce the flow of fluids from the capillaries into the interstitial space. The reduced flow of these fluids, containing nutrients and oxygen, would have serious consequences for cells; in fact, they would die.

The opposite effect of the process just described would permit an accumulation of fluid in the interstitial spaces, causing edema.

The discussion thus far may have left the impression that there are actual spaces surrounding the cells. The term *potential space* is perhaps more appropriate. The fluid from capillaries that bathes the cells is constantly being reabsorbed by the venous side of the capillaries as well as the lymph channels, leaving the cells in a relatively "dry" state compared to the fluid state of plasma in the capillaries and lymph in the lymph channels.

The amount of fluid surrounding cells is even more restricted by the fact that cells are not merely floating in a sea of fluid but rather are surrounded by a semisolid substance called a *gel*. Nutrients and water can diffuse throughout and across the gel almost as efficiently as in a fluid. The quantity of free fluid normally found in the interstitial space is quite small. When free fluids are present, they tend to be pulled by gravity into the lower extremities, producing edema.

Edema in various parts of the body is caused by a variety of causes. The most common cause is differential pressures into and out of the interstitial spaces. Another cause is blocked lymph channels, which will be discussed later.

Lymph channels differ from veins in several respects—they are thinner walled, have more valves, and in some parts of the body they have enlargements called *lymph nodes*, as shown in Figure 11.1.

Lymph nodes are made up of spaces called *sinuses*, which are filled with tissue called *lymphoid tissue*. The sinuses are lined with phagocytic cells. When fluid enters the sinuses it is filtered by the lymphoid tissue, trapping bacteria or other particulate matter. This matter is then attacked by the phagocytic cells, which destroys or deactivates it.

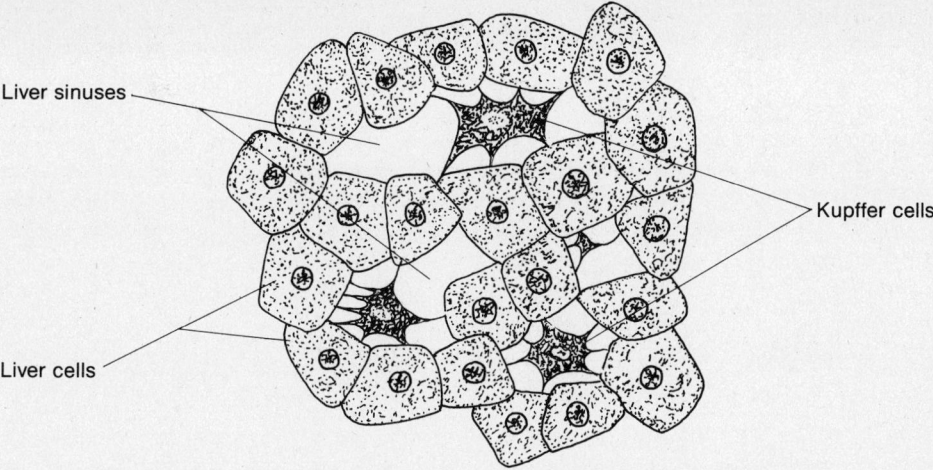

Fig. 11.3 Kupffer cells, which line liver sinuses, aid in removing foreign particulate matter from blood.

THE RETICULOENDOTHELIAL SYSTEM

The reticuloendothelial system involves tissues from many parts of the body. It includes tissues of lymph nodes, the liver, spleen, bone marrow, tonsils, and the thymus gland.

All these tissues possess specialized cells called *reticulum cells*. These cells can differentiate into a variety of forms. Depending upon where they are found, reticulum cells may differentiate into lymphocytes, red blood cells, white blood cells, or plasma cells.

The following discussion will be limited to the tissues listed above and more specifically to the functions of phagocytosis and antibody production.

Lymph nodes are producers and reservoirs of *lymphocytes* and *plasma cells*. Lymphocytes are wandering cells that are attracted to extraneous matter caught in the mesh filters of lymph nodes. The extraneous matter—bacteria, dead cells, and so on—is phagocytized (eaten) by lymphocytes, thereby cleansing the lymph fluid of potentially harmful debris.

Plasma cells synthesize chemical substances called *antibodies*, which react with and destroy bacteria or their toxins. Lymphocytes are also found wandering throughout the circulatory system, often squeezing their way into and out of arteries, capillaries, and veins. Their role here is the same as in lymph nodes—that is, to destroy by phagocytosis any potentially harmful extraneous material. Lymphocyte activity is accelerated in tissues such as the liver and spleen.

The Liver

The sinuses of the liver contain large numbers of reticulum cells called *Kupffer cells* that phagocytize invading microorganisms or other extraneous protein matter (see Figure 11.3). These substances may have escaped the action of other lymphocyte activity in lymph channels or blood vessels.

The Spleen

The *spleen* is also the site of massive lymphocyte activity. The spleen is literally a filter made up of lymphoid tissue capable of producing lymphocytes in large numbers. It is here that much cellular debris, such as dead red blood cells, is phagocytized and removed from circulation.

Red Bone Marrow

Red bone marrow produces lymphocytes and other white blood cells as well as red blood cells. The activity of reticulum cell differentiation is at its best in red bone marrow. Reticulum cells in bone marrow continually form hematoblasts. These cells have a nucleus that is normally expelled within one to two days after it enters the circulation. The cell, minus its nucleus, becomes shaped like a biconcave disc. This cell is the erythrocyte or red blood cell.

Tonsils
Tonsils are composed largely of lymphoid tissue that produces lymphocytes and plasma cells. When tonsils become infected, they provide a rather direct route for the microbial agents to enter the blood and lymph vessels.

The Thymus Gland
The *thymus gland* plays an important role in creating the immune system. Specialized cells in the newborn thymus are released into the bloodstream. These cells migrate to the lymph channels and are caught in lymph nodes, where they take up permanent residence. These specialized cells become lymphocytes responsible for the production of antibodies. Children born with a defective thymus have a deficient immune system and are susceptible to many diseases.

THE ANTIGEN-ANTIBODY REACTION

The term "plasma cell" has been mentioned previously. Its precise function, however, has not yet been discussed. Plasma cells synthesize and secrete substances called antibodies. An antibody is defined as a substance produced in response to an antigen. The chemical reaction elicited when a specific antigen comes into contact with a specific antibody is called the *antigen-antibody reaction* or, more commonly, the *immune response*.

Antigens are usually large protein molecules such as those found in bacterial cell walls, pollen grains, or in fact any foreign substance of a protein nature with high molecular weight. Antigens stimulate certain reticular cells to synthesize specific antibodies. Antibodies are usually protein in nature and can be stored in the body for varying lengths of time. *Globulins* are protein antibodies stored and carried as a normal blood component.

When certain plasma cells are first exposed to a foreign protein (antigen), they synthesize antibodies that will neutralize or destroy the antigen. This first response takes one or more weeks to accumulate an antibody sufficiently concentrated to be effective. If the same antigen is reintroduced, the secondary response is much more swift and rigorous in the production of antibodies. Once sensitized, it is as if the plasma cell can identify the antigen more easily the second time. Plasma cells are also capable of distinguishing which cells belong to the body (self) and those which are foreign (not-self).

KINDS OF IMMUNITY

The term *immunity* to most people means protection. It may be either a specific or nonspecific type of protection. *Nonspecific immunity* involves all the protective mechanisms the body maintains against general types of invaders. For instance, tears and saliva contain enzymes that destroy some types of bacteria. The low pH of some body fluids inhibits or destroys some microorganisms. If bacteria are introduced superficially, such as in a flesh wound, white blood cells rush to the wound, seal it off, and phagocytize the invading microorganisms.

Specific immunity implies a definite relationship between an antigen and the antibody induced by the presence of the antigen. Specific kinds of immunity can be further categorized as natural and acquired immunity.

Natural Immunity
Natural immunity is a resistance to disease or infection to which the body may not have been previously exposed. For example, the human body is immune to many animal pathogens. According to some authorities, what appears to be a natural form of immunity may be the result of a series of low level exposures to an antigen, none of the exposures being of sufficient duration to elicit an antigen-antibody reaction (a distinct illness). The series of exposures, however, may have succeeded in building up a relatively high level of antibody concentration.

Acquired Immunity

As its name implies, *acquired immunity* is acquired in some fashion. Two common ways to acquire immunity are to have a disease or to be innoculated with live, dead, or weakened microorganisms. Plasma cells will react even to dead bacteria, producing specific antibodies against future invasion of that bacteria. Antibodies may also be produced that counteract toxins (bacterial secretions). If diluted toxins are injected, specific antibodies are produced that will neutralize or destroy the toxin if it appears later through infection.

Passive Immunity

Passive immunity is gained from other persons or animals. If an animal is injected with bacteria normally pathogenic for humans but nonpathogenic to the animal, the animal's immune system will respond by synthesizing antibodies. These antibodies may then be isolated from the animal's blood and injected into the human bloodstream, giving a person short-term protection against the pathogen. Antibodies from a person with previous exposure to a pathogen may also be isolated in a serum and injected into another person, giving short-term protection.

Autoimmunity

Autoimmunity describes the condition in which an individual's own plasma cells produce antibodies against their own tissue. The precise chemical reactions that maintain body functions have been called by some "the wisdom of the body." Autoimmunity therefore must be labeled "the ignorance of the body." It is certainly of no advantage for some body cells to secrete substances that destroy other healthy normal cells.

An example of autoimmunity is rheumatic fever. In response to a streptococcal infection, antibodies are synthesized. The immune response would be therefore induced in a normal manner. There are tissues in the body, however, that apparently resemble the antigen so closely that the antibody attacks them. One of these tissues is cardiac muscle and valves. The severe damage done to the heart continues after the streptococcal infection is under control.

Another tissue attacked by the antibody is the *glomerulus* and *Bowman's capsule* of the kidney nephron. The glomerular membrane becomes inflamed, perhaps even ruptured, permitting large protein molecules and blood cells to enter Bowman's capsule. The glomerulus and capsule may ultimately be incapable of filtering and absorbing substances from the blood. The infection that results from the antigen-antibody reaction is called *glomerular nephritis*. Chronic streptococcal infection may ultimately lead to acute renal shutdown.

Other diseases likely to result from autoimmunity include *Addison's disease*, *rheumatoid arthritis*, and *Hashimoto's disease*.

An *allergy* represents an antigen-antibody reaction. The antigen is usually one that is not communicable and nonpathogenic. Examples of *allergens* (antigens) include such substances as house dust, animal fur, pollen, and toxins of insect bites.

The response the body makes may be immediate or prolonged. The immediate reaction may occur within a few minutes after exposure to the antigen. In these cases, the person normally has

---

Autoimmune Disease: The Enemy Within

Our immune system is so finely tuned that the slightest hint of foreign cells in the body commands an all-out attack on the invader. Leukocytes produce antibodies designed to selectively destroy foreign cells. In autoimmune disease, some of our own cells, for reasons not known, in some manner change their chemistry ever so slightly. This change may be adequate to trigger our immune system.

DNA normally is kept within the nucleus. For some reason, nuclear DNA may seep through the nuclear membrane. DNA outside the nucleus is a foreign substance susceptible to attack by the immune system. The DNA and its attacker, an antibody, form a complex molecule that becomes trapped in various parts of the body, such as the kidney, where it causes tissue damage and disease. This disease is called lupus erythematosis.

Other autoimmune diseases include rheumatoid arthritis, hay fever, rheumatic fever, multiple sclerosis, some anemias, and perhaps cancer.

a history of previous exposure to the antigen and thus has been sensitized. The person's system has over a period of time synthesized and stored high levels of antibodies. A chance exposure to the antigen, usually a large quantity, results in a severe antigen-antibody reaction called *anaphylaxis*. This condition may be severe enough to cause death within a few minutes. Individuals who are particularly sensitive to insect venom must be alert to the possibility of some degree of anaphylaxis if they are bitten.

The sensitivity to certain drugs, such as penicillin and sulfa drugs, follows a pattern similar to allergies. Some individuals are sensitized by an initial injection, then suffer some degree of anaphylaxis upon receiving subsequent injections.

DISEASES OF THE RETICULOENDOTHELIAL SYSTEM

The diseases discussed here are limited to those influencing lymphoid tissues or structures considered to be accessory to lymphoid tissue.

Hodgkin's Disease

Hodgkin's disease is cancer of the spleen and lymph nodes. Other tissues that may also be involved include the liver and bone marrow. The cause of the disease is unknown. It is characterized by the presence of giant cells called *Reed-Sternberg cells*. Lymph nodes are destroyed and replaced by fibrous tissue. Irradiation therapy is now prolonging the life of persons with the early stages of the disease.

Leukopenia

Leukopenia is a disease resulting from multiple causes, including infectious disease, chemicals, and physical factors. The term leukopenia means reduced white cell count. The normal adult white cell count ranges between 5,000 and 10,000 cells per ml of whole blood. A white cell count below 4,000 per ml is called leukopenia. Chemical substances such as benzol, antimicrobial drugs, and physical factors such as X rays may induce leukopenia.

Leukemia

Leukemia is a disease characterized by extreme proliferation of certain white blood cells. Numbers may exceed 500,000 cells per ml. The cause of leukemia is not known. However, in some types viruses appear to be implicated.

The anemia associated with leukemia becomes so severe that most body functions are affected, resulting in death. There is no known cure for leukemia. Treatment involves making the patient as comfortable as possible until the disease runs its course.

AGING OF THE IMMUNE SYSTEM

The immune system of elderly persons reacts much more slowly than the immune system of a young person. The reason for this is not clear. The immune system, like other systems, may suffer deterioration due to the impairment of other systems. Impairment of red bone marrow, heart disease affecting blood flow, and a reduced rate of specific cell production, particularly leukocyctes, may all be factors.

SUMMARY AND REVIEW

A. The lymphatic system returns fluid from interstitial spaces to the bloodstream.
   1. Lymph channels, containing one-way valves, carry lymph away from cells.
   2. Lymph nodes filter the lymph, trapping bacteria and other particulate matter.
   3. Lymphocytes are white blood cells capable of destroying bacteria or other antigens.
B. The reticuloendothelial system includes lymph nodes, liver, spleen, bone marrow, tonsils, and the thymus gland. The major role of this system is to produce white cells (lymphocytes) that synthesize and release antibodies.
C. The antigen-antibody reactions
   1. An antigen is any substance that stimulates the formation of antibodies by plasma cells.
   2. The immune response is another name for antigen-antibody reaction.

D. Types of immunity
   1. Natural immunity is a resistance to a disease or infection to which the body may not have been previously exposed.
   2. Acquired immunity is immunity gained by having the disease or through inoculation.
   3. Passive immunity is gained from other persons or animals.
   4. Autoimmunity is the condition in which an individual's own plasma cells produce antibodies that attack their own tissues.
E. Allergies represent antigen-antibody reactions.
F. Diseases
   1. Hodgkin's disease is cancer of the spleen and lymph nodes.
   2. Leukopenia is a reduced white cell count.
   3. Leukemia is a disease that results in massive increases of white blood cells.
G. Aging of the immune system
   1. Immune system reacts more slowly in the aged.
   2. Slow response may be due to other impaired systems.

12

The Endocrine System

MAJOR CONCEPTS
  ENDOCRINE GLANDS AND HORMONES
  THE PITUITARY GLAND
    The Posterior Pituitary (Neurohypophysis)
    The Anterior Pituitary (Adenohypophysis)
    The Gonadotropins
    Adrenocorticotropic Hormone
  THE THYROID GLAND
    Basal Metabolism Test
    Protein-Bound Iodine Test
    Radioactive Iodine Uptake
    Iodine Requirement for Normal Thyroid Function
    Toxic Goiter (Grave's disease)
    Thyroid Storm
    Calcitonin
  THE PARATHYROID GLANDS
  THE ADRENAL GLANDS (SUPRARENAL GLANDS)
    The Adrenal Medulla
    The Adrenal Cortex
  THE THYMUS GLAND
  THE PANCREAS
    Anatomy of the Pancreas
    Endocrine Function of the Pancreas
    The Diabetic
  AGING AND ENDOCRINE FUNCTION

Endocrine glands secrete hormones, which in turn control and integrate cellular metabolism. This chapter discusses the specific role each hormone plays in metabolic activities. In addition, the close, cooperative relationship that exists between endocrine glands and the nervous system is discussed.

The complex nature of the body working as a unit makes it imperative that some coordinating mechanisms be available. Essentially there are two coordinating systems in the body—the nervous system and the endocrine system. *Hormones*, secreted by *endocrine glands*, collectively have little in common except that they are all secreted by specialized ductless (endocrine) glands. These glands empty their contents directly into the bloodstream. Other glands have ducts or tubes that lead to a specific target organ. These glands are commonly referred to as *exocrine glands*.

ENDOCRINE GLANDS AND HORMONES

Hormones are secreted directly into the bloodstream and carried to all parts of the body, so their coordinating or regulatory functions may be at a point possibly a great distance from the point of secretion. Although hormones are diffused throughout the body by the bloodstream, only certain tissues and cells respond to each hormone. The responsive units are called target organs, tissues, or cells.

Hormones do not belong to any one particular family of chemical compounds; at least one, insulin, is a protein, while others belong to a quite diverse group of simpler compounds.

The activity of some hormones exceeds that of strictly a coordinating function. As we shall see, the complete absence of some hormones makes it impossible for the body to continue to function properly, causing illness or death. Figure 12.1 shows the general location of the major endocrine glands in the body.

THE PITUITARY GLAND

The pituitary gland is a small mass of tissue located inside the skull protruding from the base of the brain. The gland is enclosed by a bone depression in the sphenoid bone called the *sella turcica*. A stalk of tissue, the *infundibular stalk*, connects the pituitary gland with the hypothalamus. Figure 12.2 shows the location of the pituitary gland.

The pituitary gland is composed of two different kinds of tissue. The anterior portion, the *anterior pituitary* or *adenohypophysis*, is composed mainly of secretory cells interwoven with capillaries. These capillaries are connected by portal veins with capillaries within and near the base of the hypothalamus.

The *posterior pituitary* or *neurohypophysis*, an outgrowth of the hypothalamus, is made up of neuroglia-type cells plus neurons. These neurons establish a definite neural pathway from the hypothalamus to the neurohypophysis. Figure 12.2 illustrates the internal structure of the adenohypophysis and the neurohypophysis.

The hypothalamus is now regarded as the basic coordinating center in the body. This is an area where many of the signal impulses going to and from the brain and then to the tissues are coordinated for specific actions. It is also an area where there is specific coordination between neural and chemical functions.

The hypothalamus in some instances secretes a hormone that is stored in the pituitary gland. The hypothalamus controls the rate at which this secretion is emptied into the circulatory system. At least 15 different hormones are secreted by the pituitary gland. As we shall see, overproduction or inadequate secretion of pituitary hormones produces some spectacular consequences for the body.

The Posterior Pituitary (Neurohypophysis)

Secretions of the posterior pituitary are produced by neurosecretory cells in the hypothalamus. These secretions are transferred to the posterior pituitary and stored there for later release. Two major secretions of the posterior pituitary are *antidiuretic hormone* and *oxytocin*. **Antidiuretic Hormone** Known clinically as *vasopressin*, antidiuretic hormone (ADH) has an effect upon kidney tubule cells, making them more permeable to water. This process is essential for the body to regulate its water balance. If the posterior pituitary fails to secrete ADH, water entering the kidney tubules cannot be reabsorbed in adequate quantities and is excreted. *Diabetes insipidus* is the name given to this condition. Urine volume from a normal individual is about 1 to 1½ liters per day. Individuals having diabetes insipidus may excrete as

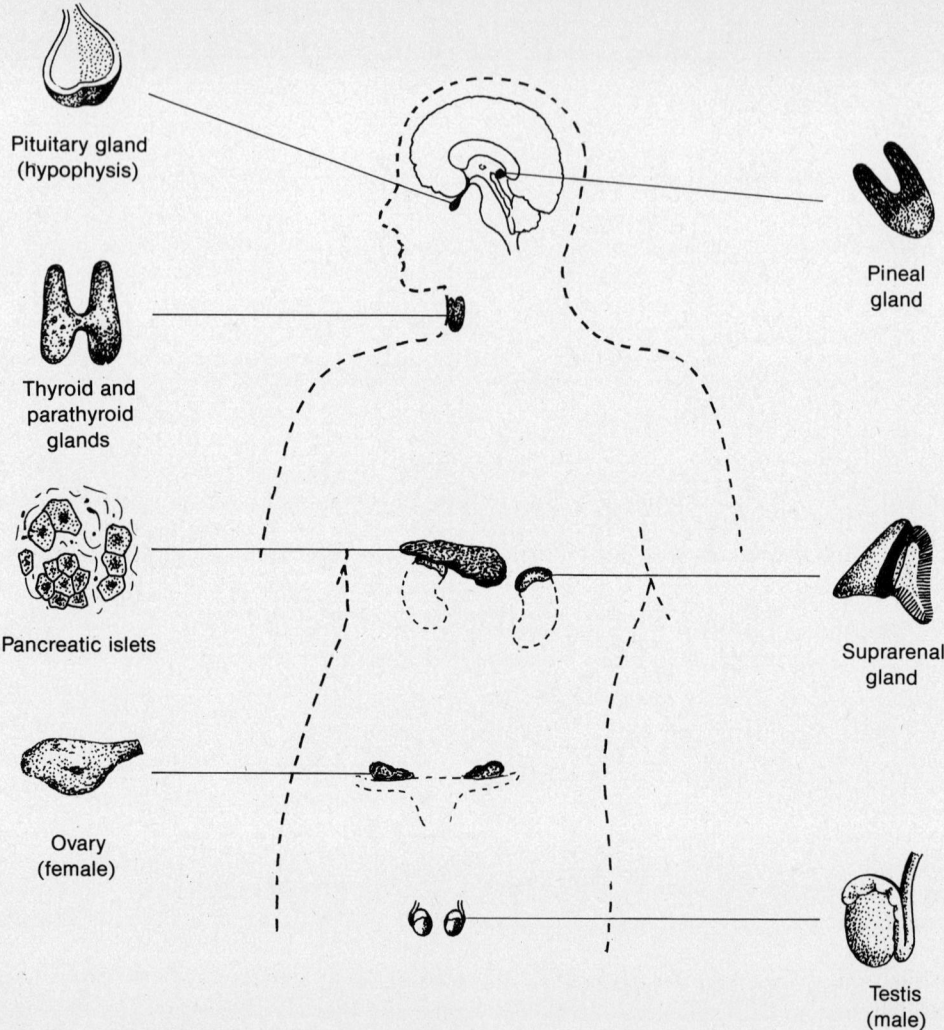

Fig. 12.1 General location of endocrine glands. Endocrine glands all secrete hormones that diffuse directly into the bloodstream. The effect produced by a specific hormone may be quite some distance from the gland that secreted it.

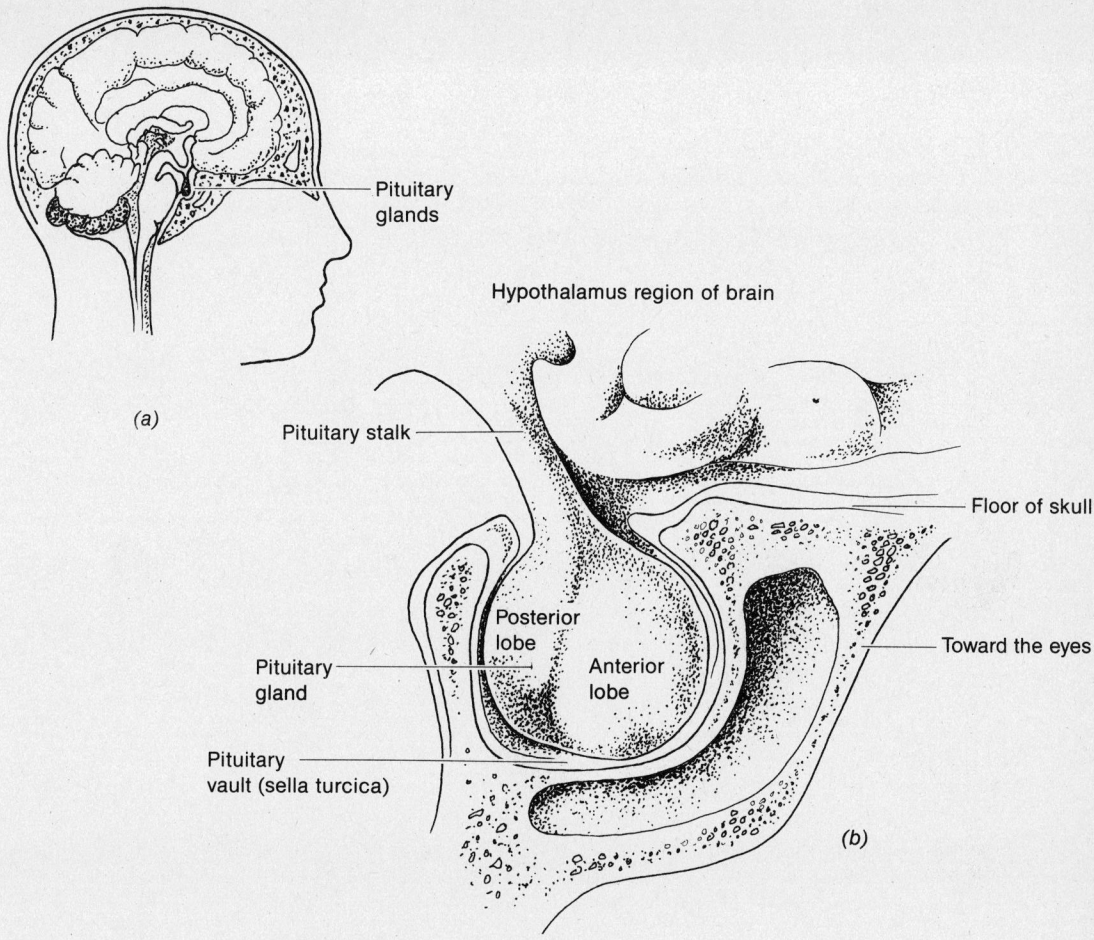

Fig. 12.2 The pituitary gland (hypophysis). (a) View of general location of the glands. (b) Enlarged view, showing the close proximity of the gland to the hypothalamus. The pituitary gland is composed of two different kinds of tissue. The anterior portion is called the adenohypophysis (anterior pituitary) and the posterior portion is called the neurohypophysis (posterior pituitary).

much as 6 to 8 liters or more per day and have a tendency to become dehydrated. They may lose many essential electrolytes along with the loss of water from the body. These individuals drink excessively large amounts of water, which tends to offset the dehydration. They also have a craving for salty substances, which decreases the tendency for electrolyte imbalance.

Stimulation for the release of ADH from the posterior pituitary includes factors that remove fluids from the body, such as severe hemorrhage or surgery. In both these instances, ADH production is stimulated by decreased blood volume. More ADH in the blood causes a greater reabsorption of water in the kidney tubules, allowing the body to maintain a homeostatic fluid balance even though fluids have been lost.

Alcohol and caffeine inhibit ADH release. This permits less water to be reabsorbed by the kidney tubules, and hence more water is excreted in the form of urine.

ADH deficiency can be treated, but not cured, by injecting extractions of animal posterior pituitary into the bloodstream. However, since ADH is a polypeptide, it is chemically degraded by protolytic (protein-digesting) enzymes in the blood and destroyed in 2 to 10 minutes. Intramuscular injections are longer lasting (15 to 30 hours). Other methods include a pill that can be dissolved under the tongue, and a nasal spray. However, these methods also have only short-term value.

Oxytocin Known clinically as *pitocin*, *oxytocin* is produced by the hypothalamus and stored in the posterior pituitary. Its major action is on smooth muscle, particularly that found in the mammary glands and the uterus. A lactating animal or human does not have milk stored and available for immediate use in the nipple area. Stimulation of the nipple by the infant's suckling produces nervous impulses that reach the hypothalamus and the posterior pituitary, causing a release of oxytocin. This release induces the contraction of smooth muscles in the mammary glands and causes the milk to be released. Oxytocin is also released during childbirth, causing the smooth muscles of the uterus to contract, which aids in expelling the fetus and the afterbirth.

The Anterior Pituitary (Adenohypophysis)
At least eight different hormones are synthesized, stored, and secreted by the anterior pituitary. Chemical messengers from the hypothalamus called *release factors* stimulate different segments of the adenohypophysis bringing about the release of specific hormones. Many of the hormones secreted by the adenohypophysis induce the secretion of hormones of other endocrine glands. These hormones in turn affect specific tissues or organs. The hypothalamus thus indirectly exercises a great deal of influence over body functions and provides a distinct link between body hormones and the nervous system.

Following is a description of the hormones secreted by the anterior pituitary.

Growth Hormone (Somatotropic Hormone)
This hormone (also called *somatotropic hormone*) is essential for normal tissue development during formative years. Chemically, *growth hormone* (GH) is a high molecular weight protein that can be synthetically produced. The major function of GH is to stimulate growth.

The exact nature of how growth hormone stimulates growth is not known. However, it is known that GH stimulates protein synthesis, increases fat metabolism, and increases blood sugar concentration. GH also appears to increase the rate of diffusion of amino acids across cell membranes. These amino acids are utilized in the formation of proteins, which constitute a major body component.

An increase in GH causes a corresponding increase in blood glucose levels. A higher level of blood glucose stimulates the islet cells of the pancreas to secrete additional insulin. GH is called a *diabetogenic* substance because a continuous hypersecretion may burn out insulin-secreting cells, causing diabetes mellitus.

Hyposecretion of GH during the normal growing years leads to stunted growth, causing a type of *dwarfism* (pituitary dwarfism).

An oversecretion of growth hormone during the formative years produces an excessive growth and elongation of long bones, causing *giantism*. Once normal growth has been attained, by age 19 or 20, a deficiency of growth

Table 12.1 Location and Function of Major Pituitary Glands

GLAND	HORMONE	FUNCTION
Adenohypophysis (anterior pituitary)	Growth hormone (GH)	Stimulates protein synthesis, accelerates growth
	Thyroid stimulating hormone (TSH)	Stimulates the thyroid to secrete thyroxine
	Adrenocorticotropic hormone (ACTH)	Stimulates the adrenal cortex
	Follicle stimulating hormone (FSH)	Stimulates the maturation of ovarian follicles and ova; induces follicle cells to secrete estrogens
	Luteinizing hormone (LH)	Induces the formation of the corpus luteum
	Prolactin	Initiates and maintains mammary gland secretion (milk)
Neurohypophysis (posterior pituitary)	Antidiuretic hormone (ADH)	Increases the reabsorption of water by kidney tubules
	Oxytocin	Stimulates smooth muscle of mammary glands and uterus
Thyroid	Thyroxin	Regulates metabolic activities
	Calcitonin	Lowers blood calcium; increases bone calcium
Parathyroid	Parathyroid hormone	Increases blood calcium; reduces bone calcium
Adrenal medulla	Epinephrine, norepinephrine	Stimulates a variety of body organs and tissues to prepare the body to withstand an emergency or stress
Adrenal cortex	Aldosterone (mineralocorticoid)	Increases the reabsorption of sodium ions by kidney tubules
	Cortisol, cortisone (glucocorticoids)	Aids in decomposition of amino acids; inhibits the immune system; acts as antiinflammatory agent
Pancreas islet cells	Insulin	Increases cell membrane permeability for glucose; aids the liver in converting glucose to glycogen
	Glucagon	Aids the liver in converting glycogen to glucose
Thymus	Thymus hormone?	Stimulates antibody production?

hormone apparently has no effect. However, an oversecretion during this period brings about an abnormal growth and elongation of bones in the hands, feet, chin, nose, and ears, producing a grotesque appearance known as *acromegly.*

Thyrotropic Stimulating Hormone (TSH) This hormone, secreted by the anterior pituitary, stimulates the thyroid gland to release the hormone *thyroxin.* Thyroxin concentration in the blood is maintained at a fairly constant level, owing to a feedback control. Once stimulated, the rate of thyroid secretion of thyroxin is increased. The increased concentration of thyroxin in the blood has an inhibiting influence upon TSH release by the anterior pituitary. Lower TSH secretion means less thyroid stimulation, and therefore less thyroxin is secreted. Feedback controls such as this are self-regulating procedures that are capable of maintaining precise control over hormone secretion. However, as the concentration of thyroxin increases, the pituitary secretion of thyrotropic stimulating hormone is inhibited, and this in turn inhibits further thyroxin secretion. A low concentration of blood thyroxin stimulates the release of thyrotropic hormone, which stimulates thyroxin secretion. All tropic hormones are involved in this typical feedback secretory control. Tropic hormones are those which stimulate another endocrine gland to secrete its hormone.

Prolactin Prolactin initiates and maintains milk production by the mammary glands. Prolactin is released only if other hormones, such as estrogen, progesterone, and other substances are released in sufficient quantities. The hypothalamus does not produce a releasing factor Instead it apparently secretes a prolactin-inhibiting factor that prevents lactation. High-level secretion of these other hormones during the late stages of pregnancy overrides the inhibitory factors and allows lactation to be initiated and maintained.

The Gonadotropins

Gonadotropic hormones secreted by the anterior pituitary stimulate the gonads (ovaries and testes) to secrete other hormones. These will be discussed in greater detail in the chapter dealing with reproductive physiology, and therefore we will limit the discussion to a few brief statements.

Follicle Stimulating Hormone (FSH) This hormone stimulates the growth of ovarian follicles, which are responsible for the normal maturation of eggs in the female. FSH released in the male is essential for normal maturation of sperm cells.

Luteinizing Hormone (LH) This hormone initiates the formation of specialized ovarian structures in the female and the subsequent production of hormones (estrogen and progesterone) that stimulate uterine tissue development. LH secretion by the male stimulates the testes to produce the hormone *testosterone*.

Adrenocorticotropic Hormone

Adrenocorticotropic hormone (ACTH) is secreted by the anterior pituitary and stimulates the adrenal cortex to release cortical hormones. The specific action of ACTH will be discussed with adrenal gland activity. The release of ACTH by the hypothalamus is regulated by release factors also located in the hypothalamus.

Table 12.1 summarizes the location and function of the major pituitary glands.

THE THYROID GLAND

The *thyroid gland* is a mass of tissue located on the anterior surface and sides of the trachea just below the larynx. The gland in a normal adult, as shown in Figure 12.3, weighs about 30 g ($\pm$ 10 g) and is approximately 5 cm long and 3 cm wide. The location of the thyroid gland has nothing to do with its function of regulating metabolic activities throughout the body.

The major secretion from the thyroid gland is called *thyroxin*. The thyroxin molecule is an amino acid combined with iodine molecules. This substance influences the rate of metabolism in all cells of the body. A decrease in the amount of thyroxin lowers metabolic activity, and an increase speeds up metabolic activity.

Insufficient thyroxin during the early growing years produces a condition called *cretinism*. This term describes a person who has not normally developed mentally or physically. If this condition is diagnosed at an early stage in the infant, it

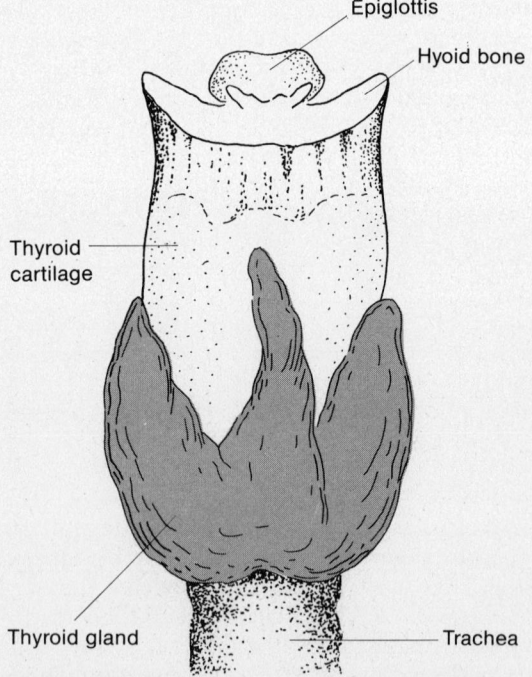

Fig. 12.3 The thyroid gland and its position relative to throat structures. It is located on the anterior surface and sides of the trachea just below the larynx. Two secretions of the thyroid are thyroxin and calcitonin.

can be corrected by injections of thyroxin, which will permit the normal mental and physical development of the child. If there is a normal thyroid function until after growth and development are complete and then a low level of thyroxin is secreted, the person fatigues easily, there is characteristic edema of the tissues, particularly around the eyes, and poor hair growth. This condition is called *myxedema.*

Efforts to establish the precise action of thyroxin have largely gone unrewarded. Excesses and deficiencies of thyroxin produce dramatic, although slowly developing, abnormalities. An injection of thyroxin is followed by a period of a week to 10 days with little or no evidence of its influence. Its effects, once observable, can last a month or more. This has led to an assumption that thyroxin affects enzyme synthesis. Other researchers have suggested that thyroxin has a specific effect upon the synthesis of ATP. Either or both of these influences (enzyme or ATP synthesis) could account for metabolic irregularities associated with abnormal thyroid activity.

Tests employed to determine the level of thyroid activity include the *Basal Metabolism Test* (BMT), *Protein Bound Iodine Test* (PBI), and *Radioactive Iodine Uptake.*

Basal Metabolism Test
BMT is a measure of the rate of oxygen utilization by the body. This rate varies according to sex, weight, age, and the body surface area of an individual. It is a rather inexact measurement, although normal ranges have been established. Exaggerated fluctuations of oxygen consumption in either direction from the norm suggest improper thyroid activity.

Protein Bound Iodine Test
Most of the iodine carried in blood plasma is chemically bound to protein molecules. In the PBI test, a sample of plasma is obtained from an individual and the protein fraction is precipitated out. By analyzing the amount of iodine present in the precipitated protein, the level of thyroid activity can be calculated. A *hypothyroid* condition exists if a less than normal quantity of iodine appears in the plasma sample. *Hyperthyroidism* produces an excessively large quantity of iodine bound to plasma protein.

Radioactive Iodine Uptake
The use of radioactive isotopes provides a simple means of determining how much ingested iodine is taken up by the thyroid gland. The rate of uptake is a measure of thyroid gland activity.

An individual is asked to drink a solution containing a known concentration of radioactive iodine. Twenty-four hours later, the person is placed in front of a radiation counter that scans the thyroid. Data obtained from the radiation counter can be used to determine how much of the originally ingested iodine has been captured by the thyroid gland.

After 24 hours, a normal thyroid will take up 15 to 45 percent of the previously ingested radioactive iodine. If less than 10 percent is taken up, hypothyroidism is indicated. A 60 percent or greater uptake indicated hyperthyroidism.

Iodine Requirement for Normal Thyroid Function
A normal thyroid can utilize a limited quantity of iodine. If a more than adequate quantity is ingested, it is secreted from the kidney tubules and carried away as waste in the urine.

The normal requirement is 30 to 50 mg per year. Ingested iodine is absorbed from the gastrointestinal tract into the bloodstream. Approximately 60 percent of it is excreted along with urine. The remaining 40 percent is taken up by the cells of the thyroid, where it is converted to a stored form called *thyroglobulin.* Thyroglobulin is stored in specialized cells of the thyroid, later to be changed to thyroxin before it is secreted.

Toxic Goiter (Grave's Disease)
Hyperthyroidism may be caused by an oversecretion of thyroxin from a *hyperplastic thyroid.* The term hyperplastic thyroid describes thyroid enlargement caused by an increase in thyroid cell numbers, which enlarges the gland. Goiter produced in this manner is called *toxic goiter* or *Grave's disease.*

Hyperthyroidism caused by toxic goiter is believed to be caused by an excessively high secretion of thyrotropin from the anterior pitui-

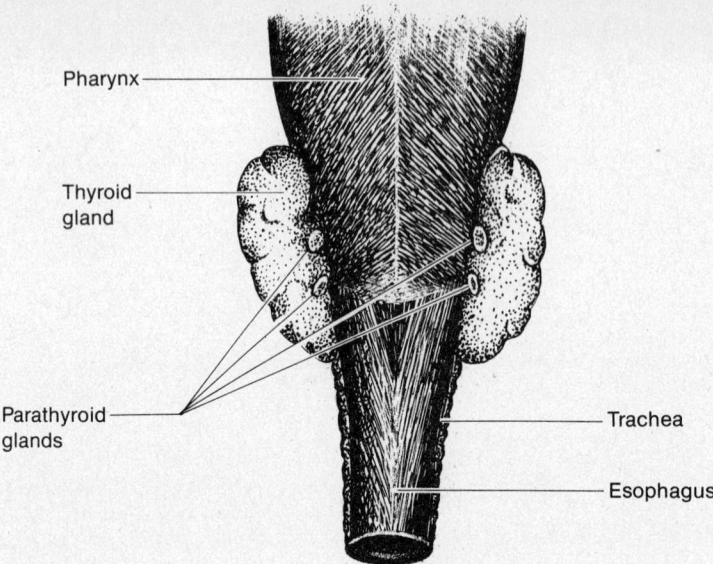

Fig. 12.4 The parathyroid glands. Embedded in the posterior surface of the thyroid glands, these four bits of tissue secrete parathyroxin (parathyroid hormone), which helps regulate blood calcium levels.

tary, which stimulates thyroid cell reproduction. The usual treatment is to surgically remove a part or all of the thyroid or destroy some of the thyroid cells with radioactive isotopes of iodine.

A person having toxic goiter suffers from an increased cardiac output, hypertension, and profuse sweating. Other symptoms include extreme nervousness, irritability, and, in some cases, bulging eyeballs *(exophthalamos)*. The eyeball may protrude to the extent that the eyelid cannot completely close, causing excessive drying of the eye membranes. Infections of the eye and blindness have resulted from this condition.

Thyroid Storm

If excessive amounts of thyroxin are suddenly emptied into one's blood system, the symptoms normally associated with hyperthyroidism are drastically (often fatally) accentuated. This condition of severe thyroid toxicity is called *thyroid storm*. It most commonly occurs a few days after surgical removal of the thyroid and is apparently due to excessive thyroxin being released during surgery. A high incidence of mortality is associated with thyroid storm. Without treatment, almost all individuals die.

Calcitonin

The hormone *calcitonin* was until recently thought to be produced by the parathyroid glands. It is now known to be secreted by thyroid cells. Calcitonin secretion is stimulated by high blood calcium. The hormone is responsible for an accelerated uptake of calcium by bones, thus lowering blood calcium concentrations. When normal blood calcium is reached, calcitonin secretion is no longer stimulated. This is another example of a self-regulating feedback process.

THE PARATHYROID GLANDS

The *parathyroid glands* are four small bits of tissue embedded in the posterior part of the thyroid gland, as shown in Figure 12.4. The location of the parathyroids close to the thyroid gland has no bearing upon their activity since the function of their secretion is entirely different from that of thyroxin. Removal of the thyroid must be done with extreme care since the parathyroids may be removed in the process.

The parathyroids secrete a hormone called

parathormone, which regulates calcium metabolism. A constant and specific level of calcium must be maintained in the blood since calcium is essential not only as a bone component but also as an ion that affects cell membrane permeability.

A low level of parathormone secretion brings about an accelerated removal of calcium from the blood and causes excess deposition of calcium in bones. A low level of blood calcium makes muscles and nerves hypersensitive. If the parathyroid is removed, the hypersensitivity is of such magnitude that only slight stimuli to nerves and muscles throws the whole body into muscular spasms (tetany). This can cause death because of violent contractions of the throat muscles shutting off the oxygen supply.

An excess of parathormone, which may be caused by a tumor of the parathyroid gland, causes calcium to be withdrawn from bones, thereby weakening them and causing an excess of calcium in the blood. This excess causes muscles and nerves to be less sensitive to stimuli. The nervous system is depressed and muscular contraction is weakened, which has a pronounced affect upon the movement of food materials through the gastrointestinal tract. This may produce a loss of appetite, bring on constipation, and accelerate ulcer formation. The specific effect of parathyroid hormone on bone cells is discussed in Chapter 5.

The parathyroids are apparently not controlled by hormones from the anterior pituitary as are the thyroids. An increased level of blood calcium has a direct inhibiting effect upon the secretion level of the parathyroids. A low blood calcium level causes accelerated absorption of calcium through the gastrointestinal tract into the circulation and there is an increase in bone demineralization until a normal blood calcium level is reached. This feedback mechanism is similar to thyroid regulation except that the parathyroids are directly influenced by the calcium ion concentration in the blood.

Calcitonin, secreted by the thyroid gland, aids in lowering blood calcium. Calcitonin and parathyroid hormone therefore have antagonistic functions. One does not directly control the other, but the fluctuations in blood calcium caused by either secretion certainly indirectly influence each one.

Hypoparathyroidism can be treated with purified parathyroid extract. *Hyperparathyroidism* is usually corrected by surgically removing portions of the parathyroid.

THE ADRENAL GLANDS (SUPRARENAL GLANDS)

The *adrenal glands* are found buried in fat tissue on top of the kidneys. Adrenal means "at the kidney." The adrenal glands represent another example of two glands physically combined into one basic structure, each having different functions. In fact, each part is derived from different embryonic tissue.

Figure 12.5 shows the location of the adrenal glands and a cross section of a gland showing the outer layer, called the *adrenal cortex,* and the inner layer or core, called the *adrenal medulla.*

The Adrenal Medulla

The central segment of the adrenal gland secretes the hormones *epinephrine* and *norepinephrine.* These hormones were formerly called *adrenalin* and *noradrenalin.* Both epinephrine and norepinephrine are released when nerve impulses arrive at the adrenal medulla from the sympathetic nervous system. The effect of these hormones upon the body is a wholesale regimentation to meet an emergency or stress. The activities associated with the influence of these hormones are the same as those initiated by the sympathetic nervous system.

Both male and female hormones are secreted by the adrenal cortex, but the secretions appear to be dominated by gonad hormones. An overproduction of the male sex hormone by the adrenal cortex produces precocious development of male genitalia. The masculinization may be obvious in very young children. If the child is a female, she will exhibit male secondary sex characteristics. The cause of oversecretion of sex hormones by the adrenal cortex may be caused by tumors. The overall effect produced by these hormones is called the *adrenogenital syndrome.*

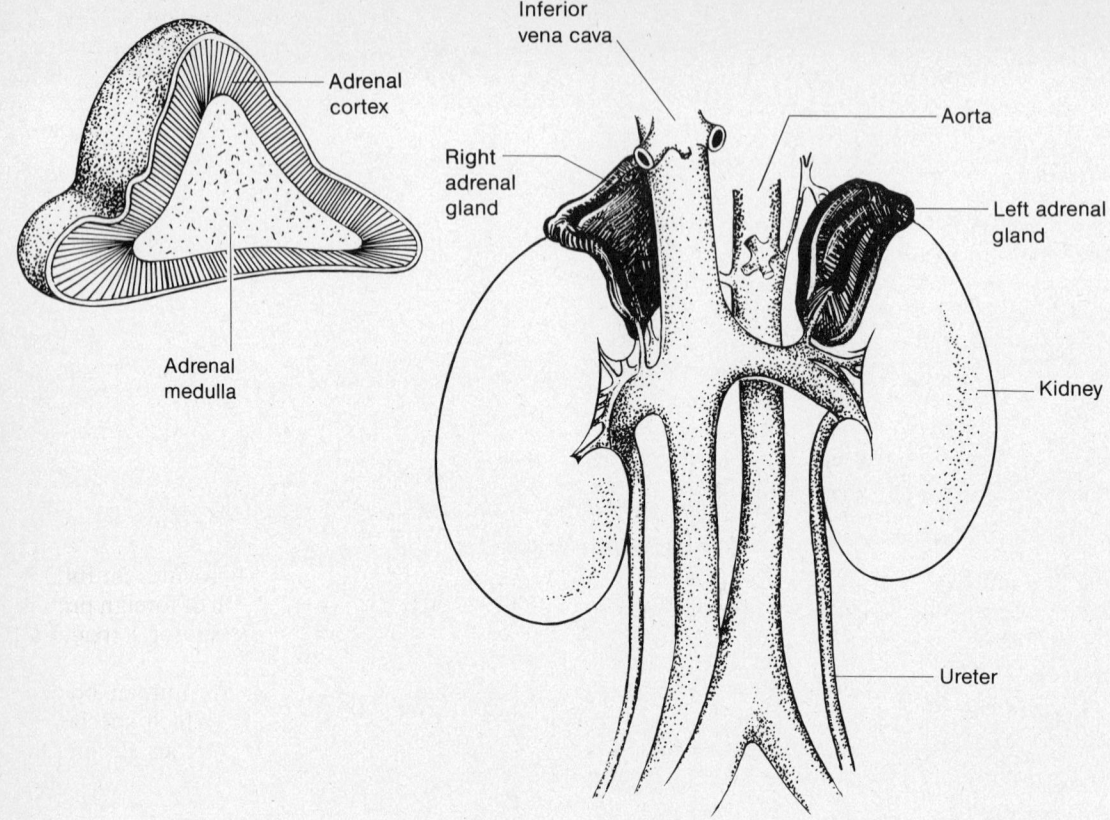

Fig. 12.5 Position and internal structure of the adrenal glands. A cross section of the gland shows that it is composed of an outer portion called the cortex and an inner portion called the medulla. These different areas secreting different hormones have no relationship except their physical contact with each other.

The Adrenal Cortex

The adrenal cortex secretes three groups of hormones—the *mineralocorticoids, glucocorticoids,* and *sex hormones.* All of these hormones are steroids derived from the base chemical substance *cholesterol.*

The term mineralocorticoid refers to those substances that affect the mineral and ion concentration of the intercellular and extracellular fluids of the body. The major hormone in this classification is *aldosterone.* It acts upon kidney tubules, causing them to reabsorb larger quantities of sodium. Along with this activity is the concurrent retention of water, and the loss of potassium and hydrogen ions.

Increased secretion of aldosterone, which in effect increases the sodium retention of the cellular fluids, also tends to remove hydrogen ions from these fluids, making them alkaline and producing the condition called *alkalosis.* A decreased secretion of aldosterone does just the opposite, in other words, less sodium is retained and more hydrogen is in the cell fluids, producing an acid condition of cell fluids called *acidosis.* The processes of alkalosis and acidosis are more completely discussed in Chapter 16.

The *glucocorticoids* include the substances *cortisol* and *cortisone.* Once again a homeostatic mechanism is involved. A high concentration of glucocorticoids thrown into the bloodstream has an inhibiting effect upon the factors that increase the overall secretion of glucocorticoids.

Cortisol functions in controlling many body activities, among which is the decomposition of amino acids and proteins and the synthesis of by-products of sugar. Also involved is the water balance of the body, certain central nervous system influences, gastric acidity, bone maintenance, and certain antiinflammatory functions.

The immunology of the body is affected by the glucocorticoids due to the fact that the lymphocyte numbers in the body are reduced; this is an important activity influencing the formation of antibodies. In transplants, the person is given large doses of glucocorticoids; this lowers the number of lymphocytes, which in turn limits the level of the antibody production in the body so that the transplanted organ is not rejected.

A hypersecretion of these substances produces a whole series of effects upon the body that are lumped into a single term called *Cushing's syndrome.* This condition is characterized by a flushed, moon-shaped face, obesity, and general weakness.

Hyperactivity of the adrenal cortex before puberty permits a greater secretion of the androgens—testosterone in the male, estrogen in the female—inducing precocious development of sex organs. A decreased secretion of these substances in the adult brings about another set of influences called *Addison's disease.* This disease is characterized by a bronzing of the skin due to excess skin pigmentation, hypoglycemia, low blood pressure, renal failure, dehydration, acidosis, increased serum potassium, and decreased serum sodium.

THE THYMUS GLAND

The *thymus gland* is located in the upper chest directly behind the tip of the sternum and anterior to the aorta and other major blood vessels. It is generally classified as an endocrine gland, but as yet there is no concrete evidence that it secretes a hormone.

Until 1960, almost nothing was known about the function of the thymus gland. Even today, scientific research is yielding only limited information, some of which is only speculative. The thymus gland is quite large (40 to 50 g) in the newborn, remaining relatively large until the age of puberty, after which it diminishes in size. In a normal adult, the atrophied thymus is scarcely distinguishable from the fatty tissue in which it is buried. The large size of the thymus in the newborn and its diminished size after puberty suggest that its major function is completed in early childhood. Further evidence for this notion is supplied by the fact that surgical removal of the adult thymus causes no apparent ill effect.

The role of the thymus in early childhood seems to be that of initiating the immune system. If the thymus is removed from newborn mice, many of them die within three or four months. Within this period of time, these mice are capable of accepting skin transplants from other species with no apparent rejection. Mice with an intact thymus will normally reject a skin transplant from other species in about 10 days. This evidence further substantiates the notion that the thymus is in some manner responsible for the immune system, which provides antibodies that guard against the invasion of foreign protein whether it is a skin transplant or a bacterial infection.

The immune system of the human body is made up of lymph nodes in which specialized cells are trapped and induced to proliferate. These cells, the *lymphocytes,* are the primary producers of antibodies that protect the body from disease and other foreign protein.

Lymphocytes are not initially produced by the lymph nodes. They are apparently produced, during the later stages of fetal development or shortly after birth, from specialized cells called *stem cells* located in the thymus. Lymphocytes produced by the thymus enter the bloodstream and are carried throughout the body. Lymph nodes, with their network of meshlike tissues, trap some of the lymphocytes, where they undergo rapid reproduction. The lymph nodes then become sites of antibody production, providing the body with an immune system.

The role of the thymus in this overall process is to provide the initial lymphocytes that seed the lymph nodes. If the thymus gland is removed after the seeding process has been accomplished, the immune system does not appear to be affected.

THE PANCREAS

The *pancreas* is composed of several kinds of specialized cells. Some of these cells have an exocrine function, while others have an endocrine function.

The exocrine function is limited to the secretion of digestive enzymes, which are carried by a duct (pancreatic duct) to the upper portions of the small intestine.

The endocrine function involves the secretion of hormones directly into the bloodstream.

Anatomy of the Pancreas

The pancreas is located just below the stomach and above the first loop of the small intestine, as shown in Figure 12.6.

The pancreas, approximately 15 cm long and 2.5 cm thick, is composed of a head, body, and tail. The head portion lies close to the duodenum, into which the digestive enzymes are emptied. The adult pancreas weighs 50 to 70 g. The exocrine cells empty their secretions into ducts which interconnect and lead toward the pancreatic duct. Interspersed among the exocrine cells are clumps of cells called the *islets of Langerhans* (islet cells). These cells secrete the hormones *insulin* and *glucagon*.

Endocrine Function of the Pancreas

Of the 50 to 70 g weight of the whole pancreas, approximately one gram of this weight is islet cells. Individual islets measure between 75 to 150 microns.

The islets are made up of a variety of cell types—*alpha cells* secrete glucagon; *beta cells* manufacture and store insulin; and *delta cells* have an unknown function. The larger portion of the islets are composed of beta cells.

Beta Cells and Insulin Production Insulin secreted by beta cells of the islets of Langerhans controls blood sugar levels throughout the body. The relationship between insulin secretion and *diabetes mellitus* (sugar diabetes) was established in 1889, when it was demonstrated that all the symptoms of diabetes mellitus could be produced by removing the pancreas from an experimental animal. Although the connection between the pancreas and diabetes mellitus was established, the control of the disease had to await the isolation of purified insulin in 1921.

Until only recently, the sole source of insulin has been from slaughterhouse animal pancreas. Insulin is a protein; therefore a rather complicated process is necessary to separate it intact from other pancreatic secretions, namely the proteolytic enzymes that would digest the insulin and deactivate it.

Insulin has been artificially synthesized. However, it is not yet available commercially since its use is still limited to experimentation.

Action of Insulin Insulin decreases blood sugar levels and therefore is referred to as a *hypoglycemic factor*. Its primary function is to increase the rate of transfer of glucose across cell membranes. Since all cells require the nutrient

Fig. 12.6 The pancreas in endocrine activity. (a) View of general location of the pancreas. (b) A bit of pancreas tissue showing the islet cells. These cells secrete two hormones, insulin and glucagon.

Diabetes Mellitus, Circa 1922

A successful treatment for diabetes mellitus was discovered by three Canadians, Banting, Best, and Macleod, in 1922. The discovery was that insulin, extracted from animal pancreases, could be injected into the human bloodstream to effectively control blood sugar levels. The following paragraph is a hypothetical clinical description of diabetes mellitus before 1922.

Diabetes mellitus (sweet water disease) is marked by a persistent excess in discharge of urine. On examination, the urine is found to have a high specific gravity and to contain sugar. The exact cause of the disease is not known. Hereditary influences are believed to play a part, and obesity and gout appear to be predisposing factors. In many cases, disease of the pancreas (sweetbread) is present. The onset of the disease is insidious, but sometimes acute cases occur, with rapid emaciation. Increase of thirst and appetite, muscular weakness, and increased necessity to pass water are generally first symptoms. Diabetic coma and death are routine expectations among adults with the disease. Children, without exception, are expected to die within a few months to a year.

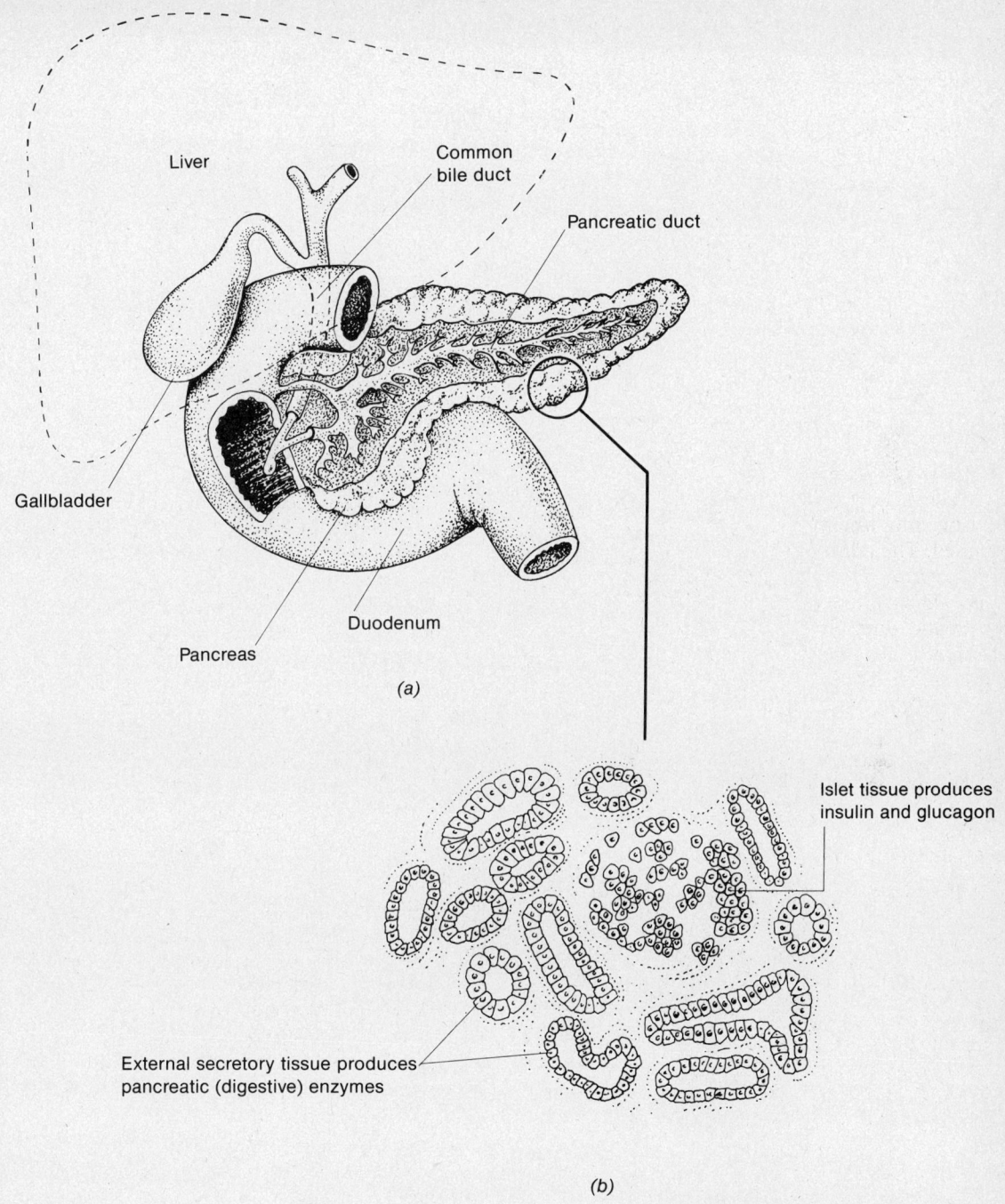

THE ENDOCRINE SYSTEM

glucose, it is essential that all cells receive an adequate supply. Some cells of the body, muscle cells for instance (including the heart), fluctuate in their need for glucose and depend upon a system to pump additional glucose into the cell as needs arise.

Red blood cells and nerve cells have a constant need for glucose and their cell membranes do not present a barrier to glucose entry. Muscle and fat cells, however, are more selective and prevent the entry of sufficient glucose without the aid of insulin.

The exact nature of the insulin-glucose transport system is not known. Insulin in some manner changes the permeability of cell membranes, permitting the entry of larger quantities of glucose. Without insulin, cells can receive adequate glucose only if the blood sugar level is extremely high.

Another organ critically involved in blood sugar levels is the liver. Glucose arriving in the intestines after the digestive processes are completed does not depend upon insulin to carry it across intestinal wall cells into the bloodstream. Most of the food nutrients are carried from the intestines via the hepatic portal vein to the liver.

The liver is not dependent upon insulin for the transfer of glucose into its cells. After a digested meal, the blood from the intestines is extremely rich in glucose. Upon arrival at the liver, the excess glucose, with the aid of insulin, is converted into a storable form called *glycogen*. If insulin is not available in sufficient quantities, excess sugar is emptied into the bloodstream, producing a condition known as *hyperglycemia*. Excess glucose in the blood is normally filtered and retained in the body by the kidney tubules. However, in the diabetic the load of glucose in the blood is so high that it overloads the kidney's capacity to remove it, and it is discharged in urine as a waste product.

In the normal body, excess glucose stored as glycogen in the liver is available for future use between meals or during stress, where there is a greater demand by the cells for glucose.

High blood sugar levels stimulate increased insulin production by the pancreas. In some manner not yet understood, the high glucose

---

> **"Eat, Drink, and Be Merry"—and Perhaps Shorten Life**
>
> *Obese individuals require larger than normal quantities of insulin to maintain normal blood glucose levels. Higher insulin levels stimulate liver cells to synthesize more triglycerides, resulting in higher blood serum triglyceride levels. A close relationship exists between high serum triglyceride levels and certain kinds of fatal heart attacks.*
>
> *Hypersecretion by pancreatic-insulin producing cells, brought about by obesity, may hasten full-blown diabetes. A potential diabetic who is also obese will tax the capabilities of a defective pancreas to the point of diminished secretion or complete inactivity. The price paid by an obese potential diabetic is the early onset of diabetes, the need for life-long insulin therapy, and the possible shortening of a normal life span.*

level triggers some mechanism, causing an increased rate of production and a release of insulin into the bloodstream. In the normal body, the increased insulin will lower the glucose level in the blood by:

1. hastening the transfer of glucose across cell membranes into cells, and
2. by storing glucose in the form of glycogen in the liver.

Alpha Cells and Glucagon Production

Alpha cells of the islets of Langerhans secrete the hormone *glucagon*. This hormone is required for the conversion of glycogen to glucose. As the blood sugar level decreases due to a sudden demand by the cells for glucose, the alpha cells are stimulated to produce and release larger quantities of glucagon, which converts the stored liver glycogen into glucose that may then be released into the bloodstream. Since glucagon increases blood sugar levels, it is called a *hyperglycemic factor*.

The Diabetic

An individual with advanced diabetes mellitus displays symptoms of hyperglycemia (sugar present in the urine), excessive loss of water and salts (NaCl), loss of weight, thirst, and hunger. Until the isolation of insulin in 1921, these individuals were doomed to a relatively short life span.

It is now certain that diabetes is a hereditary disease. Case studies of "diabetic families" point to the fact that individuals within these families, even though they show no outward sign of the disease, have certain characteristics that make them prone to the disease. Some of the characteristics of the prediabetic may include the following:

1. Young women having large babies (10 to 12 lb).
2. An abnormal pattern of blood vessel arrangement in the eye conjunctiva and retina.
3. Kidney glomerulus thickening (observed in biopsy).
4. A low blood sugar level, which may indicate excessively high insulin production by irritated beta cells.
5. A high level of insulin in the blood.

Apparently in many diabetics, the beta cells are "burned out" because they have been called upon to secrete excessively large quantities of insulin over extended periods of time. This damage is irreparable.

Since glucose retention exists, storage and utilization is impaired or lost in the diabetic. The body attempts to compensate for this loss by resorting to other means of supplying nutrition to the cells.

Protein Metabolism in the Diabetic Proteins are constantly being synthesized by some cells from amino acids. At the same time, other cells are involved in degrading proteins to amino acids. Insulin, or the lack of it, has no direct influence upon these protein metabolic cycles. However, it has a pronounced indirect effect. A body cell lacking the capability to absorb glucose must resort to other means of obtaining essential nutrients. One of these is the rapid degradation of muscle protein into amino acids. Cell proteins are broken down to amino acids faster than they are synthesized.

The amino acids thus obtained are *deaminated* (lose their amino groups), and the remaining parts of the molecules are rearranged to form glucose by a process called *gluconeogenesis*. As you may imagine, the loss of protein in this manner can have a devastating effect upon the body, bringing about weight loss, nitrogen imbalance, poor wound healing, and other consequences of protein deficiency.

Fat Metabolism in the Diabetic In a diabetic, the synthesis of fats is depressed and the degradation of fats to glycerole and fatty acids is increased. In a glucose-deficient system these free fatty acids can be converted and utilized by the cells as a source of energy in place of glucose. Some of the fatty acid molecules reach the liver, where they are converted into *acetoacetic acid*. This acid is dumped from the liver into the bloodstream, where some of it is converted into *acetone* (a ketone).

A high concentration of acetoacetic acid plus acetone (combined, called *ketone bodies*) produces *ketosis,* a specific type of acidosis. The kidneys are unable to reclaim the high concentrations of acetoacetic acid, and hence it is lost in the urine along with valuable sodium ions, which attach to the acid. The loss of sodium, a basic ion, only accentuates the acidosis. The acetone portion of the ketone bodies, being more volatile, is blown off during breathing and gives the breath of an advanced diabetic a fruity odor.

The *pineal gland,* shown in Figure 12.1, has long been considered a mystery gland. One hormone, *melatonin,* is known to be secreted. In amphibians (frogs and salamanders) melatonin causes a contraction of skin pigment cells, thus changing the color of the animal to camouflage it with respect to its environment. The function of melatonin in the human body is only speculative. It has been proposed that it inhibits ovarian function, regulates menstrual cycles, and inhibits luteinizing hormone secretion.

Hormones secreted by the ovaries and testes are discussed in detail in the chapter on reproduction. These hormones include estrogen and progesterone, secreted by the ovaries, and testosterone, secreted by the testes.

Gastrointestinal hormones, such as *secretin* and *cholecystokinin,* and other hormones, such as *chorionic gonadotropins,* secreted by the placenta, and *renin,* secreted by kidney cells, will be examined in the discussion of body systems.

AGING AND ENDOCRINE FUNCTION

There is no method at present to test how well the pituitary glands function. There are methods that indicate how other endocrine glands fare with aging. Adrenal gland activity, for example, is determined by how the glands respond to the appropriate pituitary hormone. The adrenal glands characteristically show a decline in secretory activity as age increases. Since adrenal hormones are the "stress hormones," this may result in an impaired capability to respond to stress.

The aged thyroid, except for disease, displays almost no reduction in capability to manufacture and release thyroxine during rest.

The healthy pancreas of elderly persons is capable of secreting adequate insulin, thus maintaining a constant blood sugar. During exercise or stress, the absorption of glucose from blood into cells is slower than in a young person. This follows a pattern evident in the physiology of aging persons—a progressive loss in capacities to return to normal after disturbance in equilibrium.

SUMMARY AND REVIEW

Endocrine glands secrete hormones that are emptied directly into the bloodstream.

A. The pituitary gland is composed of two different kinds of tissue. The anterior portion is called the adenohypophysis. The posterior portion is called the neurohypophysis. The neurohypophysis secretes ADH, which controls water reabsorption, and oxytocin, which stimulates smooth muscle of mammary glands and the uterus.
B. The adenohypophysis secretes growth hormone, thyroid stimulating hormone, adrenocorticotropic hormone, follicle stimulating hormone, luteinizing hormone, and prolactin. The functions of these hormones are shown in Table 12.1.
C. Pancreatic islet cells secrete insulin from beta cells and glucagon from alpha cells. Insulin aids in the transfer of glucose into cells and the conversion of glucose to glycogen in the liver. Glucagon aids the liver in converting glycogen to glucose.
D. The adrenal cortex secretes mineralocorticoids, glucocorticoids, and sex hormones.
  1. Aldosterone, a mineralocorticoid, controls sodium reabsorption in the kidneys.
  2. Cortisone and cortisol are glucocorticoids that control glucose metabolism and amino acid metabolism. They also influence the immune system, body fluids, and ion balance.
E. The adrenal medulla secretes epinephrine and norepinephrine, which act to prepare the body to meet emergencies and stress.
F. The thymus gland is involved in providing the initial cells that become the cornerstone of the immune system.
G. The effect of aging upon some endocrine glands is not known. The thyroid and pancreas show little decline in secretion. The adrenal glands tend to decline in secretory activity as age increases.

PLATE 5   POSITION OF ORGANS OF THE TRUNK

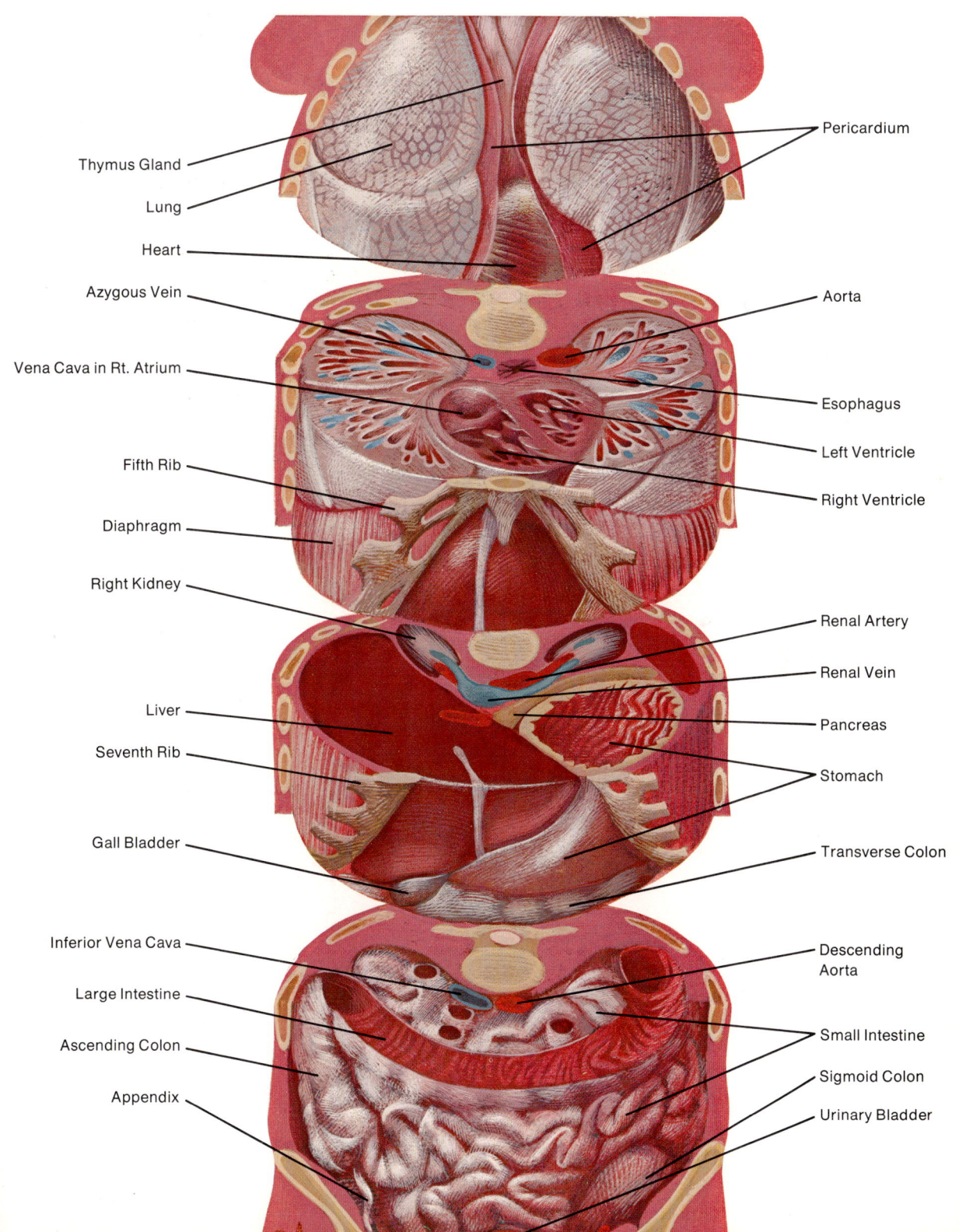

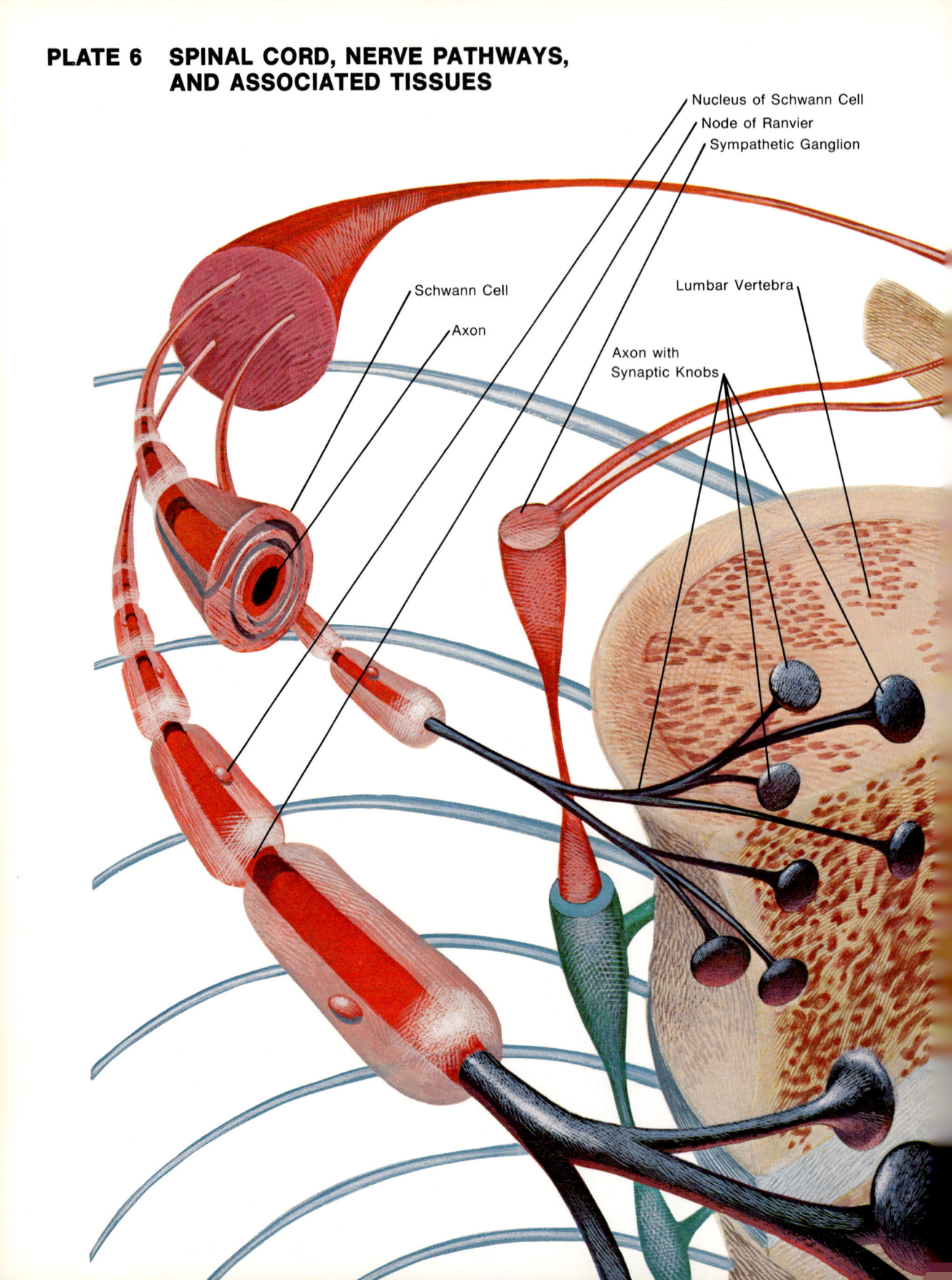

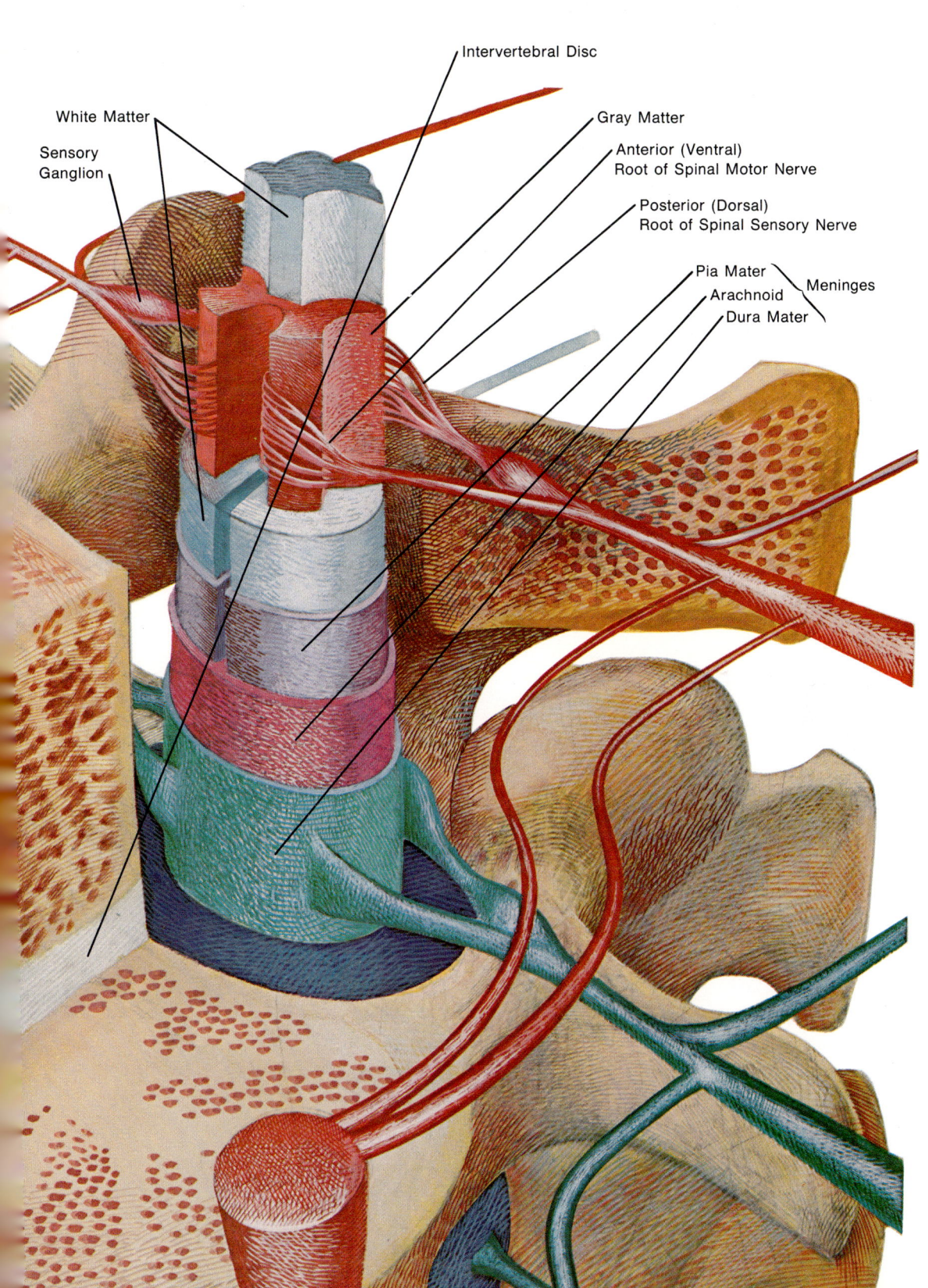

PLATE 7 CIRCULATION — THE HEART AND BLOOD VESSELS

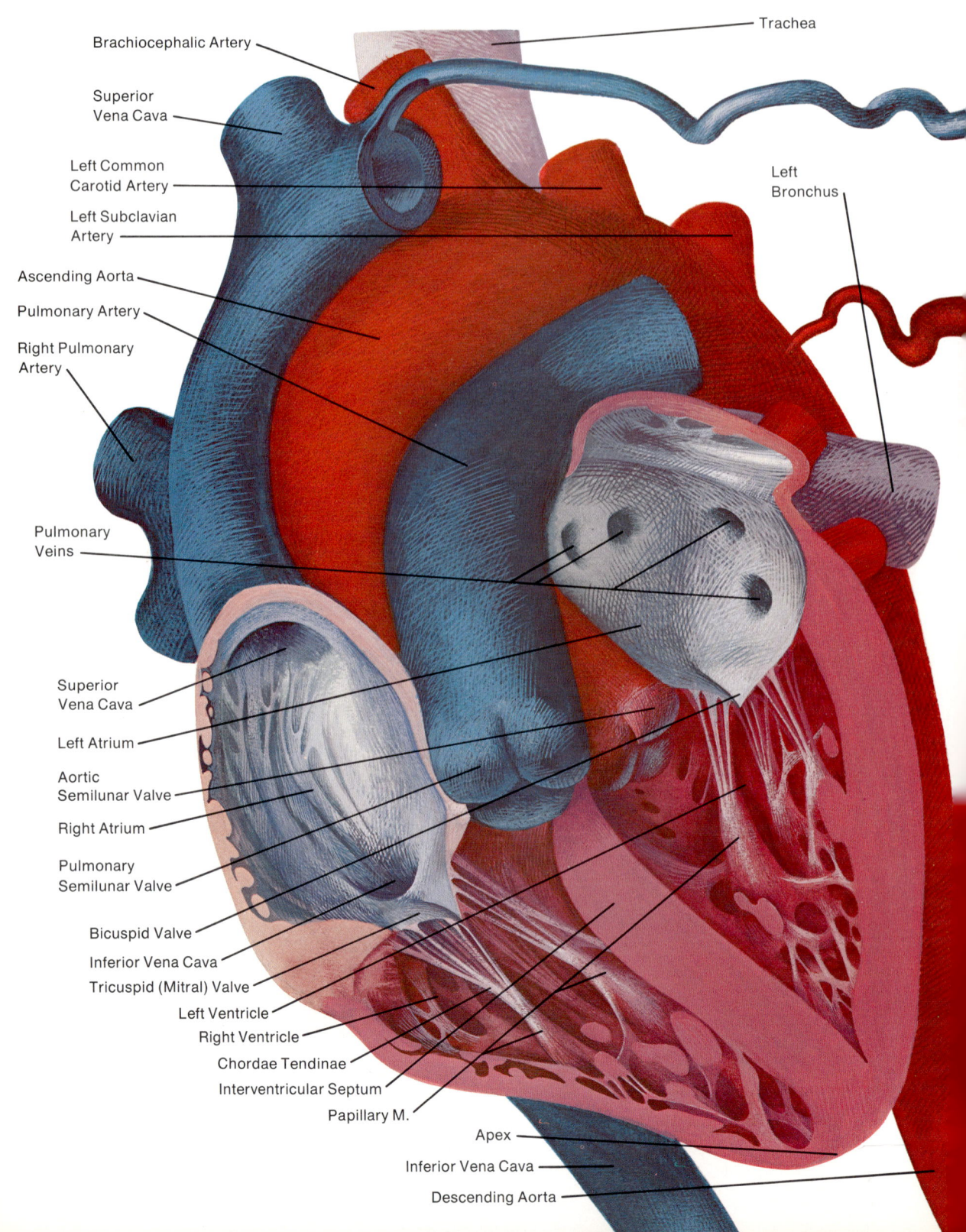

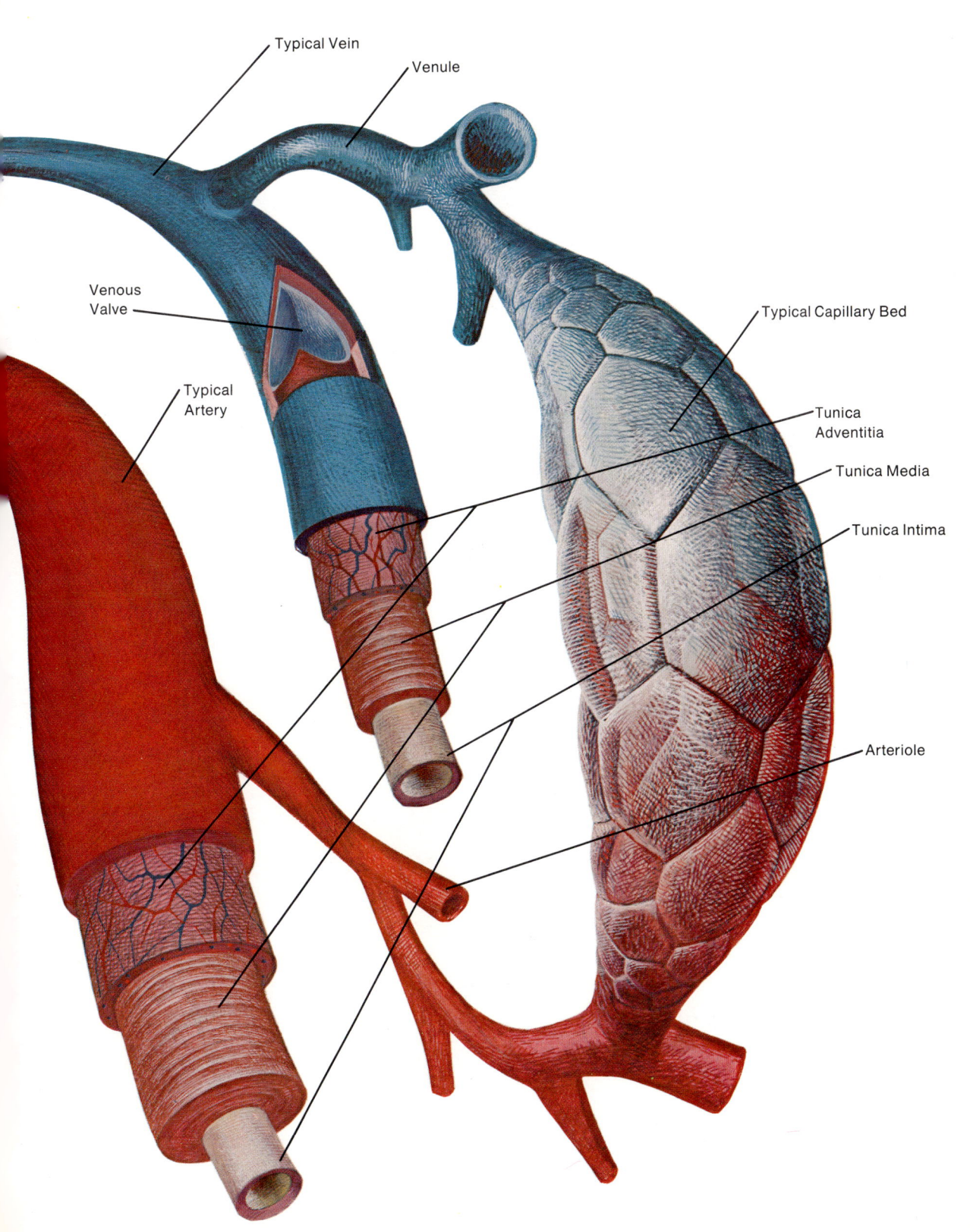

PLATE 8   DIGESTIVE SYSTEM

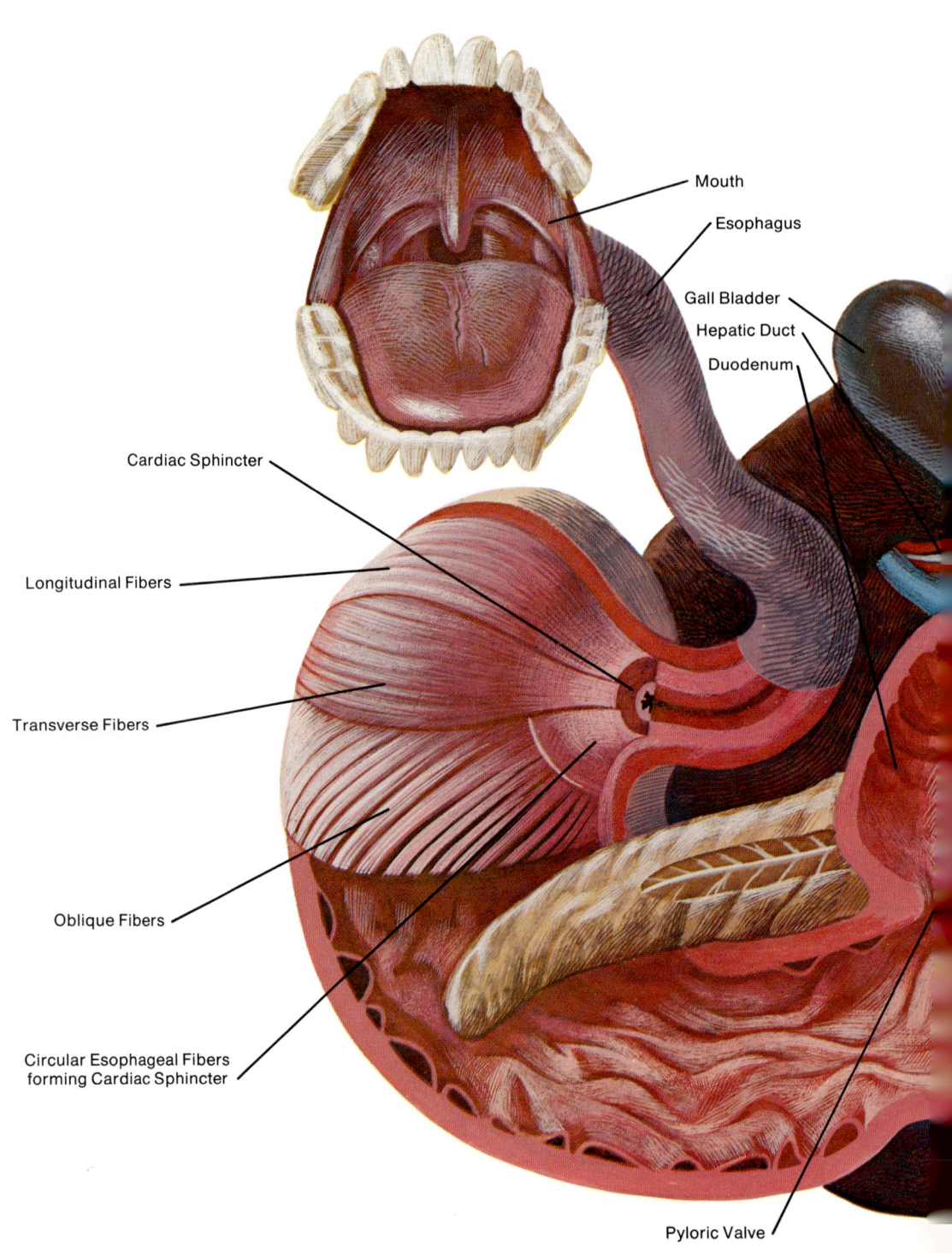

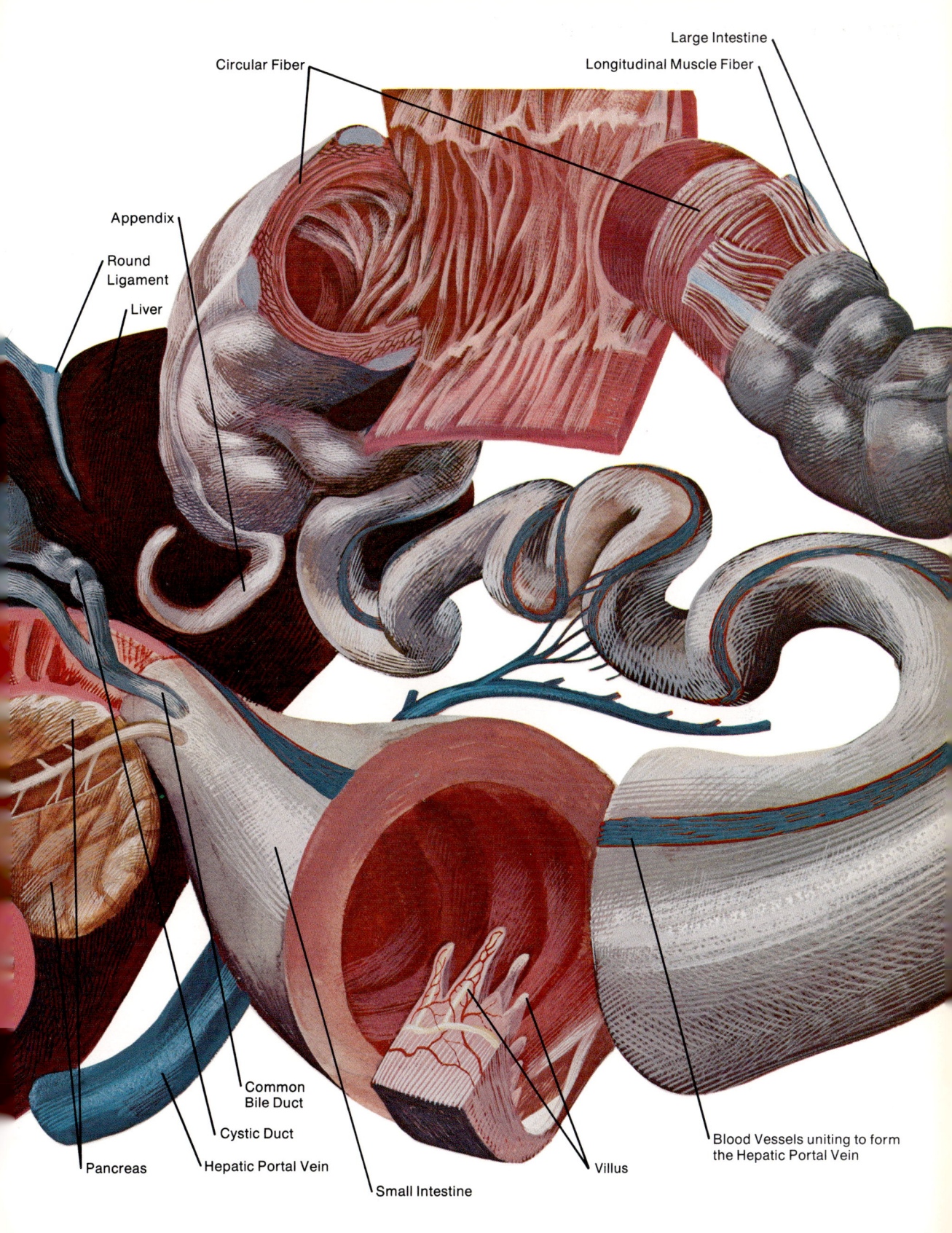

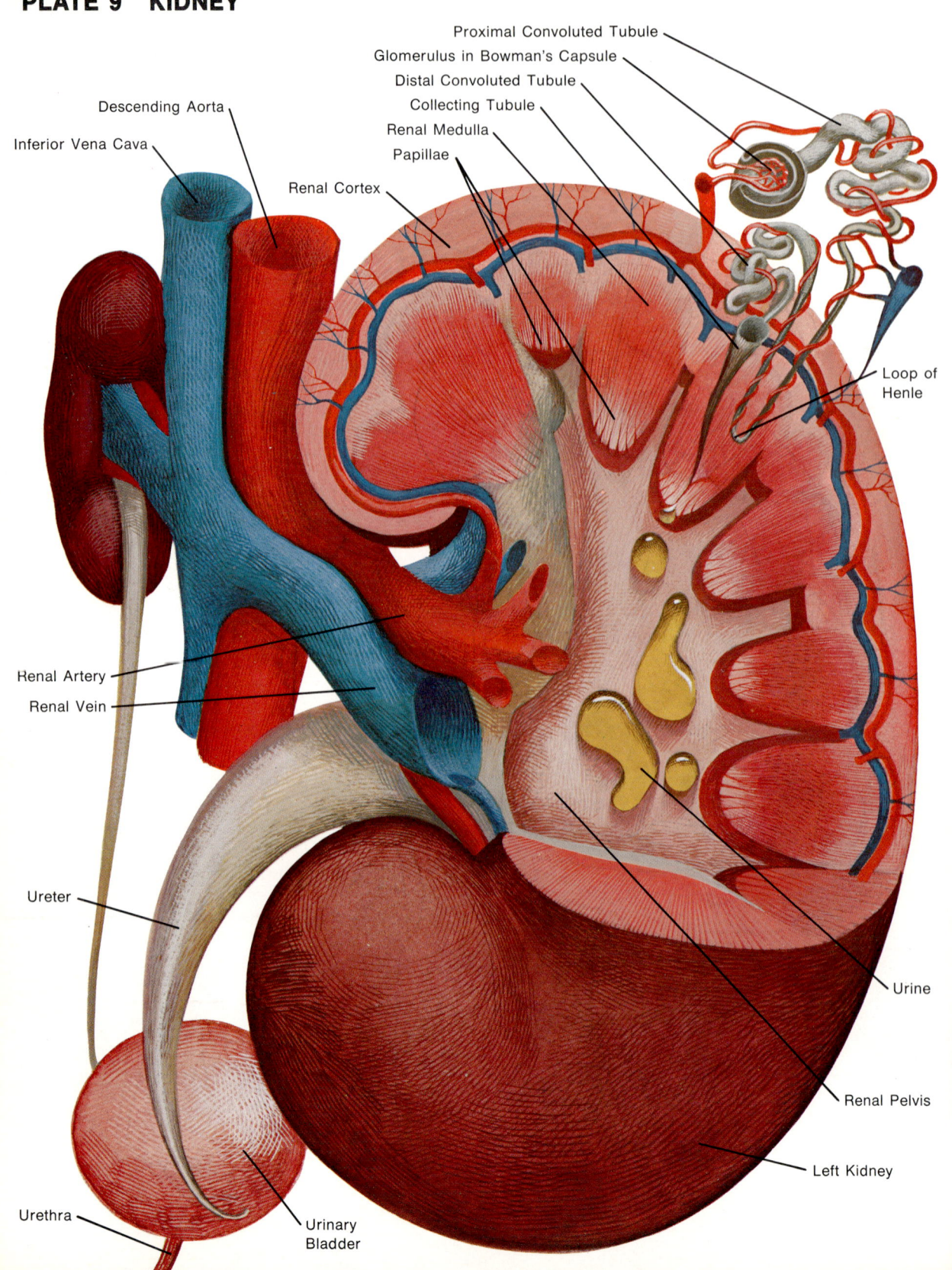

13
The Digestive System

MAJOR CONCEPTS
  TEETH
  ANATOMY OF THE MOUTH AND THROAT
  THE ESOPHAGUS
  ANATOMY OF THE STOMACH
    Functions of the Stomach
    Secretions of the Stomach
    Vomiting
  ANATOMY OF THE SMALL INTESTINE
    Secretions of the Small Intestine
    The Pancreas
    Protein Digestion in the Small Intestine
    Fat Digestion in the Small Intestine
    Absorption in the Intestines
  ANATOMY OF THE LARGE INTESTINE
  THE LIVER
  PHYSICAL DISORDERS OF THE
  GASTROINTESTINAL TRACT
  INFECTIOUS DISEASES OF THE DIGESTIVE
    SYSTEM
  AGING AND THE GASTROINTESTINAL TRACT

The digestive system provides all body cells with nutrients, water, minerals, and vitamins. The key concepts here are digestion, which is the chemical breakdown of large molecules to smaller ones, and absorption, which is the process of getting nutrients from the digestive tract into the bloodstream.

Nerves and hormones also play an important role in digestion. In addition, this chapter will discuss the location, structure, and function of each part of the gastrointestinal tract.

The human digestive system is basically a tube that runs the length of the trunk, beginning at the mouth and ending at the anus. Accessory organs that secrete digestive materials are situated along the length of the digestive system. These accessory organs include the liver, pancreas, gallbladder, and the salivary glands.

The digestive pathway is called the *alimentary canal,* or more commonly the *gastrointestinal tract.* Segments of the gastrointestinal tract, such as the *mouth,* the *esophagus, stomach,* and *intestines,* are specialized to carry out different functions. The overall function of the digestive system is to receive nutrient materials, grind them into small particles, and chemically change the food into a form that can be absorbed and used by the body. Materials that cannot be chemically broken down are disposed of through the anus, as waste materials.

In the following discussion, we will look at the overall anatomy of the gastrointestinal tract, the movement of materials through the specialized segments of the tract, and the functions of individual components of the digestive system.

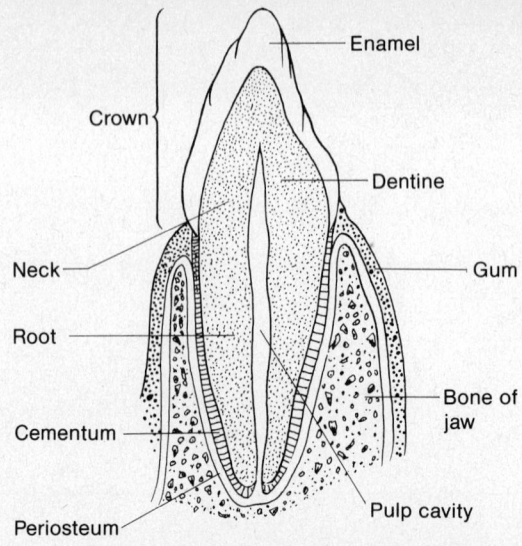

Fig. 13.1 Structure of the tooth. Basic components include the crown, composed of enamel; dentine, which makes up the bulk of the tooth; and pulp, the innermost layer containing nerves and blood vessels. Teeth are set in sockets of fibrous tissue in the gums. Fibrous connective tissue binds gums to the jawbone.

TEETH

Teeth have no function in moving the food through the gastrointestinal tract. Their essential function is to grind, tear, and shred large segments of food. The teeth are set in sockets of fibrous tissue in the gums. Fibrous connective tissue also binds the gums to the jawbone. Figure 13.1 shows the generalized structure of a tooth. The *crown* or *cap* is composed of a very hard crystalline material called *enamel.* Beneath the enamel is a softer substance called *dentine,* which makes up the largest portion of the tooth. The innermost layer, the *pulp,* contains blood vessels that carry nourishment to the tooth structure itself. Nerves are also found in the pulp.

The first set of teeth, gained during early childhood and adolescence, are called the *milk teeth* or *deciduous teeth.* Within a 24-month span, the child gains a set of 20 milk teeth. By age 11 or 12, these are all shed and are replaced by permanent teeth, which number 32 in the adult. The extra 12 teeth that adults have are called the *first, second,* and *third molars.* The molars do not appear until about age 9 or 10. The third molar, which is sometimes called the *wisdom tooth,* appears between age 18 to 25. It often becomes impacted and has to be surgically removed.

Tooth decay results from a variety of factors, among them certain kinds of sugars that ferment in the mouth. Some types of bacteria may grow in these sugars and, in the process, secrete enzymes that attack the tooth covering. Another factor that is probably significant is the chemical makeup of the tooth itself. Controlled experiments have determined that certain chemical substances tend to form a tooth surface covering that is stronger or at least resists decay better. Among these substances are salts called *fluorides,* which seem to have some effect on changing the chemical composition of the tooth enamel, making it more resistant to decay. The most effective preventative against tooth decay is proper cleansing of the teeth.

ANATOMY OF THE MOUTH AND THROAT

The mouth cavity contains the hard and soft palate, the tongue, and of course is enclosed by the cheeks. The palate separates the oral and nasal chambers, as shown in Figure 13.2. In the anterior upper part of the mouth is a bony structure called the *hard palate*. Posterior to this and connected with it is a mass of soft tissue called the *soft palate*, which extends toward the back of the mouth.

The tongue is made up of voluntary muscle and is covered by small projections called *papillae*. Chemical sensors called *taste buds* are located on or within these papillae, enabling the detection of the four primary taste qualities—sweet, sour, salty, and bitter. The tongue is useful in pushing food against the teeth, aiding the grinding and mastication process. Once the food is ground into small particles, the tongue forms the mass of masticated material into a ball-like mass called a *bolus* that is forced backward by the tongue into the *pharynx*. This initiates the swallowing process.

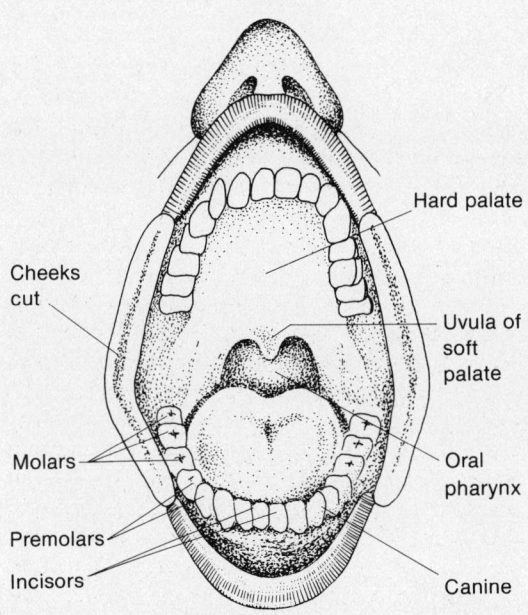

Fig. 13.2 The mouth and associated structures. The mouth cavity includes the teeth, the tongue, and the hard and soft palates, all of which are enclosed by the cheeks.

Note in Figure 13.3 the location of three pairs of salivary glands. These are the *parotid*, located just beneath the ear; the *submandibular*, located along the lower jawline; and the *sublingual*, located beneath the base of the tongue. Secretions from the parotid glands are of a thin watery type called *serous*. The sublingual secretion is largely *mucus*, and the submandibular secretes a mixture of serous and mucus. Mucus is a thick viscous substance that helps lubricate dry foods and permits a smoother passage of these substances along the digestive route. The sublingual glands that secrete mucus are stimulated when dry or rough materials, such as dry bread, are taken into the mouth.

Serous secretion by both parotid and submandibular glands contains the enzyme *salivary amylase*. This enzyme aids in converting starch to maltose. Besides mucus and salivary amylase, saliva also contains water and inorganic salts approximately in the same ratios as found in plasma. Certain organic salts are also found in saliva, along with a substance called *lysozyme*, which has a bacteriocidal effect.

Persons with very low saliva output tend to have more tooth decay than those who have a copious saliva secretion. A heavier flow of saliva aids in rinsing and cleansing the mouth, freeing it of some of the fermenting foods and bacteria.

About 1 to 1.5 liters of saliva are produced per day. Approximately two-thirds of the total saliva volume is secreted by the submandibular gland, about one-fourth by the parotid, and the remainder by the sublingual. The composition of saliva changes depending upon what type of food is taken into the mouth. For example, when acids are taken into the mouth, the parotid secretion of a serous-type saliva is increased. This helps to dilute the acid. This protects the pharynx and the upper esophagus from corrosive substances. The control of secretion is primarily under the influence of the nervous system. Chemical receptors in the taste buds at the back of the tongue promote or initiate salivation. Other factors initiating saliva flow include the entrance of the food into the mouth,

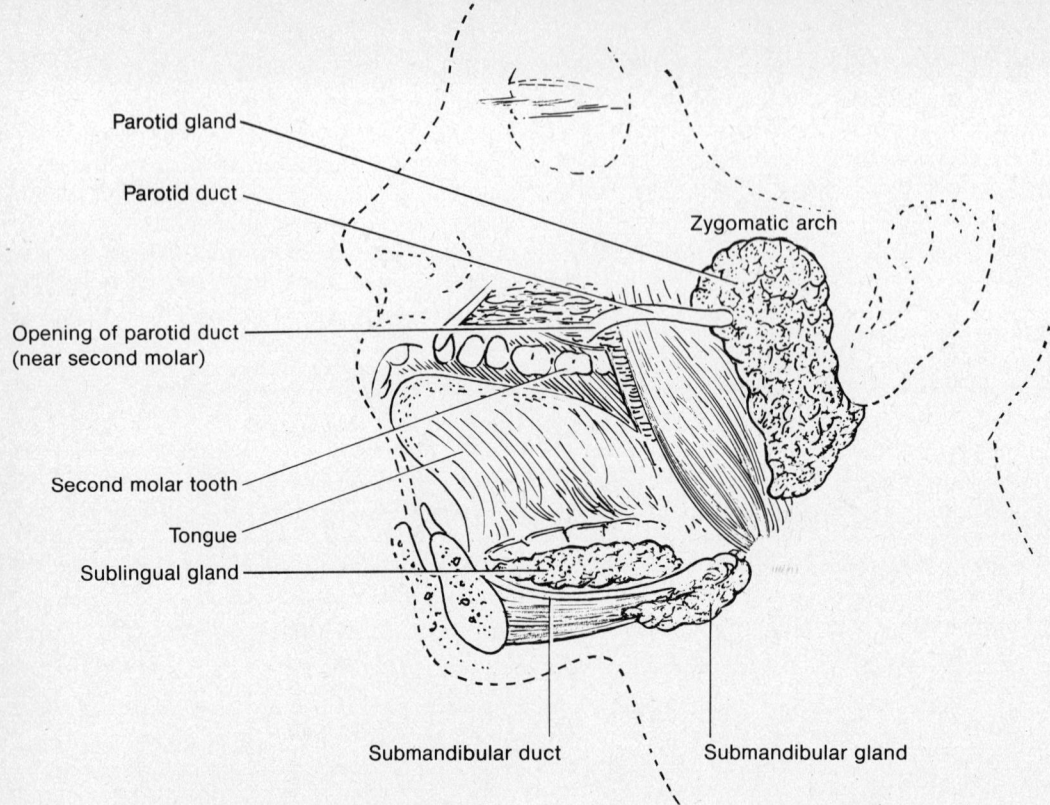

Fig. 13.3 The three pairs of salivary glands. The parotid glands are located just beneath and slightly anterior to the ear. The submandibular glands are found along the lower jaw line. The sublingual glands are located beneath the tongue.

sour or irritating substances, nausea, chewing movements, or the smell and sight of food.

In summary, saliva performs several important functions:

1. It provides salivary amylase for the conversion of starch to maltose.
2. It facilitates chewing and swallowing.
3. It rinses the mouth and moistens the mouth tissues, which facilitates speech.

Saliva is also bacteriostatic and provides protection by diluting harmful substances.

Swallowing is a complicated process involving a coordinated action of the tongue, the pharynx, part of the esophagus, the soft palate, and structures that close off the bronchi leading to the lungs. Note in Figure 13.4 the arrangement of the air and food passages. The upper portion of the air passage is called the *larynx*. Once food enters the pharynx, the voluntary phase of swallowing is ended. From here on the transfer of ingested materials throughout the system is completely involuntary.

As the food enters the pharynx, several things must happen to prevent food from entering the larynx and the bronchi. The vocal cords close; a small flap of tissue above the vocal cords, called the *epiglottis*, is pushed down over the larynx, which prevents food from entering; the larynx is also pulled up out of the way; the soft palate is elevated, which keeps food out of the nasal passages; and respiration is inhibited. If a person inhales while having food in the mouth, food particles may be dislodged and sucked into the air passages. Since the swallowing reflexes have

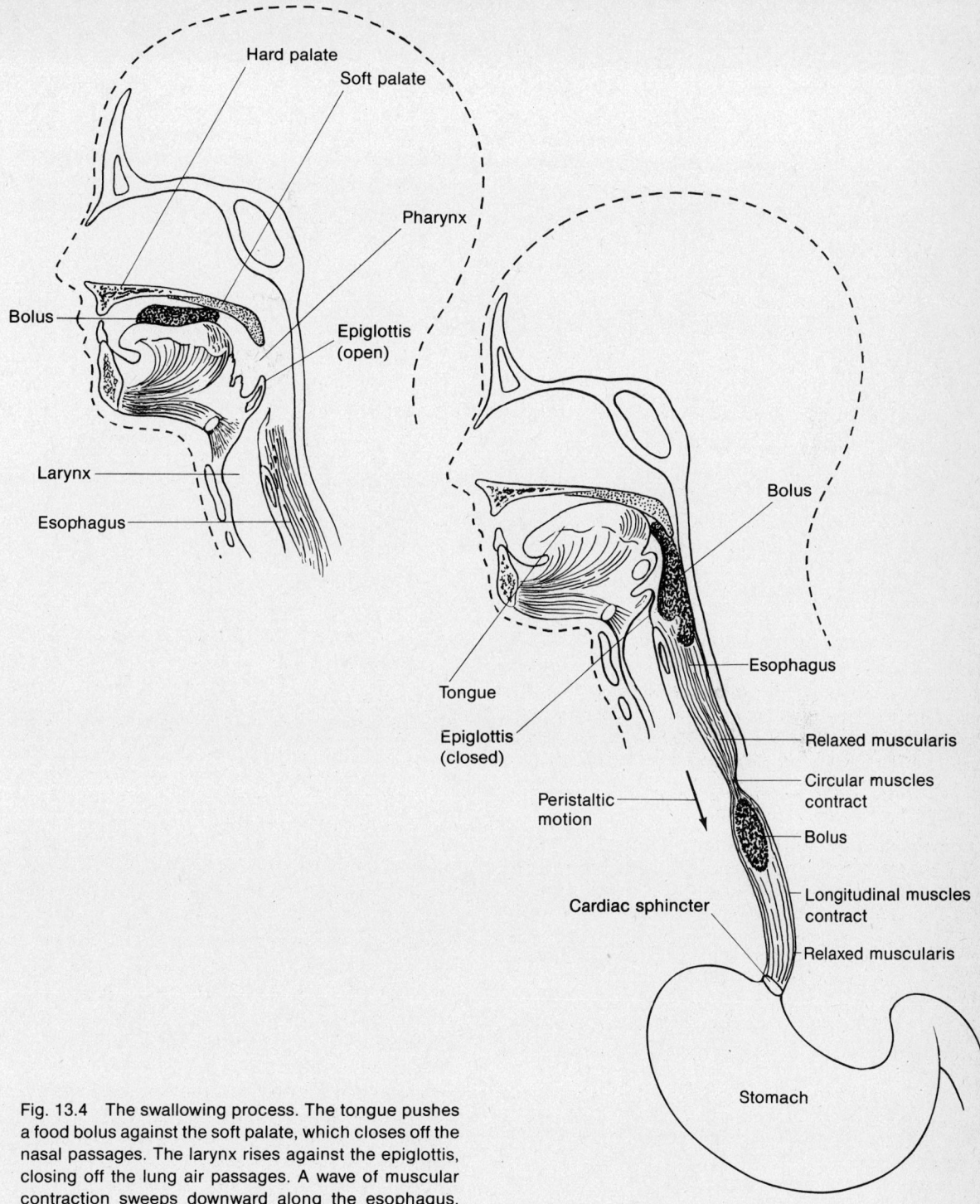

Fig. 13.4 The swallowing process. The tongue pushes a food bolus against the soft palate, which closes off the nasal passages. The larynx rises against the epiglottis, closing off the lung air passages. A wave of muscular contraction sweeps downward along the esophagus, pushing the food bolus toward the stomach.

THE DIGESTIVE SYSTEM 211

not been activated, the air passages are open and food may enter them. When this happens, an involuntary activity called *choking* is initiated. This violent action is the body's attempt to force air out of the lungs, thereby dislodging the particle of food from the air passages.

THE ESOPHAGUS

The *esophagus* is a tube that transports food from the pharynx to the stomach. This transport is accomplished by wavelike constriction above the food bolus that pushes it along toward the stomach. This particular kind of movement is called *peristalsis.*

Peristalsis is a coordinated movement that propels the food along the esophagus in one direction. The specific coordination is brought about by both the sympathetic and parasympathetic portions of the autonomic nervous system. The peristaltic wave is elicited by distension of the esophagus by a bolus. The esophagus, lined with smooth muscle, as is the rest of the gastrointestinal tract, begins its peristaltic motion in response to a food bolus, as shown in Figure 13.4.

A wave of relaxation precedes a wave of constriction. The esophagus has no function except to transport the food from the pharynx to the stomach. The lower end of the esophagus joins the stomach. Here we find a constriction called the *cardiac sphincter.* A *sphincter* is a ringlike muscle encircling an opening, capable of contraction, which closes the opening.

When the relaxed wave of peristalsis reaches the sphincter, the sphincter relaxes. Once the food passes the sphincter, it contracts forcibly as the wave of contraction reaches it, thereby preventing the return of food into the esophagus from the stomach. This particular sphincter is called a *physiological sphincter* rather than a structural one since it is capable of relaxing when the peristaltic wave reaches it. Other sphincters do not function in this manner, in other words, they must be forced open. A tight closure between the esophagus and the stomach is essential to prevent food from being forced back into the esophagus along with damaging chemical substances such as acids, which may erode the lining of the esophagus.

ANATOMY OF THE STOMACH

The area adjacent and distal to the cardiac sphincter is called the *cardiac region* of the stomach, shown in Figure 13.5. The part of the stomach bulging upward from this area is called the *fundus.* As the stomach bulges outward, the curve is referred to as the *greater curvature.* The opposite side curves inward and is referred to as the *lesser curvature.* The major portion of the stomach is called the *body.* The stomach tapers toward the small intestine. This smaller tube is called the *pyloric antrum.* The *pyloric sphincter* closes off the passageway between the stomach and the duodenum. The innermost layer of the stomach, the *mucosa,* is thrown into large folds called *rugae* when the stomach is empty. These folds smooth out when the stomach is stretched by filling. Microscopic openings in the mucosa admit glandular secretions into the stomach that aid in digestion.

Functions of the Stomach

Functions of the stomach include:

1. storage of masticated food,
2. mixing food materials,
3. controlling the release of food into the small intestine,
4. secreting digestive enzymes and acids.

A major function of the stomach is food storage. As food is emptied from the stomach, the muscular wall, being in a state of partial contraction, causes the stomach to shrink. As food enters, however, there is a wave of receptive relaxation of the stomach wall; this permits the stomach volume to increase without increasing the pressure within the stomach. The stomach of a normal adult can accommodate about 1 liter of food and liquids without any increase in pressure. If this 1 liter volume is exceeded drastically, pressure builds up within the stomach.

Another function of the stomach is that of mixing food materials with secretions from the stomach wall and saliva from the mouth. A wave of contractions begins in the fundus area and progresses toward the pyloric valve. The pyloric

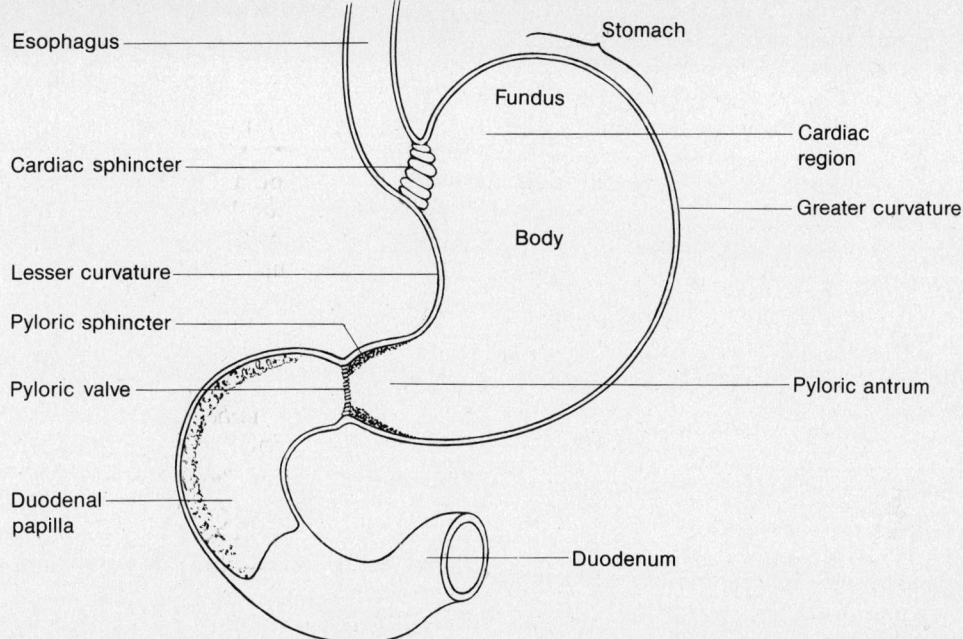

Fig. 13.5 The stomach and adjacent structures. The cardiac sphincter, located between the lower end of the esophagus and the stomach, keeps masticated food from returning to the esophagus. The pyloric sphincter restricts stomach chyme from flowing rapidly into the duodenum.

valve is a very firm structural sphincter and is not easily opened by pressure from the stomach. Food under insufficient pressure to open the pyloric valve is forced back up toward the fundus. This produces a churning motion that mixes the food thoroughly until it becomes a semiliquid called *chyme.* Chyme is a mixture of masticated food, saliva, acids, and enzymes secreted by mucosa cells of the stomach.

The contraction waves of the stomach pick up in intensity as they approach the pyloric sphincter. Eventually sufficient pressure is exerted to overcome the sphincter, allowing food to be emptied into the small intestine. This process is called the *antral* or *pyloric pump.* How much material is allowed to pass the pyloric sphincter is regulated by several factors. The volume of the contents has some effect. The greater the volume within the stomach, the more rapid is the emptying of the stomach. The fluidity of the material apparently has some effect upon how rapidly the stomach is emptied. A liquid meal empties much faster than a solid meal. The type of solid food eaten also affects the speed at which the stomach is emptied. Foods such as fats and proteins slow down the rate of stomach emptying.

Controlled emptying is the term given to the process by which the quantity of chyme released into the duodenum is regulated. The rate of release depends upon factors such as the quantity of foods and liquids in the stomach, and the types of food (protein, fats, and so on). At least two factors regulate the rate of stomach emptying. These are the *enterogastric reflex* and the secretion of the hormone *enterogasterone.*

The enterogastric reflex inhibits gastric contractions when the duodenum becomes overdistended with chyme from the stomach. Overdistension of the duodenum may be caused by large quantities of chyme entering the duodenum or may be due to some obstruction elsewhere in the small intestine. The inhibitory reflex action works this way. In response to stretching (overdistension), receptors in the muscular layer of the duodenum send impulses via the vagus nerve to the medulla. Vagus fibers

Table 13.1 Hormones of Digestion

HORMONE	WHERE FORMED	RELEASE TRIGGERED BY	ACTION
Enterogasterone	Wall of duodenum	Fat	Inhibits gastric motility
Gastrin	Stomach mucosa	Protein, alcohol, caffeine, gastric filling	Accelerates secretions of gastric juice
Secretin	Intestinal mucosa	Acid in the intestine	Stimulates pancreatic juice secretion
Pancreozymin[a]	Intestinal mucosa	Partially digested proteins and fats	Stimulates pancreatic juice secretion
Cholecystokinin	Intestinal mucosa	Fats, acids, partially digested proteins	Contraction of gallbladder, forcing bile into duodenum

[a]It has been determined that pancreozymin and cholecystokinin are the same chemical substance.

carry impulses to the stomach muscle that inhibit gastric contractions.

The type of substance entering the duodenum may also stimulate nerves, thus initiating the enterogastric reflex. These substances include protein fragments (dipeptides, polypeptides), fats, and excess stomach acids. Proteins and fats are thus permitted to enter the small intestine in small quantities so that digestion can be more complete. In the case of excess acid, some enzymes of the intestine function best in a slightly alkaline environment. The inhibited stomach action slows down the rate at which the acid chyme is admitted so that alkaline bile and pancreatic juice can neutralize it.

The hormone *enterogasterone,* described in Table 13.1, also aids in controlling stomach emptying. In response to fats emptied into the duodenum, secretory cells of the walls of the duodenum release enterogasterone. This hormone inhibits gastric motility, and thus slows down the release of fat into the duodenum.

Secretions of the Stomach

The term *gastric juice* incorporates all stomach secretions (gastric secretions). Gastric juice is made up of hydrochloric acid, mucus, water, salts, and digestive enzymes. The following discussion will pertain specifically to enzyme activity. The role of enzymes is to aid in chemical action necessary to degrade large molecules such as proteins to their component, and smaller, molecules. Small molecules may be absorbed in the intestines whereas large molecules are not.

Gastric juice is highly acid due to the secretion of hydrochloric acid (HC1) by *parietal cells* of the stomach fundic mucosa. The initial pH of freshly secreted HC1 is about pH 0.8. Food in the stomach along with mucus and fluids dilutes this to about pH 3.5.

Chief cells of the fundic portion of the stomach secrete *pepsinogen,* which will be converted to the enzyme *pepsin* when it mixes with hydrochloric acid. Pepsin aids in the chemical breakdown of proteins. Pepsin can only function in a highly acidic environment. Protein molecules under the influence of pepsin are converted to peptides. These peptides will later be acted upon by intestinal enzymes to yield amino acids, as shown in Figure 13.6.

What is occurring in the stomach with these secretions? The stomach, a muscular sac made up of protein, has specialized cells secreting protein-digesting enzymes along with very strong hydrochloric acid. An analogy would be starting a wood fire in a stove completely made of wood. What prevents the stomach secretions from digesting the stomach wall?

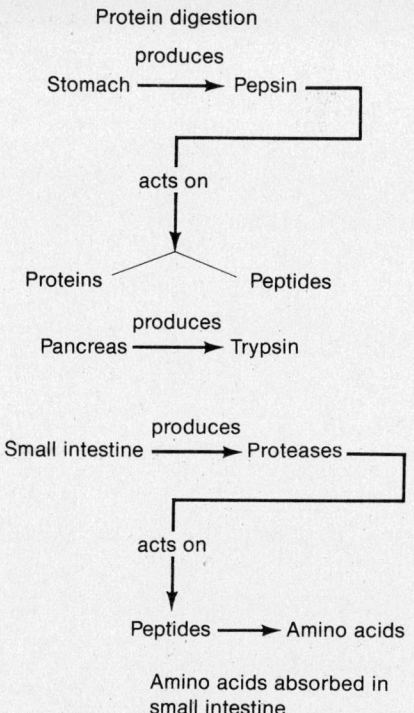

Fig. 13.6 The chemical processes involved in protein digestion. The initial chemical action on proteins is that of pepsin. Smaller molecules called peptides, formed as a result of pepsin activity, are acted upon by trypsin and proteases, yielding amino acids. Amino acids are then absorbed into the bloodstream from the small intestine.

Ulcer: An Ancient Distress Even More Prevalent in Modern Society

Ulcers have been written about since the seventh century A.D. Then, as today, part of the treatment was abstention from acidic foods and alcohol.

Ulcers afflict between 10 and 15 percent of the population. They appear more frequently in the duodenum than other parts of the digestive tract and afflict more men than women. A causal relationship appears to exist between the incidence of ulcers and society-induced stress. Such stress may include family tension, climbing the executive ladder at work, or simply dissatisfaction due to the apparent incapability to meet certain life goals, however unrealistic they may be.

The roles of the pituitary gland, adrenal glands, and the nervous system are all currently questioned in the cause of ulcers. The role of emotion, which incorporates the previously mentioned glands and systems, cannot in itself be blamed as the cause of ulcers. But neither can it be considered blameless. Other systems perhaps suffer even more from emotional upset. One need only consider other influences such as hypertension, skin rashes, headaches, and suicidal tendencies. Modern societies will no doubt be faced with ulcers for the immediate future, and perhaps far into the future.

In a normal functioning digestive system, the protective mechanisms of dilution of the hydrochloric acid and the mucus covering within the stomach are sufficient to protect the stomach muscles from digestion. If either or both of these mechanisms fail, the stomach wall is digested. The irritated and eroded portion of the stomach wall is called an *ulcer*. Ulceration may continue until the capillary walls, small arteries, and veins are eroded. This produces what is called a *bleeding ulcer*. If the erosion continues, a hole is eaten through the stomach wall. The stomach contents then are thrown into the abdominal cavity, causing serious infections. This is called a *perforated ulcer*.

Persons with ulcers are usually put on a bland diet, since increasing the acid content of food increases the irritation of the eroded lining. The imbalance of secretions apparently causes the ulcer formation, or at least the initiation of ulcers. This is due to nervous control in most instances. Persons who are hypersensitive, have a tendency to worry, and become easily overexcited tend to be more susceptible to ulcers than are more complacent individuals.

The amount of pepsin secreted by the stomach is controlled by a feedback mechanism initiated by the cells of the duodenum. Protein and peptide molecules entering the duodenum stimulate certain cells to secrete a hormone called *gastrin*. Gastrin is carried in the bloodstream back to the fundic portion of the stomach, stimulating these cells to produce more pepsinogen, which is then converted to pepsin upon contact with hydrochloric acid. This induced secretion can also be brought about by the introduction of certain meat extracts and amino acids and dilute alcohol. This may be as good a reason as any to have a glass of wine with meals.

Gastric secretion is not maintained at a contant high level. It tends to fluctuate depending upon several factors. Some of these factors include the taste, sight, or smell of food. Probably most important is the physical presence of food in the stomach or the intestine. Even during minor fasts or periods of nondigestion, a small amount of gastric secretion is maintained. Several factors tend to inhibit gastric secretion. Among these are a pH of less than 3.5, the presence of fat in the stomach or duodenum, or the presence of bile salts in the duodenum. The antral portion of the stomach secretes alkaline substances that neutralize the hydrochloric acid. If this secretion is in large amounts, it may produce a very high pH, and this in turn may have an inhibitory influence upon gastric secretion.

An infant's stomach does not produce very large amounts of pepsin and hydrochloric acid. The normal diet of infants is milk, which is acted upon by another enzyme present in the stomach called *rennin*. The function of rennin apparently is to coagulate the casein substance in milk to form a curd, which permits easier digestion of milk products and also aids in retaining the milk in the stomach and intestines for a longer period of time so that digestion may be more complete. With age and diet change, more hydrochloric acid is produced in the stomach and this alone is sufficient to curdle the milk and produce the same benefits in the adult that rennin may produce in the infant.

Vomiting

Vomiting is the forceful expulsion of gastric and/or intestinal contents from the mouth. It is a reflex brought about by several factors that stimulate vomiting (emetic) centers of the medulla. Some of the factors that will bring about vomiting are severe distension of the stomach, duo-

---

What Is Not Known About Stress

Hans Selye, of the University of Montreal, has defined stress as "sort of an omnipresent reaction to almost any stimulus." He has proposed that a certain stress could trigger any number of diseases in one person and be harmless to another. There seems to be little doubt among researchers that psychosomatic (stress-related) illnesses are very real and very damaging. However a problem arises when we try to relate a specific stress to a specific disease. For example:

1. *Who becomes ill under stress while others under a similar stress remain healthy?*
2. *What does stress do to the nervous system to specifically induce hypertension?*
3. *What types of stress are associated with gastrointestinal ulcers or hypertension?*
4. *What socioeconomic factors induce types of stress that are potentially life shortening?*

These questions and many others have been researched for years. We do not at the present have pat answers for any of them. As one researcher has stated, "Every one of the hypotheses which attempts to define a close relationship between a disease and a specific stress is contradicted by as much evidence as exists for its support."

---

denum, or the throat; impulses arriving from the inner ear, bladder, or uterus; or severe pain elsewhere in the body. Some of the activities associated with vomiting are nausea, increased salivation, increased mucus secretion, distension of the esophagus, and a relaxation of the cardiac sphincter. As pointed out earlier, materials that return from the stomach back into the esophagus contain harmful substances that may damage the lining of the esophagus. A condition we are all familiar with, *heartburn*, is caused by the return of irritating substances into the esophagus.

The lower portion of the esophagus, near the stomach, has a layer of mucus-secreting cells. This thin layer of mucus protects this part of the esophagus. However, the upper portion does not contain these protective secretory cells, and therefore the material from the stomach containing acids and protein-digesting enzymes may erode the walls of the upper esophagus. The

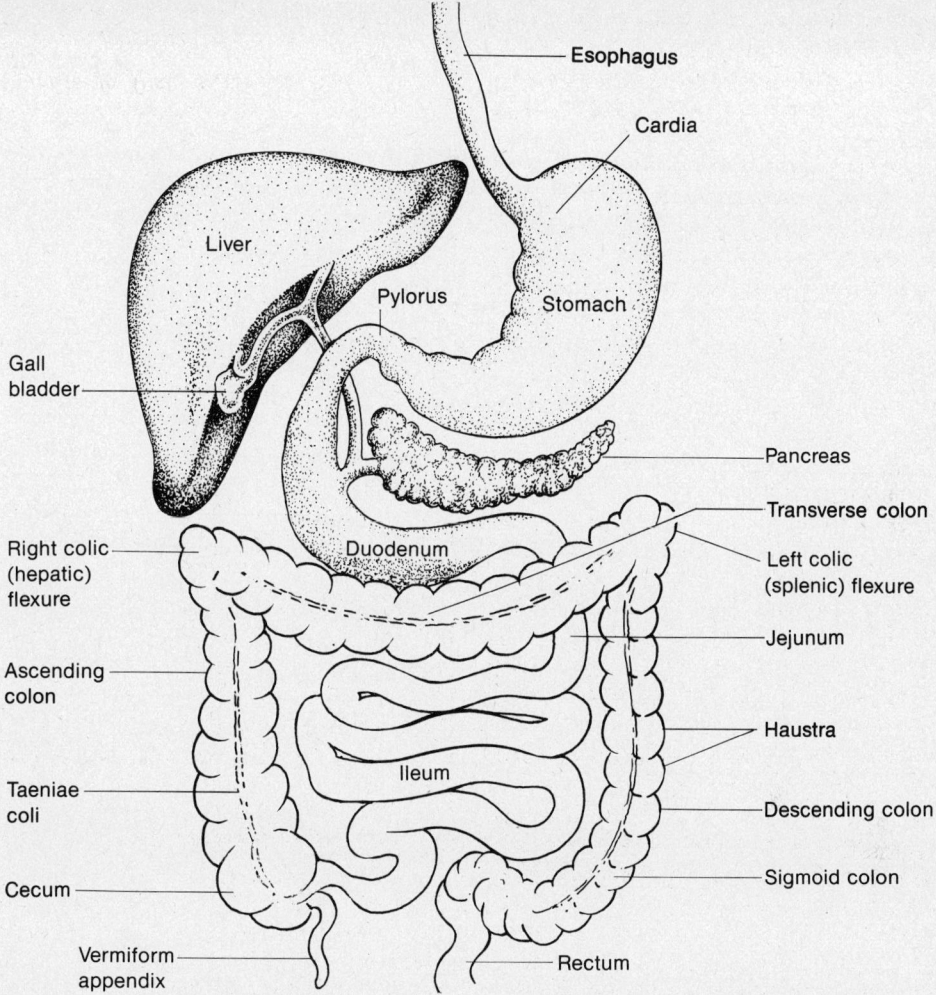

Fig. 13.7 The small intestine. The first 25 to 30 cm of the small intestine are called the duodenum. The next 100 to 150 cm are the jejunum. The remaining 2.5 to 3 meters are called the ileum. This figure also shows the anatomical relationship of the small intestine and the colon.

increased salivation that normally precedes vomiting would appear to be an attempt by the body to provide a diluting substance for these potentially harmful materials.

ANATOMY OF THE SMALL INTESTINE

The small intestine begins at the pyloric sphincter and extends about 4 m to the colon. The opening of the small intestine into the colon is guarded by a sphincter called the *ileocecal valve*.

The small intestine is normally divided into segments. The first 25 to 30 cm are called the *duodenum*. The next 100 to 150 cm are the *jejunum*, and the remaining length is the *ileum*, as shown in Figure 13.7.

The lining of the intestine is thrown into small folds, which in turn are extensively covered with very numerous microscopic projections called *villi*. Smaller projections on villi are called *microvilli*. The folds of villi and microvilli increase the surface area of the small intestine, aiding in increasing the rate of absorption of nutrients.

The duodenum is the segment of the small intestine into which *bile* from the liver and *pancreatic juice* from the pancreas are emptied.

Secretions of the Small Intestine

Digestion of food within the gastrointestinal tract occurs mainly in the small intestine. The small intestine is also responsible for the absorption of food nutrients into the bloodstream. *Villi* provide an extensive surface area for the absorption of nutrients. Specialized cells line the villi that absorb materials from the *lumen* (the inside of the intestine), which are then picked up by capillaries.

Many digestive enzymes are supplied by accessory organs such as the pancreas and liver. However, the intestinal lining itself has specialized cells that secrete some enzymes. Duodenal cells grouped into clusters called *Brunner's glands* secrete mucus. Mucus facilitates the movement of chyme through the intestine. Scattered throughout the intestine are other cells called *crypts of Lieberkuhn* that secrete copious amounts of a watery substance similar in content to interstitial cell fluid.

As much as 3,000 ml of this fluid per day are secreted. This secretion aids in the movement of the chyme through the intestine by making it more fluid. It also aids in the absorption of nutrients by the villi of the intestines. Most of this secreted substance is reabsorbed by the intestines along with food nutrients. That which is not absorbed here is later absorbed by the large intestine. This fluid also contains some weak enzymes that help digest proteins and sugars. However, the major digestive enzymes are secreted by cells within the intestinal wall or by the villi, which provide digestive enzymes just before the substances are absorbed by the intestine. Cells in the intestinal mucosa secrete a weak amylase for starch digestion. Other cells secrete the enzyme *enterokinase*, whose function we will describe in a moment.

For our discussion of small intestine activity to be complete, we must describe another accessory organ, the *pancreas*.

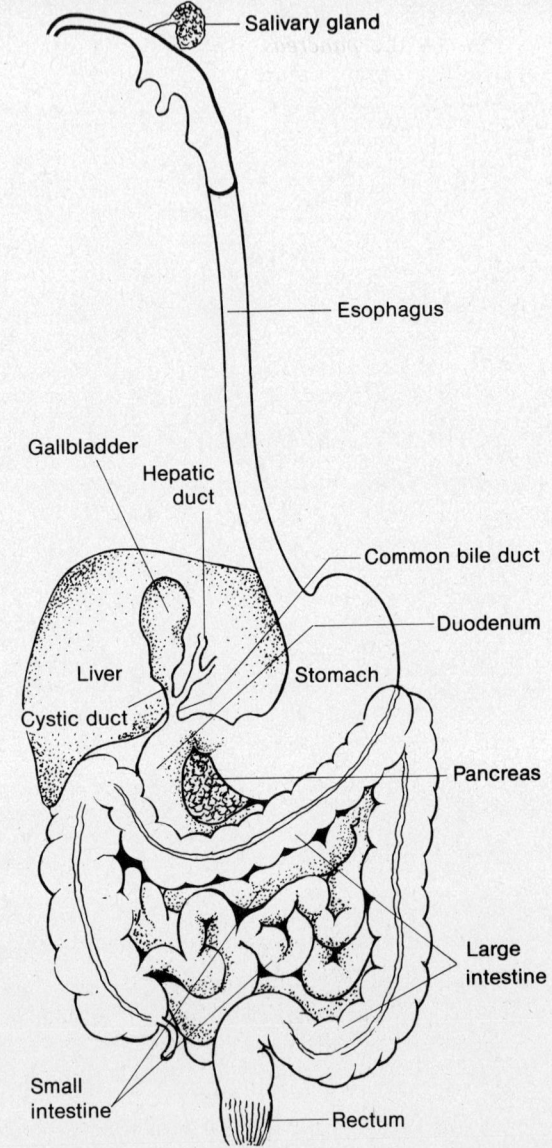

Fig. 13.8 The digestive system showing the location of the pancreas and liver. Much of the chemical activity involved with digestion is due to the action of enzymes secreted by the pancreas and transported to the duodenum by the pancreatic duct.

The Pancreas

Much of the chemical activity involved with digestion is due to the action of enzymes secreted by the *pancreas*, shown in Figure 13.8. Digestive enzymes secreted by the pancreas are carried to the duodenum by the *pancreatic duct*. The pancreas is made up of specialized cells, some having endocrine function, others having exocrine function.

The endocrine function of the pancreas was discussed along with hormones, in Chapter 12. Exocrine function of the pancreas is primarily limited to the production and secretion of digestive enzymes. The basic difference between endocrine and exocrine glands is the way in which the secretions are distributed to different parts of the body. Endocrine glands are ductless and secrete their products (hormones) directly into the bloodstream, where they are then distributed to all parts of the body. Exocrine glands are connected by ducts or tubes leading to specific body organs.

The pancreas secretes at least three essential enzymes. One is *lipase,* which aids in the splitting of the fat molecule into fatty acids and glycerole. Another enzyme, *amylase,* aids in the digestion of complex sugars, converting them to simple sugars such as glucose, galactose, or fructose. *Trypsin* is an enzyme that aids in the conversion of proteins or polypeptides to amino acids.

In addition to enzymes, the pancreas secretes a thin, alkaline fluid, which, together with the enzymes just mentioned, makes up *pancreatic juice.* The alkalinity of this fluid neutralizes the acid content of chyme emptied into the duodenum. The amount of secretion from the pancreas is controlled both by nerves and hormones. The hormone *pancreozymin* is secreted by intestinal mucosa cells in response to acid in the intestine. This hormone stimulates the secretion of pancreatic juice.

Acid substances emptied from the stomach into the duodenum stimulate certain duodenal cells to secrete a hormone called *secretin.* This hormone, carried by the blood to the pancreas, stimulates the pancreas to produce more pancreatic juice. In other words, the pancreas does not secrete large quantities of fluid continually; only when food along with acid is emptied into the intestine is it brought into greater activity.

Table 13.2 Enzymes of Digestion

ENZYME	WHERE FORMED	ACTION
Salivary amylase	Saliva	Converts starches to maltose
Pepsin	Stomach mucosa	Converts proteins to dipetides
Rennin	Stomach mucosa	Coagulate milk for slower emptying (action appears to function only in infants)
Trypsin	Pancreas	Digests proteins into polypeptides
Chymotrypsin	Pancreas	Same as trypsin
Carboxy-peptidase	Pancreas	Digests proteins to amino acids
Pancreatic amylase	Pancreas	Digests polysaccharides to disaccharides
Pancreatic lipase	Pancreas	Digests fats to glycerol and fatty acids
Dipeptidase	Intestinal mucosa	Digests dipeptides to amino acids

Protein Digestion in the Small Intestine

The stomach begins the digestion of proteins with its secretion of pepsin. However, this does not completely break down all proteins to the amino acid level. Also, some proteins in the chyme escape even partial digestion and pass whole into the duodenum. The enzyme trypsin is capable of digesting whole proteins or fragments of these protein molecules, eventually breaking them down to amino acids. Trypsin is not secreted in its active form by the pancreas, but rather in an inactive storage form called *trypsinogen*.

Trypsinogen, upon contact with some trypsin in the duodenum plus the acid content of the chyme, is converted to trypsin. Both trypsin from the pancreas and pepsin from the stomach are called *protein-hydrolyzing enzymes*. A hydrolyzing enzyme is one that aids in converting large molecules such as proteins to smaller amino acid molecules. The process of *hydrolysis* is the name given to the digestion of carbohydrates, fats, and proteins by hydrolyzing enzymes. Other enzymes secreted by the pancreas include pancreatic lipase, pancreatic amylase, chymotrypsin, and carboxypeptidase.

Pancreatic lipase hydrolyzes fats to glycerol and fatty acids that can be directly absorbed into the bloodstream from the intestinal villi.

Pancreatic amylase continues the job begun by salivary amylase—the conversion of starch to maltose. Maltose cannot be absorbed directly, it must be further altered, as will be seen.

Chymotrypsin, converted from chymotrypsinogen (an inactive storage form) in the intestines, acts upon proteins, hydrolyzing them to the polypeptides level.

Carboxypeptidases are enzymes that hydrolyze polypeptides to shorter chains, dipeptides. Dipeptidases finish the job of protein digestion, yielding amino acids.

Fat Digestion in the Small Intestine

Pancreatic lipase, a component of pancreatic juice, is a fat-hydrolyzing enzyme. In order for pancreatic lipase to do an efficient job, the surface area of the fatty substance must be increased. The greater the surface area the faster the enzyme can work. This is where bile from the liver goes to work, as shown in Figure 13.9.

Bile is secreted by the liver and stored in the gallbladder. Note in Figure 13.15 the location of the gallbladder and the common bile duct, which leads from the gallbladder to the duodenum. Bile is a mixture of several substances including *bile salts* (the only substances important in digestion), *cholesterol,* and *bilirubin*. Bilirubin is a bile pigment, largely a waste

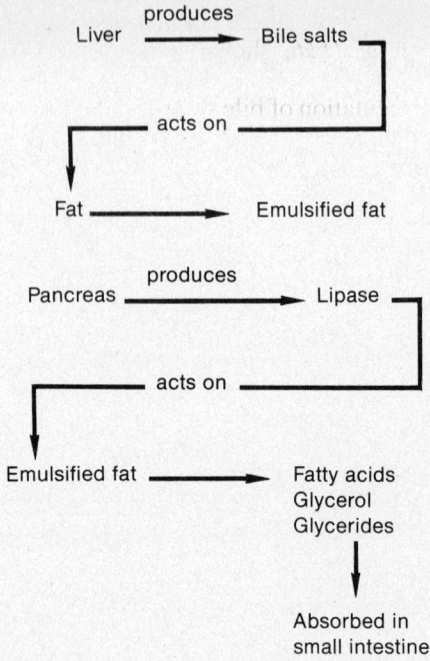

Fig. 13.9 The chemical process involved in fat digestion. Bile emulsifies fat, thus increasing the surface area of fat particles so that digestive enzymes (lyase) can degrade large fat molecules to organic acids and glycerol.

product of red blood cell destruction. Bile also contains fatty acids and electrolytes in about the same proportion as in plasma.

Bile salts do not actually digest food, since they are not enzymes. However, they are important in the digestion of fats because of their detergent action, which lowers the surface tension between water and fats and helps form emulsions. Emulsification increases the surface area of fats for faster action of lipase on them. Bile salts also aid in the absorption of fats by making them more soluble and therefore more easily absorbed.

Bile secreted by the liver is stored in the gallbladder. It is not, however, emptied into the duodenum continuously. The presence of fat in the intestine stimulates cells in the mucous layer of the duodenum to secrete the hormone *cholecystokinin*. This substance, carried by the blood, stimulates the gallbladder to contract, forcing bile into the common bile duct leading to the duodenum. Cyolecystokinin also causes the *sphincter of Oddi,* shown in Figure 13.15, to relax.

The regulation of bile secretions represents an almost classical example of a self-regulating system.

In a 24-hour period, 700 to 1,000 ml of bile may be emptied into the duodenum. Only about 20 percent of this amount is utilized in the digestive process. The remaining 80 percent is reabsorbed and returned to the liver. The reabsorbed bile has a great stimulating effect upon the liver to secrete more bile. As you might imagine, if this procedure went unchecked, large quantities of bile would be continuously secreted. After a meal containing fats is digested, the duodenal cells are no longer stimulated to secrete cholesytokinin, and therefore, the gallbladder is not stimulated to contract. With less bile entering the duodenum less is reabsorbed, and therefore the liver is not stimulated to secrete more bile. This regulation of bile secretion is illustrated in Figure 13.10.

Some of the other substances in bile, such as cholesterol, fatty acids, and electrolytes, are absorbed by the intestines. However, the bile pigment bilirubin is eliminated as a waste product, which gives feces a characteristic dark color.

In some individuals, bile may be so concentrated in the gallbladder that cholesterol crystallizes into small stonelike masses called *gallstones.* These may become so large that they block the common bile duct, restricting the flow of bile to the duodenum and restricting the flow of pancreatic juice from the pancreas. The blockage of the common bile duct creates digestive problems for the individual and is also quite painful. Gallstones have to be surgically removed if they continue to block the secretory duct.

Absorption in the Intestines

The absorbing structures of the intestine, the villi, are shown in Figure 13.11. The smaller projections on the villi themselves, microvilli, are shown in Figure 13.12. Villi are found nowhere else in the digestive tract. These structures are lined with specialized cells, some that secrete mucus, others that contain digestive enzymes, and still others that absorb materials previously digested elsewhere in the digestive tract.

Substances such as maltose, lactose (milk sugar), and sucrose cannot be absorbed due to their size and/or their molecular structure. Columnar cells lining each villus secrete enzymes to hydrolyze these substances to the monosaccharide level. Figure 13.13 illustrates carbohydrate digestion.

Monosaccharides now may be absorbed into blood vessels in each villus. Some columnar cells secrete enzymes called *dipeptidases,* which convert dipeptides to amino acids, which may then be absorbed.

Glycerol and fatty acids from the hydrolysis of fats are also absorbed by the villi into the bloodstream.

The extensive length of the small intestine along with the surface area created by the villi represents a vast absorbing area for substances

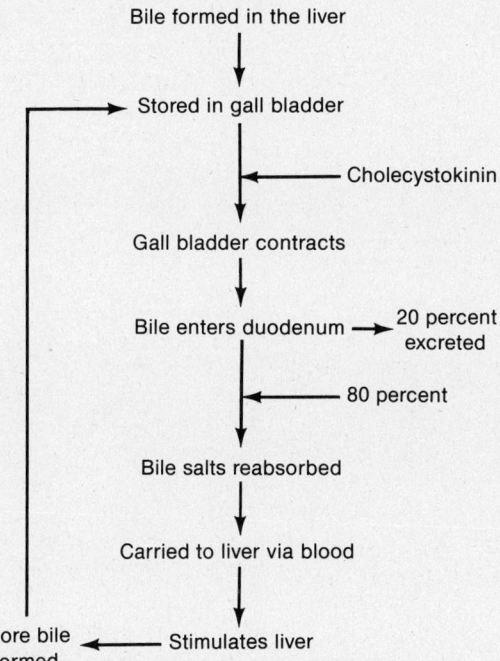

Fig. 13.10 Summary of bile secretion. The hormone cholecystokinin is secreted in response to fat entering the duodenum. This initiates a greater flow of bile from the gall bladder. The process comes to a halt if no fat enters the duodenum.

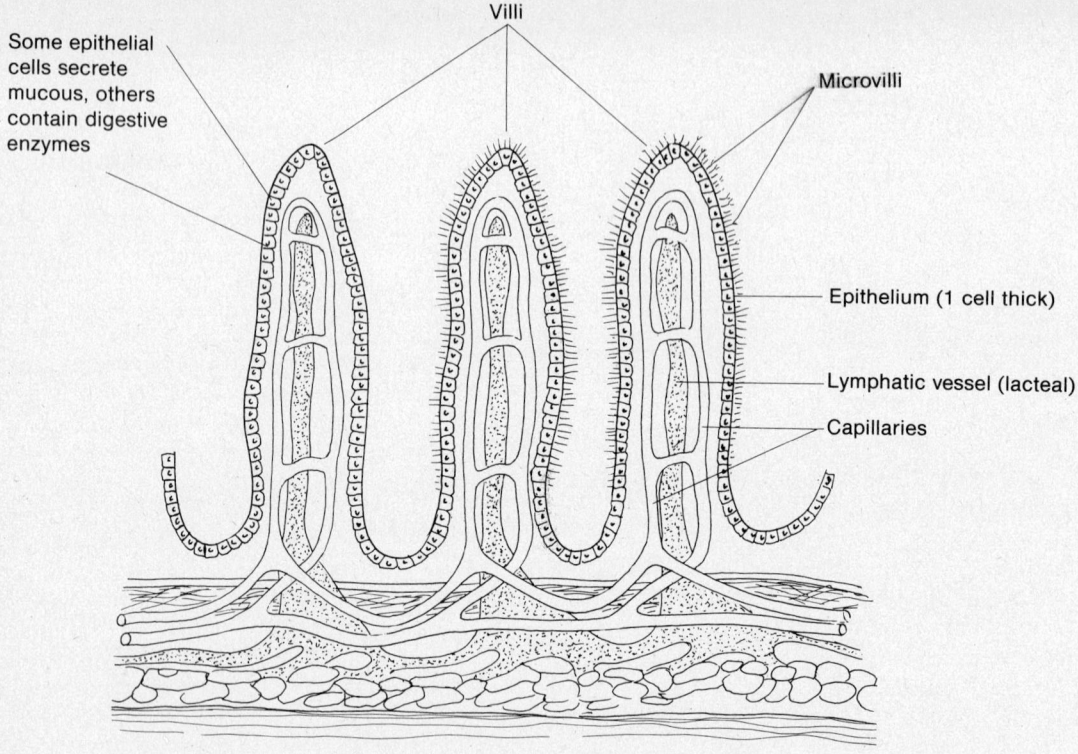

Fig. 13.11 Villi, the absorbing structures of the small intestine. These structures, found nowhere else in the gastrointestinal tract, are specialized to absorb nutrients into blood capillaries.

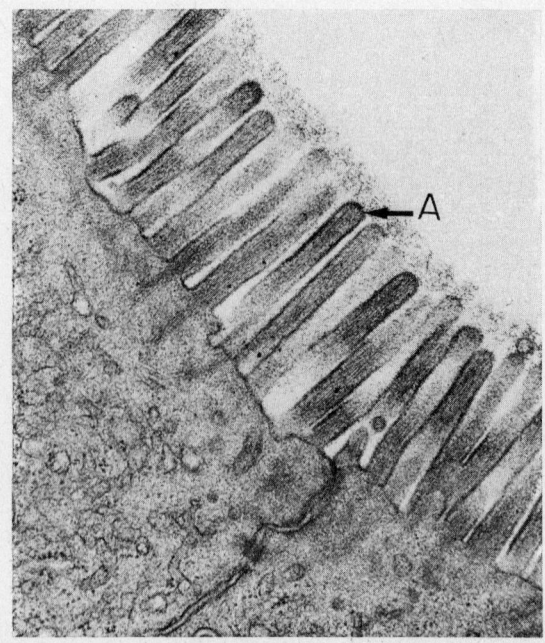

Fig. 13.12 Microvilli shown at *(A)* are extremely small projections extending from certain cells. Shown here is an intestinal cell. Microvilli increase the surface area of cells, which increases their absorbing capacity. Microvilli are too small to be seen with an ordinary light microscope. The view shown here is an electron micrograph. *(Courtesy of Santiago Plurad.)*

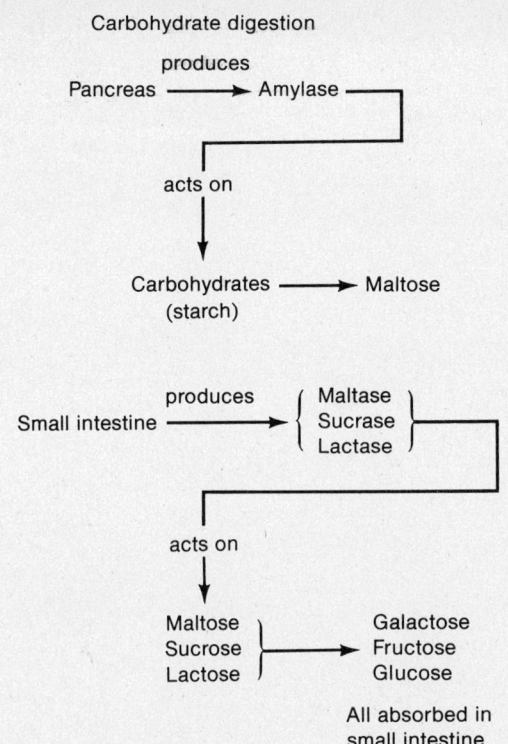

Fig. 13.13 Processes involved in carbohydrate digestion. Enzymes secreted by specialized cells of the small intestine facilitate the conversion of disaccharides (maltose, sucrose, and lactose) to monosaccharides (galactose, fructose, and glucose).

required by body tissues. All these substances, except for some fat molecules absorbed by the lacteals, are transported to the liver. Capillaries in each villus converge to form the *hepatic portal vein*, which carries its nutrient-laden blood to the liver. The term "portal" is applied to any vessel that connects capillary beds. The hepatic portal vein connects the capillaries of the intestinal villi with liver capillaries. What the liver does with these nutrients is discussed later. For now, the focus of the discussion is the small intestine.

Substances are moved through the intestines by peristalsis toward the *ileocecal junction*. At this junction the small intestine empties its contents, mostly waste material and water, into the large intestine. The *ileocecal valve* regulates the flow of materials into the large intestine. The ileocecal valve permits material to enter the large intestine. However, it resists movement back into the ileum. The distal end of the ileum, immediately in front of the ileocecal valve, is wrapped with a muscular sphincter, the *ileocecal sphincter*. This sphincter controls the rate of discharge of materials from the ileum into the large intestine. A reduced rate of flow of material from the ileum permits more time for absorption.

Neural control of the sphincter is in the form of reflexes. When the cecum is overdistended, the sphincter contracts more forcibly, thus reducing the flow. The controlling reflexes are also set in motion whenever the cecum is irritated, such as in appendicitis.

ANATOMY OF THE LARGE INTESTINE

The large intestine is about 1.3 m long and 6 cm in diameter. It extends from the ileocecal valve to the anus. Segments of the large intestine include the cecum, vermiform appendix, colon, and the rectum, as shown in Figure 13.14.

The *cecum* is a blind pouch about 5.6 cm long adjacent to and below the ileocecal valve. The *vermiform appendix* is a small wormlike appendage 2 to 3 cm long found on the cecum. The *colon* constitutes the major mass of the large intestine. It is customarily divided into four segments—the *ascending, transverse, descending,* and *sigmoid colon.*

The ascending colon runs anteriorly from the cecum to just under the liver where it makes a turn and runs horizontally across the upper abdominal cavity, forming the transverse colon. The angle formed by this turn is called the *right* or *hepatic flexure.*

The transverse colon crosses the body to the left side, turning downward near the spleen, forming the *left* or *splenic flexure.*

The descending colon progresses posteriorly to the upper level of the pelvis where it turns toward the midline of the body to become the sigmoid colon. The sigmoid (S-shaped) colon extends to the rectum.

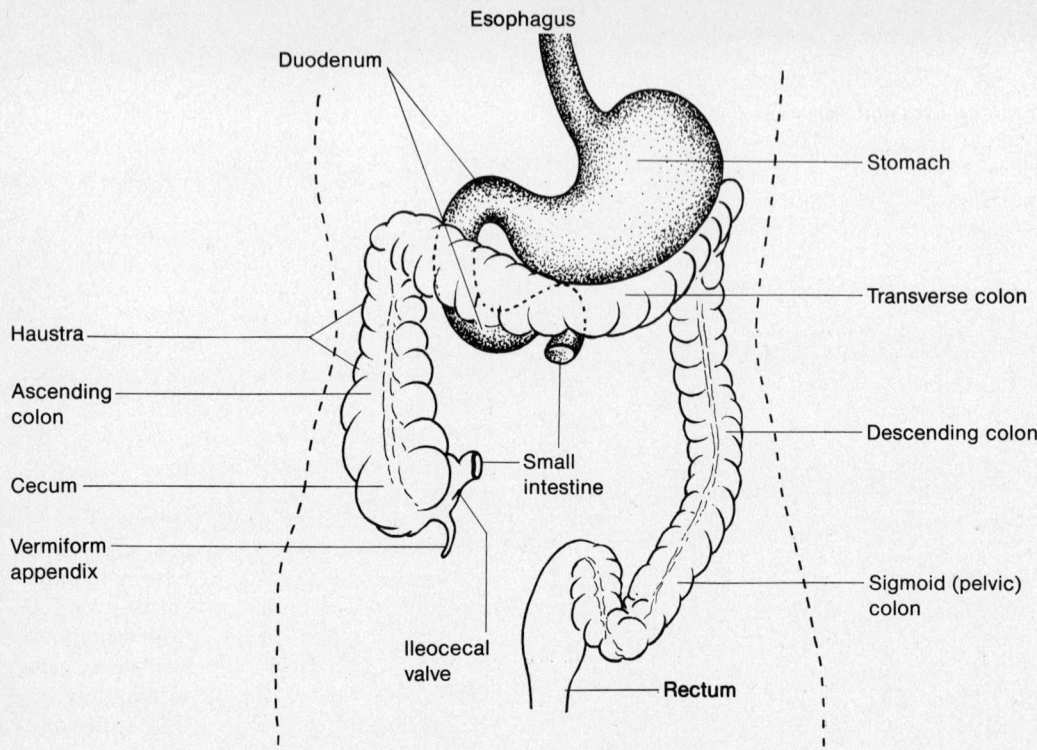

Fig. 13.14 The colon, its parts and position relative to the stomach and small intestine. Colon function is largely one of water absorption, temporary storage of undigested matter, and the elimination of wastes from the body.

Colon function is largely one of water absorption and temporary storage of undigested matter. Within the colon are segments called *haustra*, shown in Figure 13.14. These are pouchlike structures that inhibit the rapid movement of materials through the colon. Sluggish peristaltic contractions push the waste, now called *feces* or *fecal matter*, along the colon toward the anus. Circular bands of muscle around the colon contract, producing the haustra. The fluid fecal matter is thus squeezed and churned, mixing it and permitting more time for water and electrolyte absorption.

The body has a great need to conserve water thrown into the tract for the digestive process. When certain body malfunctons do not allow adequate reabsorption of water from fecal matter, the feces appear watery. This condition is called *diarrhea*. If excessive amounts of water are removed from the feces, *constipation* results. Large numbers of bacteria are normally found in the colon. These are not harmful except under extraordinary circumstances.

Peristalsis occurs in the colon; however, it is slower than that found in the small intestine. Substances carried by the colon move in one direction, from the ileocecal valve toward the anus. The rate of peristaltic contractions in the colon can be increased by food intake or by an overdistended duodenum or ileum.

Fecal matter eliminated from the body is composed largely of bacteria, mucus, bile salts, and undigested food particles such as cellulose. Excessively light or dark colored feces may be an indication of some body malfunction and can be useful in certain diagnosis. The distal end of the colon is called the rectum. The valve keeping the fecal matter within the rectum and the colon is the *anal sphincter*. Bowel movement (defecation)

> ### The Helpful Nature of Fiber in Nutrition
>
> Dr. Denis P. Burkitt, a British surgeon and epidemiologist, lived and worked for 20 years in rural Africa. He observed that the fiber diet of Africans appeared to reduce the incidence of appendicitis, hemorrhoids, diverticula, and cancer of the colon. When diets were altered, such as when an individual moved to an urban area, the incidence of these diseases increased. Westerners in Africa having different diets were also observed. Those on the rural, high-fiber diet appeared to have less gastrointestinal difficulties.
>
> Fiber is a plant product, mostly cellulose, that cannot be digested by humans. Good sources are whole grains, nuts, fruits, and vegetables. The benefit of fiber is its ability to absorb water, which adds bulk to the food eaten. More bulk, a portion of which is indigestible, helps speed the mass of materials through the digestive tract. This helps to reduce the time potentially harmful concentrates of bile salts, fermenting food stuffs, and bacteria remain in the tract.

is the result of a reflex action. Since babies have not learned the social niceties that adults have attained, the defecation reflex is sufficient to empty the bowel. In normal adults, the muscles of the anal sphincter can be voluntarily contracted and delay bowel movement.

The secretory function of the colon is limited primarily to that of mucus secretion. Mucus provides lubrication so that fecal matter can easily move through the colon. Mucus also may serve as a physical barrier between the epithelial lining of the colon and the feces.

THE LIVER

The *liver*, shown in Figure 13.15, is located in the upper portion of the abdominal cavity immediately beneath the muscular diaphragm. The largest part of the liver is concentrated on the right side of the cavity and constitutes the largest mass of glandular tissue in the body.

Functions of the liver are so diverse and involve such precise coordination with so many other body systems that it is difficult to present an overall picture of these functions.

For convenience of study we may divide liver functions into the following categories:

1. Filtration, storage, and detoxification of blood.
2. Metabolic activities involving carbohydrates, proteins, fats, and vitamins.
3. Phagocytosis.
4. Bile formation and secretion.
5. Formation of urea and uric acid.

Filtration, storage, and detoxification of blood are functions that are critical for body cells since they all depend upon a blood supply free of harmful substances. The function of the liver as a blood storage unit helps maintain a constant and adequate flow of blood throughout the body.

Blood enters the liver from the *hepatic portal vein*, which carries blood laden with nutrients from the small intestine and from hepatic arteries, which carry nourishment to liver cells.

Blood leaves the liver by the *hepatic veins*, which empty into the inferior vena cava. A defective heart incapable of adequately receiving blood from the inferior and superior vena cava may cause blood from the liver to be dammed up, creating a back pressure in the hepatic veins. When this occurs, liver sinuses become engorged with blood, causing the liver to swell. This storage function at least partially relieves pressures upon an already overtaxed heart.

The basic unit of the liver is the *lobule*. Single lobules vary from less than 1 to 2 mm in diameter and may be several millimeters long. There are an estimated 80,000 to 100,000 individual lobules in the liver. The intricate meshwork of cells, sinuses, arteries, and veins make the lobule an effective blood filter for trapping particulate matter.

Detoxification results in the removal or chemical alteration of substances that may be harmful to other body cells. It is not surprising that the liver is one of the first organs severely damaged by poisonous substances intentionally or unintentionally taken into the body. Some harmful by-products of metabolism are also removed or

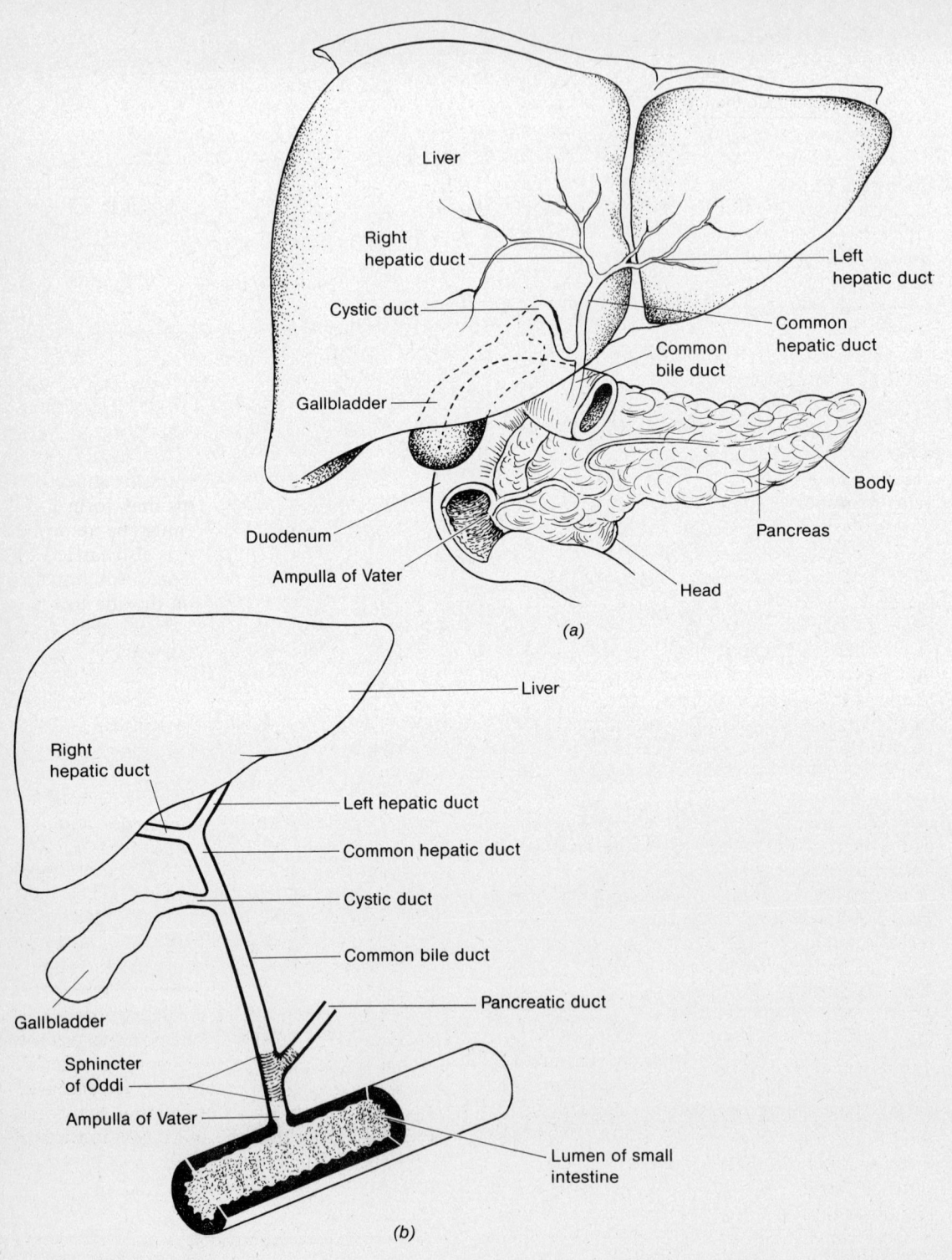

Fig. 13.15 The liver and associated viscera. (a) A representation of these structures and their normal locations. (b) The liver ducts and gallbladder and their physical relationship to the liver and duodenum.

made less toxic as will be seen in a moment in the discussion of protein metabolism.

Carbohydrate metabolism in the liver involves the conversion of glucose into glycogen (for storage), the synthesis of glucose from amino acids, and the synthesis of glucose from galactose and fructose.

Blood arriving at the liver from the hepatic portal vein is loaded with nutrients (glucose, amino acids, fatty acids, and so on) processed and absorbed from the small intestine. The liver serves as a storehouse for these substances, which arrive in large quantities after a meal is digested, and releases nutrients into the hepatic veins in response to body needs. For example, during increased physical activity, muscle cells require a greater supply of glucose.

After a meal rich in carbohydrates has been digested, the products of digestion are carried from the intestine by way of the hepatic portal vein to the liver. Large quantities of glucose are therefore available for use by the body. Since the level of blood sugar arriving at the tissues must be precisely maintained at a constant level, it is the job of the liver to store the excess glucose. Excess glucose is converted by the liver with the aid of specific enzymes and the hormone insulin into glycogen (animal starch). This process is called *glycogenesis*. The role of insulin is discussed in Chapter 12.

When, due to body demands, blood sugar becomes low, some of the liver glycogen is reconverted to glucose. This process of changing glycogen back to glucose is called *glycogenolysis*. The stimulus for glycogenolysis to occur is low blood sugar. Blood arriving at the liver low in glucose is the signal to the liver to convert the animal starch into the more usable glucose for body tissues.

The role of maintaining a constant glucose level in the blood depends upon several factors, including the degree of physical activity, the state of health of the individual, or certain glandular dysfunctions. If a person is on a very low calorie diet or is fasting, the blood sugar level becomes quite low, and the liver is stimulated to increase the activity of converting glycogen to glucose.

The stimulation of the liver to begin the process of converting glycogen to glucose is a secondary response initiated by a hormone from the pancreas. This hormone, glucagon, is released from the pancreas in response to low blood sugar. Glucagon is discussed in Chapter 12.

The liver responds to glucagon by increasing the rate of conversion of glycogen to glucose.

The pancreas secretes the hormone insulin. Insulin is also instrumental in the conversion and the utilization of glucose in the body.

Protein metabolism in the liver involves the removal of amino groups (NH_2) from amino acids. The process is called *deaminization*. The remaining parts of amino acids (minus the amino groups) are used to synthesize glucose.

The removed amino groups may form toxic substances (ammonia) that must be removed from body fluids. This process is also carried on in the liver in the formation of *urea*. Ammonia is chemically united with carbon dioxide to form urea, which is less toxic to body cells than ammonia and is removed effectively by the kidneys.

The liver also synthesizes *plasma proteins* from amino acids. Plasma proteins are instrumental in regulating the osmotic pressures of body fluid.

Lipid metabolism is initiated in the liver by converting fatty acids into substances required by cells to release energy from nutrients. One of the products is cholesterol, which is a component of bile salts.

PHYSICAL DISORDERS OF THE GASTROINTESTINAL TRACT

The gastrointestinal tract is quite literally an extension of the external body surface. All ingested materials must pass into and through this tubular pathway along with toxins, bacteria, and indigestable substances. The fact that the mouth and anus are in direct contact with the surface environments makes it understandable why there are so many disruptions—chemical, physical, and microbial—along the length of the gastrointestinal tract.

Proper function of the gastrointestinal tract depends upon free movement of materials within it. Any factor that impedes this movement may cause serious consequences. Factors that constrict or block the movement of substances in the tract include improper closing of sphincters, sphincters that do not open adequately, and physical blockage due to cancer, hernia, or ulcers. Physical blockage may also result from muscles not responding to nerve impulses due to disease or damaged nerve centers.

Achalasia is a condition in which the cardiac sphincter, located between the esophagus and the stomach, fails to respond to the wave of relaxation that precedes a bolus of food. Since peristalsis requires nerve-muscle coordination, the cause of the failure of the cardiac sphincter apparently is due to some dysfunction of nerve impulses normally supplied to the lower esophagus.

The esophageal constriction prevents a normal flow of foods into the stomach, resulting in a damming up of food in front of the cardiac sphincter.

Hiatus hernia is a condition in which the upper portion of the stomach protrudes through and above the diaphragm, as shown in Figure 13.16. The opening in the diaphragm through which the esophagus passes normally fits tightly around the esophagus. If this passageway is enlarged, pressures due to stomach wall contraction may cause the upper segment of the stomach to balloon out above the diaphragm.

The added back pressure on the cardiac sphincter may force gastric contents into the esophagus. This highly acidic material may bring about deterioration and ulceration of the lower esophageal lining.

Pyloric stenosis is an early childhood disorder in which the muscular pyloris (a sphincter between the stomach and the duodenum) becomes

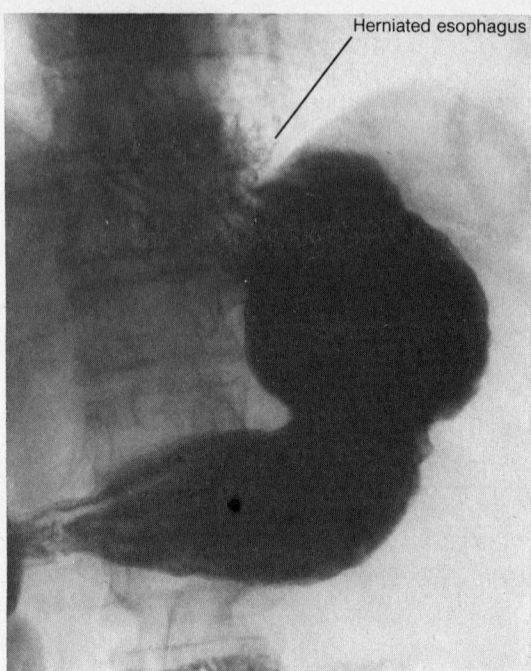

Fig. 13.16 X ray showing a hiatus hernia. The upper segment of the stomach protrudes above the muscular diaphragm. *(Courtesy Radiology Department, Normandy Osteopathic Hospital, St. Louis, Mo.)*

hypertrophied to the extent that the lumen of the pyloris is drastically decreased in diameter. It is generally classified as a congenital disorder. It occurs more commonly in male children than female, and is more prevalent in certain families than others. Pyloric stenosis is a physical obstruction that inhibits the movement of food from the stomach into the duodenum.

Intestinal obstruction may involve a variety of factors that interfere with the normal flow of material through the intestines. The ileum, having the smallest lumen, is more susceptible to blockage than the rest of the small intestine. The colon, with its very large lumen, is much less prone to blockage.

Diverticulosis is the presence of *diverticula*, which are evaginated (out-pocketed) segments of the intestine. Diverticula usually occur along the intestine where pressure is greater and the muscular wall and mucosa are weakened.

Diverticulitis is an inflammation of a diverticulum. Intestinal materials may be forced into the pocket where they are not easily expelled.

leading to fermentation and eventual erosion of the diverticulum lining. The diverticulum may rupture, causing *peritonitis,* an inflammation of the lining of the abdominal wall and viscera of the abdomen.

The treatment for diverticulitis includes antibiotic therapy, diet management, or, in extreme conditions, surgery.

Tumors of the gastrointestinal tract, both *benign* and *malignant,* constitute a major health hazard. Benign tumors, although rare, may cause pain, bleeding, and blockage.

Malignant tumors cause severe tissue destruction. The destruction of tissue is progressive and is usually only halted by surgical removal of the affected part. When a segment of the intestine must be surgically removed, it is often necessary to make an opening in the abdominal wall to permit elimination of fecal matter. The name of this surgical procedure is *colostomy.*

INFECTIOUS DISEASES OF THE DIGESTIVE SYSTEM

The digestive tract and its accompanying accessory organs are particularly prone to microbial infection. The major cause of these infections is inadvertent and often careless ingestion of microbes along with food and drink. Microbial agents include bacteria, viruses, protozoa, fungi, and parasitic worms.

The digestive tract is the natural home for a great number and variety of bacteria. Most of these resident bacteria cause no serious medical problems unless one species gains some advantage over others. An advantage may occur if a certain medication a person takes eliminates or reduces the population of one or more species of bacteria. Other unaffected bacteria take advantage of the lack of competition and multiply rapidly. These opportunistic agents and their toxic wastes may cause intestinal disorders.

Gastroenteritis is an inflammatory disorder of the stomach and intestines. Gastroenteritis may be caused by bacteria, viruses, or chemical agents. Irritation of the mucosa of the stomach and intestines may produce nausea, abdominal cramps, vomiting, and loss of appetite. When vomiting of gastric and/or upper intestinal contents occurs, acid-base balances are disturbed and essential electrolytes may be lost. Food poisoning is a leading cause of gastroenteritis. Table 13.3 lists the agents commonly associated with food poisoning.

Peritonitis is an inflammation of the peritoneum, the membrane that lines the abdominal cavity and covers the viscera. Bacteria released into the abdominal cavity due to a ruptured appendix, ruptured diverticulum, or perforated ulcer, are a common cause of peritonitis.

Ulcerative colitis is ulceration of the colon mucosa and underlying tissues. The lower colon is a common site of infection, although the entire colon may be affected. The specific cause is not known. However, bacteria, viruses, food allergies, and the autoimmune system are all suspected.

Parasitic worms may cause severe intestinal and liver damage. Some parasitic worms, such as tapeworms and roundworms, may reproduce in sufficient numbers to physically block portions of the intestinal tract.

Viral hepatitis is a viral infection of the liver. *Infectious hepatitis* is usually transmitted by fecal-contaminated food or water. Nausea, fever, enlarged and sore liver, and jaundice are all symptoms. *Serum hepatitis* produces similar symptoms, but it is usually transmitted through viral-infected blood used in transfusions or by contaminated hypodermic needles.

Appendicitis is an inflammation of the appendix. The inflammation caused by trapped bacteria may cause the mucosa and underlying tissues of the appendix to rupture. The subsequent contamination of the peritoneum causes peritonitis.

Pancreatitis is an inflammation of the pancreas caused by the rupture of small branches of the pancreatic duct. These ruptures are caused by blockages of the pancreatic duct such as gallstones. The dammed-up pressure bursts the vessels, releasing proteolytic (protein-digesting) enzymes among pancreas cells. Pancreas tissue is digested, which may set the stage for subsequent microbial infection of the pancreas.

Table 13.3 Food Poisoning

AGENTS CAUSING FOOD POISONING	FOODS THAT MAY BE CONTAMINATED	SYMPTOMS RELATED TO BACTERIAL TOXIN	ONSET OF ILLNESS	DURATION OF SYMPTOMS	SERIOUSNESS OF THE DISEASE
Staphylococcus aureus	Poultry stuffing, ham or chicken salads, dairy products, cream-filled pastries, egg dishes	Nausea, vomiting, diarrhea	1–8 hours	1–2 days	Rarely fatal
Clostridium perfringens	Improperly stored stews and gravies, rare meats	Diarrhea, abdominal cramps	8–24 hours	2 days or less	Rarely fatal
Clostridium botulinum	Improperly canned foods, such as green beans, corn, meats, fish, and mushrooms	Blurred or double vision, respiratory paralysis	12–36 hours	1–10 days	Fatal in approximately one-half of the victims
Ergot (fungal)	Grains contaminated with the fungi	Convulsions, sometimes gangrene of the hands and feet	2–3 days	3–4 days	Rarely fatal

AGING AND THE GASTROINTESTINAL TRACT

The gastrointestinal tract suffers less from the aging process than most other organ systems. There tends to be a progressive loss of numbers of taste buds, and the gag reflex is lessened.

Approximately 10 percent of the population suffers from diverticulosis. These individuals are generally middle-aged and older. Specific causes of diverticulosis are not known. However, diet and the loss of tissue elasticity of the colon are at least implicated.

SUMMARY AND REVIEW

The digestive system includes the mouth, esophagus, stomach, small intestine, large intestine, and accessory organs. The pancreas and liver have an indirect role in digestion, and thus are called accessory organs.

A. Teeth grind and shred foods.
  1. First set of 20 teeth of early childhood are called the milk teeth (deciduous teeth).
  2. Second set of 28 teeth replace the milk teeth by age 11 or 12.
  3. Third molars appear between ages 18 and 25. The final number of adult teeth is 32.
B. Mouth and throat.
  1. The hard and soft palate separate food and air passages.
  2. Three pairs of salivary glands empty into the mouth cavity. These are the parotid, submandibular, and sublingual glands.
  3. The back of the mouth is continuous with the pharynx, which is continuous with the esophagus.
  4. Swallowing is a reflex act.
C. The esophagus is a muscular tube that transports food and water to the stomach.
D. The stomach is a muscular sac that serves as a reservoir and a mixing container where digestive substances are secreted.
  1. The cardiac sphincter, located at the terminal end of the esophagus, keeps food in the stomach.

2. The pyloric valve and sphincter regulate the flow of chyme out of the stomach.
3. Parts of the stomach anatomy include:
   a. fundus
   b. greater curvature
   c. lesser curvature
   d. body
   e. pyloric antrum
   f. pyloric valve and sphincter
4. Functions of the stomach
   a. Mixes food and secretions to produce chyme.
   b. Secretes mucus.
   c. Secretes hydrochloric acid.
   d. Secretes pepsinogen, which is converted to pepsin in the stomach.
5. Protein digestion begins in the stomach by the action of hydrochloric acid and the enzyme pepsin.
   a. Protein molecules are hydrolyzed to polypeptides.
   b. Mucus secreted by the stomach acts as a barrier between digestive substances and stomach tissue.
   c. The hormone gastrin, secreted in response to protein appearing in the stomach, induces a larger secretion of pepsinogen.

E. The small intestine is divided into:
   1. Duodenum, the first 20 to 30 cm just beyond the pyloric valve. It receives secretions from the liver (bile) and pancreas (pancreatic juice).
   2. Jejunum, the next 100 to 150 cm.
   3. Ileum, the remaining length of the intestine, ending at the ileocecal valve.
F. Functions of the small intestine include:
   1. Accepting secretions from the liver and pancreas.
   2. Secreting digestive enzymes (Table 13.2).
   3. By peristalsis, transporting food and water toward the large intestine.
   4. Absorbing nutrients, water, ions, and minerals into the bloodstream.
G. The large intestine is divided into:
   1. cecum and appendix
   2. ascending colon
   3. transverse colon
   4. descending colon
   5. sigmoid colon
   6. rectum

H. Functions of the large intestine include:
   1. Absorption of water and ions from material emptied from the small intestine.
   2. Temporary storage of waste (feces).
   3. Secretion of mucus.
I. The liver is an accessory organ of the digestive system. It:
   1. Produces and secretes bile to aid in fat digestion.
   2. Receives nutrients absorbed from the small intestine via the hepatic portal vein.
   3. Converts glucose to glycogen.
   4. Converts glycogen to glucose.
   5. Detoxifies the blood.
   6. Functions in metabolic activities involving carbohydrates, proteins, fats, and vitamins.
   7. Filters and stores blood.
J. The pancreas is an accessory organ that secretes numerous digestive enzymes (Table 13.2).
   1. Empties pancreatic juice into duodenum.
   2. If the pancreatic duct is blocked by gallstones, a condition called pancreatitis is produced.
K. Physical disorders of the gastrointestinal tract
   1. achalasia
   2. hiatus hernia
   3. pyloric stenosis
   4. intestinal obstruction
   5. diverticulosis
   6. tumors
L. Infectious diseases of the digestive system
   1. gastroenteritis
   2. peritonitis
   3. ulcerative colitis
   4. parasitic worms
   5. viral hepatitis (infectious, serum)
   6. appendicitis
   7. pancreatitis
M. Aging and the gastrointestinal tract. Major changes in the gastrointestinal tract are:
   1. Loss of taste buds.
   2. Gag reflex lessened.
   3. Increase in diverticulosis.

14

Nutrition and Metabolism

MAJOR CONCEPTS
CARBOHYDRATE METABOLISM
   Glycolysis
   The Kreb's Cycle
CALORIC VALUE OF FOODS
FAT METABOLISM
PROTEIN METABOLISM
ENZYMES
VITAMINS
MINERALS
WATER
MEDICAL PROBLEMS RELATED TO NUTRITION
   Phenylketonuria
   Galactosemia
   Kwashiorkor
   Obesity
   Underweight

Nutrition is a process of supplying all the substances required to sustain life. Metabolism is a process that utilizes nutrients. This chapter discusses some of the chemical processes involved with cellular metabolism. The important roles of enzymes, vitamins, minerals, and water are also included.

A nutrient is any substance from which energy is derived by a living organism or any substance essential for normal metabolism. Energy-rich substances include carbohydrates, fats, and proteins. These substances are called foods. Substances that do not release energy include minerals, vitamins, water, and oxygen. This latter group of substances is required in synthesis and decomposition reactions that occur throughout the body, mainly at the cellular level.

Nutrition is the process of supplying all the substances required to sustain life.

CARBOHYDRATE METABOLISM

Glucose, a carbohydrate, is considered energy rich in the sense that the molecule can be systematically degraded, with energy released in small packets. Fats, proteins, and other carbohydrates are all considered energy-rich substances. The simplest molecule we can use to illustrate energy release is glucose.

Glucose, a product of carbohydrate hydrolysis, is transported away from the digestive system and eventually carried to all body cells. When glucose arrives at the cell level, it diffuses out of the capillaries and across the cell membrane into the cytoplasm. Here a variety of enzymes begins to act upon it, initiating the decomposition of the molecules and the subsequent release of energy tied up in chemical bonds.

Glycolysis

Glycolysis is the process by which a glucose molecule ($C_6H_{12}O_6$) is acted upon by enzymes, splitting it into two molecules of pyruvic acid, as shown in Figure 14.1

During this molecule-splitting process high energy electrons are thrown off. Special molecules called *electron acceptors* capture the electrons, momentarily removing some of the energy. Some of this energy is used directly. The rest is used to create a new molecule called adenosine triphosphate (ATP). The electron acceptors operate as a stepwise process, each removing some energy from the electrons. The final electron acceptor, after all the energy has been removed, is an oxygen atom. This oxygen atom combines with hydrogen to form a water molecule.

The cytoplasm contains molecules called *adenosine diphosphate* (ADP) along with some phosphate molecules. If energy is available (from the hydrogen acceptors), ADP will combine with phosphorus, yielding a new molecule called *adenosine triphosphate* (ATP). This molecule now contains the energy released from glucose. Furthermore, this molecule can be accumulated and stored throughout the body. When the body has a need for extra energy, ATP can quickly be changed to ADP + P + E. Figure 14.2 shows a summary of ATP synthesis and decomposition.

The Kreb's Cycle

A small amount of energy is released in the chemical processes that yield acetyl coenzyme A from glucose. Acetyl coenzyme A is a two-carbon molecule (see Figure 14.3) that will enter and become a functional part of the Kreb's cycle. A great deal more energy is released in the final splitting of carbon and hydrogen atoms from compounds yielding CO_2 and H_2O. The *Kreb's cycle* is a series of chemical reactions that begins with the introduction of acetyl coenzyme A and ends with the final products of energy, carbon dioxide, and water. Figure 14.1 summarizes the process. This process occurs on the cell organelles called mitochondria. Located here are specific enzymes that aid in the final decomposition and release of energy.

When the C_2 (acetyl coenzyme A) molecule arrives at the mitochondria it unites with a C_4 molecule. This new molecule, citric acid, now has six carbons ($C_6H_8O_7$). Due to enzyme action, one carbon, two oxygens, and two hydrogens are lost, forming a five-carbon molecule called *ketoglutaric acid* ($C_5H_6O_5$). The carbon atom and two oxygen atoms combine to form the molecule CO_2 (carbon dioxide). The high-energy hydrogen atoms are passed along the acceptor series previously described, releasing energy for ATP synthesis. Consequently, oxygen from respiration unites with the spent hydrogen yielding water molecules.

The next chemical reaction is similar to the one just described. Another carbon atom, two

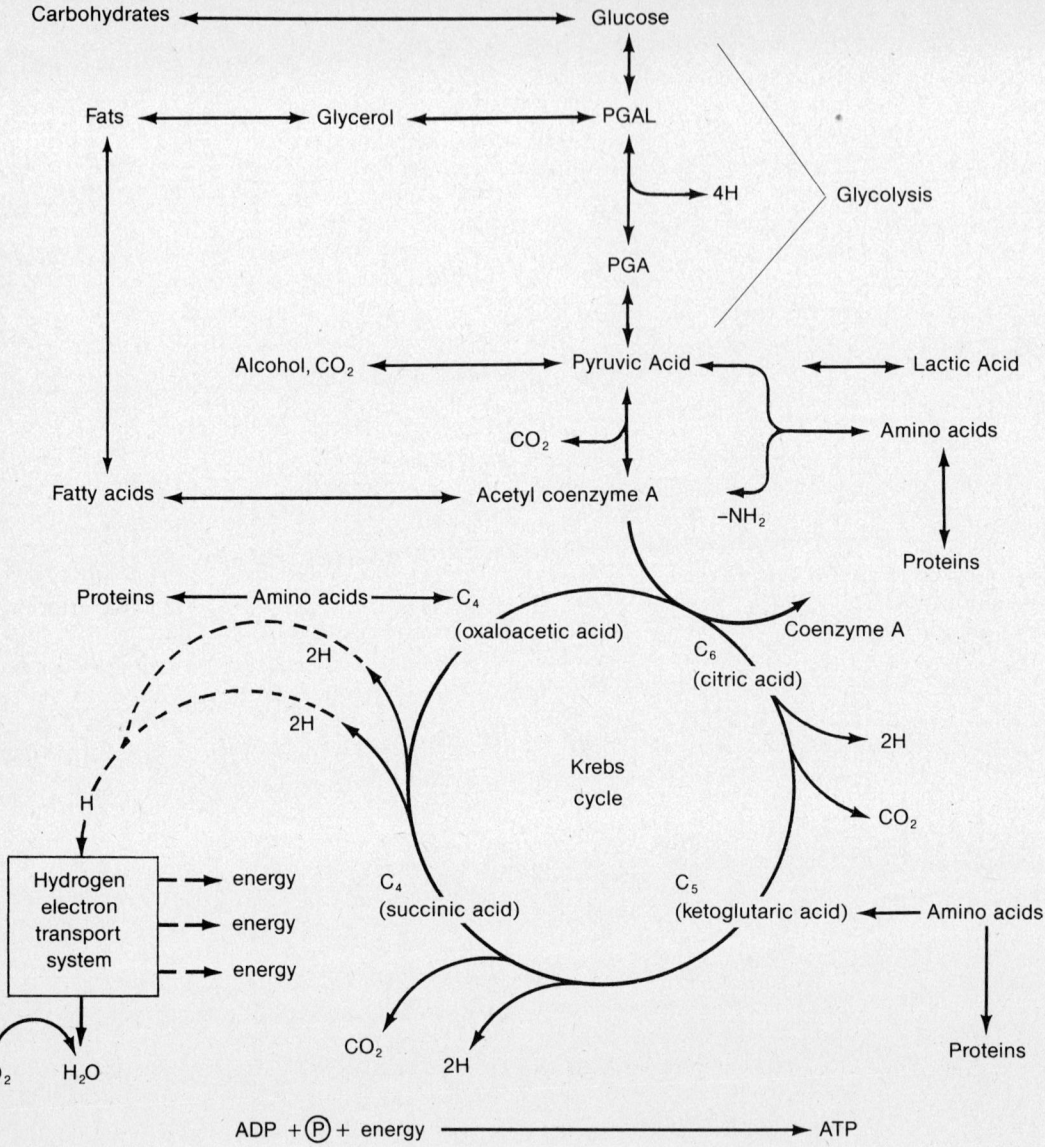

Fig. 14.1 Metabolic pathways involved with cellular respiration. The glucose molecule is gradually degraded to eventually produce a two-carbon molecule called acetyl coenzyme A. This molecule combines with a four-carbon molecule (oxaloacetic acid) to produce a six-carbon molecule called citric acid. In succeeding steps two carbon atoms, oxygen, and hydrogen are removed and energy is released. Once again the four-carbon molecule oxaloacetic acid joins the two-carbon acetyl coenzyme A. The process is repeated again and again as long as the cell lives.

more oxygen atoms, and two high-energy atoms are lost from $C_5H_6O_5$ (ketoglutaric acid) after rearrangement and the addition of a water molecule. This C_4 molecule (oxaloacetic acid) now can unite with the C_2 molecule (from glycolysis) making citric acid (C_6), starting the cycle all over again.

The process just described occurs in every living cell. The final product, ATP, is the basic energy-storage molecule for all living organisms. An overall simplified illustration of glycolysis and the Kreb's cycle is shown in Figure 14.3. Later in the chapter is a discussion of this process related to fat and protein metabolism.

Anaerobic respiration is the release of energy in the absence of oxygen. This occurs during the first steps of glycolysis. If oxygen is not available, pyruvic acid cannot change to acetyl coenzyme A. A new molecule, *lactic acid*, becomes the final hydrogen acceptor in place of oxygen. The accumulation of lactic acid in tissues, muscles in particular, is called oxygen debt.

The *calorie*, a unit of energy measurement, is the amount of heat required to raise the temperature of 1 kg of water 1 degree C. Most of the energy released from foods ends up as heat. Some energy released during the Kreb's cycle is released directly as heat, which helps maintain normal body temperature.

CALORIC VALUE OF FOODS

Fats yield at least twice the amount of energy per gram as carbohydrates and proteins. A gram each of carbohydrate and protein yield approximately the same amount of energy. An individual on a self-prescribed low fat diet may not be aware of fat energy levels and consequently may continue to gain weight.

The American diet with regard to carbohydrates, fats, and proteins has changed since the early 1900s. We are consuming fewer carbohydrates, slightly more protein, and considerably more fat.

This trend, along with a higher consumption of salt, may be the forerunner of cardiovascular problems such as atherosclerosis, arteriosclerosis, hypertension, and so on.

Several methods have been devised to determine the caloric intake required to perform specific kinds of work. The number of calories needed depends upon age, sex, weight, and the type of work performed.

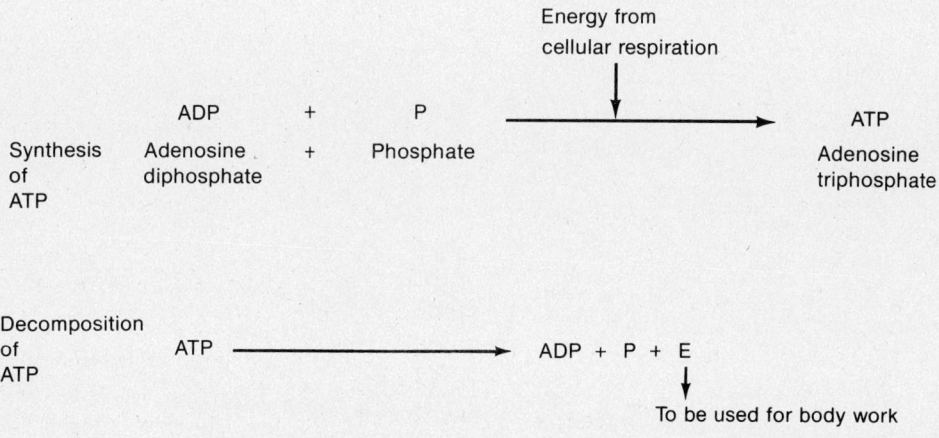

Fig. 14.2 Excess energy from foods is stored in molecules of ATP. This energy is easily reclaimed as cells require it.

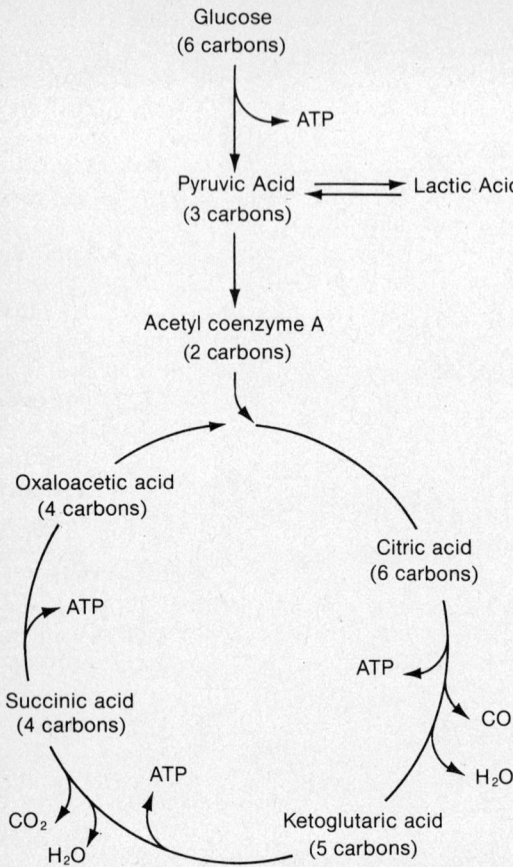

Fig. 14.3 Generalized diagram of glycolysis and the Kreb's cycle. This process occurs in all living cells. All living organisms that carry on aerobic respiration obtain their energy through the mechanisms shown here.

FAT METABOLISM

Although fats yield more energy per gram than do carbohydrates, the body prefers carbohydrates for energy. The hydrolysis of fats and the chemical rearrangement required to release energy is much more complicated than carbohydrate metabolism.

Fats are hydrolyzed to glycerol and fatty acids. The liver can convert glycerol to glucose, which can be metabolized directly by cells.

Energy can be obtained from fatty acids, but the process is more complicated than with glycerol. Some fatty acids are broken into two-carbon fragments and converted to acetyl coenzyme A. This substance then enters the Kreb's cycle. Other fatty acid fragments are converted to ketone bodies that may enter the Kreb's cycle directly or be converted to acetyl coenzyme A, and then enter the Kreb's cycle.

The liver is also capable of converting glucose to fat, which is then stored throughout the body.

Dietary Goals for the United States

In 1977, the United States Select Committee on Nutrition and Human Needs prepared a list of goals regarding nutrition which, if followed, would alter the eating habits of many Americans and would certainly, if followed, affect many levels of the American economy. The committee, chaired by Senator George McGovern, recommended the following goals:

1. Increase carbohydrate consumption to account for approximately 55 to 60 percent of the energy (caloric) intake.

2. Reduce overall fat consumption from approximately 40 percent to 30 percent of energy intake.

3. Reduce saturated fat intake to account for about 10 percent of total energy (caloric) intake, and balance that with polyunsaturated and monounsaturated fats.

4. Reduce cholesterol consumption to about 300 mg a day.

5. Reduce sugar consumption to about 40 percent to account for about 15 percent of total energy intake.

6. Reduce salt consumption by about 50 to 85 percent to about 3 g per day.

This study is based on the assumption that a corrected diet, within the limits of the proposed goals, can lead to better health, a longer life span, and greater satisfaction from work, family, and leisure time.

PROTEIN METABOLISM

Most body protein is utilized in chemical reactions involving the synthesis of specialized body components. These include muscle, enzymes, antibodies, and blood proteins.

Ingested protein is hydrolyzed to amino acids. Some amino acids are further degraded and energy derived from them. This involves removing the amino group (deaminization). The amino group, containing nitrogen, is a waste substance that must be eliminated from the body. A carbon dioxide molecule combines with two amino groups yielding a molecule of urea. Urea is carried by the blood to the kidneys where it is filtered out and excreted as a component of urine. The reaction shown below describes urea formation.

$$CO_2 + (NH_2)_2 \longrightarrow NH_2 \cdot CO \cdot NH_2$$
carbon dioxide + amino groups ⟶ urea

ENZYMES

Enzymes are substances that initiate and control biological chemical reactions. More specifically, an enzyme may be defined as a protein that acts as a catalyst. A *catalyst* is any substance that initiates, speeds up, or otherwise controls a chemical reaction.

The vast majority of chemical reactions occurring in the normal body is controlled by enzyme activity. This activity may include such things as aiding the breakdown of large molecules into small molecules during digestion, the synthesis of small molecules into large molecules for storage and tissue building, or mediating chemical reactions in which energy is released from foods. Other specialized functions of enzymes will be discussed with specific systems. Since enzymes are protein molecules, they are destroyed or deactivated, as are other proteins, by excessive heat, strong alkali, or other enzymes.

VITAMINS

Vitamins are organic substances needed by the body in very small quantities. The amount needed each day ranges from 1 to 20 mg. Most vitamins cannot be synthesized by the body, and therefore must be obtained through foods or by taking supplements. Vitamins cannot be categorized as a single class of substance since their molecular structures are widely variable. Some are acids, some are alcohols, others are amines, and others are polypeptides. Vitamins furnish no calories, and they are not pep pills, as many persons assume.

The daily requirement of each particular vitamin varies considerably with several factors. Among these are the size and age of the individual, sex, physical activity, the amount of food eaten, disease, and pregnancy. It is still not known precisely how vitamins work. However, it is known with some degree of precision where vitamin activity takes place. It is known, for instance, that vitamins are essential for the transformation of chemical energy found in foods into chemical energy that cells can use. This transformation involves the process of changing food material such as fats and proteins into smaller molecules from which energy can be released. It is also known that many enzymes cannot work without forming a cooperative association with some vitamins. These vitamins have been given the name *coenzyme*.

Excess vitamins are treated as waste by the body and are excreted as such. There is also documented evidence that vitamins A and D stored in the liver of animals can be toxic if large amounts of the liver of these animals are eaten.

Vitamins are referred to as *trace substances* since the optimum need is measured in milligrams or micrograms. In daily diets, one microgram, or 1/23,000,000 of an ounce, of vitamin B_{12} is sufficient to prevent certain types of anemia. If you consumed one microgram of this vitamin a day until the total consumption equaled one ounce, you would be 76,000 years old when you took the last dose.

Even though the requirements for vitamins seem infinitesimally small, they are absolutely essential for many metabolic functions, which are in turn essential for maintenance of life.

To make ATP, the cells must change one compound to another in a long series of reac-

Table 14.1. Vitamins, Their Source and Deficiency Effect

	VITAMIN	SOURCE	DEFICIENCY EFFECT
Fat soluble	A	Eggs, liver, cheese, carrots	Night blindness, lowered resistance to disease
	D	Fish, liver, dairy products	Rickets, osteomalacia
	E	Wheat germ oil, leafy vegetables	Reduced capability to prevent nemolysis of red blood cells
	K	Leafy vegetables, produced by intestinal bacteria	Inhibited blood clotting capability
Water soluble	C (B-complex)	Citrus fruits, vegetables	Lowered resistance to disease, scurvy
	B_1 (Thiamine)	Whole grains, yeast, pork liver	Beriberi, impaired carbohydrate metabolism, stunted growth
	B_2 (Riboflavin)	Vegetables, whole wheat, yeast, liver	Eye cataracts, impaired cellular metabolism
	Niacin	Yeast, fish, whole grains	Pellagra
	B_6 (Pyridoxine)	Liver, yeast, cereals	Dermatitis, nausea
	Folic acid	Yeast, wheat germ, liver	Anemia due to malformed red blood cells
	B_{12} (Cyanocobalamin)	Intestinal bacteria, liver, milk, cheese	Pernicious anemia
	Pantothenic acid	Eggs, vegetables	Impaired food metabolism
	Biotin	Intestinal bacteria, eggs, liver	Fatigue, sensitive skin, nausea

tions, each of which requires a special molecule—an enzyme—that catalyzes it. Along with the necessary enzyme, often a vitamin is necessary, working as a coenzyme.

When chemists learned how to isolate pure enzymes, it became possible to study these molecules. One procedure used was analogous to putting enzymes into a plastic bag and then submerging the bag in a solution to see what soaked out of it. The membrane or wall of the bag contained pores too small to let the big protein molecules of an enzyme slip through. Yet when scientists removed certain pure enzymes from the bag and put them into the proper environment to do their work, nothing happened. The protein molecule of the enzyme was "dead" or useless.

A very small and vital molecular part of an enzyme had escaped through the small membrane pores. Reattachment of the missing part (coenzyme) restored the enzyme to full activity. Vitamins probably work also in other ways, but the essential partnership of vitamins with enzymes in catalyzing many chemical reactions indicates their importance.

Vitamins A, D, E, and K are fat soluble. These are obtained directly from fatty foods or indirectly from provitamins, which are not fatty. These vitamins are also stored to some degree in body fat. A *provitamin* is a substance from which a vitamin is produced. For example, *carotene*, the orange pigment of carrots, egg yolk, and so on, is the provitamin of animal tissue. All provitamins are produced by plants. Animal tissue may store the provitamins from which vitamin A is synthesized.

Vitamin C and the B-complex vitamins are water soluble. These vitamins are often lost if foods containing them are cooked in water, with the water then discarded.

Table 14.1 lists the common vitamins, some common sources, and specific disorders associated with deficiencies of each vitamin.

MINERALS

Minerals are inorganic substances that make up approximately 4 percent of the adult body. The other 96 percent of body weight is of organic composition. If a mixture of organic and inorganic substances is burned completely, the minerals would constitute the ash. Some minerals found in fairly large amounts in the body are called *macrominerals*. These include sodium, calcium, phosphorus, chlorine, sulphur, and magnesium. Minerals found in lesser quantities in the body, called *microminerals*, include iron, iodine, manganese, copper, cobalt, zinc, and fluorine.

Most minerals are supplied by normal diets. However, some may not be absorbed by the intestinal mucosa due to some malfunction or inability to secrete the proper carrier molecule. An example of this latter process is the lack of a substance called *intrinsic factor* normally secreted by the intestinal mucosa. If intrinsic factor is not secreted, vitamin B_{12} cannot be absorbed. Vitamin B_{12} is essential for the proper maturation of red blood cells. A type of anemia (pernicious anemia) results from deficient vitamin B_{12}. Table 14.2 lists common minerals needed by the body for normal growth, development, and maintenance.

Table 14.2 Some Important Minerals Required by the Human Body

MINERAL	FUNCTION
Sodium	Osmotic pressure
Potassium	Osmotic pressure, growth
Calcium	Bone and teeth development, muscle contraction
Phosphorus	Bone and teeth development, carbohydrate metabolism
Iodine	Production of thyroxine by thyroid gland
Iron	Synthesis of hemoglobin
Sulfur	Synthesis of proteins and hormones
Chlorine	Osmotic pressure, component of HCl in gastric juice
Manganese	Hemoglobin synthesis, enzyme activator

WATER

The role of *water* is of such importance that a major portion of Chapter 16 is devoted to it. The short discussion here will be limited to general functions.

Body metabolism depends upon chemical reactions between substances in solution. Water is the major solvent for these substances. Water aids in heat distribution, heat regulation, hydrolysis, and synthesis of body substances. It also acts with other substances as a lubricant between organs. The need for water is greater than any other nutrient. Mineral, vitamin, and food deficiency can be tolerated for a much greater length of time than water deficiency.

MEDICAL PROBLEMS RELATED TO NUTRITION

Everyone is familiar with recorded accounts of starvation and malnutrition. To some degree we are also familiar with excess nutrition resulting in obesity.

Some medical problems involved with these and other nutrition deficits and metabolic dysfunctions are discussed in this segment.

Phenylketonuria

Phenylketonuria is a genetic disease in which a liver enzyme is not properly synthesized. A specific enzyme is required to convert an excess of the amino-acid *phenylalanine* into another amino acid, *tyrosine*. Phenylalanine is a component of many ordinary foods such as whole wheat cereals and meats. An excess of phenylalanine in the blood causes the disease phenylketonuria (PKU). This disease in young children is characterized by malformed leg bones and severe mental retardation. Excess phenylalanine inhibits the proper maturation of nerve cells.

A simple screening test provided by most hospitals can determine whether or not a baby is afflicted. If so, medication and a restricted diet will prevent the occurrence of PKU symptoms.

The child must remain on the restricted diet until age 9 to 12. By this time the nerve cells are fully mature. Excess phenylalanine does not affect mature nerve cells nor apparently any other body tissue.

Galactosemia

Galactosemia, another genetic disease, is similar to PKU in that an enzyme is involved in a metabolic process. In galactosemia, the child's metabolism is incapable of converting the monosaccharide galactose to glucose. Galactose and glucose are products of the hydrolysis of lactose (milk sugar). Human milk and cow's milk both contain lactose. Galactose concentration in the child's blood reaches high levels and has, like PKU, an adverse effect upon maturing nerve cells. Treatment is to restrict lactose in the diet.

Drug-Induced Nutritional Deficiency

The interaction of drugs and the gastrointestinal tract may in some instances lead to nutritional deficiencies. The following examples point up the need for a certain level of expertise in intestinal physiology when drugs to be ingested are prescribed:

1. Mineral oil used daily as a mild laxative or salad dressing may lead to the loss of fat soluble vitamins A, D, and K. These vitamins are dissolved in and carried out of the gastrointestinal tract with mineral oil.

2. Glucocorticoids delay calcium transport and absorption. Osteoporosis may result from decreased calcium absorption.

3. Tetracyclines and antacids containing calcium form insoluble compounds. These two substances therefore should not be given together.

Kwashiorkor

Kwashiorkor is a disease caused by severe protein deficit. The disease makes its appearance in children between the ages of six months and three years and typically occurs after a child is weaned from mother's milk and is placed on largely a carbohydrate diet of cereals.

The symptoms of kwashiorkor are edema, muscle degeneration, stunted growth, and behavior change.

Kwashiorkor is treated by including high quality protein in the diet. This includes skim milk or powdered milk protein or vegetable protein.

Obesity

The term *obesity* is usually equated with overweight. However, overweight and "fatness" may not be the same. The problem of definition is further confused due to the fact that we have no single standard of what overweight means. Some

Table 14.3 Minimum Skin Fold Thickness as Indicator of Obesity[a]

AGE	MALES (mm)	FEMALES (mm)
5	12	14
6	12	15
7	13	16
8	14	17
9	15	18
10	16	20
11	17	21
12	18	22
13	18	23
14	17	23
15	16	24
16	15	25
17	14	26
18	15	27
19	15	27
20	16	28
21	17	28
22	18	28
23	18	28
24	19	28
25	20	29
26	20	29
27	21	29
28	22	29
29	22	29
30–50	23	30

[a]Table is for normal thickness, American Caucasians. Standards are not available for children under 5 years of age.

authorities indicate overweight as a weight 20 percent in excess of normal weights shown in weight tables for age and sex. Some use a 5 to 10 percent excess as an overweight criterion. More recently, the skin fold test has gained recognition. The skin fold test measures the thickness of a fold of skin along the triceps muscle. The measurement is made in mm. Table 14.3 shows skin fold norms for male and female Caucasian Americans.

Excessive weight gain may be attributed to many factors. The major factor is simply explained—an overweight person is usually one who eats more food than can be metabolized. Obesity is a major health problem. It is a specific risk factor in surgery, coronary artery disease, athersclerosis, and arteriosclerosis.

Overeating may be due to glandular disturbance, emotion, peer standards, or social influence. Evidence exists that should assure the obese individual that his or her life expectancy will probably be shortened.

Underweight

Like obesity, *underweight* is difficult to define. Guidelines are anywhere from 5 to 10 percent under normal weight charts. The condition may be due to disease, glandular malfunction, digestive disorders, or psychological disorders. The latter factor is most often observed in teenage girls who impose on themselves a restrictive—and often starvation—diet. The desire to lose weight may result from previously being overweight. Ridicule by parents or friends may be a contributing factor in the self-imposition of a starvation diet. Underweight persons generally have a low resistance to infection and may suffer from vitamin deficiency.

SUMMARY AND REVIEW

A. Carbohydrate metabolism involves a stepwise process of chemically removing carbon and hydrogen atoms.
1. Glycolysis represents the first phase of cellular respiration, in which a glucose molecule is split into two molecules, each having three carbon atoms. These molecules are pyruvic acid.
2. The Kreb's cycle is a series of chemical reactions that further degrades carbon molecules, eventually releasing the energy held by them.
B. Fat metabolism involves hydrolysis that yields glycerol and fatty acids. Molecules of these substances may then be altered and enter the Kreb's cycle reactions so that energy may be released from them.
C. Protein metabolism includes hydrolysis, which yields amino acid. Deaminated amino acid fragments may then be altered and enter the Kreb's cycle.
D. Enzymes are organic catalysts involved in all phases of chemical reactions occurring in the body.
E. Vitamins are coenzymes. Most vitamins are obtained by eating foods containing them. Those synthesized within the body include K, B_{12}, D, and folic acid.
F. Minerals are inorganic substances required for many body functions. Some are required in larger quantities than others. These include:
1. sodium
2. calcium
3. phosphorus
4. chlorine
5. sulfur
6. magnesium

Smaller quantities of the following are required:
1. iron
2. iodine
3. manganese
4. copper
5. sulfur
6. zinc
7. flourine
G. Medical problems related to nutrition
1. phenylketonuria
2. galactosemia
3. kwashiorkor
4. obesity
5. underweight

15

The Excretory System

MAJOR CONCEPTS
  BASIC ANATOMY OF THE URINARY SYSTEM
    Internal Structure of the Kidney
    Nephrons: The Functional Units of Excretion
  STRUCTURE OF THE NEPHRON
  HOW MATERIALS GET INTO BOWMAN'S CAPSULE
  FORMATION OF URINE
    Glomerular Filtrate
    Tubular Reabsorption
    Control of Sodium Ions
    Regulation of Body Water
  DIURETICS
  MAINTAINING ACID-BASE BALANCES
  MAINTAINING FLUID BALANCES
  COMPOSITION OF NORMAL URINE
  COMPOSITION OF ABNORMAL URINE
  MICTURATION
  RENAL DISEASES
  THE AGING KIDNEY

Few other body systems can compare with the excretory system when it comes to maintaining homeostatic balances. These balances include body fluid volume, acid-base balance, and electrolyte balance. A major emphasis of this chapter is the anatomical detail of the excretory system and the role each segment of the system plays in homeostatic balances and the excretion of wastes.

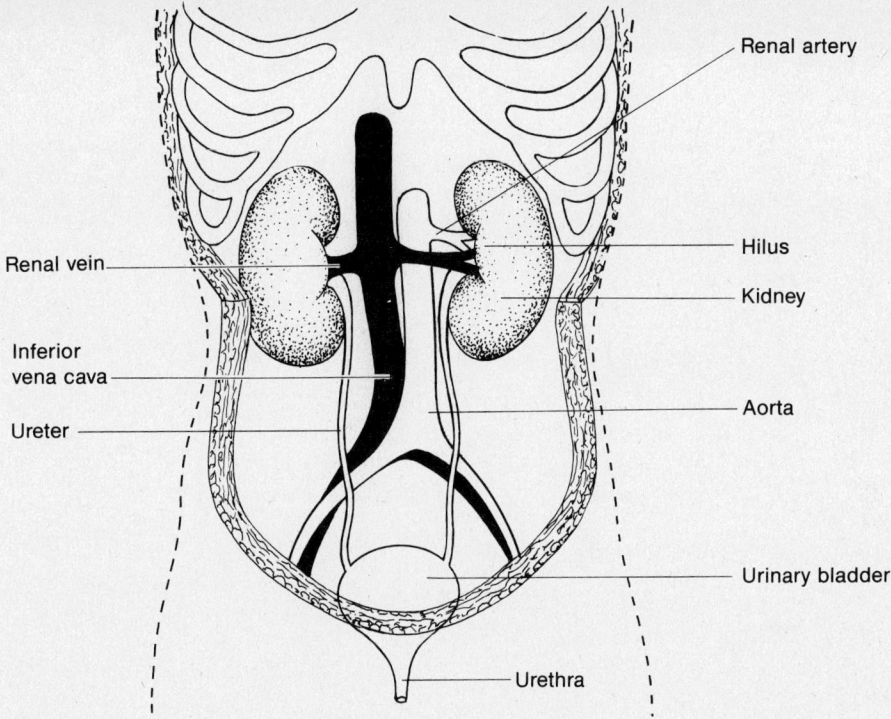

Fig. 15.1 The location of the kidneys in relation to the abdominal cavity and major blood vessels. The kidneys lie tightly against the posterior wall of the abdominal cavity. The upper part of each kidney is near the lower level of the rib cage. The kidneys are separated from the abdominal cavity by the parietal layer of the peritoneum.

If we had to choose one organ that best exemplifies the homeostatic process, the kidneys would easily win the contest. One of the main functions of the excretory system, of which the kidneys are a part, is to maintain the constant internal environment for the body cells, getting rid of substances that are harmful and conserving those that are helpful for normal body functioning. Except for its cellular content, the constancy of the chemical makeup of blood, which surrounds cells and provides them with nutrients, is largely dependent upon normal kidney function. Although the discussion in this chapter relates only to kidney function, it is important to point out that other body systems and tissues excrete wastes. Included among these are the skin (Chapter 4), the lungs (Chapter 16), and the colon (Chapter 13).

Waste removed by the kidneys is a mixture called *urine*. An analysis of urine can often yield valuable information for diagnosing a body malfunction. In a healthy person, the composition of urine is surprisingly constant. Disease or illness may disrupt one or more physiological systems, which in turn do not perform normally. If a substance such as glucose cannot be utilized by the body, it is treated as a waste product and is removed from the blood by the kidneys.

Other major functions of the kidneys include the maintenance of body fluid levels and normal blood pH levels, and the control of ion concentration in body fluids. The overall function of the kidney is to regulate the volume and composition of body fluids.

BASIC ANATOMY OF THE URINARY SYSTEM

The kidneys are paired, bean-shaped organs about 12 cm long, 7 cm wide, and 3 cm thick.

They weigh about 150 g each. The kidneys lie tightly against the posterior wall of the abdominal cavity, as shown in Figure 15.1. The upper part of each kidney is near the lower level of the rib cage. The kidneys are separated from the abdominal cavity by the parietal layer of the peritoneum. The concave portion of each kidney faces the medial line of the body.

Near the center of the concave side is an area called the *hilus* that serves as an entry for an artery and nerves and as an exit for veins, nerves, and the ureter.

Internal Structure of the Kidney

A kidney cut in longitudinal section reveals its internal structure, as shown in Figure 15.2. A tough fibrous membrane called the *capsule* covers the kidney. Just internal to the capsule is a dense area called the *cortex*. The inner part of the kidney, called the *medulla,* is composed of triangular-shaped masses of tissues called *renal pyramids*. The apex of each triangle is called a *renal papilla*.

A large portion of the kidney, including the cortex and medulla, is composed of tubules, some of which are microscopic in size. These tubules form a network that accumulates fluids and other substances, then empties them through the renal papilla into the part of the kidney called the *pelvis*. The kidney pelvis drains into and is continuous with the *ureter,* which in turn drains into the urinary bladder.

A portion of the cortex dips in between the renal pyramids forming columns of tissue called the *renal columns,* made up of microscopic tubules.

Nephrons: The Functional Units of Excretion

A *nephron* is a microscopic structure located in the cortex of the kidney. Each kidney contains about a million nephrons. Nephrons remove wastes such as urea, salts, and excess substances such as water, thereby maintaining a constant fluid volume and a chemical balance for all body cells.

About 1.2 liters of blood per minute flow into both kidneys through renal arteries. In the

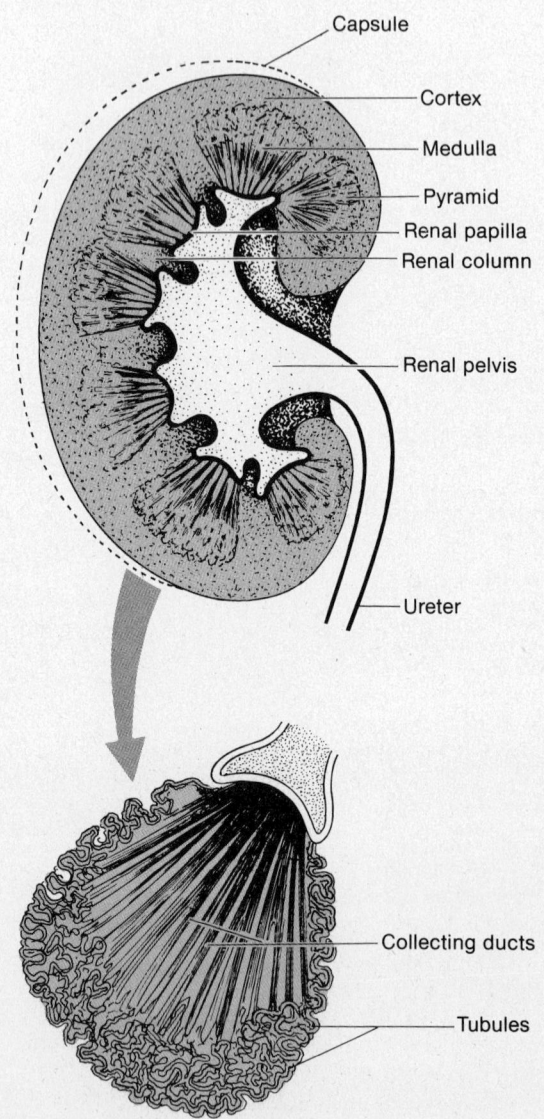

Fig. 15.2 A longitudinal section through the kidney, showing internal structure. The lower figure is an enlarged view of a renal pyramid. A large portion of the kidney is composed of microscopic tubules that filter and drain wastes into the kidney pelvis, then into ureters.

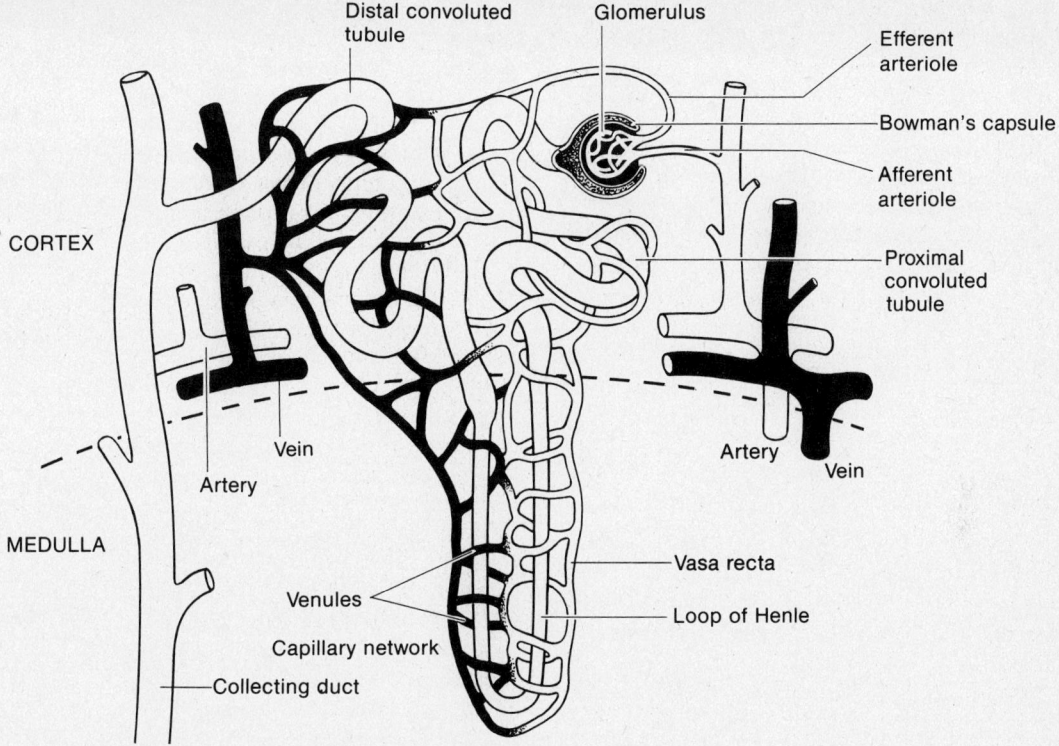

Fig. 15.3 A kidney nephron, showing the blood supply entering and leaving. The tubule drains into collecting ducts, which in turn empty into the renal pelvis.

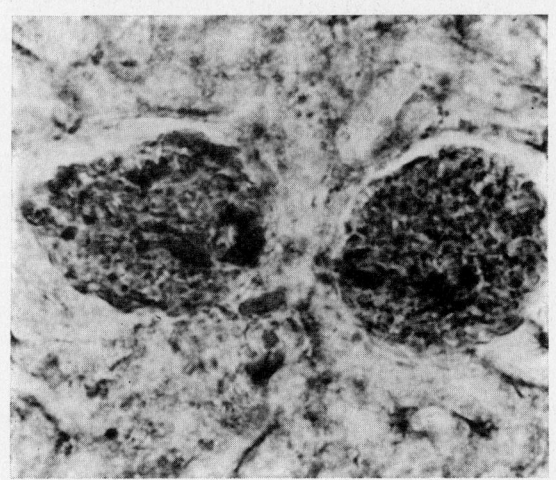

Fig. 15.4 Photomicrograph of nephron glomerulus and Bowman's capsule. *(Author's collection.)*

kidney nephrons, blood is filtered, removing wastes and excess substances. From the 1.2 liters of blood per minute both kidneys remove about 125 ml per minute of filtrate. Of this, only about 1 percent is finally discarded as urine.

STRUCTURE OF THE NEPHRON

The nephron is the functional unit of the kidney. Figures 15.3 and 15.4 show the physical relationship of nephrons to other kidney structures. Each kidney is actually an accumulation of nephrons connected to collecting tubules. These structures control the fluid volume and chemical balance of all body fluids.

The *renal artery*, a branch of the dorsal aorta, enters the kidney, then branches into smaller vessels called the *afferent arterioles*. Afferent arterioles pass into a capillary bed called the *glomerulus*. The glomerulus is the site of filtration of water and other materials from the blood into the *kidney tubules.*

The capillaries of the glomerulus reunite into vessels called the *efferent arterioles,* which leave the glomerulus. The efferent arterioles branch and form a meshwork of small vessels called the *peritubular capillaries,* which surround the tubules of each nephron. The tubular area of the nephron is the site of the precise control of body fluids. Blood vessels leading from the peritubule capillaries converge into *venules. Renal veins* collect blood from the venules and empty into the inferior vena cava, which carries blood back to the heart.

Refer again to Figure 15.3. The proximal convoluted tubule dips downward (toward the pelvis) and is reduced in diameter. This portion of the tubule is called the *loop of Henle* and is made up of a descending and an ascending segment. The tubule again enlarges, forming the *distal convoluted tubule,* which empties into *collecting ducts.*

When we refer to the renal tubule we are speaking in general terms that include all the segments. It is often necessary to refer specifically to a precise segment of the tubule since different segments may be involved with different reabsorption capabilities. Although the renal tubule is a continuous vessel beginning at *Bowman's capsule* and ending at the collecting ducts, each segment removes or adds to the filtrate. The final filtrate entering the collecting duct is urine.

Many tubules drain urine continuously into collecting ducts that empty into the renal pelvis. The quantity of urine that drains into the renal pelvis, then into the urinary bladder depends upon how much excess fluid is ingested over a specified length of time.

HOW MATERIALS GET INTO BOWMAN'S CAPSULE

The glomerulus is a knot of capillaries whose membranes are permeable to water and under certain conditions to other substances. These substances then filter across the inner membrane of Bowman's capsule into the *proximal convoluted tubule.* Formed elements of blood and large protein molecules cannot filter across the membrane.

The driving force that pushes substances across these membranes is blood pressure. Blood entering the renal artery from the aorta initially has a pressure of approximately 100 mm of mercury. As the renal artery divides into smaller vessels, the arterioles, the resistance to blood flow reduces the pressure at the glomerulus to approximately 70 mm of mercury. The efferent arteriole leaving the glomerulus is smaller than the afferent arteriole entering the glomerulus. As blood enters the efferent arteriole, a reverse pressure is thus created in the glomerulus. This same mechanism is responsible for a relatively low pressure in the peritubular capillaries.

If it were not for the differences in arteriole size, as water and other substances left the glomerulus the pressure would drop. Blood pressure in the glomerulus maintained at about 70 mm of mercury is the major force pushing substances out of the blood, into the glomerulus, and across Bowman's capsule membranes. This is the pressure pushing blood into Bowman's capsule. There are, however, pressures within Bowman's capsule that resist the glomerular pressure, thereby significantly reducing the inward pressure.

The osmotic pressure of blood due to the proteins it contains offers a negative pressure. Since proteins cannot filter through the glomerular membranes and the capsule wall, they remain in the glomerulus. If there were no pressure other than the osmotic pressure, which is due to the protein molecules, the flow of fluids would be reversed. In other words, the flow would be from the capsule into the glomerulus. The glomerular pressure due to the high protein content is equal to about 30 mm Hg (mercury). Another opposing influence is *hydrostatic pressure,* which is the pressure of fluids within Bowman's capsule as it becomes filled with filtrate. This pressure resists the incoming fluids. *Capsular hydrostatic pressure* is equal to about 12 mm Hg. If we look at the incoming pressures and the opposing pressures, we find 70 mm Hg pressure pushing liquids into the capsule and 30 + 12 or 42 mm Hg resisting the incoming pressure, with a net pressure of 28 mm Hg pushing liquids from

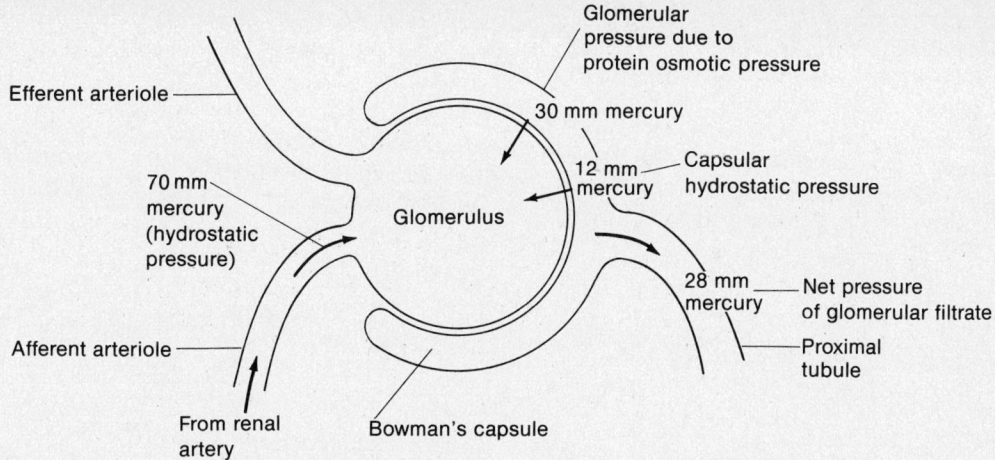

Fig. 15.5 How substances get into Bowman's capsule. Blood hydrostatic pressure into the glomerulus is 70 mm of mercury. Resisting this pressure is osmotic pressure of 30 mm of mercury, owing to capsular proteins. Also resisting the incoming pressure is 12 mm of mercury, owing to hydrostatic pressures created in the capsule as it fills. The effective inward pressure is thus equal to 70 mm Hg − (30 mm Hg + 12 mm Hg) = 28 mm Hg. This is called the effective filtration pressure.

the glomerulus into the capsule. This pressure is called the *effective filtration pressure*, as shown in Figure 15.5.

Most authorities place the effective filtration pressure between 23 and 30 mm Hg. This pressure is highly variable depending upon factors such as quantity of liquids taken into the body, and constriction or dilation of the glomerulus due to nervous mechanisms, hormones, drugs, hypotension, hypertension, and stress.

FORMATION OF URINE

All substances that enter the glomerulus make up what is called the *glomerular filtrate*. As the filtrate enters the proximal tubule and continues along the tubular pathway, it is significantly altered. Two processes that alter the filtrate are *tubular reabsorption* and *tubular secretion*. The final altered material that enters the collecting ducts is urine.

Glomerular Filtrate

Substances that filter into Bowman's capsule are collectively called the *glomerular filtrate*. The composition of this substance is almost the same as blood plasma minus large protein molecules.

Glomerular filtrate contains glucose, urea, creatinine (a by-product of muscle contraction and the utilization of chemical energy), and a trace of protein, and ions of sodium, chloride, potassium, and sulfates. The small quantity of protein in normal filtrate remaining in the renal tubule is reabsorbed by the proximal tubule or is excreted in the urine where it is not detected by usual methods.

The composition of glomerular filtrate changes suddenly and radically upon entering the proximal convoluted tubule. More alteration of the filtrate continues as it passes along the other segments of the renal tubule.

The process by which substances are absorbed out of the renal tubule back into the bloodstream is called *tubular reabsorption*.

Tubular Reabsorption

The glomerular filtrate accumulates at the rate of about 125 ml per minute by both kidneys. As this fluid passes along the winding tubule route, only about 1 ml per minute is emptied into the bladder as urine. This means that 124 ml per minute of the original 125 must have been lost somewhere along the tubular length. Refer to Figure 15.3 and observe the anatomical relationship between the efferent arteriole and nephron

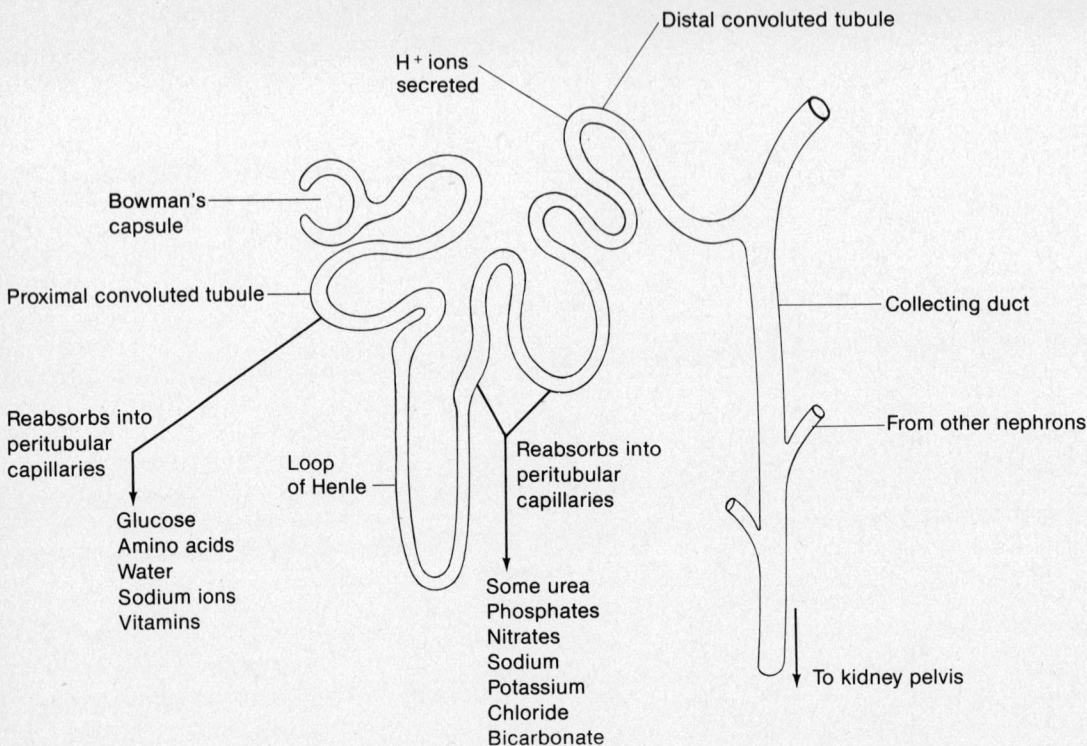

Fig. 15.6 Absorption and secretion in different parts of nephron tubules. All along the tubule, substances are selectively reabsorbed (put back into the blood). It is this process of saving desirable substances and letting excesses and wastes continue along the tubule to be excreted that maintains the constancy of body fluids.

tubules. The efferent arteriole subdivides many times, forming a capillary bed called the *peritubular capillaries*. The close proximity and the nature of the renal tubule and peritubular capillaries permit the transfer of substances from one to the other. This is accomplished by both passive diffusion and active transport.

The process of tubular reabsorption therefore restores substances to the blood that were removed by the glomerulus and Bowman's capsule.

Ions reabsorbed into the blood capillaries from the proximal tubules include sodium, chlorine, and potassium. Other substances reabsorbed in the proximal tubule include all the glucose, amino acids, vitamins, and any protein that may have been filtered into the glomerulus. Hydrogen ions, phosphates, and urea are reabsorbed to a small extent, also.

Figure 15.6 summarizes the tubular reabsorption process. This diagram illustrates the reabsorbing activity along the entire nephron tube. Some substances such as sodium, in the loop of Henle, and ammonia, in the distal tubule, enter the tubule from the capillary. This process, just the reverse of reabsorption, is called *secretion*. Tubular secretion adds substances to the filtrate. These substances may include hydrogen and potassium ions, creatinine, ammonia, and certain drugs such as penicillin. The purpose of tubular secretion is to aid in controlling pH and to expel certain substances from the blood.

Control of Sodium Ions

The sodium ion is the most abundant electrolyte in the body and the one most necessary to control with respect to maintaining homeostasis.

The glomerular filtrate has about the same concentration of sodium ions as blood plasma. (Table 15.1 shows relative concentrations of electrolytes in blood plasma and glomerular filtrate.) Sodium reabsorbed into the peritubular capillaries increases the concentration of water molecules in the tubule. Consequently, water diffuses out of the tubule. Simply stated, when sodium is lost or retained, it is accompanied by water. This is why the control of sodium is so critical for the body. If excess sodium is retained in the tubule then excess water is also retained, both eventually being excreted in urine.

Table 15.1 Relative Concentrations of Substances and Ions in Plasma and Glomerular Filtrate[a]

CATIONS	mEq/B	ANIONS	mEq/L
Sodium	142	Bicarbonate	24
Potassium	5	Chloride	105
Calcium	5	Biphosphate	2
Magnesium	2	Sulfate	1

OTHER SUBSTANCES	mg/100ml
Glucose	0.01
Urea	0.03
Creatinine	0.001
Protein	7,000–9,000 (plasma)
	10–20 (filtrate)

*Ion concentration in body fluids is commonly expressed in milliequivalents per liter (mEq/l).

$$mEq/l = \frac{\text{milligrams of ion per liter of solution} \times \text{The valence of an ion}}{\text{Atomic weight of the ion}}$$

Example: Na^+/liter of plasma = 3,270 mg/l

$$mEq/l = \frac{3{,}270 \times 1}{23} = 142.2$$

The mechanics of sodium and water regulation are closely tied to nephron anatomy. Some of the peritubular capillaries dip into the medullary tissues along with the loop of Henle. This segment of the capillaries is called the *vasa recta*. Collecting tubules leading toward the renal pelvis are closely associated anatomically with the loop of Henle and the vasa recta. This relationship is illustrated in Figure 15.7. As the glomerular filtrate passes into the descending arm of the loop of Henle, much of the sodium has been reabsorbed higher up in the proximate tubule. The descending loop, however, is capable of absorbing large quantities of sodium. The ascending arm of the loop, on the other hand, cannot absorb sodium. In fact, sodium is removed from the tubule by active transport. As you can see in Figure 15.7, sodium removed from the ascending arm of the loop is absorbed by the descending arm of the loop. The concentration of sodium in the base of the loop therefore rises to much higher levels than found elsewhere in the nephron system. The process just described is called the *countercurrent mechanism*. The high concentration of sodium in the base of the loop thus is further increased by more sodium entering the descending loop from filtrate constantly entering the glomeruli.

The interstitial fluid surrounding the loop of Henle increases in sodium concentration. This will then increase the sodium levels within the vasa recta, which has its own countercurrent mechanism.

Countercurrent mechanisms prevent the loss of sodium from the tissues. The loss of large quantities of water from the blood and tissues is called *dehydration*. Since normal body function depends upon specific quantities of water, dehydration causes severe physiological disturbances.

When excess sodium is reabsorbed, excess water is also reabsorbed. This means the blood and tissues hold more water than normal, producing edema, which also disturbs some body functions.

Aldosterone, a hormone secreted by the adrenal cortex, acts upon the renal tubules, increasing the rate of sodium reabsorption. An insufficient level of aldosterone secretion reduces sodium reabsorption, thereby increasing the quantity of sodium excreted in urine. Since water follows sodium, it also is excreted in abnormal amounts. (See the discussion of osmosis in Chapter 2.)

The control of aldosterone secretion is regulated by the concentration of sodium in extracellular fluids. Blood low in sodium arriving at the adrenal cortex will stimulate the

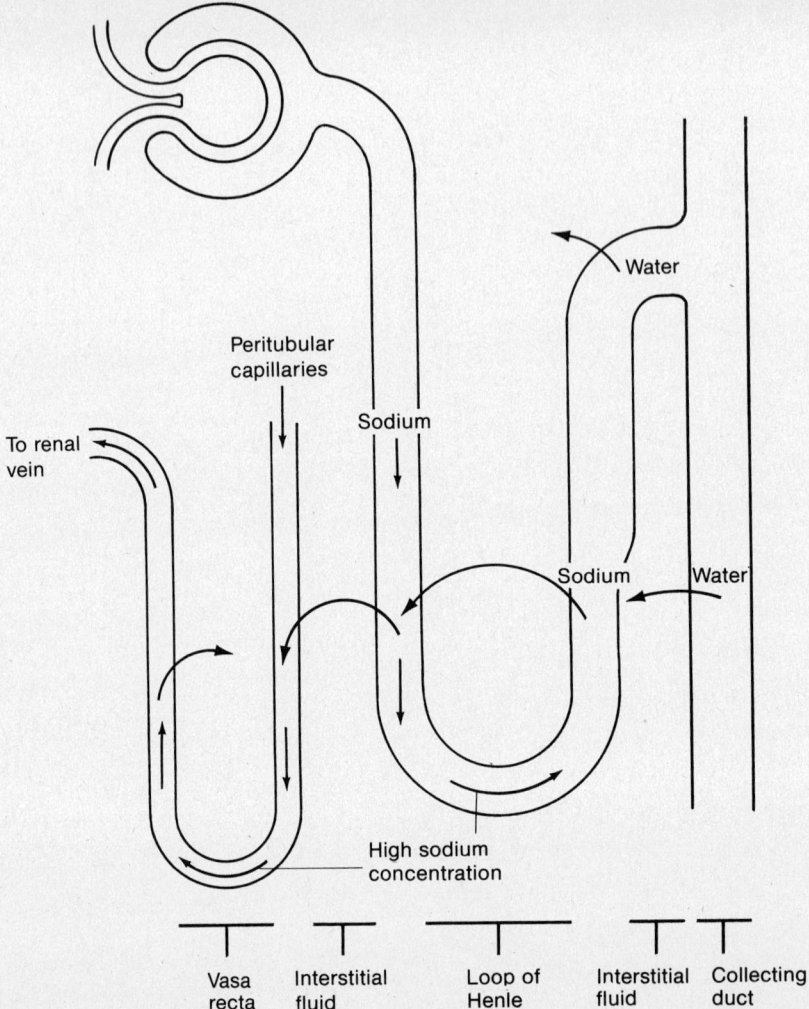

Fig. 15.7 The countercurrent mechanism. Sodium concentrations increase in the loop of Henle, medullary interstitial fluids, and the vasa recta. This process helps preserve sodium in the body.

adrenal cortex to secrete higher levels of aldosterone. The increased reabsorption of sodium in response to aldosterone restores the proper sodium level in the tissues. The adrenal glands, no longer stimulated, stop secreting aldosterone.

The process just described is an excellent example of how different parts of the body work together to maintain homeostasis.

Regulation of Body Water

The amount of water taken into the body by the ingestion of water, water in foods, and the synthesis of water by metabolic processes equals about 2.4 liters a day in the normal adult. The daily loss of water must equal the amount of water ingested. About 1.4 liters per day are lost by excretion, 100 ml per day are lost in sweat, 200 ml per day are lost in feces, and about 700 ml per day are lost owing to evaporation from the lungs. If an excess of about 2.4 liters per day are lost by the body, it becomes dehydrated. If more than 2.4 liters per day are retained, the fluid collects and the result is edema.

Water reabsorption rate in the tubules is controlled by the action of *antidiuretic hormone* (ADH) upon nephron collecting ducts. Antidiuretic hormone is secreted by the hypothalamus and released by the neurohypophysis (posterior lobe of the pituitary). In the absence of ADH the collecting duct membranes become impervious to water, reducing the quantity of water that can be returned to the interstitial fluids. In other words, water continues into the renal pelvis and is excreted. This produces a very dilute urine. In the presence of ADH the collecting duct membranes easily permit water to leave the tubule, thus a concentrated urine is produced. The action of ADH upon the collecting tubules represents a major mechanism for water retention for the body. A complete lack of ADH causes the disease called *diabetes insipidus*. Persons suffering from this condition have a urine volume of 5 to 15 liters per day as opposed to the normal 1 to 2 liters per day. They suffer from constant thirst and excrete a very dilute urine.

DIURETICS

Any substance that tends to increase the output of urine is called a *diuretic*. The action of diuretics either increases the glomerular filtration rate or decreases the tubular reabsorption rate. These increasing or decreasing actions may be produced by some of the following:

1. Norepinephrine increases the arterial pressure. When the blood pressure increases, the pressure within the glomerulus is also increased, forcing more fluids into the capsule, and hence producing a greater urine output.
2. Digitalis, which is used for congestive heart failure, also increases glomerular pressure.
3. Caffeine dilates the afferent arterioles. This dilation permits a larger volume of blood to flow into the glomerulus, therefore increasing the hydrostatic pressure.
4. Substances such as urea, sucrose, or manitol, when injected into the bloodstream are filtered from the glomerulus into the capsule, but these substances are not reabsorbed into the peritubular capillaries. Since they remain in the tubule they increase the osmotic pressure within the tubule, preventing water reabsorption and adding to the fluid that will be excreted as urine.

MAINTAINING ACID-BASE BALANCES

As we have seen earlier, the pH of urine may vary. The changes in pH are much more variable in disease. (Chapter 20 discusses pH, and acids and bases.)

The kidneys play a major role in maintaining proper acid-base balances largely by controlling the bicarbonate ion. Excess CO_2 in interstitial fluids is absorbed by the epithelial cells of the tubule (proximal, distal, and collecting). Carbon dioxide reacts with cellular water, forming carbonic acid, H_2CO_3, which dissociates into hydrogen (H^+) ions and bicarbonate (HCO_3^-) ions. Hydrogen ions are then secreted into the tubule, as illustrated in Figure 15.8.

Sodium bicarbonate in the tubule, a normal glomerular filtrate component, upon decomposition yields sodium ions plus bicarbonate ions. Hydrogen ions secreted into the tubule then join bicarbonate ions producing carbonic acid (H_2CO_3). An increase in H_2CO_3 lowers the pH (more acid).

Sodium ions left in the tubule when sodium bicarbonate ($NaHCO_3$) dissociates are reabsorbed by the epithelial cells and transferred to interstitial fluids. This may be due to an electrical "pay back" for the secreted hydrogen ions.

When carbonic acid in the epithelial cell dissociates to yield hydrogen ions, bicarbonate ions also are produced. These ions are reabsorbed into the interstitial fluid where they join the reabsorbed sodium ions, producing sodium bicarbonate ($NaHCO_3$).

The rate of hydrogen ion secretion depends upon the concentration of CO_2 in the interstitial fluids. Increased metabolic rate increases CO_2 levels, thus increasing hydrogen ion secretion. The lungs also play a vital role in controlling

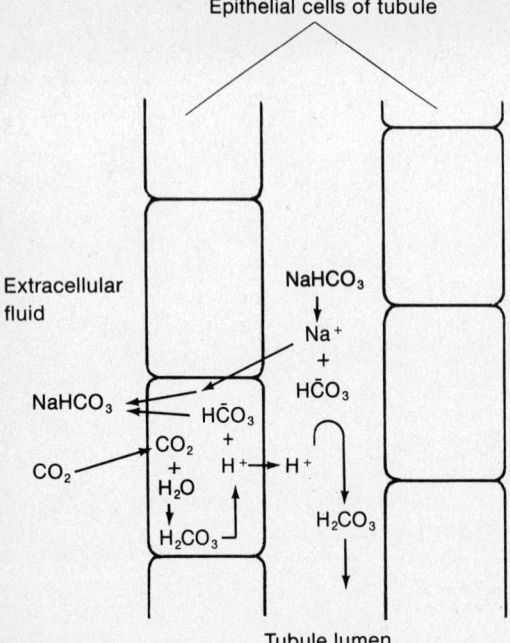

Fig. 15.8 The chemical reactions necessary for hydrogen ion secretion. Carbon dioxide absorbed into tubule cells reacts with H_2O forming H_2CO_3 (carbonic acid). H_2CO_3 promptly dissolves into H^+ ions and HCO_3 ions. The hydrogen ion enters the tubule (secretion), where it joins a bicarbonate ion producing H_2CO_3, which is then excreted as waste.

acid-base balances (see Chapter 17). Decreased CO_2 concentrations due to excess removal of CO_2 from interstitial fluids (for example, hyperventilation) reduce hydrogen ion secretion.

MAINTAINING FLUID BALANCES

The amount of fluid in the body must be maintained at a constant level if cells are to survive. There must be a balance between the volume of fluids entering the body and the volume of fluids leaving the body. Under normal conditions, fluid loss is adequately balanced by fluid intake. However, this balance is drastically and often fatally altered due to illness, injury, or organ malfunction. An ill or disabled patient may be incapable of drinking or eating. Body fluids, along with the essential substances in the proper concentration must be maintained by other procedures—intravenous feeding, for example. The seriously ill patient may also suffer additional loss of fluids through vomiting, diarrhea, hemorrhage, or seepage from traumatized tissue.

Body fluid is contained in two different types of reservoirs, referred to as intracellular and extracellular spaces. *Intracellular fluid* includes all fluids within all cells and represents the largest space. By weight the body is roughly 60 percent water. Of this 60 percent, 45 percent is due to the weight of cellular fluid. The remaining 15 percent is the weight of *extracellular fluid*, made up of blood, lymph, and interstitial fluids in the space between cells. These percentages are variable and dependent upon several factors such as age, sex, and obesity, as shown in Tables 15.2 and 15.3. Women generally have a smaller percentage of body fluids than do men. Children and babies have a very high percentage of body water. As we grow older, we progressively lose body fluids. Obese persons have a much lower percentage of body fluids, owing to the fact that many of the body spaces that normally would be filled with fluid are occupied by stored fat.

One reason why elderly and obese persons are relatively poor surgery risks is the fact that the percentage of body fluids is too low to begin with. The loss of more fluids during surgery can only increase the fluid imbalance at a time when it is most important to be maintained.

Cellular fluid is more chemically constant than extracellular fluid. It is surprising that the chemical and physical qualities of the extracellular fluid remain as constant as they do when we consider the diversity of foods eaten and the different amounts of fluids ingested each day.

If the kidney tubules cannot, for some reason, remove excess water, the extracellular space be-

Table 15.2 Relation of Body Weight to Body Fluids

	PERCENTAGE OF BODY WEIGHT		
	MALE ADULT	FEMALE ADULT	INFANT
Water	60	50	77
Solids and fats	40	50	23

Table 15.3 Variation of Water Content Related to Fat Content

	PERCENTAGE OF BODY WEIGHT		
	LEAN	NORMAL	OBESE
Males			
Water	70	60	50
Fat	4	18	32
Solids	26	22	18
Females			
Water	60	50	42
Fat	18	32	42
Solids	22	18	16

Table 15.4 Normal Urine Components and Characteristics

SUBSTANCE	AMOUNT EXCRETED DAILY
Water	600–2,000 ml
Urea	25–30 g
Creatinine	1.4 g
Ammonia	0.7 g
Uric acid	0.7 g
Ascorbic acid	15–50 mg
Chloride ions	10 g
Sodium ions	4 g (varies with intake)
Potassium ions	2 g (varies with intake)
Calcium	0.2 g
Magnesium	0.15 g
Phosphate ions	trace
Sulfate ions	trace
Hippuric ions	trace
Total volume	0.6–2.5 l (varies with H_2O intake)
Specific gravity	1.003–1.030
pH	4.7–8.0

comes overfilled with fluid, causing edema. A certain amount of edema may result in some tissues, even though the kidney function is adequate. For example, if one stands quietly for a long period of time, hydrostatic pressure builds up in the blood vessels of the legs and feet due to the pull of gravity on these fluids, forcing fluids into the interstitial spaces, causing edema.

COMPOSITION OF NORMAL URINE

Although urine is a mixture of water, wastes, and excess substances, its composition is surprisingly consistent within ranges established from healthy individuals.

Fresh, normal urine is a light amber color with little odor. If urine is permitted to stand, it takes on a strong odor of ammonia, owing to bacterial action and changes in pH.

The *specific gravity* of normal urine is from 1.010 to 1.025. (The specific gravity of water is 1.000. See the discussion of specific gravity in Chapter 1.)

Normal urine has a pH ranging from 5.0 to 7.8, with an average of 6.0 (slightly acid). The pH varies with the health and nutrition of the individual. Table 15.4 indicates components of normal urine.

COMPOSITION OF ABNORMAL URINE

Abnormal urine may vary considerably from the normal ranges. *Urinalysis*, the physical, chemical, and microscopic examination of urine, is an important medical procedure.

Dark-colored urine may be due to a variety of causes, including inadequate water intake and the presence of red blood cells. Although it has nothing to do with body malfunction—in fact, quite the contrary—red blood cells may be normally found in the urine of menstruating women. Concentrated, dark urine may also result from excessive loss of water owing to fever or profuse perspiration.

The presence of glucose in urine may indicate diabetes mellitus. However, a single test is not conclusive. Normal urine usually contains no glucose.

Albumin (protein) found in urine usually indicates one of several types of kidney infections. The presence of pus may also indicate kidney infection or disease.

Red blood cells normally cannot enter Bowman's capsule. When they are found in urine, there is a strong indication of acute damage to glomerular and capsular membranes or infection of parts of the kidney beyond the tubules, including the urinary tract. Table 15.5 lists some abnormal components of urine.

Table 15.5 Abnormal Components of Urine

1. Serum albumin—a normal constituent of plasma. Its presence in urine is usually due to increased permeability of the glomerulus membrane.
2. Glucose—normally absent from urine unless blood glucose level exceeds 180 mg percent.
3. Indican (indoxyl potassium sulfate)—a salt derived from indole. Indole is produced when protein food putrifies in the intestine.
4. Ketone bodies—substances that result as by products of excessive fatty-acid metabolism, e.g., diabetes mellitus.
5. Casts—molds of kidney tubule cells that detach and are washed away with urine. The type of material composing the cast may indicate specific malfunctions of the kidney.
6. Calculi—mineral salts in urine that may solidify.
7. Pus cells—cells that appear in urine due to infection of urinary organs.
8. Bile pigments—due to massive destruction of red blood cells caused by hemorrhage or hemolytic disease.

MICTURATION

The voiding of urine, or bladder emptying, is called *micturation (urination)*. This basic function of the urinary bladder is to store urine that has been accumulated from the ureters, one from each kidney.

The urinary bladder is essentially a muscular sac, as shown in Figure 15.9. The main muscle of the bladder is the *detrusor muscle*, which functions to squeeze the urine out of the bladder.

In children the bladder is somewhat cone shaped. In adults it is almost spherical when filled and has a triangular segment called the *trigone*. It is in this area that the ureters empty into the bladder. The bladder empties its contents into the *urethra*, which also originates at the distal end of the trigone.

As urine accumulates in the pelvis of each kidney it drains into the ureters. The ureters, however, are more than just passive drainage

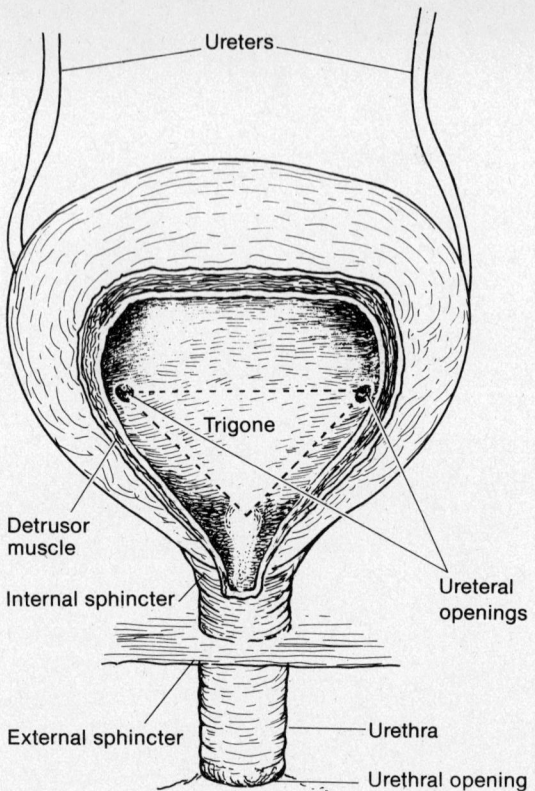

Fig. 15.9 The urinary bladder. As urine accumulates in the pelvis of each kidney, it drains into ureters. A pair of ureters (one from each kidney) transports urine to the urinary bladder. When the volume of urine in the adult bladder reaches 300 to 350 ml, the urge to urinate is initiated by nerve reflexes.

ducts for urine. Each ureter is lined with smooth muscle that is capable of peristaltic contractions that constantly squeeze and push the urine along in the direction of the urinary bladder.

The muscular bladder itself maintains a certain level of contraction called *tonus* and is capable of expanding without appreciably increasing the internal pressure. For example, as urine enters the bladder from the ureters the quantity increases. However, the bladder is capable of filling and becoming larger due to the extra urine entering without increasing the pressure within the bladder. When the volume of urine in the bladder reaches 300 to 350 ml the pressure does start increasing and triggers reflex actions that initiate micturation. The quantity of urine in the bladder that initiates micturation may deviate considerably from the quantity shown. Disease, medication, and emotional stress may all influence the quantity that stimulates reflex centers.

The female urethra is very short—about 3.8 cm. Since the urinary bladder is located directly behind the pubic symphysis and attached to the wall of the vagina, the distance between this and the external surfaces is very short. The urethra opens into the genital area between the clitoris and the opening of the vagina.

The male urethra is much longer—about 20 cm. Part of this extra length as compared to the female urethra is due to the fact that the male urethra is enclosed by the penis.

At the emptying end of the bladder, the trigone, a mass of glandular tissue called the *prostate gland* surrounds the male urethra. As we shall see later, the prostate gland is instrumental in supplying certain substances to the reproductive fluid. This structure, however, causes some complications in male urination, particularly in elderly males, when the prostate gland enlarges. Prostate gland enlargement causes a constriction of the urethra, making urination difficult. In extreme cases the prostate gland must be removed surgically or reduced in size by hormone treatment so the urination can occur normally.

RENAL DISEASES

Renal diseases are classified according to which portion of the nephron is affected. Two important dysfunctions of nephron activities are *acute renal shutdown* and *chronic renal insufficiency*.

Acute renal shutdown is a condition in which the kidneys stop functioning. One such instance involves an antigen-antibody reaction as a complication or side effect of streptococcal infection. The condition known as *glomerular nephritis* may result from this type of infection.

During severe streptococcal infection the microorganisms release a toxin that stimulates antibody production. These antibodies not only react with and neutralize the toxin but also, for some unknown reason, may react with the glomerular membrane of nephrons. This type of antigen-antibody reaction is known as *autoimmunity*. A similar reaction occurs with heart valve inflammation in rheumatic fever. When the glomerular membrane becomes inflamed, white blood cells collect in large numbers in the glomerulus, causing serious blocking of the filtering mechanisms of many of the glomeruli. The glomeruli that are not blocked become extremely permeable, allowing large protein molecules to filter into the capsule. The membrane may rupture, allowing red and white blood cells to enter the glomerular filtrate. If acute glomerular nephritis goes unattended the individual will die.

Chronic renal insufficiency results from a reduced number of nephrons and may be caused by periodic attacks of glomerular nephritis. Each occurrence of the disease destroys more and more nephrons.

Renal insufficiency may also result from arteriosclerosis, where the blood supply to the kidney is insufficient due to fibrous and calcified deposits collecting in arteries leading to the kidneys. Kidney cells, like all other cells in the body, need a constant supply of oxygen and nutrients. When these are lacking, nephron cells die. Renal failure requires immediate attention if the individual's life is to be maintained. The artificial kidney can be used to reduce the level of wastes and toxins.

Fluid and electrolyte balances must be restored and special precautions are necessary to prevent or slow infection. Antibiotic therapy, blood transfusions, dialysis, diet supervision, and asepsis are all crucial in the treatment of renal failure.

Pyelonephritis, the most common type of renal disease, is an infection of the renal pelvis and surrounding tissues. The infection results from bacterial invasion of the kidney pelvis from the blood, lymphatic system, or from the urinary bladder via the ureters.

The most common causative agent is *E. coli*, a common bacteria of the intestinal tract. *E. coli* is ordinarily a harmless microbe if it stays in the intestine. However, its invasion of the urinary tract results in a serious medical problem.

Other bacteria often associated with pyelonephritis are *Proteus vulgaris, Pseudomonas aeruginosa,* and *Aerobacter aerogenes*.

Pyelonephritis occurs more commonly in infants and women. The unhygienic practice of using infant diapers, although there are few alternatives, provides for the contamination of urinary tracts with fecal *E. coli.*

The female urethra, owing to its short length, can easily transmit bacteria to the rest of the urinary tract. Once again, unhygienic personal habits tend to increase the incidence of urinary infection.

Symptoms of pyelonephritis include chills and fever, nausea, vomiting, and abdominal and lower back pain.

Pyelonephritis treatment involves antimicrobial treatment, regulation of diet, and the maintenance of fluid and electrolyte balances.

Cystitis is an inflammation of the urinary bladder. It may precede or follow other urinary infections, such as pyelonephritis. Cystitis occurs more frequently in women than men and often is caused by microbial contamination of the urinary tract. Except in chronic conditions, cystitis may be controlled with antibiotic and chemical therapy in conjunction with large fluid intake. Excess fluids (3 or more liters per day) taken orally dilute the urine and make it less of an irritant to the inflamed tissue.

Renal calculi are crystalline bodies that form in the renal pelvis or renal tubes. When similar formations localize in the bladder they are called *bladder calculi.* Calculi are often small enough to pass through renal structures along with urine. Occasionally the crystals are needle-shaped and may physically damage renal tissue or in some instances become large enough to block renal passages, as shown in Figure 15.10.

Some probable causes of calculi include *hypercalcemia* (high blood calcium), which may be induced by a hyperparathyroid condition, gross excess of calcium in the diet, and bone deterioration due to the lack of physical activity (bedridden patients).

Renal tumors may be induced by a variety of factors including ingested chemicals. Carcinomas of the kidney and bladder may cause blockage of renal tubes and cause pain due to pressure exerted on nerves. Surgical removal of the tumor is often required.

Two major advances in maintaining the life of

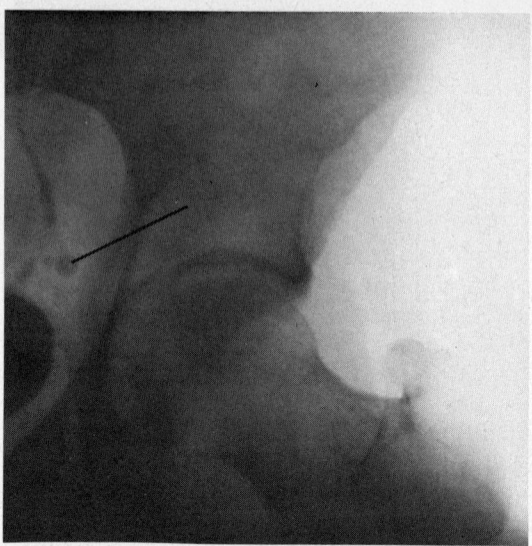

Fig. 15.10 X ray of very large bladder stone (arrow). *(Courtesy Normandy Osteopathic Hospital Radiology Department, St. Louis, Mo.)*

Kidneys and Calcium Absorption

The kidneys play so many important physiological roles it is difficult to list them all in a single text. Here is one more function from the long list.

Vitamin D is required to aid in the absorption of calcium from the intestine into the bloodstream. The vitamin in its raw form, however, must be given special treatment by the liver and kidneys before it is capable of performing its proper role. The liver secretes an enzyme that converts vitamin D to 25-hydroxy-vitamin D. This substance is then further modified in the kidney by another enzyme resulting in a compound called 1, 25 hydroxy-vitamin D. It is this substance that is directly responsible for the absorption of calcium in the intestine. If the kidneys are diseased, the normal utilization of vitamin D is altered. Individuals suffering from kidney disease therefore may show symptoms similar to rickets, a disease of vitamin deficiency.

---

patients who have lost the function of both kidneys are dialysis and kidney transplants. *Hemodialysis* is a procedure that passes the patient's blood through a coil of cellophanelike tube surrounded by a specifically designed fluid. as shown in Figure 15.11. Body wastes are

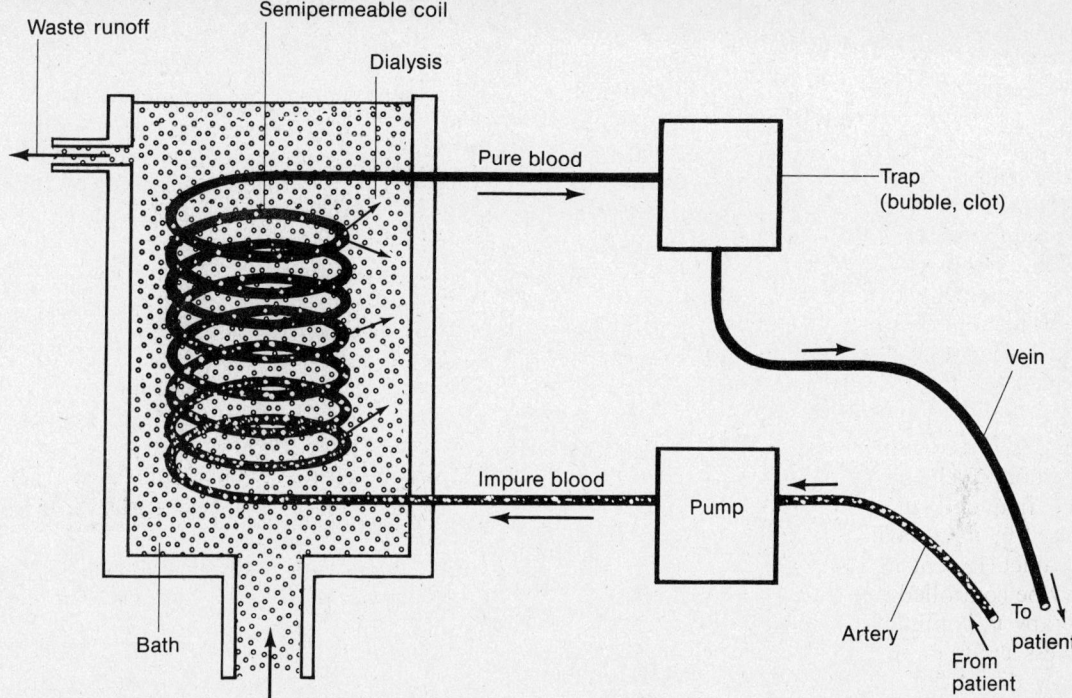

Fig. 15.11 The kidney machine, used for hemodialysis. In this procedure, the patient's blood is pumped through a tube of cellophanelike material surrounded by a fluid bath. Wastes are selectively removed from the blood, which is, after cleansing, returned to the patient.

filtered out of the tube into the surrounding, circulating bath by osmosis.

Hemodialysis requires that the patient be hooked up to the apparatus once or twice a week for 3 to 6 hours. Hemodialysis is at present an expensive process using expensive equipment. However, for the patient requiring such treatment the alternative may be death. It is hoped that the design of hemodialysis apparatus in the future will lead to more compact units less expensive to maintain.

Peritoneal dialysis utilizes *peritoneal* (abdominal lining) *membranes* to remove wastes. A special fluid is introduced into the peritoneal cavity where it may remain for a short time and then be removed and discarded. Osmosis filters some substances from the blood across the peritoneal membranes into the cavity.

Kidney transplants are relatively new in medicine, the first successful one occurring in 1954. Some of the same problems plague the procedure today as they did then, namely the problems of locating a suitable donor and the tendency of one person's tissues rejecting those of another.

The more closely related the donor and host are, the less likelihood that the tissues are incompatible. The ideal donor-host relationship is identical twins. Beyond the problem of tissue incompatibility lies another problem, that of infection. Surgical procedures are never completely aseptic. The recipient of a transplant is given medication to reduce the effectiveness of the person's *immunological system*. This system builds antibodies that not only fight bacterial infection, but also produce antibodies in response to foreign protein (the transplanted kidney). The reaction between these antibodies and the transplant results in its rejection. To prevent this, the antibody-producing machinery is inhibited before the surgery. This inhibition reduces the body's defenses against bacteria, and therefore there is a greater risk of infection following transplant surgery compared to other surgical procedures.

THE AGING KIDNEY

Many so-called kidney problems of the aged are the result of faulty blood (renal) circulation. The kidney cannot properly do its job if the blood to be purified is not available in adequate quantities. It has been estimated that by age 80, the flow of blood through the kidney has been reduced by 55 percent. There is some evidence that the total number of glomeruli is reduced. However, the blood supply remains the more important health problem.

SUMMARY AND REVIEW

A. Functions of the excretory system
  1. Removes wastes from the blood.
  2. Aids in maintaining normal fluid balances.
  3. Aids in maintaining normal blood pH.
  4. Controls sodium ion concentration.
B. Basic anatomy of the urinary system
  1. The kidney removes wastes from the blood.
  2. The ureter carries urine to urinary bladder.
  3. The urinary bladder stores urine.
  4. The urethra is a structure that drains the urinary bladder.
C. Internal structure of the kidney
  1. The capsule is the outer covering of the kidney.
  2. The cortex is the outer layer of kidney tissue.
  3. The medulla is kidney tissue internal to the cortex.
  4. The pelvis is a cavity into which collecting tubes empty urine.
D. Structure of a nephron, the basic kidney unit
  1. The renal artery branches into afferent arterioles that enter a glomerulus.
  2. The glomerulus is a knot of capillaries enclosed by Bowman's capsule.
  3. Bowman's capsule is the structure into which is absorbed most of the plasma contents.
  4. Efferent arterioles leave the glomerulus and branch into the peritubular capillaries.
  5. Nephron tubules continuous with Bowman's capsule accept the glomerular filtrate. Most of the absorbed plasma contents are reabsorbed into the peritubular capillaries from nephron tubules.
E. Substances get into Bowman's capsule mainly by passive diffusion due to hydrostatic pressure.
F. Glomerular filtrate, the substance absorbed into Bowman's capsule, is similar to plasma except for large blood proteins.
G. Tubular reabsorption is the process whereby most of the glomerular filtrate is reabsorbed into the blood.
H. Sodium ions are controlled by two mechanisms:
  1. Aldosterone secreted by the adrenal cortex makes the tubules more permeable to Na^+, permitting its reabsorption.
  2. The countercurrent mechanism, which involves the loop of Henle, the vasa recta, and collecting tubules.
I. Acid-base balance of the blood is maintained by the tubular secretion of H^+ ions into the lumen of the tubule and thus excreted.
J. Water balance is maintained:
  1. indirectly by the action of aldosterone (water follows sodium).
  2. directly by the action of ADH upon nephron tubules, making them more permeable to water molecules.
K. Normal urine is a mixture of water, salts, and waste products. Urinalysis is the microscopic and chemical inspection of these substances.
L. Abnormal urine may contain red and white blood cells, bacteria, pus, and glucose.
M. Renal diseases
  1. glomerular nephritis
  2. pyelonephritis
  3. cystitis
  4. renal tumors
N. Aging kidney. Most kidney problems in aging are the result of inadequate renal circulation rather than the loss of or defective nephrons.

16

Body Fluids and Electrolytes

MAJOR CONCEPTS
  FUNCTIONS OF BODY FLUIDS
  FLUIDS AND BODY WEIGHT
  ORGANS THAT REGULATE FLUIDS
    The Kidney
    The Posterior Pituitary
    The Adrenal Cortex
  ELECTROLYTES
    Sodium
    Potassium
    Potassium Deficit (Hypokalemia)
    Potassium Excess (Hyperkalemia)
    Calcium
    Magnesium
    The Chloride Ion
    The Phosphate Ion
  ACID-BASE BALANCE
  RESPIRATORY AND METABOLIC ACIDOSIS
  RESPIRATORY AND METABOLIC ALKALOSIS

The major function of body fluids is to transport all essential substances to cells and carry away metabolic wastes. Although much of this has been discussed in previous chapters, some of these functions and where they occur in the body will be reviewed here. Also included in this chapter is a discussion of the major electrolytes found in body fluids, their normal concentrations, what they do, and what happens if they are excessive or deficient.

Body fluids are housed in several compartments. The two major compartments are the *extracellular compartment* and the *intracellular compartment*. The extracellular compartment contains fluids outside cell membranes that include *plasma* and *interstitial fluids* (fluids surrounding cells). The intracellular compartment contains all the fluids within cell membranes. Approximately 60 percent of an adult male's body weight is due to fluids. Of this 60 percent, about 45 percent is intracellular and 15 percent is extracellular. The body of a healthy individual maintains these percentages within a fairly narrow range.

With only membranes separating the compartments, how is the rather constant quantity of water in each compartment maintained? As we will see, there is a direct relationship between fluid balance among the compartments and dissolved substances in the fluids, creating osmotic pressures.

FUNCTIONS OF BODY FLUIDS

The major function of body fluids is to transport all essential substances to cells and carry away metabolic wastes. Other functions include heat distribution, hormone and enzyme transport, and transport of substances called *electrolytes*. Electrolytes are substances that dissociate in solution, yielding charged particles called ions.

Fluids move from one fluid compartment to another. The exchange constantly bathes cells with nourishment and oxygen. The processes involved in fluid movement include osmosis and active transport. These processes have been defined and discussed elsewhere in the text. However, a summary of each will be given here.

Osmosis is the process by which water molecules move across semipermeable membranes. The direction of movement is always from an area of the highest concentration of water molecules (a dilute solution) to an area of lower water concentration (a concentrated solution). If pure water is placed on one side of a semipermeable membrane and a salt solution is placed on the other side, water molecules will move from the pure water side across the membrane into the salt solution.

Active transport is the movement of materials through a membrane from an area of low concentration toward an area of higher concentration. This process produces a flow of materials opposite to that of osmosis. Active transport requires an energy expenditure whereas osmosis is a passive process requiring no energy. The process of active transport is believed to involve specialized carrier molecules. These molecules are capable of forming a loose chemical attachment with the substance to be transported across a membrane. Upon the arrival of these substances at the other side of the membrane, the carrier molecule dissociates from the transported substance.

Diffusion, like osmosis, is a physical process. Diffusion may be defined as a scattering of molecules from an area of high concentration to an area of low concentration. Diffusion may occur across membranes. The passage of oxygen from lung alveoli into the blood capillaries and carbon dioxide from blood capillaries into lung alveoli is accomplished by diffusion.

FLUIDS AND BODY WEIGHT

The percentage of body weight assigned to fluids depends upon several factors including age, sex, and fat content.

Compartments that normally contain fluids when filled with fat reduce the total fluids in the body. This makes obese persons poor surgery risks, since fluid balances are difficult to maintain. The loss of plasma fluids during surgery reduces overall fluid levels. After surgery, the individual may also drastically reduce fluid intake. If a person enters surgery with a fluid deficit (owing to obesity or other factors), the problem of maintaining adequate fluids is intensified.

ORGANS THAT REGULATE FLUIDS

Some body organs directly regulate fluid quantities, while others have an indirect effect. The major direct controlling organ is the kidney. Indirect control is maintained by secretions of the posterior pituitary and adrenal cortex.

The Kidney

The *kidney* regulates not only the quantity of fluids but also substances dissolved in the fluids. In a 24-hour period, the kidneys filter out approximately 180 liters of fluid. All but about 1 percent of this (1.8 liters) is restored to the bloodstream. The quantities used here represent an average since the quantity of fluids ingested may be highly variable, but the normal body needs a certain amount of fluid at all times. When excess fluid is ingested, the kidneys collect and discard more. When small quantities of fluids are ingested, the kidneys collect and discard smaller amounts.

Water may readily shift from one fluid compartment to another. Intracellular fluid content is more stable than extracellular fluid content. There is a constant interaction between plasma and interstitial fluids in normal individuals.

Fluid levels in these compartments may fluctuate drastically due to diseases, burns, or hemorrhages. When large quantities of fluids are lost from extracellular spaces, fluids from the intracellular spaces will also be reduced in time. The term *hypovolemia* is used to describe extracellular fluid deficit. In addition to the causes listed above, other factors may include diarrhea, vomiting, and excess perspiration.

Hypovolemia presents potentially dangerous problems for the body. Since plasma is the main fluid affected, blood pressure drops and the heart rate increases in an attempt to compensate for the reduced blood volume. This puts an additional workload on the heart. In addition, a measurable weight loss may be recorded, as high as 5 to 8 percent of the normal body weight. Substances dissolved in these fluids are also lost. In fact, the loss of electrolytes often has a more profound influence upon the body than the water loss. The loss of electrolytes may initially go unnoticed, since the electrolyte-water ratio may remain relatively constant. As water is lost, the electrolytes are also lost, in approximately the same ratio as the remaining fluids.

Hypervolemia describes an excess of extracellular fluid. This condition may result from kidney malfunction in which excess fluids are retained rather than excreted. In hypervolemia, the ratio between water and electrolytes is altered.

The Posterior Pituitary

The *posterior pituitary* secretes antidiuretic hormone (ADH) that influences kidney tubules to reabsorb much of the water filtered from plasma. A reduction in the secretion of ADH reduces reabsorption, and, therefore, water is excreted in large quantities. The controlled secretion of ADH thus controls plasma water as body needs change.

The Adrenal Cortex

The *adrenal cortex* secretes the hormone aldosterone, which controls the reabsorption of sodium from the kidney tubule. If sodium is retained in the tubule, water is also retained, both being excreted as a component of urine.

Thus far our discussion has been primarily about fluid balance. The term "electrolyte" has been referred to several times. It is nearly impossible to discuss fluid balance without discussing electrolyte balance. The shift in electrolyte balance in one fluid compartment may produce both an electrolyte shift and a fluid shift in another compartment.

ELECTROLYTES

Substances whose molecules dissociate in solution yielding ions are called *electrolytes*. Important electrolytes include sodium, potassium, calcium, and bicarbonate. Substances that do not dissociate (nonelectrolytes) may also be dissolved in body fluids and may also affect fluid balances. These include glucose and urea. For the moment we will concentrate on the electrolytes.

Electrolytes serve a variety of functions. Some are required for the synthesis of large molecules. Others, such as the hydrogen and bicarbonate ions, are involved in acid-base balance of body fluids. One of the most important functions is the production of osmotic pressures. Substances, by osmosis, cross semipermeable membranes. The direction of flow of substances is from an area of high concentration to an area of lower concentration. The ease with which substances cross membranes depends upon the number of particles in solution. Particles may be atoms, molecules, or ions.

The supreme test of the force created across the membrane by a dissolved substance would be to place pure water on the opposite side of the membrane. Water molecules will rush across the membrane into the solution while, if the membrane permits, substances in solution will rush into the water. The maximum force created is called the *potential osmotic pressure* (P.O.P.)

The formula for deriving the P.O.P. is as follows:

P.O.P. = molar concentration of solution × number of ions × 19,300

The number 19,300 is a constant. The answer will be in mm of Hg. In order to determine the P.O.P., one need only know what the molarity of the solution is and what ions, if any, are dissociated. A substance with two dissociated ions would create a greater pressure than would one with no ions.

Table 16.1 illustrates P.O.P. of 5 percent glucose and 5 percent NaCl. A substance such as Na_3PO_4 would yield three sodium ions and one phosphate, for a total of four ions. Na_3PO_4 would therefore create a P.O.P. twice that of NaCl (2 ions) if the molarity of both solutions were initially the same.

Table 16.1 Potential Osmotic Pressure of Glucose and Sodium Chloride

SUBSTANCE (5% SOLUTION)	AMOUNT	P.O.P. (mm Hg)
Glucose =	0.27 molar =	0.27 × 19,300 = 5,130
Sodium chloride =	0.86 molar =	0.86 × 2 × 19,300 = 32,680

Electrolyte balance means that the number of negatively charged ions, *anions*, in a solution must equal the positively charged ions, *cations*.

Since we cannot actually count ions, a system has been devised to equate numbers. This system is called the *milliequivalent system* (mEq). With this system the quantities of electrolytes are expressed in mEq/liter. Table 16.2 shows how mEq is derived within a fluid compartment such as plasma. The total number of cation mEq/liter should equal the total number of anion mEq/liter. The three fluid compartments differ in their composition with respect to electrolytes. Table 16.3 compares the electrolyte composition of the different fluid compartments' substances. Glucose and urea are not normally considered to be electrolytes. However, both of these substances specifically influence osmotic pressures. Their effect is accomplished by increasing the number of particles (molecules) in solution.

Table 16.2 Milliequivalents of Some Common Substances

SUBSTANCE	MOLE g	MILLI-MOLE g	mEq (MILLIMOLE VALENCE) g
Sodium	23	0.023	0.023
Calcium	40	0.040	0.020
Potassium	39.1	0.0391	0.0391
Magnesium	24.3	0.0243	0.0121

Note in Table 16.1 the concentration of sodium and potassium in each of the three fluid compartments. Plasma electrolytes and interstitial fluid electrolytes are roughly the same concentrations with respect to both cations and anions. One major deviation is the lack of proteinate in interstitial fluid. Intracellular fluid concentration of electrolytes deviates radically from the other two fluid compartments. The sodium ion is greatly reduced, while potassium is increased.

Of the three fluid compartments, plasma is the most readily accessible for analysis. The constant flow of electrolytes from one compartment to another alters the concentration in each, although if only one of these fluids is available for study, some prediction can be made concerning the other spaces. Disease, trauma, drugs, or physical disorders may alter the permeability of membranes separating the fluid spaces. This altered permeability may involve many ions at once or individual ions may be singled out with a

Table 16.3 Comparison (in mEq/l) of Electrolyte Composition of Different Fluid Compartment Substance

CATIONS	mEq/l	ANIONS	mEq/l
Plasma electrolytes			
Na^+	142	HCO_3^-	24
K^+	5	Cl^-	105
Ca^{++}	5	HPO_4^-	2
Mg^{++}	2	SO_4^-	1
		Organic acid-	6
		Proteinate-	16
Total	154		154
Interstitial fluid electrolytes			
Na^+	142	HCO_3^-	30
K^+	5	Cl^-	114
Ca^{++}	5	HPO_4^-	2
Mg^{++}	2	SO_4^-	1
		Organic acid-	7.5
		Proteinate-	16
Total	154.5		154.5
Intracellular fluid electrolytes			
Na^+	142	HCO_3^-	10
K^+	5	Cl^-	1
Ca^{++}	5	HPO_4^-	100
Mg^{++}	2	SO_4^-	20
		Proteinate	63
		Proteinate-	16
Total	194		194

special pass through the membrane. If the fluid can be analyzed when this occurs, the increased or decreased electrolyte concentration can be detected. The analysis of fluids, plasma in particular, with regard to electrolyte concentration has in the past 25 years added a new dimension to medical practice.

Each of the electrolytes has precise, specific functions. The following discussion will elaborate on these functions.

Sodium

Sodium, a cation, is the single most abundant electrolyte in plasma and interstitial fluids. Approximately 90 percent of all cations in these spaces is sodium. Sodium cannot easily cross membranes. It is this characteristic that makes sodium an important factor in determining where water is held and where it is released.

The kidney tubules have an intimate relationship with the sodium ion. If one can determine where the highest sodium concentration is, there will also be found the largest quantity of water. In other words, water follows sodium across membranes.

A classical example of homeostasis referred to in the chapter on excretion warrants repeating here. The hormone ADH, secreted by the anterior pituitary, is inhibited by a decreased sodium concentration in blood plasma. ADH increases the reabsorption of water into the peritubular capillaries, thus conserving body water. Plasma sodium decrease below normal causes an inhibition of ADH. In response to the reduced ADH, the kidney tubule permits more water to be retained, therefore returning a smaller than normal quantity of water to plasma. The retained water is excreted in urine. Plasma with its decreased sodium now has a decreased water concentration, and the proper ratio of 142 mEq/liter has been reestablished. Once the sodium deficit is corrected, ADH is no longer inhibited. This homeostatic control also is in effect if plasma sodium levels are higher than normal. In this case, ADH secretion is stimulated. The kidney tubules become more permeable to water, thereby diluting sodium in plasma back to the normal (142 mEq/liter).

A severe sodium loss due to any cause leads to severe water loss for the body. This loss leads to hypovolemia and predictable, undesirable, and often fatal effects. The precise daily requirement of sodium per kg of body weight is not known, but the average requirement is approximately 6 g per day for adults. The daily requirement of sodium for children is about 3 g and for infants about 1 g. The daily exchange (ingested-excreted) is approximately the same regardless of age. The major source of sodium in the human diet is sodium chloride.

Salt-free or low-salt diets are often recommended for persons suffering from hypertension. When excess salt is ingested, excess water may be retained. Excess retention of water contributes to hypervolemia, which may contribute to greater hypertension. Persons wishing to lose weight are often placed on low-salt diets. Less salt means less water retained, which means less weight.

The role of sodium in this brief narrative has been only superficially explored. The precise medical control of this electrolyte may mean the difference between health and illness, even life and death.

Are You a Salt Addict?

The average adult needs about 150 to 300 mg of salt (NaCl) daily. This is about 1/10 teaspoon. Most Americans use 6 to 40 times this amount. When at home, they may routinely salt food before tasting it. So much salt is available in processed foods that the average person far exceeds the daily requirement. Beef or chicken bouillon cubes may be 50 percent or more salt.

A high-salt diet produces a gradual rise in blood pressure in susceptible individuals. Children of parents susceptible to hypertension are excellent targets for salt craving. These children, with a potential strike already against them, are unknowingly creating additional hazards for middle-age hypertension and a potentially shortened life.

Potassium

Potassium is a cation essential for protein synthesis, nerve impulse transmission, carbohydrate metabolism, and many other normal body functions. The human body lacks specific hormones that directly regulate potassium concentrations in body fluids. The kidneys may continue to excrete potassium even though a potassium deficit exists elsewhere in the body.

Referring again to Table 16.1, note that the potassium ion is the dominant cation in intracellular fluids. In other words, most of the body's potassium is located in cells with a concentration of approximately 150 mEq/liter. Potassium, like sodium, does not easily cross cell membranes. A potassium deficit may exist in cells and an excess simultaneously may occur in extracellular fluids.

An exact reverse may also occur. Either of these conditions may have profound, even fatal, consequences.

A close relationship apparently exists between sodium and potassium. An excess of one tends to reduce the concentration of the other. Potassium concentration is therefore controlled to some degree by sodium concentration. Either a deficit or a surplus of potassium may result in unpleasant circumstances. The body lacks the capability to conserve potassium. The daily loss of potassium is somewhere around 40 to 50 mEq. This rate of loss continues irrespective of potassium deficit in the body.

Potassium Deficit (Hypokalemia)

Potassium deficit, or *hypokalemia*, is seldom due to inadequate intake. The deficit usually is associated with fluid loss from such causes as diarrhea, diseases of the intestinal tract, surgical operations of the intestinal tract, excessive use of diuretics, fever, and excessive sweating.

Extracellular fluid normally has a potassium concentration of 3.5 to 5.5 mEq/liter. Less than 3.5 mEq/liter is considered a potassium deficit. A deficit can be corrected by ingesting potassium salt or including high-potassium foods in the diet.

Low potassium affects the heart, causing reduced cardiac output. Skeletal muscles weaken to such a degree that peristalsis is inhibited and respiratory function is affected.

Potassium Excess (Hyperkalemia)

Potassium excess, or *hyperkalemia*, is more rare than hypokalemia. Most of the body's potassium is kept within cell membranes. If large numbers of cells are crushed through injury, large quantities of potassium may be released into the extracellular spaces, creating an excess. Other excesses may be found in persons whose kidney function is restricted and therefore incapable of excreting the normal 40 to 50 mEq of potassium per day. Hyperkalemia is indicated with a laboratory finding in excess of 5.6 mEq/liter. An excess of 7 mEq/liter may affect heart muscles sufficiently to produce cardiac arrest.

Calcium

Calcium is an important component in teeth and bones. It is also essential for muscle contraction and nerve transmission. Calcium supplied in normal diets is absorbed from the intestines into the blood with the help of vitamin D. The normal extracellular fluid concentration of calcium is 5 mEq/liter. A reduced calcium level, *hypocalcemia*, may result from inadequate calcium in the diet, excess loss due to diarrhea, pregnancy, or malfunction of the parathyroid gland. The symptoms of hypocalcemia include muscular cramps, tetany, and convulsions.

Hypercalcemia occurs when the plasma concentration exceeds 5.8 mEq/liter. The condition may be due to excessive intake (too much milk), excess vitamin D, or a hyperactive parathyroid gland. (See Chapter 5.) An excess of parathyroid hormone causes bone calcium to become soluble and be reabsorbed into the bloodstream. This, in addition to raising blood calcium, leaves the bones porous (spongy), permitting fractures to occur with relative ease. Hypercalcemia also depresses muscle metabolism, including the heart.

Magnesium

Magnesium is a cation essential for enzyme activity. Its role as an enzyme activator is important in carbohydrate and protein metabolism and neuromuscular chemistry. Normal plasma levels range between 1.7 to 2.3 mEq/liter. Magnesium is found in smaller concentrations in bones and within cells.

The anions all have a negative charge. Under the influence of a direct electric current anions migrate to the positive pole. The following discussion will deal with the most abundant and significant anions in the body.

The Chloride Ion

The *chloride ion* is usually associated with sodium, so a loss of sodium usually also indicates a loss of the chloride ion. Sodium and chloride loss are not 100 percent correlated since another anion, the bicarbonate ion, may replace chloride. In other words, chloride may be lost with an equivalent sodium loss. As we will see in a moment, the shift of the bicarbonate ion alters the pH of fluids. The concentration of plasma chloride is 100 to 106 mEq/liter, making it the most abundant anion in the body.

The Phosphate Ion

The *phosphate ion* is the most abundant anion in intracellular fluid, with a concentration of about 100 mEq/liter of cell water. Plasma concentrations are much lower—about 1.7 to 2.3 mEq/liter. Phosphate function is controlled metabolically in a way similar to calcium. Parathyroid hormone inhibits reabsorption of phosphates by the kidney tubules. An increase in parathyroid hormone therefore increases the excretion of phosphates.

Other anions found in body fluids include *sulfates, organic acids,* and *protein.* Although the latter two substances may be present as undissociated molecules, they react chemically as if they carried negative charges. The most important organic acid is *lactic acid,* which plays an essential role in metabolism during strenuous exercise. This role was discussed in Chapter 6.

Proteins as anions play an important role in determining the direction fluids flow from one fluid compartment to another. Plasma protein concentration exceeds interstitial fluid concentration, therefore "pulling" water into the capillaries from interstitial spaces. Although this concept was discussed in Chapter 15, a review of protein influence on fluid direction will be useful.

Fluids entering capillary beds by way of arterioles are under a pressure created by the pumping action of the heart. This pressure is adequate to force fluids into the capillaries and through the capillary walls into the interstitial spaces. This pressure is called *hydrostatic pressure.*

Another pressure, *osmotic pressure,* due mainly to protein concentration differences on either side of the capillary wall, tends to pull fluids back into the capillary. As fluids enter the minute capillary network, the hydrostatic pressure is reduced. Fluids can only be pushed into the interstitial spaces when the hydrostatic pressure exceeds the opposing osmotic pressure. Once the hydrostatic pressure is reduced further along the capillary, osmotic pressure gains the upper hand, pushing fluids back into the capillary. Fluids from this end of the capillary enter venules, then veins, and return to the heart.

ACID-BASE BALANCE

Many body systems play important roles in maintaining an equilibrium between acids and bases. The normal pH of extracellular fluids falls between 7.35 and 7.45 (slightly alkaline). A shift of the pH to as small an amount as 7.0 may be fatal. A variety of causes may alter acid and base concentrations in body fluids. Excess acids and bases may be ingested or be due to respiratory or metabolic processes. Normally, these excesses are properly disposed of by kidney excretion, the respiratory system, or chemical buffers. Occasions may arise however, in which these safeguards are inadequate, owing either to malfunction or simply to overload. When this occurs, the increased acid or base content of body fluids presents some rather serious consequences for tissues.

The anion particularly important in acid-base balance is the *bicarbonate ion*. Carbon dioxide, dissolved in body fluids during respiration, reacts with water, forming carbonic acid (H_2CO_3). This molecule, when dissociated into ions, yields hydrogen ions (H^+) and bicarbonate ions (HCO_3^-). Excess hydrogen ions are normally eliminated from fluids by buffer systems. The bicarbonate ion is also normally buffered by its union with the cations of sodium or potassium. This molecule, composed of a cation plus the bicarbonate ion, is called *base bicarbonate* ($B.HCO_3$). The ratio of base bicarbonate to carbonic acid in plasma is normally about 20 to 1. As long as this ratio is maintained there is no problem. In other words, if both base bicarbonate and carbonic acid levels increase or decrease in the same proportions, the acid-base balance is maintained. It is only when this ratio is disrupted that the body finds itself in trouble.

RESPIRATORY AND METABOLIC ACIDOSIS

If a person holds his breath or is a very shallow breather, insufficient carbon dioxide is exhaled. This increases the carbonic acid concentration in tissue fluids, upsetting the 20-to-1 ratio of base bicarbonate to carbonic acid. The ratio may be decreased to, say, 13 to 1. When this occurs plasma pH may be reduced to 7.21 (from the normal 7.4). This condition is called *respiratory acidosis*. Respiratory acidosis may result from a variety of causes. In addition to those just given, these include pneumonia, respiratory tract obstruction, or reduced alveolar surface area. All these factors plus any others that inhibit the exchange of gases may cause respiratory acidosis.

Metabolic acidosis results from either an increase of the hydrogen ion in body fluids or a decrease in base bicarbonate. Hydrogen ion increases may be due to excessive ingestion of acids, the formation of organic acids by body cells, or the inadequate secretion of hydrogen ions by renal tubules. Metabolic acidosis can also result from diarrhea. Colon fluids are rich in base bicarbonate. When this is excessively flushed out, the acid-base balance is disturbed. Acidosis as a consequence of either respiratory or metabolic causes can have serious, even fatal, effects. The nervous system is severely depressed, which may lead to coma and death.

RESPIRATORY AND METABOLIC ALKALOSIS

Respiratory alkalosis results when excess carbon dioxide is exhaled *(hyperventilation)*. The 20-to-1 ratio of base bicarbonate to acid is disturbed, resulting in an increased plasma pH from the normal 7.4 to 7.70 or higher.

Metabolic alkalosis, which occurs much less often than respiratory alkalosis, usually is caused by ingesting large quantities of alkaline substances. Another cause may be the vomiting of stomach contents without vomiting the upper intestinal contents.

Alkalosis from any cause results in hyperexcitement of the nervous system. The individual suffering from alkalosis may be extremely nervous and excitable, often with muscles of the arms, abdomen, and face going into tetany.

Convulsions are not uncommon, particularly in individuals with a history of epilepsy.

SUMMARY AND REVIEW

A. Body fluids are located in fluid compartments. These include:
1. Extracellular fluid compartment (fluid outside cell membranes). Two such fluids are plasma and interstitial fluid.
2. Intracellular fluid compartment (fluid within cell membranes).

B. Fluid movement from one compartment to another includes the processes of:
1. osmosis
2. active transport
3. diffusion

C. Organs that regulate fluids include:
1. The kidney. Kidney nephron tubules are capable of reabsorbing most of the fluid filtered from plasma and returning it to plasma.
2. The posterior pituitary gland secretes ADH, which increases the permeability of kidney tubules for water.

D. The adrenal cortex secretes the hormone aldosterone, which increases the reabsorption of sodium ions by kidney tubules. Water molecules follow sodium ions across membranes. Thus water reabsorption is increased.

E. Electrolytes are substances whose molecules dissociate, yielding positive and negative ions. Ions influence the rate of flow of fluids across membranes due to their number and their electrical charge. The major body electrolytes include:
1. Sodium, which makes up about 90 percent of all cations in the body.
2. Potassium, a cation essential for protein synthesis, nerve impulse, and carbohydrate metabolism. Potassium deficit (hypokalemia) and potassium excess (hyperkalemia) affect cardiac tissue.
3. Calcium, a cation important in teeth, bones, muscular contraction, and nerve impulses.
4. Magnesium, a cation essential to initiate some enzyme activity.
5. Other ions are:
   a. chloride
   b. phosphate
   c. sulfates
   d. organic acids
   e. proteins (proteinate)

E. Acid-base balance
1. Normal pH of extracellular fluids is 7.35 to 7.45.
2. Excess acidity of plasma may be brought about by respiratory or metabolic acidosis.
3. Excess alkalinity (higher pH) may be brought about by respiratory alkalosis or metabolic alkalosis.

17

The Respiratory System

MAJOR CONCEPTS
ANATOMY OF THE BREATHING APPARATUS
THE MECHANISMS OF BREATHING
THE ALVEOLI
REGULATION OF BREATHING
  Inspiratory Centers in the Medulla
  The Hering-Breuer Reflex
  Reflex Centers in the Pons
  Chemical Control of Respiration
TYPES OF BREATHING
PHYSICAL OBSTRUCTION AND RESISTANCE TO AIR FLOW
GASEOUS TRANSPORT
  Transport of Oxygen
  Transport of Carbon Dioxide
  Lung Capacity and Air Volume
  Carbon Monoxide Poisoning
MICROBIAL DISEASES OF THE RESPIRATORY SYSTEM
RESPIRATORY CHANGES IN THE AGED

Respiration is the exchange of gases (oxygen and carbon dioxide) between the external atmosphere and cells. The processes discussed in this chapter include breathing and structures associated with it, the transfer of gases across lung membranes, and the transfer of gases from body fluids into and from cells.

The primary function of the respiratory system is to furnish oxygen for the tissues and carry away carbon dioxide. The respiratory system includes the *lungs, bronchial* and *bronchiole tubes,* the *pharynx, nose,* and *mouth.* The biochemical transactions occurring within cells that are responsible for the release and storage of chemical energy are referred to as *cellular respiration,* that is, the exchange of gases between the atmosphere and cells. This process includes: (1) the inspiration and expiration of gases; (2) the transfer of gases across lung membranes into and out of the blood; (3) the transfer of gases from the blood through the capillaries to cells and from the cells to the blood. The gases referred to here are oxygen (O_2) and carbon dioxide (CO_2). Living cells require a constant supply of oxygen. If it is withheld for only a few minutes, death is the consequence.

The need to remove carbon dioxide is also essential, although not as immediately critical as oxygen intake.

ANATOMY OF THE BREATHING APPARATUS

The structures normally associated with breathing are shown in Figure 17.1. You will find yourself referring to this diagram often as we proceed with the discussion of the respiratory system.

The normal pathway for airflow into and out of the body is through the nasal passages. However, due to obstructions or habit, many persons are mouth breathers. There are some disadvantages, even dangers, in mouth breathing. The external nostril openings, the *external nares,* have fine hairs protruding into their channels that catch and hold particulate matter, keeping it out of the lungs. The entire nasal passage is covered with mucous membranes that secrete mucus. The more posterior air passages, the *internal nares,* contain cilia that collect even smaller particles missed by the hair. Cilia, by a wavelike motion, move the debris into the throat, where it is swallowed.

The mucous membranes are also responsible for warming the air and adding moisture to it. Air entering the mouth is not as carefully conditioned before it enters the delicate lung membranes.

Moisturized air is critical not only for the proper function of lung tissues but also for the activity of the cilia. Dry air inhibits cilia activity and there is a tendency to cough more when dry air is breathed. Dry cilia are less active, permitting extraneous material to be lodged in the lower pharynx and tracheal levels, where it stimulates nerves, setting off the cough reflex.

The mouth and nasal passages join at the back of the throat into the funnel-shaped *pharynx.* This tube accepts not only air but also food and water. It later divides into two branches, the *esophagus,* for the water and nutrients, and the *trachea,* which carries gases to and from the lungs. The trachea is reinforced its entire length by cartilage rings that prevent its collapse during the mechanics of breathing. The upper part of the trachea, called the *larynx,* contains the vocal cords. Just above the larynx is a flap of tissue called the *epiglottis.* When food or water is swallowed, the epiglottis is pushed down to cover the upper part of the larynx called the *glottis.* This, along with a muscular elevation of the glottis during swallowing, prevents food from entering the trachea. These preventive measures have not been set in motion if a person inhales air through a mouthful of food. The food is pushed along with the air into the trachea or the lower breathing tubes, causing a blocking of the air passages and initiating the choking reflex. This is an uncontrolled violent expulsion of air from the lungs attempting to dislodge the foreign particle.

The lower trachea divides into the right and left *bronchi,* which also contain cartilagenous rings for support. Each bronchus divides into small tubules called *bronchioles,* which subdivide into smaller and smaller branches. There are 23 to 25 divisions of the bronchi and bronchioles, beginning at the trachea and ending with the alveoli. The alveoli, shown in Figure 17.1, will be discussed later in the chapter.

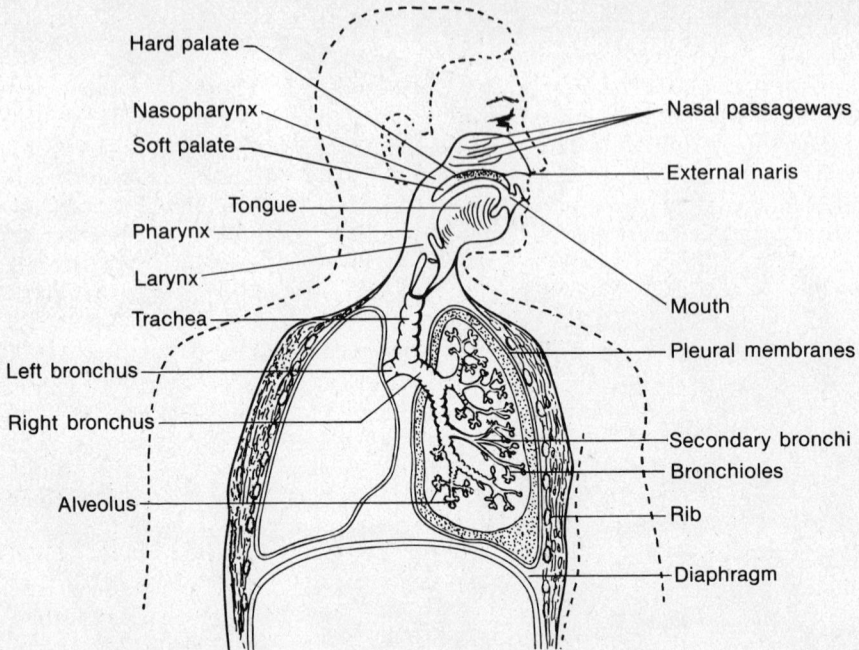

Fig. 17.1 Respiratory structures. The mouth and nasal passages join at the back of the throat into the funnel-shaped pharynx. The pharynx divides into the larynx and the esophagus. The trachea, an extension of the air passage beyond the larynx, divides into two branches, the right and left bronchi.

THE MECHANISMS OF BREATHING

The lungs are suspended in a space called the *thoracic cavity*. This cavity is enclosed by the sternum, ribs, vertebral column, and the muscular diaphragm. Breathing involves *inspiration*, taking air into the lungs, and *expiration*, the removal of gases from the lungs. The lungs are suspended in the closed thoracic cavity. The lungs are open to the atmosphere. Figure 17.2 illustrates the action of the lungs.

A double-walled membrane, the *pleura*, which covers the lungs and lines the thoracic cavity, plays an important role in breathing. The layer of the pleura adhering to the lungs is called the *visceral pleura*. The pleural layer adhering to the thoracic wall is the *parietal pleura*. These two layers are held tightly together due to the surface tension created by a fluid secreted between them. The two membranes, shown in Figure 17.3, can slide against each other but cannot easily be separated.

The *muscular diaphragm* is a dome-shaped structure that separates the thoracic and abdominal cavities and functions in changing the volume of the thoracic cavity. When the diaphragm muscles contract, the shortening makes the diaphragm less dome shaped. In other words, it flattens out, as illustrated in Figure 17.2.

When the diaphragm lowers, the volume of the cavity increases. Since this is a closed space, as the volume increases the pressure decreases. When the diaphragm muscles contract, air rushes into the air passages and into the lungs toward the reduced pressure area. The term *inspiration* or *inhalation* is given to this process. In order to expel air from the lungs, called *expiration* or *exhalation*, the diaphragm passively relaxes and assumes the dome shape that in effect reduces the volume of the lung cavity, creating more pressure, which forces the gases out of the lungs into the air passages.

PLATE 10 RESPIRATORY SYSTEM

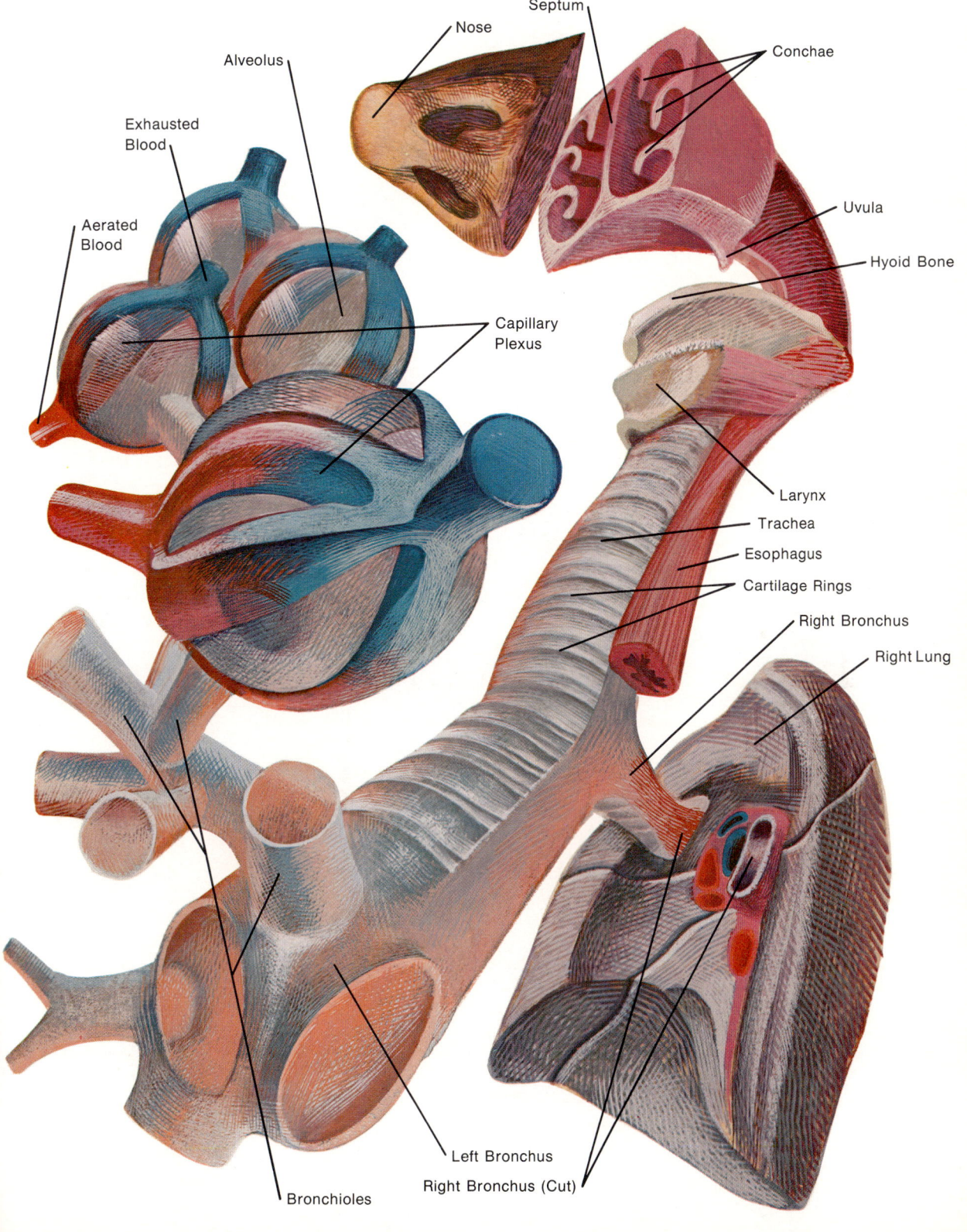

PLATE 11 MALE REPRODUCTIVE SYSTEM

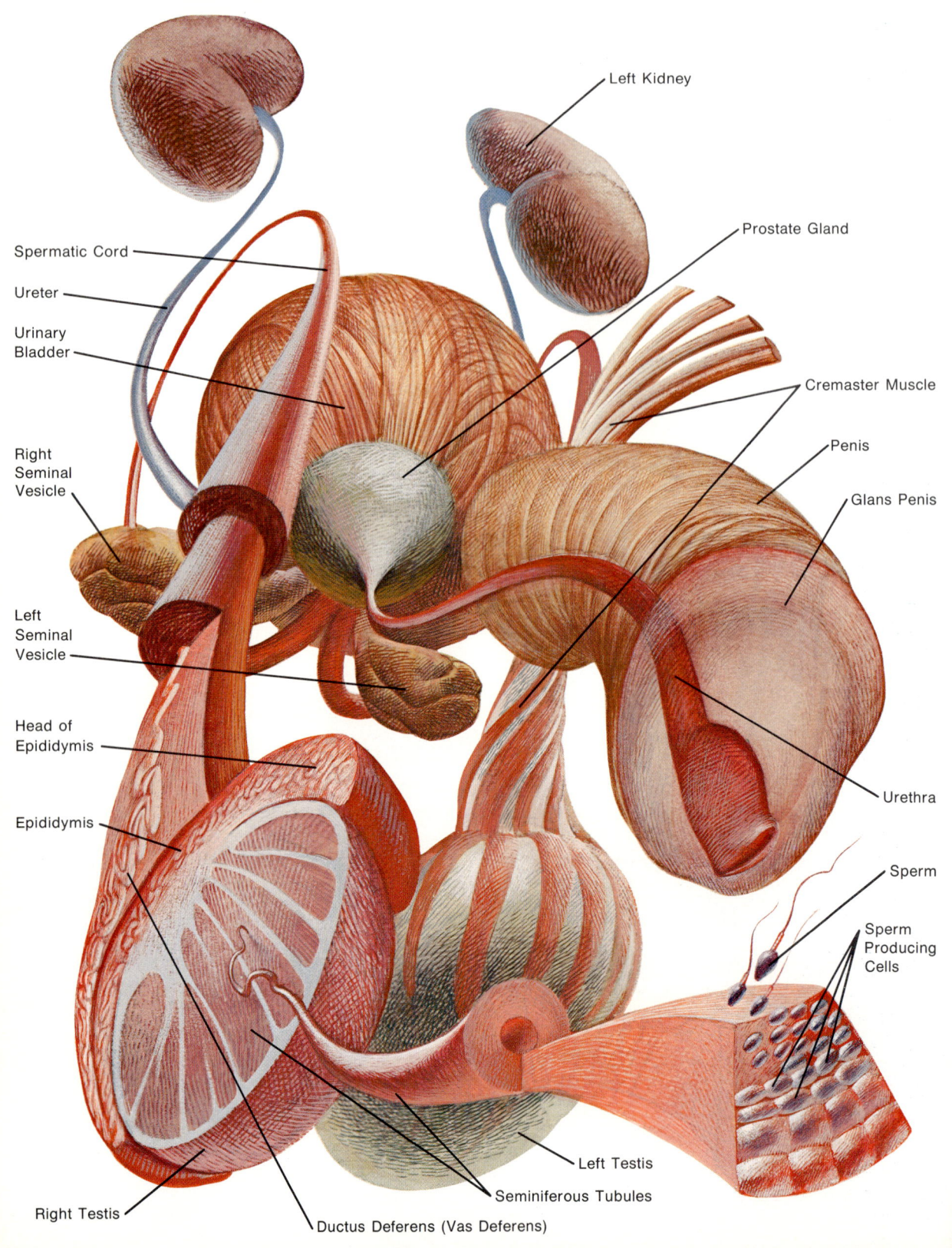

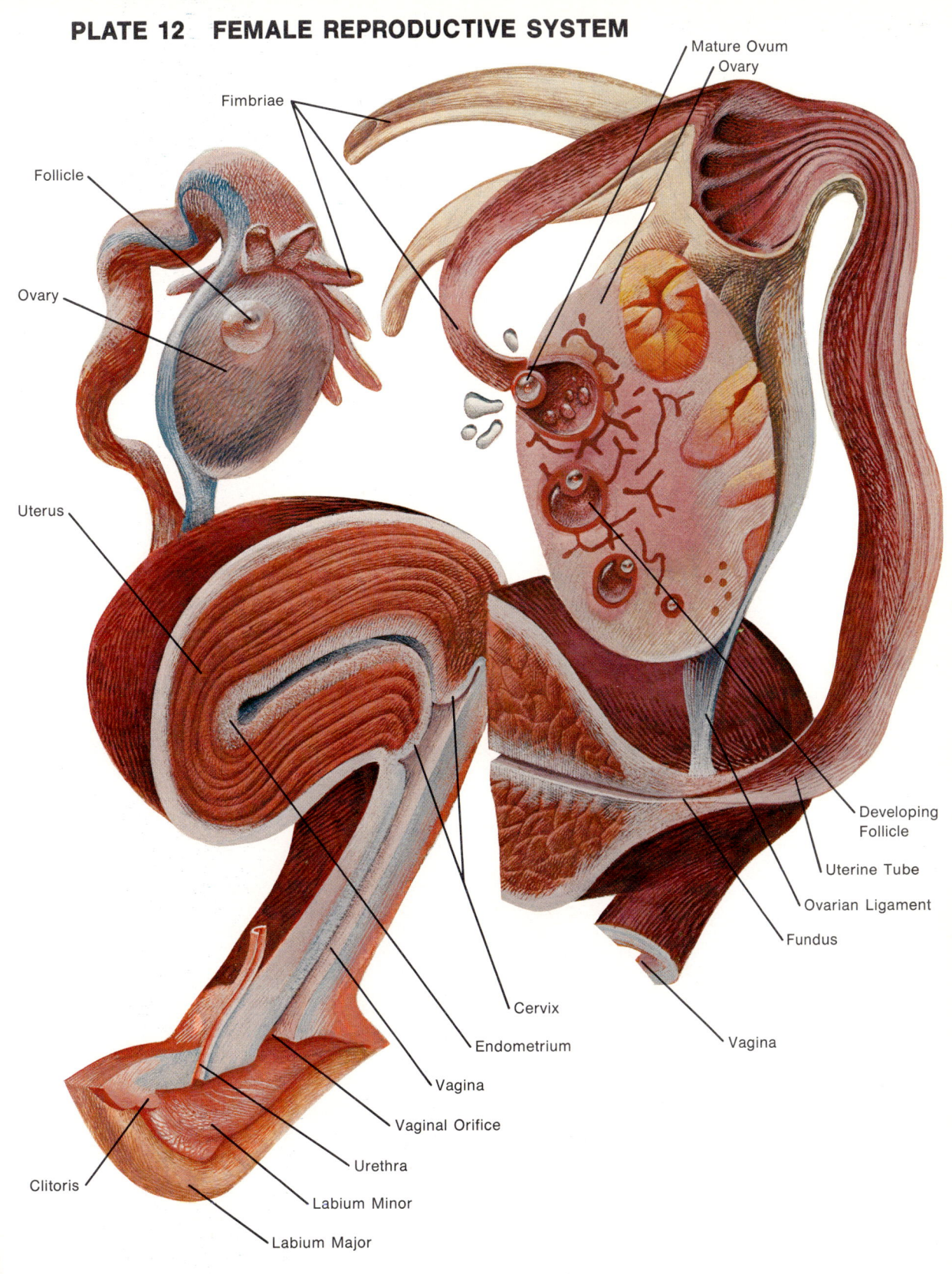

PLATE 13 CONCEPTION AND IMPLANTATION

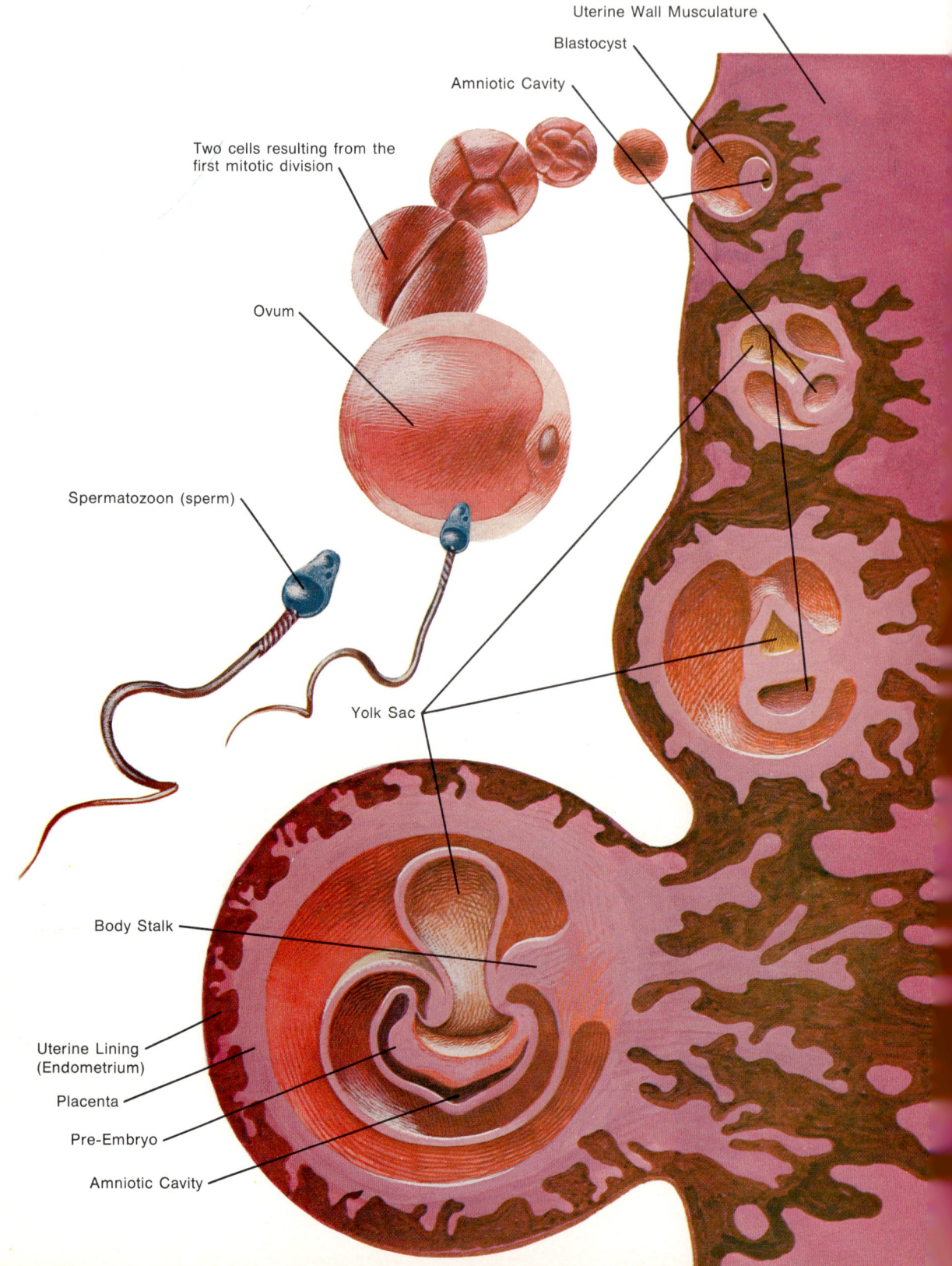

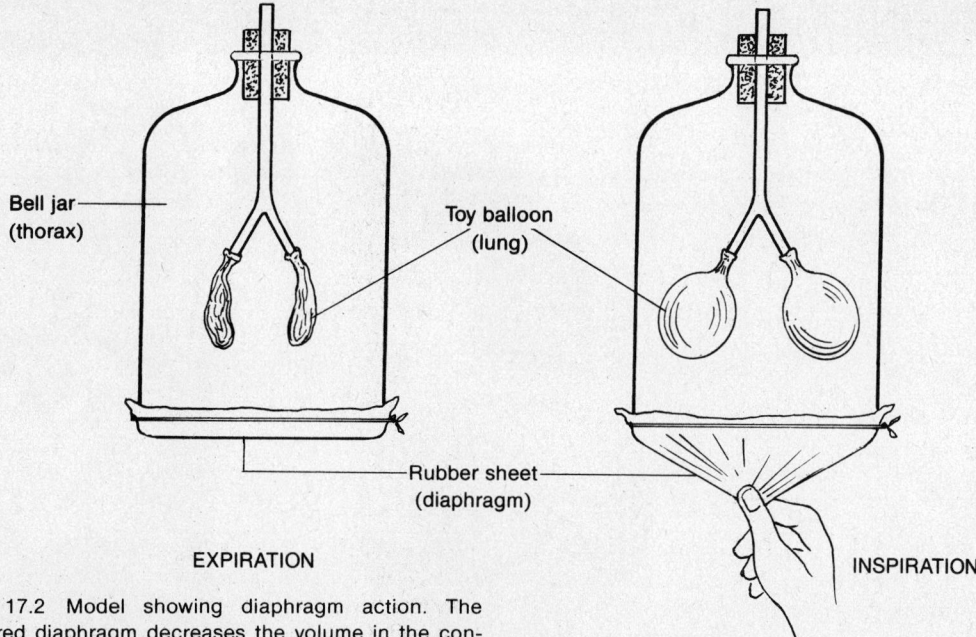

Fig. 17.2 Model showing diaphragm action. The lowered diaphragm decreases the volume in the container, permitting air to enter the balloons. The raised diaphragm increases air pressure around the balloons, forcing the air out.

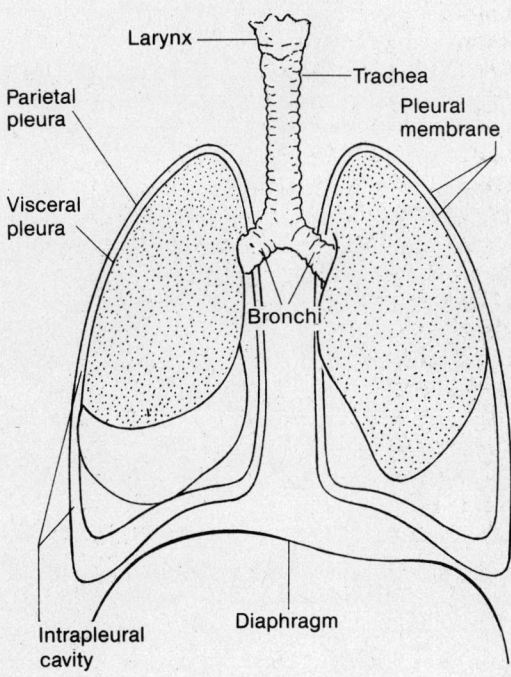

Fig. 17.3 A double-walled membrane, the pleura, covers the lungs and lines the thoracic cavity. The layer of the pleura adhering tightly to the lungs is called the visceral pleura. The layer of the pleura adhering to the thoracic wall is the parietal pleura. A fluid is secreted between the two layers to facilitate movement during breathing.

THE RESPIRATORY SYSTEM

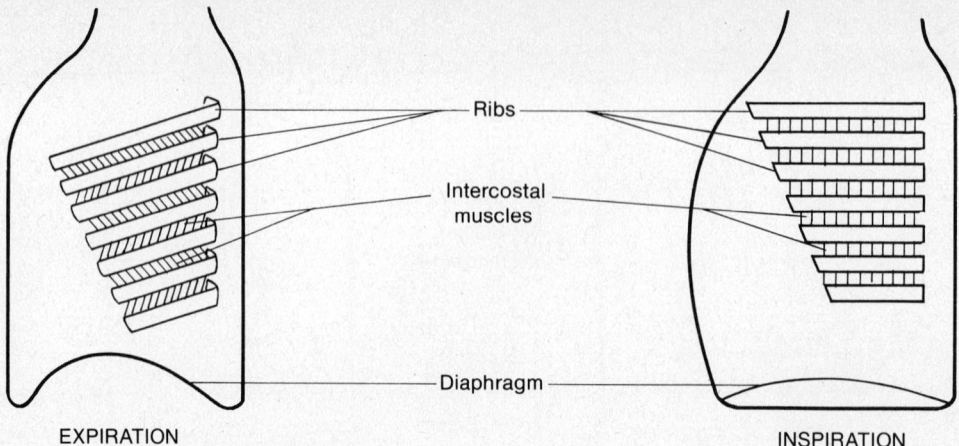

Fig. 17.4 Illustration of rib position changes during breathing. The rib cage alters its shape due to the pull of intercostal muscles.

When the diaphragm moves downward during inspiration, pressure against the underlying organs, the stomach and intestines, causes them to protrude forward and downward. Persons who breathe mainly with diaphragm action are called stomach breathers because of the movement of the stomach with each breath. Persons who constrict the stomach by clothing or by other restraining devices have to rely on other chest muscles (the intercostals) for breathing since the diaphragm movement is restricted. Intercostal muscles are shown in Figure 17.4.

The *intercostal muscles* cause the rib cage to alter its shape, thereby increasing or decreasing the volume of the thoracic cavity. Persons relying mainly on rib cage movement for breathing are called chest breathers. Factors other than tight clothing may alter the pattern of breathing. These include diseased diaphragm muscles, posture, position of the body, and cardiac influences. The heart lies in the thoracic cavity between the major lung lobes. If the heart and vessels increase in size due to an increased cardiac output (blood volume per minute), the total volume of the thoracic cavity is reduced, which therefore influences breathing patterns.

Normal-functioning lungs are at all times compressed tightly against the inner wall of the thoracic cavity. This is where the pleura, with its double-walled membrane and contained fluid, plays an important role in the breathing process. Due to cohesion of fluid molecules and adhesion of these molecules to pleural membranes, a pressure less than that of the atmosphere is created between pleural membranes. This pressure, called *intrapleural pressure,* is about 756 mm Hg (4 mm Hg less than atmospheric pressure at sea level).

When the thoracic cavity enlarges, the lungs are forced to enlarge, thus creating a reduced pressure in the alveolar spaces. The higher atmospheric pressure then rushes into this reduced pressure area. When the thoracic cavity becomes smaller due to relaxation of the diaphragm and intercostal muscles, the lungs are forced to occupy a smaller space, and thus air is expelled from them. The pressure of gases within the lungs is at this point greater than atmospheric pressure. The internal volume of the lungs fluctuates as the volume of the thoracic cavity fluctuates.

The lubricating fluid between the visceral and parietal pleuras prevents their adhesion to each other and allows free movement of the lungs within the thoracic cavity. If this lubricating fluid is not present, the two membranes cannot slide past each other easily and the resulting friction produces an excruciating type of pain associated with the disease called *pleurisy.* Since deep breathing is a painful process for individuals with pleurisy, they tend to be very shallow breathers.

THE ALVEOLI

The *alveoli*, the terminal lung tissues, are responsible for the exchange of gases between the environment and the blood. To do their job, these structures must be thin (only one cell thick), they must be kept moist, and they must be extensive enough to provide an adequate supply of oxygen and remove carbon dioxide. Figure 17.5 shows the relation of the bronchioles and alveoli and a segment of an alveolus illustrating its relative position to blood capillaries.

A substance called *surfactant* normally lines all inner alveolar tissues. This substance lowers the surface tension between the gases and the moist membranes. Without surfactant, the cohesive attraction of water molecules for each other becomes so strong that the alveolar walls are pulled together and collapse. In the collapsed state they cannot function as respiratory membranes. This condition in newborn infants is called *hyaline membrane disease*. If surfactant is completely absent, the infant will die. Infants having less than an adequate supply of surfactant usually have severe respiratory problems and die in early infancy. (See the discussion of surface tension in Chapter 21.)

The direction of diffusion of gases across the alveolar membrane and across the blood capillary membrane is due to the relative concentrations of gases on either side of these membranes.

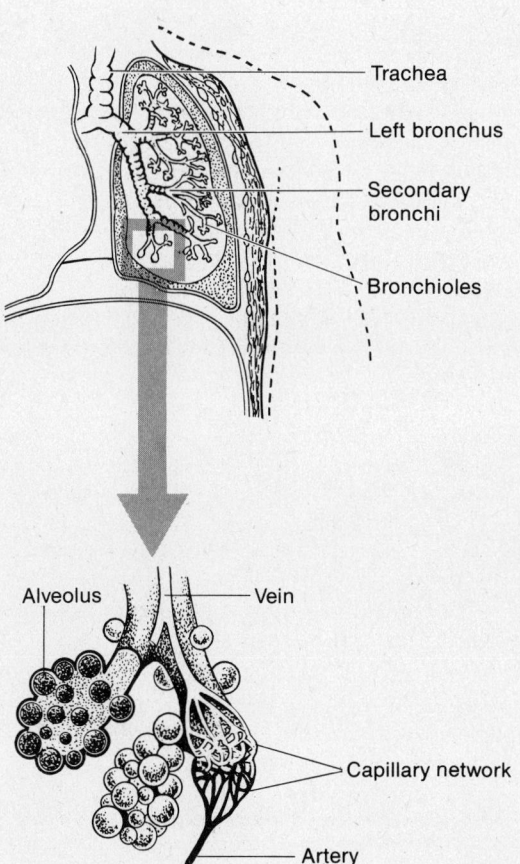

Fig. 17.5 Illustration of the location of alveoli, with respect to other respiratory structures.

Respiratory Distress Syndrome

A premature infant at birth or within two hours develops an increased respiratory rate of more than 60 respirations per minute. The infant may grunt or moan on expiration and have chest wall retractions during inspiration. Cyanosis is a major sign, and hypothermia complicates the problem. This infant has respiratory disease syndrome, or hyaline membrane disease, which is responsible for as many as 40,000 infant deaths each year.

The cause of the disease is still debated. However, it is believed to be caused by the absence of a substance called surfactant, which is necessary to lower the surface tension of alveolar membranes. When inadequate surfactant is supplied the alveoli collapse, resulting in collapse of the lung.

If the infant survives the first 72 hours, chances of survival are excellent. Without immediate and effective medical care, the infant dies.

When speaking of gas concentrations, care must be taken to designate which gas (oxygen or carbon dioxide) is being considered and furthermore what is meant by gas concentration.

We will digress for a moment to consider *Dalton's Law of Partial Pressures*, one of many physical laws explaining physical factors acting on all gases. Dalton's Law states that in any mixture of gases, each gas contributes its own pressure irrespective of other gas pressures.

For example atmospheric pressure at sea level is 760 mm Hg. Air is composed of 79.02 percent nitrogen, 20.96 percent oxygen, and 0.04 percent carbon dioxide. Therefore, nitrogen contributes 79.02 percent × 760 mm Hg, or 600.6 mm Hg toward the total atmospheric pressure of 760 mm Hg. A summary of the pressure is given below.

Nitrogen:
79.02% × 760 mm Hg =   600.6 mm Hg
Oxygen:
20.96% × 760 mm Hg =   159.1 mm Hg
Carbon dioxide:
0.04% × 760 mm Hg =   0.3 mm Hg
                            760   mm Hg

The portion of the pressure that each gas contributes is called the *partial pressure* of the gas. (Partial pressures of gases are usually designated as PCO_2, PO_2, PN_2, and so on.)

In the alveoli, the partial pressure of oxygen is greater than that of the blood capillaries, and therefore the oxygen diffuses from the alveolus into the capillaries. In other words, the oxygen diffuses from a high pressure to a lower pressure. The partial pressure of carbon dioxide is greater in the blood capillary side. Hence it will diffuse toward the lower partial pressure in the alveolus.

Figure 17.6 illustrates some of the partial pressure differences involved in the exchange of gases into and from blood. Note that partial pressures of gases in the alveoli are different from those in the atmosphere. Water vapor (a gas) contributes its own partial pressure.

If the PCO_2 in the lung alveolus is 40 mm Hg, and the PCO_2 in the capillaries bringing venous blood to the lungs is 46 mm Hg, CO_2 would diffuse from the capillaries into the alveolus under a pressure of 6 mm Hg. On the other hand, oxygen, with a partial pressure of 100 mm Hg in the alveolus, would diffuse into the capillary, which has a PO_2 of 37 mm Hg. As you can see, it is the differential partial pressures that determine the direction gases move by diffusion. The direction is always from a high partial pressure to a lower one.

At the cellular level, a cell with a PO_2 of 37 mm Hg and an arterial blood PO_2 of 100 mm Hg would cause a shift of O_2 from the capillary toward the cell. A PCO_2 within the cell of 46 mm Hg with the capillary maintaining a PCO_2 of 40 mm Hg would result in a shift of CO_2 from the cell into the capillary.

It may be helpful at this point to distinguish between gas concentration and gas pressure. *Gas pressure*, measured in mm Hg, is the driving force that pushes gases (O_2 and CO_2) across membranes. *Gas concentration* is a measure of the quantity of a specific gas in a mixture of gases. Partial pressure is the force created by the gas concentration.

REGULATION OF BREATHING

Normal breathing rates are controlled by reflex centers in the lower brain. The function of these reflex centers is to adjust the rate of inspiration and expiration to the demands of the body. The body at rest has less need for oxygen intake and carbon dioxide elimination than during exercise.

The normal rate of air intake during quiet respiration is about 16 inhalations per minute. The amount of air breathed in (tidal volume) with each breath is approximately 500 ml. By multiplying 16 inhalations per minute times 500 ml we obtain a figure of 8,000 ml, or 8 liters, of air taken in during normal respiration per minute. Of the 8 liters of air taken in, only about 20 percent is oxygen. Therefore 8 liters times 20 percent yields about 1.6 liters of oxygen taken in per minute.

The rate of air intake is drastically increased when activity is increased. Up to 100 liters of air can be taken into the body during vigorous exercise. Of this 100 liters about 20 liters are oxygen.

During vigorous exercise, the cardiac output is also increased from 5 liters up to as much as 35 liters per minute.

Also, with increased exercise we find vasodilation (enlarged blood vessels) that increases the flow of blood to the tissues. If all of these factors influence the amount of air taken into the lungs per minute, there must be some control

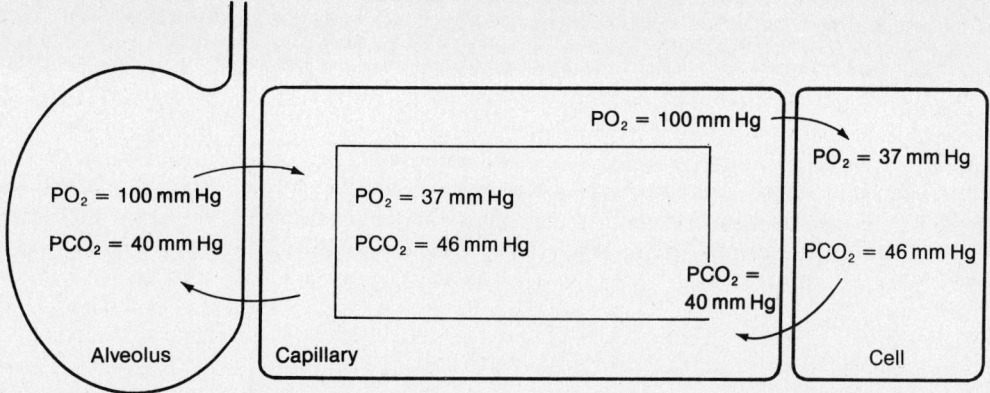

Fig. 17.6 Gas exchange depends upon differential pressures. The direction of flow is from an area of high partial pressure to one of low partial pressure. Note the 100 mm Hg PO_2 in the alveolus. Since this is greater than the 37 mm Hg PO_2 in the capillary, oxygen will enter the capillary. Also in the alveolus, a PCO_2 of 40 mm Hg is less than the PCO_2 of 47 mm Hg in the capillary, and so CO_2 will thus enter the alveolus from the capillary. This process is reversed at the cellular level, on the right side of illustration.

system that coordinates the blood flow, the air intake, and the carbon dioxide that is given off as a waste product.

When the spinal cord of experimental animals is cut above the pons no influence is evident on the normal breathing rate. When the spinal cord is cut lower in the medulla, the breathing rate is either reduced or halted. It is believed that nerve centers that control the rate of breathing are located in the medulla oblongata and the pons.

Inspiratory Centers in the Medulla

Neural control of breathing is believed to involve at least three sets of control mechanisms. It is believed that there are two clusters of nerve cells in the medulla. One cluster regulates expiration, the other regulates inspiration, with connecting neurons between the two cell-cluster centers. The nerve center in the medulla that regulates breathing is called the *respiratory center*.

The major concern with the cell-cluster hypothesis is that there is difficulty in explaining the rhythmic pattern of breathing, since there is no positive evidence of a "pacemaker" in the medulla. It is believed that the rhythm of breathing is one of self reexcitation between the inspiratory and the expiratory neurons in the cell clusters.

The cell-cluster hypothesis is diagrammed in Figure 17.7. In this figure note that the two centers are connected by a neuron. When the inspiratory center is stimulated, impulses are sent to the diaphragm and other chest muscles, which brings about the inhalation of air. At the same time, impulses travel along connecting neurons to the expiratory center. These neurons apparently inhibit the expiratory center while inspiration is occurring.

If both centers stimulated at the same time, they would counteract each other. Figure 17.7 shows the process that may occur during expiration. The neurons in the expiratory center "fire." This brings about the exhaling of air. At the same time, a connecting neuron carries an inhibiting impulse to the inspiratory center. The procedure is a theoretical explanation of how the rhythm of normal breathing is maintained.

The Hering-Breuer Reflex

A second factor controlling breathing involves a reflex called the *Hering-Breuer reflex*. This reflex not only halts further inflation of the lungs, but also prevents overinflation, which may rupture the epithelium of the alveoli.

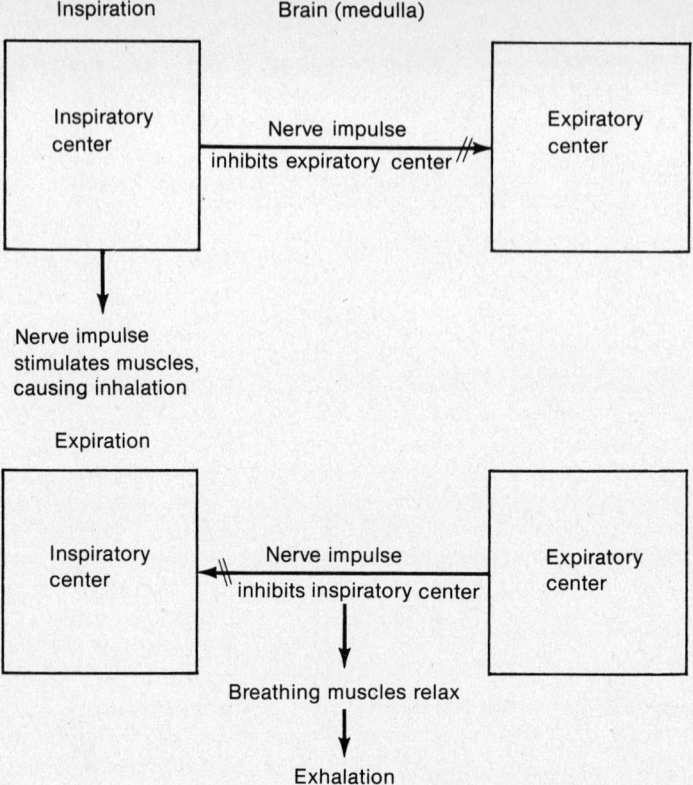

Fig. 17.7 Cell-cluster hypothesis explaining the breathing cycle.

Visceral pleura, the thin membrane covering the lungs, has nerve centers scattered among its cells that respond to stretch. During inspiration, the lung alveoli become expanded and stretched. This stimulates the sensitive stretch receptors to send impulses to the medulla reflex centers where they have an inhibiting effect on the inspiratory center, bringing about a cessation of inspiration.

Reflex Centers in the Pons

A third factor controlling breathing involves nerve centers located in the pons. If both of the previously mentioned nerve centers fail—for example, if nerves to and from them are damaged or severed—breathing may still be maintained. When the inspiratory center in the medulla is stimulated, impulses are sent not only to the diaphragm and intercostal muscles but also to the pons. Nerve centers in the pons, after a short delay, send impulses to activate inspiration.

In normal, unlabored breathing, expiration is a passive function brought about by the relaxation of the diaphragm and intercostal muscles. The elastic nature of lung tissue permits it to regain its original shape.

Impulses coming from the brain cortex can override the other nerve centers, allowing conscious control of breathing. In other words, you can hold your breath or breathe deeper at an accelerated rate. Voluntary holding of the breath cannot be extended by the average person for more than about a minute. When the breath is held, carbon dioxide levels build up rapidly in the blood, which chemically influences expiratory nerve centers. Therefore no great concern need be given to children voluntarily holding their breath.

Proprioceptor centers located in the skin layers can also be stimulated, sending impulses to the inspiratory center. A dash of cold water or pain, such as a slap or an electric shock, will cause a sudden intake of air.

Chemical Control of Respiration

At the end of a normal expiration of air from the lungs, the concentration of carbon dioxide rapidly builds up in the blood. The inspiratory center in the medulla is extremely sensitive to carbon dioxide and only minor concentration differences will affect it. An excess of carbon dioxide in the blood is called *hypercapnia*.

Carbon dioxide normally is about 0.04 percent of the total inspired air. There is no increase in the breathing rate if the carbon dioxide increases up to about 1 percent of the air. A level of 4 percent carbon dioxide in the air breathed in can double the normal breathing rate, with higher concentrations increasing the rate sharply. The normal individual can tolerate up to about a 10 percent level of carbon dioxide. However, a 20 percent carbon dioxide level causes discomfort, depression, or unconsciousness. A 40 percent level of carbon dioxide usually causes death.

As the concentration of carbon dioxide increases, the inspiratory centers send impulses to the diaphragm muscles, bringing about a more rapid breathing rate, thereby blowing off the excess carbon dioxide. If the carbon dioxide is not rapidly removed, the blood pH is lowered (becoming more acid). If the body cannot, for some reason, get rid of the surplus carbon dioxide, the pH becomes lower and a condition called *acidosis* results. At a normal pH (7.42), the ventilation rate is about 6 liters per minute. A pH of 7.1 increases the breathing rate to about 11 liters per minute. In summary, the lower the pH, the greater the increase in breathing rate. (See the discussion of pH in Appendix A.

Hyperventilation is a rapid loss of carbon dioxide from the blood due to an increased breathing rate. Rapid breathing has little effect upon building up oxygen reserves. However, if carbon dioxide is too rapidly exhausted, a condition called *alkalosis* results. In this case, the pH of the blood is elevated (made more alkaline) due to the lack of sufficient carbon dioxide and hydrogen ions. A return to normal breathing patterns or taking long deep breaths usually counteracts the effects of hyperventilation.

Oxygen concentration levels in the blood have only a limited effect upon the ventilation rate. The level of oxygen concentration has to be decreased considerably, to about 10 percent or lower, before any significant influence is seen on the rate of ventilation.

Chemoreceptors Certain nerve cells called *chemoreceptors*, located in the aorta and carotid arteries, are sensitive to carbon dioxide concentrations and pH levels. When the blood carbon dioxide concentrations increase (pH decreases), these nerve centers are stimulated to send rapid impulses to the inspiratory centers, which in turn speed up the rate of breathing. Decreased oxygen concentration levels in the blood also have a strong effect upon the chemoreceptive centers in the aorta and carotid arteries. However, the oxygen concentration has little if any effect upon the inspiratory centers directly.

Abnormal Respiratory Function We have all at some time experienced the sensation produced due to an inadequate oxygen supply. An example is the desire to take deep breaths after a short period of shallow breathing. This type of activity is normal and is the body's way of reestablishing certain chemical balances. The respiratory tract, with its door open to the environment, is vulnerable to attack by many factors. In earlier times, almost all breathing problems were lumped under the term respiratory inefficiency. Today, diagnosis can be more precise, since more is known about respiration physiology.

TYPES OF BREATHING

Each part of the body has its own vocabulary, and the respiratory system is no exception. In the physiology of respiration, normal breathing is called *eupnea*. Deep breathing or high alveolar ventilation rate is called *hypernea*. Shallow breathing is called *hypopnea*. Complete absence of breathing is called *apnea*.

Dyspnea is a physical awareness of breathing insufficiency. A person experiences "air hunger" and attempts to compensate for this by taking deep breaths. This condition is commonly due to an excess of carbon dioxide or a low pH in body fluids; to reduced flow of air through the respiratory tract due to partial blockage, for example, asthmatic conditions; and to psychological fear of suffocation.

PHYSICAL OBSTRUCTION AND RESISTANCE TO AIR FLOW

If a deep breath is taken with food in the mouth, the swallowing reflex does not occur. Therefore the protective mechanisms that would prevent food particles from entering the respiratory tract are lacking and we choke.

Asthma is a condition that reduces the diameter of the bronchial tubes and nasal passages by edema or copious secretion of mucus. Dyspnea due to asthma is generally limited to forced expiration. Inspiration seems to be less of a problem. Expiration is a rather passive act in normal individuals. However, the asthmatic individual has to actively force air through the reduced air passages during expiration.

Atelectasis is the partial or total collapse of a lung. This may be caused by a deficiency of surfactant in the premature newborn, the blockage of an air tract, or simply due to a rupture in the alveolar surfaces of the lung. When a blockage occurs, the air behind the obstruction is gradually absorbed out of the alveoli into the capillaries or the interstitial fluids, leaving the alveoli without sufficient pressure to keep them expanded. Therefore they collapse. If a whole lung is collapsed, the blood capillaries in the collapsed tissue become constricted, forcing a greater blood volume flow to the other lung, which can supply sufficient oxygen for normal activities. If a lung is collapsed due to an obstruction, the first problem is to get rid of the obstruction. This may be accomplished by some type of suction device that will pull the obstruction toward the mouth. Once the obstruction is removed, the lung will normally reinflate after a period of time.

Diffusion problems are essentially physical in nature rather than chemical. However, they differ from the previously discussed physical blockages. Diffusion of oxygen and carbon dioxide occurs at the respiratory membrane. Any factor that reduces the diffusion rate through this membrane will result in either (or both) reduced oxygen to the tissue or accumulated carbon dioxide in the tissue. *Emphysema* results in reduced alveolar surface area, which therefore reduces the diffusion rate.

Pneumonia is a general term applied to any condition that causes the alveoli to become filled with fluid and, in some instances, red and white blood cells. The reduced diffusion rate that results is proportional to the amount of accumulated fluid. If a patient with pneumonia is kept in an upright position, the lower lung alveoli become filled with fluid due to gravitational pull on the fluids. This makes the alveoli useless for gas diffusion. Placing the patient in a supine position will still permit some of the dorsal alveoli to fill up, but the total reduced alveolar surface area of the lung is lessened. The causes of pneumonia will be discussed later in the chapter.

The alveoli may become more fibrous due to an attempt by the tissue to wall off fine particulate matter such as asbestos fibers or carbon particles. This thickens the respiratory membrane, providing greater resistance for gaseous diffusion. The condition is commonly called *fibrosis*. Figure 17.8 illustrates the different alveolar structural peculiarities due to emphysema, pneumonia, and fibrosis.

GASEOUS TRANSPORT

The process of getting oxygen to the cells and carbon dioxide away is affected by factors in addition to those just mentioned. Any condition that lowers the minimum requirement of oxygen needed by cells is called *anoxia*, which means "without oxygen." The term is commonly used to designate different levels of oxygen insufficiency. Anemia is a condition where there is an insufficient number of red blood cells or a low hemoglobin content. In either instance, an adequate supply of oxygen may be available at the respiratory membrane, but the transporting

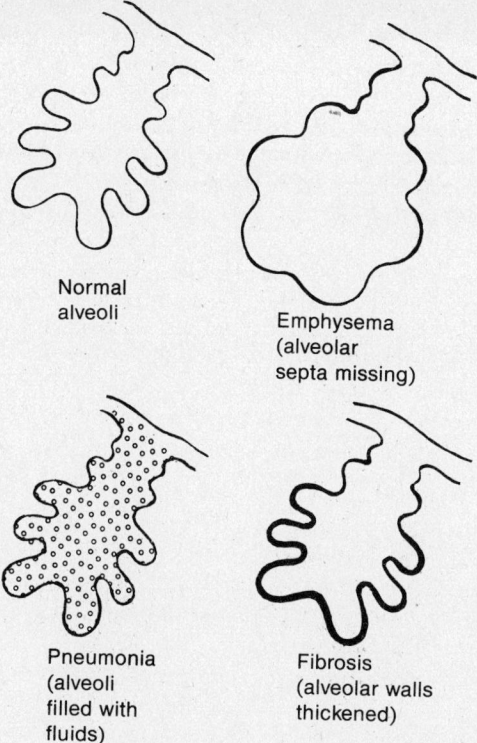

Fig. 17.8 Alveolar structure abnormalities that may cause gas-exchange problems.

mechanism is inadequate, causing *hypoxia* (low oxygen availability). Blood stagnation due to poor circulation may delay the arrival of oxygen at the cell, also causing hypoxia. A condition may exist wherein cells may be unable to utilize oxygen even though it is supplied to them in adequate quantities. This is called *histotoxic hypoxia*.

The respiratory membranes need a pumping mechanism to move gases constantly along the surface of the membrane. This is accomplished by the diaphragm and other muscles. When a mixture of gases arrives at the alveolar side of the respiratory membrane, some molecules of gas pass on through, while others do not. This same process occurs when the gases diffuse from the tissue side toward the alveoli.

Transport of Oxygen
Oxygen transport depends upon a higher pressure of oxygen on one side of a membrane than the other. The pressure difference on either side of a membrane is called a *pressure gradient*. The pressure gradient is the force that drives gases across the respiratory membrane. Partial pressures of gases remain the same whether the gases are free in a container or dissolved in fluids. In other words, we still refer to the partial pressure of oxygen in blood as if it were not dissolved or attached to the hemoglobin molecule. Once oxygen crosses the respiratory membrane, it can be chemically bound to the hemoglobin molecule or it may be dissolved in the blood plasma. Hemoglobin carries the major portion of the dissolved oxygen and, since it is capable of normally transporting more oxygen than cells need, the small amount of oxygen dissolved and carried by the plasma is insignificant.

Severely anemic individuals (lacking sufficient hemoglobin) placed in a pressurized vessel called a *hyperbaric chamber* containing 100 percent oxygen apparently can benefit from the oxygen dissolved in the blood plasma. The hemoglobin in a normal individual, at 760 mm Hg as it leaves the capillaries surrounding the lungs, is about 95 percent saturated with oxygen. As it approaches cells in the tissues, it will give up some of this oxygen to the cells because of the partial pressure gradient between the cells and the blood. When the cells have taken all the oxygen they need, the hemoglobin is still about 70 to 75 percent saturated, hence the blood carries much more oxygen than is normally needed. In vigorous exercise, this reserve is rapidly depleted.

Transport of Carbon Dioxide
Carbon dioxide transport also depends upon a pressure gradient. As cells metabolize food, carbon dioxide is released as a by-product and must be removed from the body. Arterial blood normally has a carbon dioxide partial pressure of 40 mm Hg. Carbon dioxide partial pressure builds up in the cell equivalent to 46 mm Hg. Since the pressure of carbon dioxide inside the cell is greater than that outside, it will diffuse out of the cell into the venous capillaries. Carbon dioxide

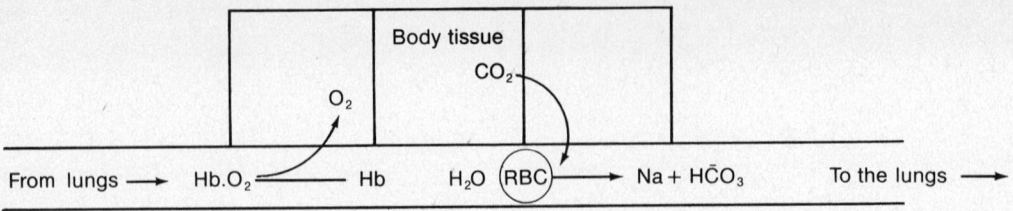

Fig. 17.9 Gas exchange at tissues. Hemoglobin gives up its oxygen and picks up CO_2, converting it to $NaHCO_3$. The carbon dioxide in this form is transported by plasma to the lungs.

is carried by the blood back to the lungs where it diffuses into the lung alveoli and is exhaled along with the other gases.

The mechanism for the transportation of carbon dioxide by the blood to the lungs is more involved than oxygen transport. The following discussion, which explains this process, is summarized in Figure 17.9.

As carbon dioxide leaves a cell, it enters the venous capillaries, which empty into veins carrying blood back to the lungs. In the capillaries, some of the carbon dioxide is simply absorbed into the blood plasma. A small amount of it becomes chemically bound to the protein part of hemoglobin molecules within the red blood cells.

Hemoglobin therefore plays a dual role. Oxygen is carried to the cell attached to the hemoglobin iron, and carbon dioxide is carried away attached to hemoglobin protein. Hemoglobin plus carbon dioxide yields a compound called *carbaminohemoglobin* ($HBCO_2$).

By far the greater portion of the carbon dioxide is carried in the plasma as a bicarbonate. Refer again to Figure 17.9. As carbon dioxide diffuses into a red blood cell it chemically unites with water, forming carbonic acid, H_2CO_3. The enzyme *carbonic anhydrase* is essential for this chemical reaction. *Carbonic acid* is an unstable molecule and breaks down to form a carbonate ion (HCO_3^-) and a hydrogen ion (H^+). Bicarbonate ions diffuse out of the red blood cells into the plasma, which carries them in this form to the lungs.

Because the bicarbonate ion is negatively charged as it leaves the red blood cell, the electrical balance of the red blood cell is disturbed. This electrical imbalance is corrected by chloride ions diffusing from the plasma into the red blood cell. These chloride ions are carried into the cell along with water molecules. This process is called *chloride shift*.

Hydrogen that results from the dissociation of carbonic acid is picked up by hemoglobin molecules, forming the compound *HHB*, called *reduced hemoglobin*. If the hydrogen ions were allowed to accumulate, the pH within the red blood cells would be lowered. Hemoglobin, in this instance, serves as a buffer, helping to maintain a constant pH.

To summarize what has occurred thus far, carbon dioxide has left the cell. A small portion of it has been absorbed into the blood plasma, a small portion has become attached to the protein fraction of hemoglobin to form carbaminohemoglobin, and the rest has combined with water to form carbonic acid. Carbonic acid breaks down to form hydrogen ions and carbonate ions. The hydrogen ions become bound to hemoglobin, and the carbonate ions diffuse into the plasma to be carried back to the lungs in this form.

When the venous blood carrying the dissolved and combined carbon dioxide arrives at the lung capillaries, the whole process is reversed, as shown in Figure 17.10. The carbonate ion reenters the red blood cell from the plasma, picks up the hydrogen held by hemoglobin, and becomes carbonic acid, H_2CO_3. The enzyme carbonic anhydrase immediately causes the dissociation of carbonic acid into water and carbon dioxide. Carbon dioxide then diffuses into the alveoli.

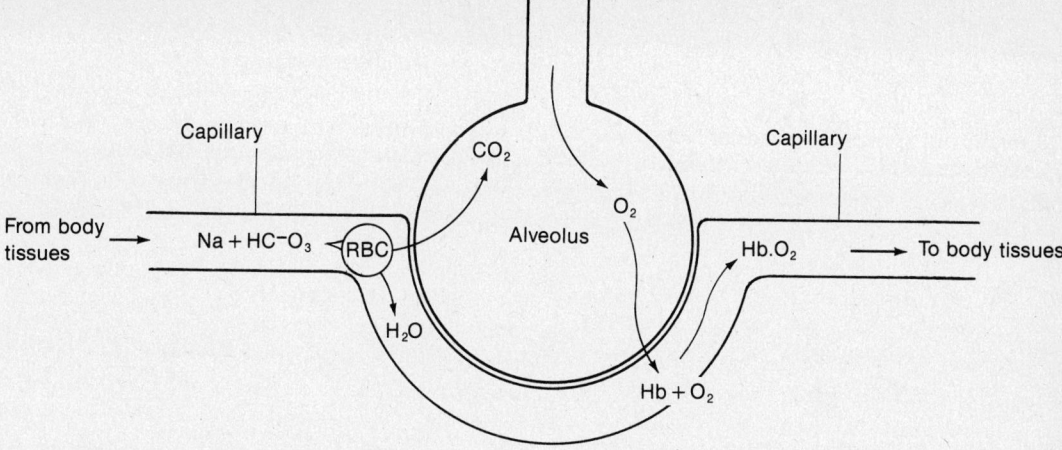

Fig. 17.10 Gas exchange between the capillaries and the alveoli. When the $NaHCO_3$ arrives at the alveoli, it is reconverted into NaCl and H_2CO_3. The carbonic acid decomposes into H_2O and CO_2. Carbon dioxide diffuses into the alveoli and is exhaled.

Carbon dioxide, including that bound to hemoglobin, diffuses out of the lung capillaries into the alveoli due to differential partial pressures of the carbon dioxide. At this point, the partial pressure of carbon dioxide on the venous side of the capillary is 46 mm Hg, compared to the partial pressure of carbon dioxide in the alveoli of 40 mm Hg. Therefore, carbon dioxide will diffuse from the venous capillary into the alveoli. Expiration by the lungs forces the accumulated carbon dioxide out of the body.

Since the partial pressure of carbon dioxide at normal atmospheric pressure is less than the partial pressure of carbon dioxide in the capillaries, carbon dioxide will not diffuse into the capillaries from the lungs, unless it is inhaled under considerable pressure. There is danger in breathing a higher concentration of carbon dioxide since it displaces needed oxygen, resulting in suffocation.

Lung Capacity and Air Volume

During normal breathing, a person will inhale and exhale about 500 ml of air with each breath. This particular quantity of air is called the *tidal volume*. After a normal inspiration, about 3,000 ml of additional air may be moved into the lungs by taking a very deep breath. This is called the *inspiratory reserve volume*.

At the end of a normal expiration, an additional 1,000 ml of air may be forcibly expelled from the lungs. This is called the *expiratory reserve volume*. Even after forcefully expelling all the possible air from the lungs, about 1,500 ml of air remain in the lungs. This is called *residual volume*. The total capacity of the lungs is the sum of all these volumes, which equals about 6,000 ml.

A more useful and more easily measured lung capacity is the *vital capacity*, which is the sum of inspiratory reserve volume, tidal volume, and the expiratory reserve volume. Vital capacity is equal to about 4,500 ml. These air volumes are summarized in Table 17.1.

Table 17.1 Respiratory Air Volumes. Total lung capacity is inspiratory reserve volume plus tidal volume plus expiratory reserve volume, equal to 4,500 ml.

	VOLUMES (ml)
Inspiratory reserve volume	3,000
Tidal volume	500
Expiratory reserve volume	1,000
Residual volume	1,500

Carbon Monoxide Poisoning

Carbon monoxide presents a different problem. Unlike carbon dioxide, which attaches to the protein fraction of hemoglobin, carbon monoxide becomes attached at the same location on the hemoglobin molecule that oxygen does. It therefore competes with oxygen for the attracting site. Even more serious is the fact that carbon monoxide has 210 times greater attraction than does oxygen for the hemoglobin binding site. This excludes oxygen, and an anoxic state results. A concentration of 0.1 percent of carbon monoxide is considered lethal. Carbon monoxide poisoning can be alleviated by administering 100 percent oxygen under a higher pressure than the atmospheric pressure. This high oxygen concentration tends to displace the carbon monoxide molecule from the hemoglobin molecule. Small quantities of carbon dioxide may also be administered that will stimulate the respiratory centers to increase the breathing rate, which helps expel carbon monoxide from the lungs.

MICROBIAL DISEASES OF THE RESPIRATORY SYSTEM

The nose and throat tissues are in almost constant contact with the external environment and hence are readily susceptible to infection caused by airborne microorganisms. Bronchi and lung tissue are further removed from the direct source of infection, but they are also susceptible.

The upper respiratory tract includes the pharynx, larynx, trachea, and upper bronchi. The lower respiratory tract includes the lungs and their associated tissues, such as alveoli, bronchioles, and pleura.

Diseases of the upper respiratory tract include the *common cold, pharyngitis, bronchitis, croup, whooping cough, diphtheria*, and *influenza*.

The *common cold* is a viral disease caused by at least 55 different types of rhinovirus (nose virus) plus other unknown viruses. The seriousness of the common cold is not so much with the discomfort it causes but rather as a forerunner of some secondary infection.

The disease is transmitted by direct contact, indirectly by airborne droplets from an infected individual, or by objects contaminated by an infected individual. Artificial immunization is not available.

Pharyngitis is an inflammation of the pharynx that may be caused by rhinoviruses or bacteria including streptococci and staphylococci. A sore throat with different degrees of severity is generally a sign of pharyngitis.

Bronchitis is an inflammation of the bronchi normally associated with upper respiratory tract infections.

Croup is a childhood disease that is an inflammation of the larynx (voice box). The distinct crowing sound made is caused by spasms of the larynx. Crouplike bronchitis is normally associated with other upper respiratory tract infections.

Whooping cough (pertussis) is a bacterial infection of the trachea, bronchi, and bronchioles. The bacteria causing the disease, *Bordetella pertussis*, are transmitted directly or indirectly by infected persons.

Whooping cough is predominately an early childhood disease. The disease leaves the individual with a definite immunity to further attacks.

Artificial immunity is available and immunization of all susceptible preschool children is recommended.

Diphtheria is an acute disease of tonsils, pharynx, and larynx caused by the bacterium *Corynebacterium diphtheriae*. Diphtheria is predominantly a childhood disease (under 5 years). However, it does occur in adults. Artificial immunity is available and is the only effective control.

Influenza is a disease of the respiratory tract caused by viruses (A, B, and C types). It is characterized by fever, chills, and headache, with later stages inducing coughing and sore throat.

The disease is transmitted by direct or indirect contact with an infected person. Although humans are the principle reservoir for the viruses, animals (swine, horses, birds) are suspected.

Diseases of the lower respiratory tract include *pleurisy, pneumonia, tuberculosis*, and *emphysema*.

Pleurisy is an inflammation of the pleural

membranes causing a reduction of the lubricating fluid between the parietal and visceral pleura.

Pneumonia is a disease attacking alveoli caused by bacteria, viruses, and organisms of uncertain classification. The most common form, caused by the bacterium *Diplococcus pneumoniae*, is spread by direct or indirect contact with an infected person.

In pneumonia, the alveoli fill with fluids, thereby inhibiting the transfer of gases across the membranes into the bloodstream. The total amount of air space (vital capacity) is reduced.

Active immunization is not available. Treatment involves antibiotic therapy. Preventive measures include avoiding crowded living and sleeping quarters, particularly in institutions, military barracks, and the like.

Tuberculosis is a chronic bacterial disease affecting the lung tissues. The bacterium responsible, *Mycobacterium tuberculosis*, is transmitted by direct or indirect contact with an infected person.

COPD Cannot Be Cured

Chronic obstructive pulmonary disease (COPD) is due to a group of diseases that includes emphysema, chronic bronchitis, and asthma. Of these three diseases emphysema ranks second to heart disease among the elderly. Although emphysema is much more common in males 50 to 70 years old, the incidence of the disease is increasing among elderly women.

The cause of the disease is not known. However, factors that antagonize the condition include cigarette smoking, air pollution, other pulmonary diseases, and genetic predisposition.

The major objective of health care directed toward the COPD patient, since no cure is known, is to provide an environment that enables the person to lead as near a normal life as possible. Some of these objectives are met by relieving bronchial obstruction mechanically and with drug therapy, teaching proper breathing techniques to ensure that lung alveoli do not collapse, oxygen therapy, and proper nutrition.

Diseased lung tissue becomes fibrous and loses its capability to transfer oxygen to the blood. Tuberculosis is progressive (chronic), with different stages recognizable by the extent of damage done to lung tissue. Periodic X ray examination and tuberculin tests are important first steps in preventing the spread of tuberculosis.

Emphysema is a disease that reduces the surface area of alveoli, thereby reducing the transfer of oxygen into the blood. The specific cause of the disease is not certain. The high incidence of emphysema among heavy smokers or industrial workers exposed to pollutants certainly implicates irritants inhaled voluntarily or involuntarily.

RESPIRATORY CHANGES IN THE AGED

The respiratory system shows definite changes with aging. The elasticity of alveolar septa is reduced, leading to a reduction of oxygen transfer into capillaries. The total amount of air the lungs can hold (total capacity) is not reduced. There are, however, reductions in vital capacity and maximum ventilation during exercise. The aged person is more vulnerable to diseases that affect the respiratory system. These include emphysema, pneumonia, bronchitis, colds, and influenza.

With increased age, tissues of the rib cage become less elastic and more rigid. This rigidity may cause reduced lung expansion, thus becoming a factor in lung congestion.

SUMMARY AND REVIEW

A. The primary function of the respiratory system is to provide oxygen for tissues and carry carbon dioxide, a waste product, away from the tissues.
B. The respiratory system includes the lungs, bronchial and bronchiole tubes, pharynx, larynx, mouth, and nose.
C. Mechanisms of breathing include:
   1. muscular diaphragm
   2. intercostal muscles

D. Regulation of breathing
  1. Neural regulation is controlled by:
     a. respiratory nerve centers in the medulla
     b. Hering-Breuer reflex
     c. nerve centers in the pons
  2. Chemical regulation is maintained due to the concentrations of carbon dioxide in arterial blood and its influence upon the respiratory nerve centers in the medulla.
E. Abnormal respiratory function may be due to:
  1. physical obstruction such as asthma or atelectasis
  2. diffusion problems, as in the case of pneumonia or emphysema
F. Oxygen transport depends upon a pressure gradient, that is, a higher PO_2 on one side of a diffusing membrane compared to the other side.
G. Carbon dioxide transport follows the same basic pattern that oxygen transport does except for the chemical reactions involved.
H. Carbon monoxide poisoning results from the fact that carbon monoxide has a higher affinity for hemoglobin attachment than oxygen has, and thus oxygen is excluded.
I. Microbial diseases of the respiratory system include:
  1. pharyngitis
  2. bronchitis
  3. croup
  4. whooping cough
  5. diphtheria
  6. influenza
  7. pleurisy
  8. pneumonia
  9. tuberculosis
  10. emphysema
J. Respiratory changes in the aged
  1. reduction in alveolar membrane elasticity
  2. reduction in vital capacity
  3. greater vulnerability to respiratory diseases

18

Reproduction

MAJOR CONCEPTS
  THE MALE REPRODUCTIVE SYSTEM
    Accessory Structures
    The Penis
    Spermatogenesis
    Male Sex Hormones
  THE FEMALE REPRODUCTIVE SYSTEM
    The Ovaries
    Uterus
    Fallopian Tubes
    Vagina
    External Genitalia (Vulva)
    Hormone Control of Female Reproduction
    Menstrual Cycle
  PREGNANCY
  FETAL DEVELOPMENT
  DISEASES OF THE REPRODUCTIVE SYSTEM
    Syphilis
    Gonorrhea
    Vulvovaginitis
    Trichimoniasis
    Puerperal Fever
  NONMICROBIAL DISEASES OF THE
  REPRODUCTIVE SYSTEM
    Malignant Tumors
    Prostate Cancer
  THE AGING REPRODUCTIVE SYSTEM

Biologically speaking, a reproductive system is nature's way of making another reproductive system. In other words, the species is perpetuated and the population is maintained. This chapter will describe the anatomy and physiology of male and female reproductive systems and the process that perpetuates life. Also included here is a discussion of abnormalities of the systems and diseases caused by pathogenic microorganisms.

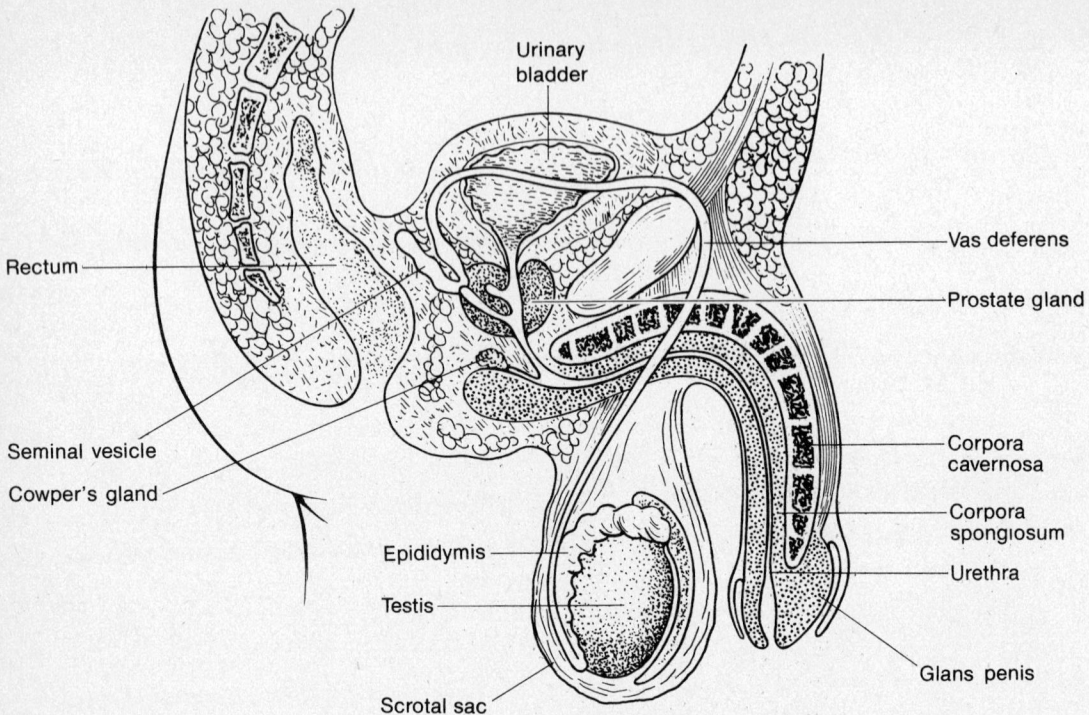

Fig. 18.1 A sagittal section through the male lower trunk showing the male reproductive organs. The testes are suspended in a saclike structure called the scrotum. A duct, the vas deferens, leads from each testis and unites with the urethra at the base of the urinary bladder.

Human reproduction promotes and maintains the continuity of life. Reproduction itself is not a requirement for the life of an individual. It is, however, a requirement for the maintenance and perpetuation of the life of the species.

Human reproduction may be described as a series of chemical and physical processes that ultimately results in the creation of another human. The continuity of life and the perpetuation of characteristics that permit each generation of humans to produce another generation of humans depends upon a transfer of genetic material from one generation to the next. The combination of genetic material from one sex with that of the opposite sex produces a specialized cell that contains all the information needed to produce another human.

This chapter will describe the anatomy and the physical and chemical processes necessary to perpetuate life. In addition, the chapter includes a discussion of abnormalities of the reproductive systems and how the systems are affected by pathogenic microorganisms.

THE MALE REPRODUCTIVE SYSTEM

The male reproductive role is basically limited to providing a specialized cell called a *spermatozoon*, commonly called a *sperm*. A secondary function is to introduce the sperm into the female vagina.

The male reproductive organs include *paired testes*, which produce sperm; a system of ducts, which carries the sperm out of the body; and the *penis*. Associated with the sperm ducts are accessory structures that contribute secretions, which together with sperm constitute a thick fluid called *semen*. Figure 18.1 shows a sagittal section of the male reproductive organs.

The testes are suspended in a saclike structure called the *scrotum*. A duct, called the *vas deferens* (also called *ductus deferens* or *sperm duct*), leads from each testis and unites with the urethra at the base of the urinary bladder.

The testes are formed in the male fetus within the *peritoneal* (abdominal) *cavity* near the level of the kidneys. As the fetus matures, the testes migrate downward and move through an opening at the base of the abdominal cavity called the *inguinal canal* into the scrotal sac.

The testes normally descend into the scrotum shortly before birth. If the testes do not descend, the condition is called *cryptorchidism*. Undescended testes after the age of puberty are inhibited in their function of producing viable sperm, apparently due to the slightly higher temperatures of the abdominal cavity. The condition can be corrected with male hormones (testosterone), therapy, or surgery.

The internal structure of a testis is shown in Figure 18.2. Each testis is covered with a tough connective tissue called *tunica albuginea*. Internally, the tissue is divided into compartments separated by *septa*. Each compartment is filled with small coiled tubules called *seminiferous tubules*, which produce sperm.

The ducts that drain the seminiferous tubules, called *straight tubules*, are in turn connected with other tubules called *rete testes*, which empty into *efferent ducts*. A mass of highly coiled ducts that collect sperm from the efferent ductules, located somewhat like a cap on each testis, is called the *epididymis*. Continuous with the epididymis and leading toward the urethra is the *vas deferens*.

Accessory Structures
Included in the accessory structures are those that secrete substances into the urethra. These include the seminal vesicles, the prostate gland, and the bulbourethral glands.

The *seminal vesicles*, located just below the urinary bladder, empty secretions into the *ejaculatory duct*. The ejaculatory duct is a muscular tube that, during sexual excitement, ejects sperm and accompanying fluids (semen) into the urethra. See Figure 18.1. The seminal vesicles produce secretions that contain simple sugars, amino acid, and mucus. It is assumed that the substances secreted by the seminal vesicles provide nutrition for the sperm until they unite with an egg.

The *prostate gland* lies in a position surrounding the upper portion of the urethra at the base of the urinary bladder. This structure secretes more fluids, which combine with sperm as they are emptied into the urethra. The substance secreted by the prostate is slightly alkaline, which changes the pH of the fluids sufficiently to provide motility to the spermatozoa.

Many aged males suffer from an enlargement of the prostate gland. When the gland enlarges, it obstructs the flow of urine from the urinary bladder. The enlarged gland squeezes the urethra, thereby inhibiting normal urination.

The *bulbourethral glands* (Cowper's gland) are two small glands located further along the urethra distal to the prostate gland. These glands secrete an alkaline, mucuslike substance that is added to the contents that have been emptied further up in the ducts. The substance expelled from the penis during sexual excitement, then, consists of a mixture of substances.

The Penis
The *penis*, the male organ for copulation, serves to introduce sperm into the female vagina and also serves as part of the male urinary system, since the distal urethral opening is at the tip of the penis.

The unstimulated penis is small and flaccid. Upon erotic stimulation, it becomes firm, enlarged, and lengthened. The structure of the penis is shown in Figure 18.1.

The penis is made up of three cylindrical bodies. Two of these cylindrical bodies lie side by side and are called the *corpora cavernosa*. The cylinder lying beneath and between the corpora cavernosa and enclosing the urethra is called the *corpus spongiosum*. These three cylinders are made up internally of spongy tissue surrounded by tough fibrous tissue. Each cylinder is highly vascularized.

During sexual excitement, blood rapidly enters the corpora cavernosa and the corpus spongiosum, filling the spongy spaces and causing the penis to become erect. The flow of blood

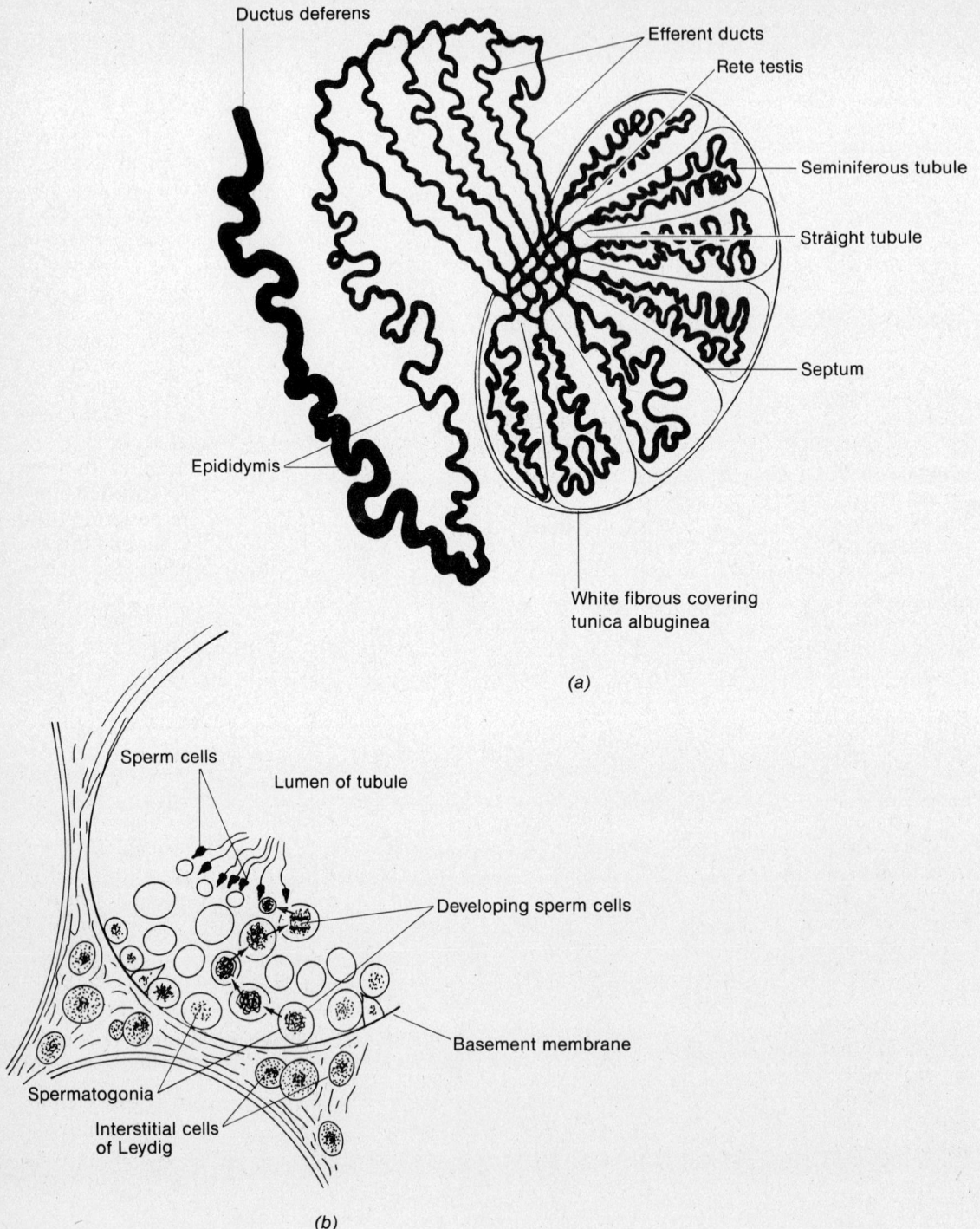

Fig. 18.2 Internal structure of testes. (a) Sagittal section. (b) Cross section of a seminiferous tubule. Each testis is covered with a tough connective tissue called tunica albuginea. Internally, each testis is divided into compartments. Each compartment is filled with small coiled tubules (seminiferous tubules), which produce sperm.

into the spaces creates a pressure upon the veins that drains the penis, reducing or shutting off the outward flow of blood. The penis remains erect until stimulation ceases, causing the arteries to constrict. Less blood enters the cylinders, and therefore less pressure is exerted on the veins, allowing the blood to leave the penis, and permitting it to regain its unstimulated, flaccid condition.

At its distal end the corpus spongiosum enlarges to become the *glans penis*. The cylinders are enclosed in muscular tissue and exteriorly are bound by skin that is continuous with that of the abdominal wall and of the scrotal sac. At the distal end of the penis the skin forms a fold or a cuff around the glans. This fold of skin is called the *foreskin* or *prepuce*. It is a common practice to remove surgically the foreskin by circumcision shortly after birth.

Spermatogenesis

Of all the structures listed above, only cells of the seminiferous tubules are capable of producing sperm. The process by which sperm are produced is called *spermatogenesis*.

The process of spermatogenesis occurs in the seminiferous tubules beginning at puberty and continuing until death, unless interrupted by disease or injury.

Figure 18.3 illustrates the process of spermatogenesis. Figure 18.2 shows a cross section of a seminiferous tubule and the location of specialized cells within the tubule that mature the sperm. Spermatogenesis is meiosis occurring in the male. (See Chapter 3.)

Around the perimeter of each seminiferous tubule are rows of specialized cells called *spermatogonia*. These cells are similar to other body cells in that each of them has 46 chromosomes, which is normal for human cells. These cells reproduce themselves by mitosis, thereby supplying large and continuous numbers of potential sperm.

Spermatogonia at different stages of maturity move toward the center of the seminiferous tubule and undergo a reduction division of chromosomes. The first division, however, is a mitotic division, which produces a cell called a *primary spermatocyte*. The primary spermatocyte has 46 chromosomes. The next division of the primary spermatocyte produces two cells, each of which have 23 chromosomes. These cells are called *secondary spermatocytes*. The secondary spermatocytes undergo an alteration in structure and are consequently called *spermatids*. Spermatids undergo no further reduction in chromosome number and maintain the normal number of chromosomes in sperm at 23. These cells now undergo a structural change and become sperm cells. See Figure 18.3. During this change the cytoplasm is greatly reduced.

A sperm is essentially a nucleus with a *tail* attached. The nuclear material is crowded into a structure called the *head*. More posteriorly is a thickened mass called the *neck*. Beyond this is a thin whiplike tail, which moves the sperm through genital structures.

Mature sperm collect in the lumen of the seminiferous tubules and are transported out of the tubules by smooth muscle contraction of the seminiferous tubules. The motility of the sperm is not great at this time. Other chemical factors are necessary to make them motile.

Smooth muscle around the tubules and cilia lining the lumen propel the sperm out of the seminiferous tubules, where they enter the series of ducts previously mentioned, eventually entering the vas deferens, leading toward the urethra.

Note in Figure 18.1 that the vas deferens rises upward out of the scrotal sac into the abdominal cavity to about the level of the urinary bladder, then heads downward again into the pelvic region toward the urethra. Just before it reaches the urethra it joins the ejaculatory duct.

Male Sex Hormones

The major male sex hormone is testosterone, secreted by cells located between seminiferous tubules. These cells, shown in Figure 18.2, are called the *interstitial cells of Leydig*. These cells

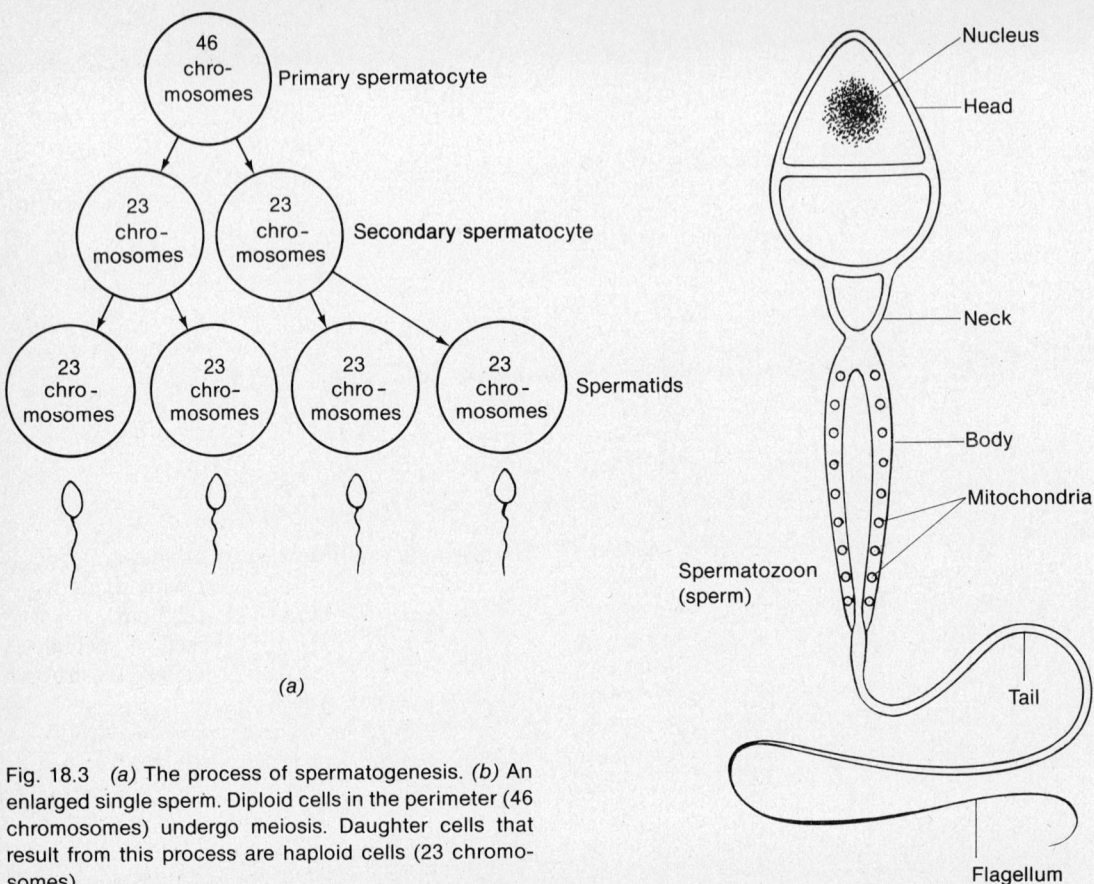

Fig. 18.3 (a) The process of spermatogenesis. (b) An enlarged single sperm. Diploid cells in the perimeter (46 chromosomes) undergo meiosis. Daughter cells that result from this process are haploid cells (23 chromosomes).

actively secrete testosterone during fetal development, then become inactive until puberty. Testosterone is actively secreted continuously after puberty.

The function of testosterone during fetal development is to aid in the differentiation of tissues that result in male characteristics. These tissues include the penis, scrotum, testes, and sperm ducts. After puberty, testosterone helps to bring about and maintain male secondary sex characteristics. These characteristics include body hair distribution, enlargement of the genitalia, fat distribution, voice change, and so on.

Other hormones stimulate the interstitial cells of Leydig to secrete testosterone. *Chorionic gonadotropin*, secreted by the placenta, stimulates the formation of fetal interstitial cells that secrete testosterone in concentrations directly proportional to their number. At birth the stimulation ceases and the interstitial cells disappear until the beginning of puberty. At puberty, two hormones—*luteinizing hormone* (LH) and *follicle stimulating hormone* (FSH)—stimulate interstitial cells to secrete testosterone. LH and FSH are gonadotropic homones secreted by the anterior pituitary.

Other male sex hormones are secreted by various body tissues. Those that influence male characteristics are called *androgens*. The adrenal gland secretes several androgens that normally have little influence except when adrenal tumors develop. Tumors of the adrenal cortex secrete large quantities of androgens. If this occurs in early childhood before puberty, there is a precocious development of male secondary sex

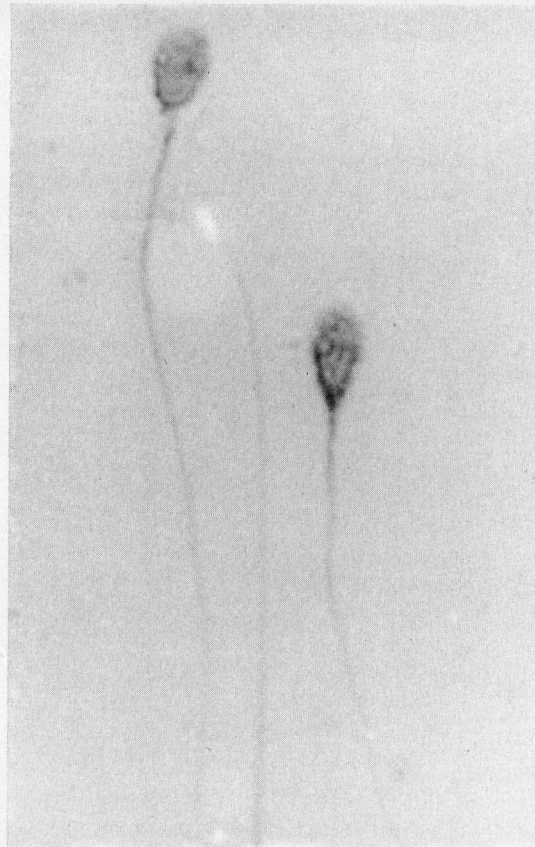

Fig. 18.4 Photomicrograph of human sperm. (photo from author's collection).

characteristics. Adrenal cortical tumors occurring in female children produce masculine characteristics.

Estrogens are secreted by male tissues. The precise location of the secretion is not known, nor is it known what effect estrogens have on the male body.

THE FEMALE REPRODUCTIVE SYSTEM

The female role in human reproduction is decidedly more complicated than that of the male since the female must not only produce a viable ovum (egg) but must also serve as a host for the fetus during the term of pregnancy.

The female reproductive organs, shown in Figure 18.5 include *paired ovaries, uterus, fallopian tubes* (uterine tubes), *vagina,* and *external genitalia.*

The Ovaries

Ovaries carry on several different kinds of activities simultaneously. These activities include the development and release of ova, and the production and release of female hormones, *progesterone* and *estrogen.* The ovaries are also acted upon and influenced by pituitary hormones.

Ovaries lie on either side of the uterus deep in the pelvis cavity, as shown in Figure 18.5. Each ovary is almond shaped and weighs approximately 6 g. At the time of birth, each ovary contains several thousand structures called *primary follicles.* Each primary follicle is a mass of epithelial cells surrounding an immature ovum. Until puberty all ova are immature.

Uterus

The *uterus* is a muscular, pear-shaped structure about 7.5 cm long and 5 cm wide at its widest part, tapering down to about 2.5 cm.

The inner lining is composed of special epithelial cells called the *endometrium,* as shown in Figure 18.5.

The uterus lies in the abdominal cavity, with a portion projecting slightly above the urinary bladder. The upper portions of the uterus are continuous with the fallopian (uterine) tubes. The lower portion of the uterus, the *cervix,* which is shaped like a small doughnut, projects slightly into the vagina.

Fallopian Tubes

The "horns" of the uterus are hollow extensions of the uterine cavity called *fallopian tubes.* These tubes at their distal ends surround a large portion of each ovary. Fallopian tubes receive mature ova released from the ovary and by muscular contraction carry the ova in the direction of the uterus.

Vagina

The *vagina,* or birth canal, is a muscular tube lined with epithelial tissue. It serves as a receptacle into which sperm is deposited by the male during sexual intercourse and gives passage to the fetus during birth. The distal end of the vagina is partially closed by the *hymen.*

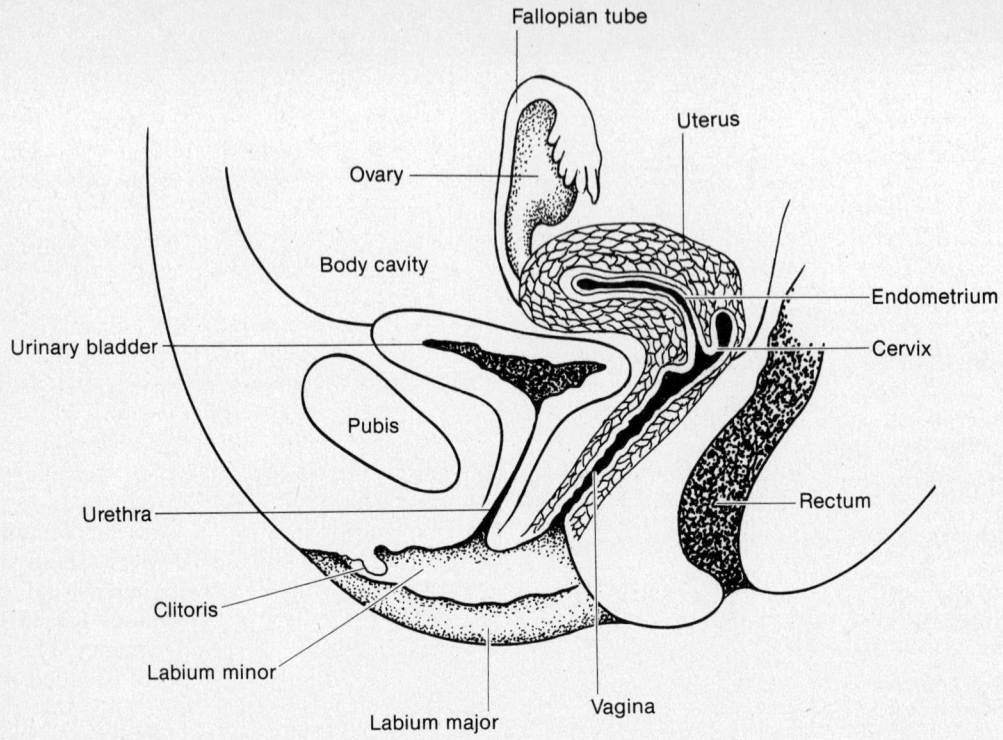

Fig. 18.5 The female reproductive system in sagittal view. The female system includes paired ovaries, the uterus, paired fallopian tubes, vagina, and external genitalia.

External Genitalia (Vulva)

Structures located external to the vaginal opening include the *labia major, labia minor, prepuce,* and *clitoris*.

The Labia Major These are the large folds of tissue that constitute the most exterior part of the genitalia. They are composed of connective tissue, muscle, fat, and glands. Superficially they are covered with skin and hair.

The Labia Minor This structure is made up of fleshy, vascular folds of tissue that surround the vaginal opening. Between the folds are located the vaginal opening and the urethral orifice.

The Prepuce The prepuce, formed by the joined upper edges of the labia minor, is a fold of tissue that partially covers the clitoris.

The Clitoris This small mass of erectile tissue contains numerous sensory nerve endings. Physical stimulation of the clitoris during sexual intercourse plays an important part in sexual arousal of the woman.

Hormone Control of Female Reproduction

At the beginning of puberty, primary follicles in the ovaries begin to mature. The maturing process is initiated by *follicle stimulating hormone* (FSH) secreted by the anterior pituitary. FSH induces one or more follicles to mature, producing one or more ova. FSH also induces follicle cells to secrete hormones called *estrogens*. Estrogens in turn influence the uterine lining, the *endometrium,* causing it to become a thick, vascular tissue. This conditioned tissue is capable of accepting and eventually nourishing the fertilized ovum.

As soon as the ovum is released from the follicle (ovulation), the follicle takes on a new and quite different function from that it originally had in the maturation of the ovum. It becomes a large yellow body called the *corpus luteum*. The corpus luteum is an induced structure produced under the influence of another hormone from the anterior pituitary called *luteinizing hormone* (LH). The corpus luteum secretes the hormone *progesterone*. Progesterone continues the preparation and the maintenance of the endometrium in anticipation of pregnancy.

FSH and LH are called *gonadotropic hormones* since both influence gonads (in this case the ovaries), inducing them to secrete their own hormones.

In summary, the anterior pituitary secretes two gonadotrophic hormones. Each of them influences the ovary or parts of the ovary in some fashion and induces it to secrete hormones that in turn stimulate the endometrium to thicken and to prepare for the acceptance of a fertilized egg.

These two hormones, as is the case with all hormones, are carried in the bloodstream. Therefore shortly after the female has ovulated there are large concentrations of these two hormones in the blood. This high concentration tends to have an inhibitory effect upon the anterior pituitary, causing it to reduce its secretion of FSH. This feedback mechanism prevents the FSH stimulation of the ovary to start preparing another egg. If pregnancy occurs, the progesterone and estrogen levels remain at a high level. This effectively inhibits the anterior pituitary and its secretion of FSH.

Hormones Secreted by the Placenta These include estrogens and progesterone. The normal quantities of these hormones secreted by the follicle cells and the corpus luteum are adequate to initiate implantation (the zygote attachment to the endometrium). Once implantation has occurred, however, there is a need for these hormones in much higher concentrations. The placenta, once formed, begins secreting large quantities of estrogens, as much as 300 times the quantity supplied by follicle cells. This hormone continues to influence the endometrium but, more importantly, also aids in the enlargement of the breasts and the external genitalia.

Progesterone Secretion of *progesterone* by the placenta is also carried on at an elevated rate. In addition to continuing endometrium development during the early stages of pregnancy, progesterone also inhibits uterine contractions that might otherwise expel the fetus.

Prolactin This hormone, secreted by the anterior pituitary, stimulates the secretion of milk by the mammary glands of the breasts. The extremely high levels of estrogen and progesterone in the maternal bloodstream during pregnancy inhibit the release of *prolactin* by the pituitary. At birth, the sudden drop in estrogens and progesterone concentrations removes the inhibition, permitting the secretion of prolactin.

Menstrual Cycle

The fluctuations of individual hormone concentrations in the bloodstream produce structural and functional changes in female reproductive organs, referred to as the *menstrual cycle*. The normal series of menstrual cycles begins at puberty and continues for 35 to 40 years, interrupted only by pregnancy. The cessation of reproductive capability is called *menopause*.

The uterus is not in perpetual readiness to receive a fertilized egg. It must undergo structural changes so that it can effectively accept the fertilized egg and nourish the fetus.

A menstrual cycle is of approximately 28 days duration. The first day of the cycle is usually timed with the beginning of menstruation and continues until the beginning of the next menstrual period. The term *menstruation* refers to endometrium deterioration and the slight bleeding that occurs.

The essential purpose of the menstrual cycle is to prepare the uterus for its acceptance of a fertilized egg. If a fertilized egg is not available after the preparation phase, reduced hormone levels cause the endometrium to degrade and then begin the preparation stages all over again. If pregnancy occurs, the endometrium is influenced by hormones to maintain its state of preparedness, culminating in the formation of placental tissues that nourish the fetus.

To clarify the anatomical and physiological changes that occur during a complete menstrual

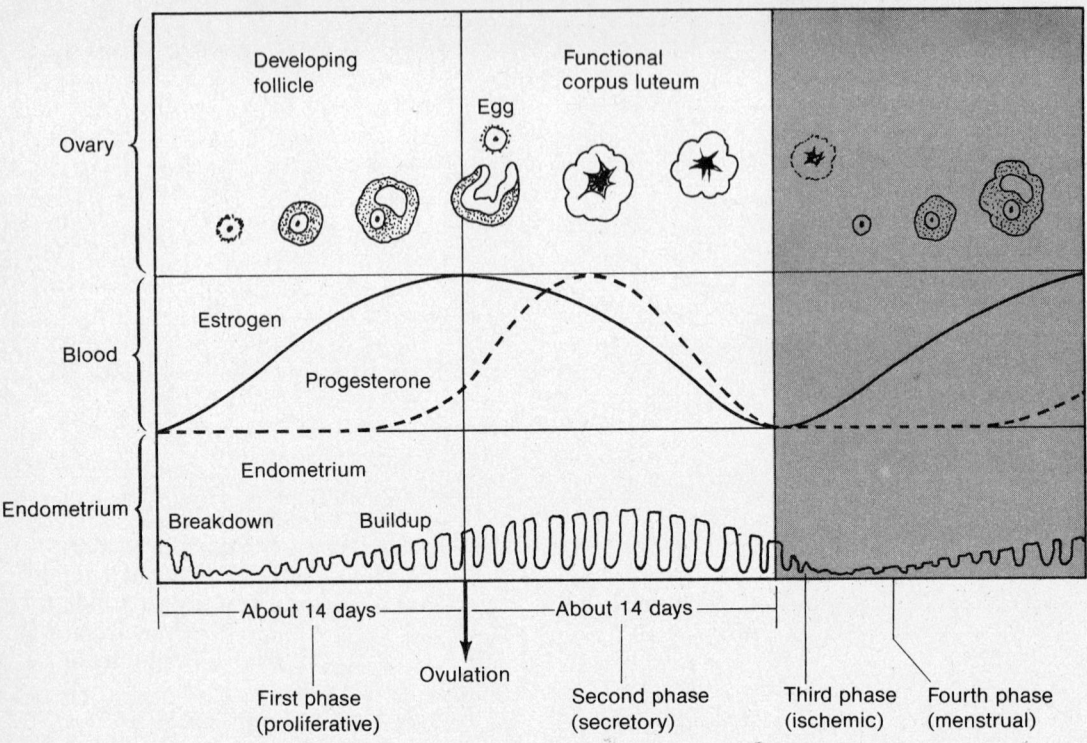

Fig. 18.6 Phases of menstruation. The first phase (proliferative) coincides with the maturation of the follicle and the subsequent ovulation. The second phase (secretory) begins with the formation of the corpus luteum and the secretion of progesterone. The third (ischemic) phase results in a rapid degeneration of endometrium cells. The fourth (menstrual) phase is marked by the sloughing off of dead endometrial cells and small quantities of blood from ruptured capillaries.

cycle, the cycle can be divided into four phases, as shown in Figure 18.6.

The First Phase (Proliferative) The first phase begins about the fifth day after the cessation of menstruation and coincides with the maturation of the follicle and ovulation. Cells surrounding maturing follicles secrete the hormone *estrogen*, which induces the endometrium to become highly vascular and thickened. The end of the proliferative phase coincides with the release of the egg from the follicle (ovulation) at about the fourteenth day of the cycle.

The Second Phase (Secretory) The second phase begins with the formation of the corpus luteum and its secretion of the hormone *progesterone*. Progesterone continues and accelerates the work begun by estrogen. The endometrium continues to thicken, becoming more vascular and developing specialized glandular cells that provide nutrient substances in anticipation of a fertilized egg. The stage is now set for a fertilized egg, upon arriving at the uterus, to be implanted in the endometrium.

If a fertilized egg is not available, the endometrium begins its ischemic phase.

The Third Phase (Ischemic) The third phase begins at about the twenty-fourth day of the cycle. The small arterioles and capillaries that have invaded the endometrium constrict, thus shutting off the blood supply to the proliferated cells of the endometrium. These cells, deprived of oxygen and nutrients, die.

The Fourth Phase (Menstrual) The fourth phase lasts from three to five days. It is marked by the sloughing off of dead endometrial cells and small amounts of blood from ruptured capillaries. When the small capillaries rupture, the blood immediately clots. The clots are liquified by enzymes. Thus the fluid that flows into the vagina is a mixture of unclotted blood, dead endometrium cells, mucus, and the microscopic unfertilized (now dead) ovum.

PREGNANCY

Ovulation coincides with the secretory phase of the menstrual cycle. A mature ovum is released from the ovary and is picked up by the ends of the fallopian tubes. The smooth muscle contractions and cilia movement along the length of the fallopian tubes carry the ovum in the direction of the uterus. This occurs at approximately the fourteenth day of the cycle. If, during this time, sperm is deposited in the vagina, fertilization of the egg may occur.

Sperm are motile cells that "swim" into the uterus and then into the fallopian tubes where a single sperm, out of several million deposited during male ejaculation, may unite with the egg. This union is called *fertilization*. Once a sperm has united with the egg, chemical and physical changes occur in the egg membrane prohibiting the entry of other sperm. Of the millions of sperm available, only one is destined to unite with the egg.

Once fertilization has been accomplished, mitosis is initiated. The fertilized egg, now with 46 chromosomes (23 contributed by the egg and 23 contributed by the sperm), divides into two cells; these two cells divide into four cells, and so on. Once cell division begins, the dividing structure is referred to as the *zygote*.

The zygote, having no motile power of its own, is carried along the fallopian tube by muscular contraction and cilia toward the uterus. The trip takes one to three days. By the time the zygote arrives at the uterine endometrium, it has undergone a series of mitotic divisions. The zygote is capable of attaching itself to the prepared and receptive endometrium. This process, called *implantation*, sets the stage for rapid changes in the endometrium, culminating in the formation of the *placenta*.

The placenta receives nutrients and oxygen from maternal blood coursing through the uterus and transmits these substances by way of the umbilical cord to the embryo.

The progesterone level of maternal blood is quite high at this time and must remain at a high concentration level throughout the gestation period. *Gestation* is the period of time dating from fertilization to birth (approximately 270 days).

The complicated processes that result in the formation of the placenta and the formation of fetal tissues belong in an area of scientific study called *embryology,* which will not be discussed in this text. The developmental processes involved in the formation of the fetus are also largely beyond the scope of this text. However, the following is a synopsis of the events culminating with the birth of the fetus.

FETAL DEVELOPMENT

The secretion of hormones and their influence upon the female body are only a prelude to the union of an ovum and a sperm. This union, called *conception,* marks the beginning of a new organism.

The single cell, now containing nuclear material from both parents, begins dividing by mitosis. Each new generation of cells doubles the number of the previous generation. Mitosis proceeds very rapidly at first, producing a mass of undifferentiated cells called a *morula.*

As cells of the morula continue to divide, they arrange themselves into layers. From these layers—*ectoderm, mesoderm,* and *endoderm*—arise all body tissues.

Cells from the ectoderm differentiate into outer body tissues such as skin, hair, nails, glands, and the nervous system.

The endoderm gives rise to the lining of the gastrointestinal tract, lungs, liver, and glands.

The mesoderm cells differentiate into muscle, bones, the cardiovascular system, lungs, and other internal organs.

Mitosis and differentiation proceed at an accelerated rate during the first month of human fetal development. Tissues, organs, and systems are formed to such an extent that by the end of the first month of pregnancy the fetus has recognizable human features.

Drugs Given During Pregnancy May Harm the Fetus

Drugs and processes often given to pregnant women may cause serious, often life threatening, conditions for the fetus and newborn. These include:

1. *Analgesic agents*
2. *Anesthetic agents*
3. *Belladonna derivatives*
4. *Antihypertensives*
5. *Intravenous therapy*
6. *Psychotropic agents*
7. *Anticonvulsive agents*
8. *Anticoagulants*
9. *Cholinesterase inhibitors*
10. *Hypoglycemic agents*
11. *Sedatives*
12. *Diagnostic procedures (X ray, radium)*
13. *Thyroid medication*
14. *Uterine stimulants*
15. *Vitamins (vitamins K and B_6)*

One or more of these drugs or processes may cause congenital malformation, miscarriage, abnormal growth, permanent physical and/or mental dysfunction, or lead to the development of cancer in later years.

DISEASES OF THE REPRODUCTIVE SYSTEM

Venereal disease is the term applied to those reproductive system diseases that are transmitted by sexual contact. The two major venereal diseases are *syphilis* and *gonorrhea*. Other diseases of less serious importance include those that produce local infections, such as *vulvovaginitis, trichimoniasis,* and *puerperal fever*.

Syphilis

Syphilis is an infectious disease caused by *Treponema pallidum*, a spirochete. Its transmission is by direct contact during sexual activity. An infected person may transmit the spirochete to another person by way of moist membrane transfer. This includes the genitals and mucous membranes of the mouth. The onset of syphilis is generally recognized by an eruption involving the skin and mucous membranes. Microscopic and serologic tests are conclusive.

The early stages of syphilis cause disturbances that resemble other infections, such as influenza. The disease may become latent and remain so for several years, then reappear causing disabling malfunctions of the cardiovascular, nervous, and other systems.

Syphilis can be controlled with antibiotic therapy and can be prevented by the use of proper prophylactics. The most effective prevention is to avoid sexual contact with infected persons. The difficulty with this last statement is that an infected person may not know he or she has the disease or, on the other hand, may not care.

Gonorrhea

Gonorrhea is a disease caused by the bacterium *Neisseria gonorrhoeae*. It is transmitted almost entirely by sexual contact with an infected person. Although gonorrhea is seldom fatal, it may spread throughout the body, carried by the blood, and cause disabling conditions in major body organs.

Gonorrhea can be controlled with antibiotic therapy. Interviewing infected persons and tracing their contacts can be effective.

Vulvovaginitis

Vulvovaginitis is an inflammation of the vulva. The agents causing the condition may, in addition to venereal diseases, include bacteria normally found in otherwise healthy females. Some bacteria become opportunists when favorable conditions occur, increasing their populations drastically.

Symptoms of vulvovaginitis include itching and burning sensations, vaginal discharge, and pain.

The protozoan usually associated with vulvovaginitis is *Trichimonas vaginalis*. The fungi *Candida albicans* is also often identified.

Trichimoniasis

Trichimoniasis is caused by an invasion of the vagina by the protozoan *Trichimonas vaginalis*. The condition can be spread by sexual contact. Trichimonas is so commonly found in the adult female vagina that some researchers believe it to be a normal parasite that causes problems mainly when the vaginal environment changes to permit its growth. The vaginal environment may change as the normal body chemistry changes, or may change due to the introduction of fluids (douches) that alter the normal pH of vaginal secretions.

Puerperal Fever

Puerperal fever is a disease of the genital tract of a woman who has recently given birth to a child. The bacterium associated with this disease is most often of the genus *Streptococcus*. The source of infection is often the lack of aseptic technique followed by medical personnel during birth.

NONMICROBIAL DISEASES OF THE REPRODUCTIVE SYSTEM

Malignant Tumors

Malignant tumors of the uterus and cervix can be detected in their early stages by the *pap smear test*. In this test, a sample of cervix tissue scraping is observed under the microscope for the presence of malignant cells.

Prostate Cancer

Tumors of the prostate, which may be either benign or malignant, occur rather commonly in elderly males. The enlarged prostate puts pressure upon the urethra, which it surrounds at its origin at the base of the urinary bladder. The flow of urine into the urethra is inhibited, making urination painful and difficult.

Prostate cancer is one of the more easily detected cancers (rectal examination) and also one of the simplest to treat surgically.

THE AGING REPRODUCTIVE SYSTEM

Less is known about the influence of aging upon the reproductive system than is known about any other body system. In the female, cessation of menstruation (menopause) leads to the atrophy of the ovaries and reduced estrogen secretion. This may have pronounced influence upon many body systems. However, many women go through this interval of time with little or no awareness of any change.

Cancer of the cervix and endometrium in women and of the prostate gland in males tends to increase in incidence as the aging process continues.

Healthy aged males are capable of producing viable sperm well into their 80s and beyond.

SUMMARY AND REVIEW

A. The male reproductive system includes:
   1. Paired testes, which produce spermatozoa and secrete the male hormone testosterone.
   2. Vas deferens, which conducts sperm from the testes to the urethra.
   3. The penis, which serves to introduce sperm into the female vagina.
   4. Accessory structures include:
      a. seminal vesicles
      b. prostate gland
      c. bulbourethral glands
      d. ejaculatory duct
B. Spermatogenesis, the process by which mature sperm are produced, involves a reduction-division of chromosomes to 23 per sperm.
C. Male sex hormones. Testosterone is the major male sex hormone. Its influence, beginning with puberty, brings about male secondary sex characteristics.
D. The female reproductive system includes:
   1. paired ovaries
   2. fallopian tubes
   3. uterus
   4. vagina
   5. external genitalia
      a. labia major
      b. labia minor
      c. prepuce
      d. clitoris

E. Hormone control of female reproduction. Influences of anterior pituitary gland
   1. FSH (follicle stimulating hormone) stimulates ovarian follicles to begin the egg maturation process and stimulates the production and secretion of estrogen.
   2. LH (luteinizing hormone) induces the formation of the corpus luteum, which secretes progesterone.
   3. Estrogens and progesterone influence the vascularization of the endometrium.
   4. Prolactin, a hormone secreted by the anterior pituitary, stimulates the mammary glands to secrete milk.
F. The menstual cycle phases
   1. First phase (proliferative) includes ova maturity and release, and endometrium vascularization.
   2. Second phase (secretory) includes the formation of the corpus luteum and the secretion of progesterone.
   3. Third phase (ischemic) results in the reduction of blood supply to endometrium cells, resulting in the death of the proliferated edometrical cells.
   4. Fourth phase (menstrual) is marked by the menstrual flow.
G. Pregnancy is the result of a sperm fertilizing an egg and the subsequent implantation of the zygote in the uterine wall.
H. Fetal development is the process involved in the formation of an individual beginning with the fertilized egg.
I. Diseases of the reproductive system
   1. Syphilis, caused by *Treponema pallidum*.
   2. Gonorrhea, caused by *Neisseria gonorrhoeae*.
   3. Vulvovaginitis, caused by opportunistic bacteria or protozoans such as *Trichimonas vaginalis*. The fungi *Candida albicans* also may be implicated.
   4. Trichimoniasis, an invasion of the vagina by the protozoan *Trichimonas vaginalis*.
   5. Puerperal fever, a streptococcal disease of the genital tract of a woman who recently gave birth.
J. Nonmicrobial diseases
   1. malignant tumors
   2. benign tumors
   3. prostate cancer

19

Human Genetics

MAJOR CONCEPTS
  CHROMOSOMES AND KARYOTYPES
  HOMOLOGOUS PAIRS OF CHROMOSOMES
  SINGLE-GENE DEFECTS
    Huntington's Disease
    Achondroplasia
    Sickle Cell Anemia
    Tay-Sachs Disease
    Phenylketonuria
    Galactosemia
  SEX DETERMINATION
  SEX-LINKED TRAITS
    Color-Blindness
    Hemophilia
    Agammaglobullinemia
  MULTIFACTORAL INHERITANCE
  CHROMOSOMAL ABNORMALITIES
    Down's Syndrome
    Structural Defects
  SEX CHROMOSOME DEFECTS
    Turner's Syndrome
    Kleinfelter's Syndrome
  TESTS FOR GENETIC ABNORMALITIES
    Nuclear Staining
    Amniocentesis
    Blood Tests
    Family Pedigree

Normal and abnormal human traits are passed along each generation. This chapter presents a brief discussion of basic genetics and an overview of genetic diseases and tests for these diseases where known.

Genetics is the study of hereditary traits. The expression of traits in an individual is the end result of the combination of sex cell DNA from two parents. Each parent contributes half of the DNA found in a fertilized egg. Structural components of cell nuclei called *chromosomes* are made up of DNA molecules. Normal human cells (excluding sex cells) have 46 chromosomes, or 23 pairs. One chromosome of each pair is contributed by each parent. Sex cells (egg and sperm) contain 23 chromosomes each. When the egg is fertilized, the sperm and egg nuclei fuse, each contributing 23 chromosomes.

The segment of a chromosome responsible for transmitting information that determines a specific trait is called a *gene*. How many genes are there on the 23 pairs of chromosomes? We would have to know how many traits humans have. The question cannot be answered.

The exploration of normal human genetics is a rewarding and worthwhile pursuit. However, the emphasis in this chapter is directed toward the abnormal aspects of human genetics.

CHROMOSOMES AND KARYOTYPES

The problem of separating the microscopic chromosones so that they may be counted and observed has only recently been solved. Living human body cells are grown in test tubes. A chemical is then added that arrests all division at a phase when the chromosomes are short, thick, and easily observed, as shown in Figure 19.1.

Chromosomes vary not only in size but also in arrangements of their components. A chromosome is made up of two threadlike strands called *chromatids*, which are held together by a minute bit of matter called the *centromere*, as shown in Figure 19.2. The centromere may be near the middle of the two chromatids, near the ends, or somewhere between these two locations. The location of the centromere helps an observer locate pairs of chromosomes. Each sex cell contributes an identically shaped chromosome, including the location of the centromere. By locating all the pairs, they may be arranged in

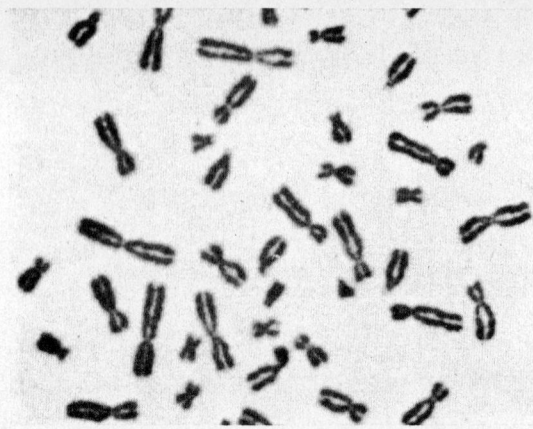

Fig. 19.1 Microscopic view of normal human chromosomes. *(From Redding, A. and Hirschhorn, K. "Guide to Human Chromosome Defects" in Bergsma, D. (ed.) B, D, O, A, S IV (4) published by The National Foundation–March of Dimes, White Plains, New York, 1968.)*

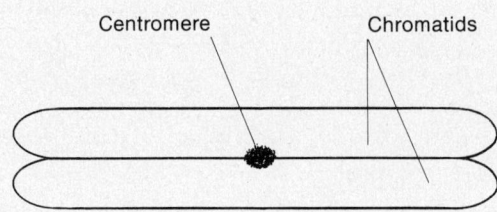

Fig. 19.2 A chromosome is made up of two chromatids held together by a centromere. The centromere may be near the middle of the two chromatids, near the ends, or somewhere in between.

order from the largest chromosome pairs down to the smallest. This arrangement, called a *karyotype*, is shown in Figure 19.3.

HOMOLOGOUS PAIRS OF CHROMOSOMES

Each member of a pair of chromosomes is structurally similar to the other. The pairs, therefore, are referred to as *homologous pairs*. If a specific segment of one chromosome carries information for a certain trait, the other chromosome of the pair will carry information for the same trait at precisely the same location (called a *gene locus*). Both genes at the same locus on each of the homologous pair are capable of expressing themselves with respect to a trait.

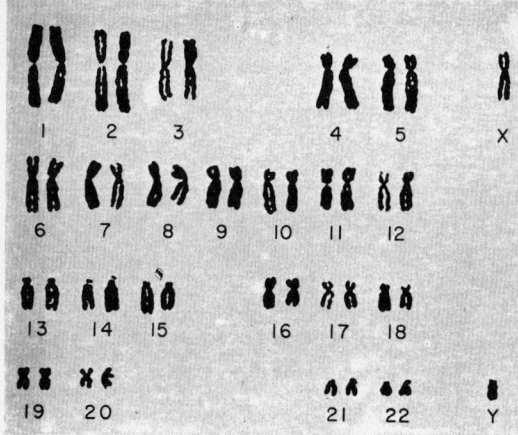

Fig. 19.3 Normal male chromosomes from Fig. 19.1 arranged in order of size and centromere location. This arrangement is called a karyotype.

The human trait of stature (height) can be used as an example. If one chromosome of a pair carries information calling for short stature, and the other chromosome carries information dictating tall stature, the individual will have an intermediate stature. Shortness dominates tallness; that is, the gene for shortness is a *dominant gene*. The gene for tallness is inhibited (not permitted to be fully expressed) and is called the *recessive gene*.

Homologous genes both carrying dominant genes, or one dominant and one recessive, will produce the dominant trait. A recessive gene can express itself only when united with another recessive gene. If both genes of a homologous pair are dominant, they are *homozygous dominant*. If one gene is dominant and the other recessive, they are said to be *heterozygous*. If both genes are recessive, they are said to be *homozygous recessive*.

A major concern in human genetics is to identify faulty genes and to predict their probable occurrence in future generations. A faulty gene is one that expresses itself in an individual causing physical, physiological, or mental deformities, and sometimes death. It has been estimated that at least 40 percent of all infant mortality results from faulty genetic factors. Each married couple stands a 3 percent risk of having a genetically defective child. The faulty gene may be either dominant or recessive.

If the genetic makeup of parents (the mother's or father's *genotype*) is known with respect to a given trait, genetic laws help predict the probability of the trait occurring in their offspring. Dominant genes are symbolized by capital letters while recessive genes are recorded in lowercase letters.

For example, suppose that a certain dominant gene, D, is faulty and is carried by the male in the heterozygous state, Dd. If the female is normal, dd (homozygous recessive), some predictions can be made concerning the offspring. The genotype Dd indicates that one of the chromosomes of the homologous pair carries the D gene while the other carries the d gene. When sex cells (gametes) arise from a cell of type Dd, the chromosome carrying the D gene will show up in one gamete and the d gene will show up in another. In other words, two different kinds of gametes are produced, one carrying a dominant gene and another kind carrying a recessive gene. This is illustrated in Figure 19.4.

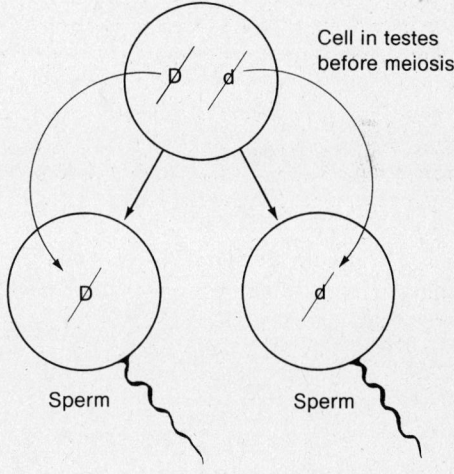

Fig. 19.4 Sperm from cells carrying the Dd combination of chromosomes will be of two types. One type will carry the D gene and another the d gene.

In this example, the female is normal (dd). Any eggs produced by her will carry the recessive trait. If we examine all possible combinations of sperm and eggs, we can predict possible genotypes of the offspring. A simple way to show this is to plot all possible male gametes along one axis and all possible female gametes along the other, then make all possible combinations, as illustrated in Figure 19.5. From this parental cross, each child has a 50 percent chance of inheriting the faulty gene and a 50 percent chance of being normal.

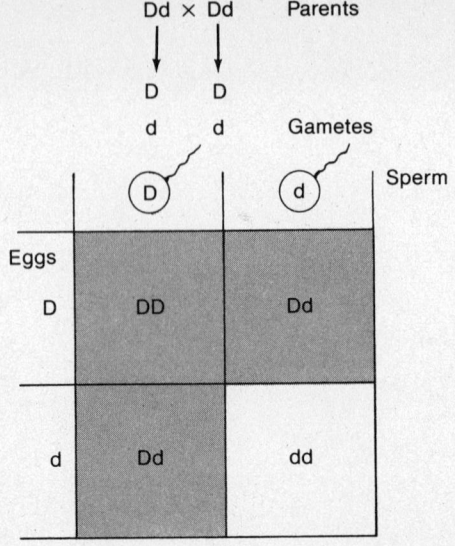

Fig. 19.5 All probable offspring from the cross between Dd and dd. From this cross each child has a 50 percent chance of inheriting the faulty gene combination and a 50 percent chance of having the dominant gene, thus being normal.

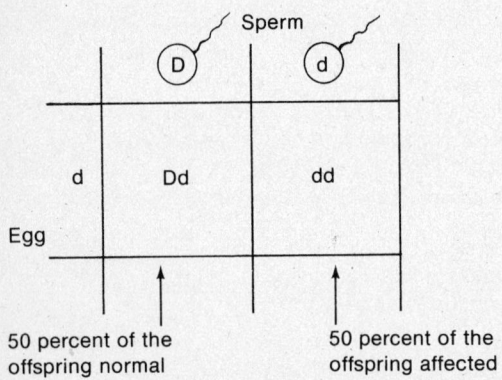

Fig. 19.6 If both parents are heterozygous, the probable ratio of offspring types will be the shaded squares, which represent the abnormal trait. Unshaded is normal. In this cross, 75 percent of the offspring may be affected and 25 percent may be normal.

The term *phenotype* describes the appearance of the individual with respect to the combination of genes. For example, if we use the term "affected" for faulty gene combinations, then the phenotype of 50 percent of the offspring from this cross is affected. The phenotype of the other 50 percent is normal.

If both parents are heterozygous (Dd), both are faulty. The probable offspring are shown in Figure 19.6.

By showing all possible combinations, we can predict a 50 percent chance of the offspring being a *carrier* and faulty like the parents; a 25 percent chance of getting a double dose (DD) of the faulty gene, and only a 25 percent chance of being normal. Since both DD and Dd carry the faulty gene, 75 percent of the offspring would be predicted to be defective and only 25 percent normal.

This description of gene action is all that is needed to describe many genetic diseases known to affect the human race. To date, more than 900 disorders involving single-gene defects have been identified. In these diseases, if one parent is carrying a faulty gene and if the gene is dominant, in each pregnancy there is a 50 percent risk of passing along the disease to offspring. If both parents carry a single, dominant faulty gene, the risk of transmitting the disease rises to 75 percent. If one parent has faulty genes on both chromosomes (of a homologous pair), all offspring will be affected regardless of what genes the other parent possesses.

If the trait is recessive, then each parent must contribute a recessive gene. The parents may be only carriers (heterozygous) of a faulty gene. Parents heterozygous for a faulty recessive gene run a 25 percent risk of passing the disease to their offspring.

SINGLE-GENE DEFECTS

Genes carried on all chromosomes other than the sex chromosomes are called *autosomal genes*. Some single-gene autosomal defects are *Huntington's disease, achondroplasia, sickle cell anemia, Tay-Sachs disease, phenylketonuria,* and *galactosemia*.

Huntington's Disease

Huntington's disease (chorea) is caused by autosomal dominant gene. The disease is a serious nervous disorder that progressively worsens with age. Hospitalization and death are the eventual consequences. The disease often is not identified in an individual until middle age. However, the onset may occur between ages 10 and 60. At middle age, a parent, unknowingly carrying the defective gene, may have transmitted the gene to one or more of his or her offspring. If signs of the disease become apparent, then all children from this parent would know they have at least a 50 percent chance of carrying the defective gene. Not only may they themselves have the disease, but they also take a chance of passing the defect along to their children.

Achondroplasia

Achondroplasia is an inherited form of dwarfism transmitted as an autosomal dominant single-gene defect. The defect results in shortened arms and legs. However, the rest of the body may be normally proportioned. There appears to be no negative effect upon the nervous system.

Sickle Cell Anemia

Sickle cell anemia is transmitted as an autosomal recessive gene. Both parents must contribute to the defect by each contributing a recessive gene. The combined recessive genes (homozygous recessive) alter red blood cell hemoglobin molecules resulting in sickle-shaped cells rather than the normal, biconcave disc-shaped cell. These cells are more fragile than normal cells. They have a very short life, hence producing an anemic condition. Sickle cell anemia is predominantly a gene defect occurring among black Americans. About 10 percent carry the recessive trait, which results in approximately 1 in every 400 to 600 black children having sickle cell anemia. The disease is not limited to the black population. It also occurs in other ethnic groups, including Greeks, Arabs, Sicilians, and Turks.

There is some reason to believe that the disease originated in tropical countries. Individuals heterozygous for sickle cell (Ss) are carriers of the gene that produces the disease. These individuals appear to have a greater resistance to malaria than persons who do not carry the gene. If this is true, these heterozygous individuals have an adaptive advantage over others without such protection. This may explain how the disease has been perpetuated for centuries.

Sickle cell anemia results in the reduced oxygen-carrying capacity of red blood cells. The individual usually appears to be short of breath and tires easily. Other symptoms may include jaundice, low immune response, and retarded physical growth.

Sickle cell anemia cannot be cured. Research today is directed toward creating an environment in which affected persons can live normal, productive lives.

Tay-Sachs Disease

Tay-Sachs disease, also known as *infantile amaurotic idiocy,* is transmitted as an autosomal recessive gene. Each parent must contribute a recessive gene. This defect begins in the early infant and always results in death, usually before age 4. This condition exemplifies what are called *lethal genes*. A carrier of the recessive gene is not affected. Therefore a more proper term than lethal gene may be *lethal gene combinations*. Tay-Sachs disease also follows ethnic lines, being more prevalent in families of Askenazi Jewish ancestry traceable back to Central and Eastern Europe. About 90 percent of American Jews trace their ancestry to this part of Europe.

Tay-Sachs disease is an enzyme defect, inhibiting chemical reactions involving lipids, particularly of the brain. Toxic accumulations of substances called *sphingolipids* affect the nervous system, progressively worsening until death results.

Phenylketonuria

Phenylketonuria (PKU) is a single-gene autosomal recessive defect. This disease is routinely tested for in hospitals shortly after birth. The affected child lacks an enzyme to complete chemical reactions involving certain proteins. Toxic residues accumulate and inhibit proper maturation of nerve cells. Different levels of mental retardation result if the condition is left untreated. A blood test at birth will reveal the condition. The child's diet may then be controlled (to age 7 to 12), omitting foods that cannot be metabolized.

Galactosemia

Galactosemia is similar to PKU except the monosaccharide galactose cannot, due to an enzyme deficit, be converted to glucose. Galactosemia is also caused by a single-gene recessive defect. The consequences of the disease as well as the treatment are similar to PKU. The child's diet must not contain milk or milk products because lactose—milk sugar—is hydrolized to glucose plus galactose.

SEX DETERMINATION

In a karyotype of female chromosomes, the 23rd and smallest pair, the sex chromosomes, are also called the X chromosomes. Females have two X chromosomes. When these separate in the production of eggs, an X chromosome will appear in each egg.

The 23rd pair of chromosomes in males is made up of an X chromosome like the female, plus another small chromosome called the Y chromosome. Males therefore produce two kinds of sperm. Fifty percent will contain the X chromosome, and 50 percent will have the Y chromosome. Note in Figure 19.7 the combination of X and Y chromosomes; the ratio of male offspring to female offspring is 1:1 or 50 percent for each. The male parent, then, carries the determiner for male or female offspring, depending upon which of his two kinds of sperm fertilize the egg.

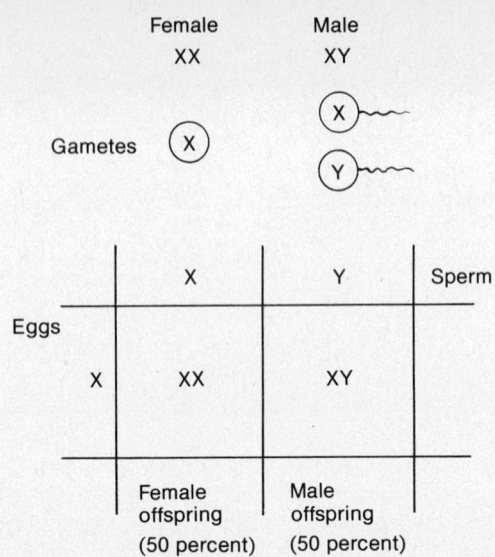

Fig. 19.7 Sex determination depends upon which kind of sperm (X bearing or Y bearing) fertilizes the egg. The male carries the determiner for male or female offspring, depending upon which of his two kinds of sperm fertilizes the egg.

SEX-LINKED TRAITS

X and Y chromosomes carry other traits besides sex determination. Over 150 known disorders are known to be carried on the X chromosome, as well as a few on the Y chromosome. These disorders are called *sex-linked traits*.

Color Blindness

We can use color blindness as an example of how sex-linked traits work. The X chromosomes of the female and the X chromosome of the male may be normal or affected. In color blindness, a common sex-linked disorder, the individual cannot distinguish between the colors red and green. Given the following genotypes of parents, we can predict what percent of male and female children from these parents may be color-blind.

In this example, the father is normal. The mother is normal but heterozygous (a carrier). If we let C stand for the normal gene and c for the defective gene, then the male genotype would be $X^C Y$. The Y chromosome carries nothing with respect to this disorder. The female genotype

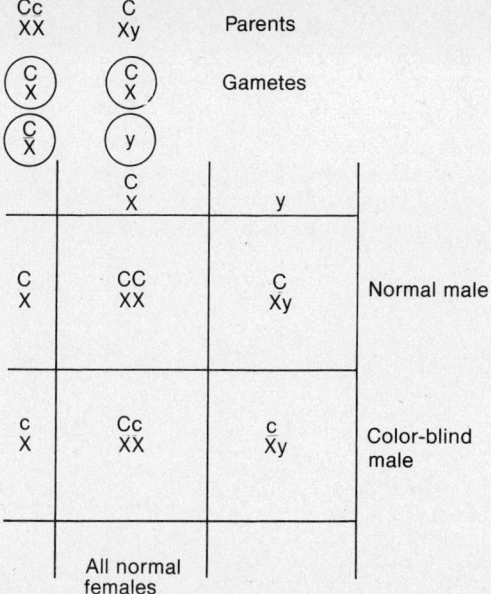

Fig. 19.8 Sex-linked trait of color blindness. From this cross, all females carry the dominant gene and are thus normal (50 percent carry the recessive gene). Half of the male offspring can be predicted to be normal and half color blind.

would be X^CX^c. One of the female's X chromosomes is carrying the dominant (normal) gene, while the other is carrying the gene for color blindness. All possible gametes and all possible combinations are shown in Figure 19.8.

All female offspring would be normal (50 percent are carriers), and 50 percent of the males would be predicted to be color-blind. Note that males can only receive their X chromosome from the mother. Therefore if either of the mother's X chromosomes is carrying a defective gene, 50 percent of her male offspring will risk having the defect. If both of the mother's X chromosomes carry the defect, all her male offspring will have the defect.

Hemophilia

Hemophilia is sex linked and recessive. Each parent must contribute a recessive gene. Only the X chromosomes carry the defective gene. Since males get their X chromosomes from the mother, the trait may be passed from mother to son.

Hemophilia, also called bleeder's disease, is the incapability of blood to clot. The incidence of sex-linked traits is much greater in the male population than the female.

Because females possess two X chromosomes, even if a female carries a defective recessive gene on one of her X chromosomes, the other X may be normal (dominant). Therefore she does not show the defect. She is a carrier and may transmit the defect to a son. Males have only one X chromosome and one Y chromosome. A recessive gene on a male's only X chromosome makes him appear with the defect. The Y chromosome can do nothing to lessen the defect.

Agammaglobullinemia

Agammaglobullinemia is a sex-linked trait that inhibits the formation of serum antibodies. Individuals with this trait often have little or no active immunity against microorganisms.

Figure 19.9 indicates age of onset and duration of some single gene defects, including some not discussed in the text.

MULTIFACTORIAL INHERITANCE

Some genetic defects result from the interaction of several genes. Predicting the incidence of transmittal from parent to offspring is not as easy as with single-gene defects. Some defects thought to be transmitted by gene interaction (*multifactorial inheritance*) are spina bifida (open spine), hydrocephalus (water on the brain), cleft lip, and cleft palate.

CHROMOSOMAL ABNORMALITIES

Thus far, a consideration has been made only with respect to specific genes. Another area of human genetics, called *cytogenetics*, is concerned with structural defects of chromosomes and their deviation in number from the normal autosomal number (46 in humans).

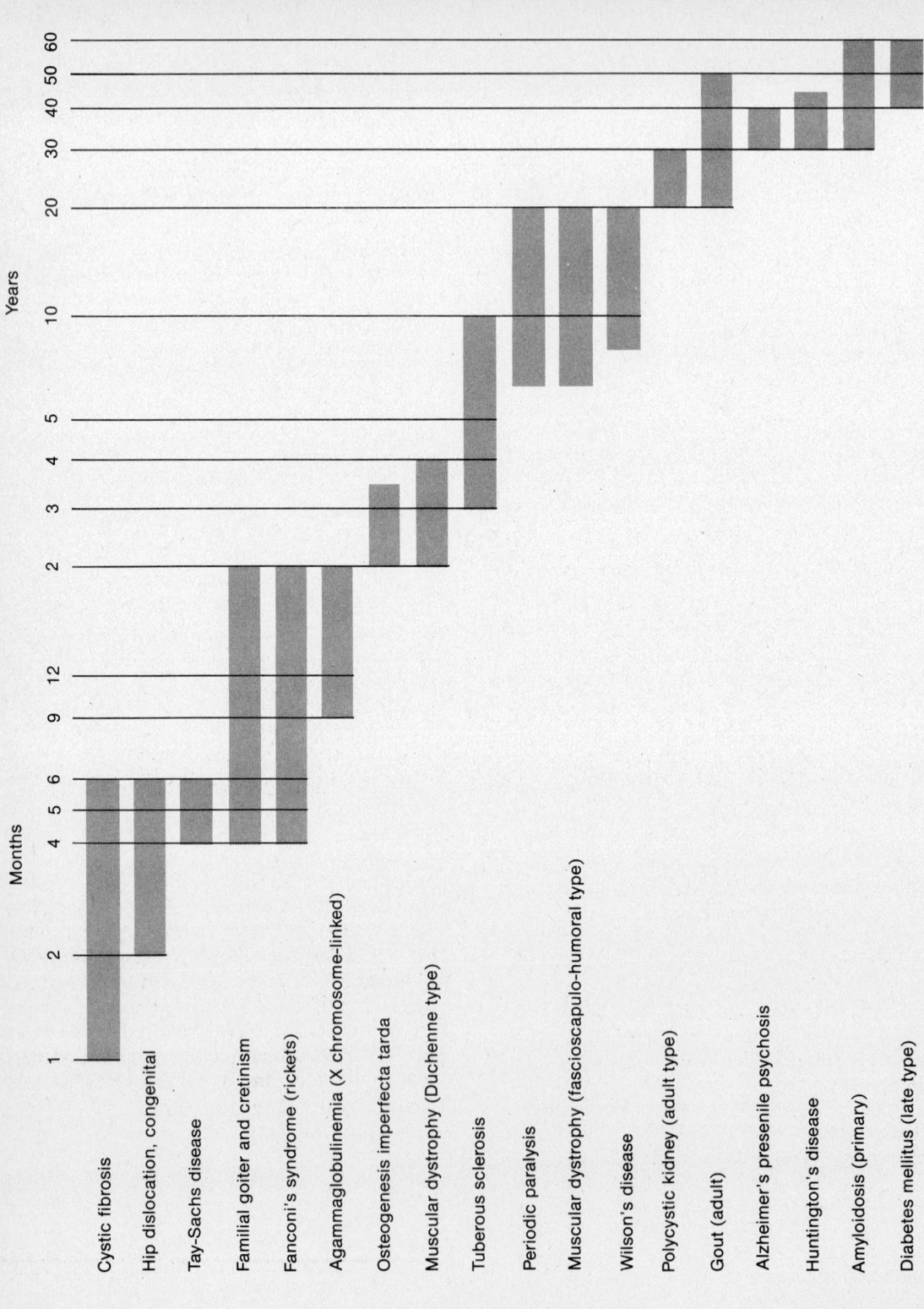

Fig. 19.9 Late-appearing birth defects. *(Courtesy National Foundation/March of Dimes, 1975.)*

Down's Syndrome

Down's syndrome, or mongolism, is technically called *trisomy 21.* This indicates that instead of the two chromosomes at a specific location in a karyotype, there are three. This increase in chromosomes is caused by *nondisjunction* (failure to separate) of the 21st pair of chromosomes during meiosis, sex cell (ova) production. The homologous pair at 21 fail to separate; thus an egg results with 24 chromosomes instead of the normal 23. If this egg is fertilized by a normal sperm (23 chromosomes), the offspring from this mating will have cells with 47 chromosomes rather than 46.

The child with Down's syndrome always suffers some degree of mental retardation; has distinctive flattened features; inner eyelid fold; a short, broad neck; and a host of other characteristics that result in a distinctive deviation from normal human physical and mental characteristics. The incidence of Down's syndrome increases as the maternal age increases. Approximately 60 percent of the mothers having children with this defect are 35 years of age or older. The risk of a mother between the ages of 15 and 29 having a child with Down's syndrome is 1 in 500. After age 44, the risk is 1 in 50. There is no medical treatment available for Down's syndrome. Later in the chapter is a discussion of prenatal diagnosis and tests for Down's syndrome and other genetic defects.

The incidence, cause, detection, treatment, and prevention for some common genetic diseases are shown in Table 19.1.

Other trisomy conditions have been identified in other chromosome pairs. *18 trisomy (Edwards's syndrome)* is much more rare than Down's syndrome (about 1 in 4,500 live births). Trisomy syndromes of progressively larger chromosomes produce more severe defects. Babies born with 18 trisomy have severe physical and mental defects, and many only live a few months; *13 trisomy (Patau's syndrome)* produces even more severe physical and mental defects than the other trisomy syndromes occurring on smaller chromosomes.

Structural Defects

Chromosomes may be broken and parts of chromosomes deleted. These missing parts carry with them specific information essential for normal development during fetal growth.

These structural defects invariably result in serious physical and mental defects in the child.

SEX CHROMOSOME DEFECTS

The term "defect" used here is not entirely proper; perhaps "genetic accident" is more appropriate. The two defects discussed here are due to sex chromosome nondisjunction.

Turner's Syndrome

Ovary cells contain cells having 46 chromosomes, including two X chromosomes. During meiosis, the two X chromosomes may fail to separate (nondisjunction), thus yielding eggs with either two X chromosomes or no X chromosomes (monosomy). If fertilized, an egg with no X chromosome (22 total chromosomes) would produce a female with 45 chromosomes, with the sex chromosomes shown as XO. This condition is called *Turner's syndrome.* Individuals with Turner's syndrome universally have very short stature, are sterile, and have a long list of physical abnormalities. Mental capabilities are less affected.

Kleinfelter's Syndrome

Kleinfelter's syndrome results from an egg (XX) fertilized by sperm carrying the Y chromosome. The individual, a male, has 47 chromosomes—the normal 22 autosomes plus XXY. The syndrome carries with it a wide variety of physical and mental defects, but usually less severe than trisomy or structural defects of autosomes. Males with the syndrome are usually taller than normal, and usually sterile. Approximately one-fourth are mentally retarded.

Table 19.1 Selected Birth Defects, U.S.A.

BIRTH DEFECT	TYPE	ANNUAL INCIDENCE[a]	PREVALENCE[b]	CAUSE
Down's syndrome (mongolism)	Functional/structural: retardation often associated with physical defects	5,100	44,000	Chromosomal abnormality
Markedly low birthweight[d]/ prematurity	Structural/functional: organs often immature	50,000	NA	Hereditary and/or environmental: maternal disorder or malnutrition
Muscular dystrophy	Functional: impaired voluntary muscular function	Unknown (late-appearing)	200,000	Hereditary: often recessive inheritance
Congenital heart malformations	Structural	24,800	248,000	Hereditary and/or environmental
Clubfoot	Structural: misshapen foot	9,300	149,000	Hereditary and/or environmental
Polydactyly	Structural: multiple fingers or toes	9,300	184,000	Hereditary: dominant inheritance
Spina bifida and/or hydrocephalus	Structural/functional: incompletely formed spinal canal; "water on the brain"	6,200	53,000	Hereditary and environmental
Cleft lip and/or cleft palate	Structural	4,300	71,000	Hereditary and/or environmental
Diabetes mellitus	Metabolic: inability to metabolize carbohydrates	Unknown (late-appearing)	90,000	Hereditary and/or environmental
Cystic fibrosis	Functional: respiratory and digestive system malfunction	2,000	10,000	Hereditary: recessive inheritance
Sickle cell anemia	Blood disease: malformed red blood cells	1,200	16,000	Hereditary: incomplete recessive—most frequent among blacks
Hemophilia (classic)	Blood disease: poor clotting ability	1,200	12,400	Hereditary: sex-linked recessive inheritance
Congenital syphilis	Structural: multiple abnormalities	(newborn only) 180	NA	Environmental: acquired from infected mother

[a] Incidence: the number of new cases diagnosed within a specific time period.
[b] Prevalence: total number living who have been diagnosed as having defect. Above statistics based on number less than 20 years of age.

DETECTION[c]	TREATMENT[c]	PREVENTION[c]
Amniocentesis, chromosome analysis	Corrective surgery, special physical training and schooling	Genetic services
Prenatal monitoring, visual inspection at birth	Intensive care of newborn, high nutrient diet	Proper prenatal care, genetic services, maternal nutrition
Apparent at onset	Physical therapy	Genetic services
Examination at birth and later	Corrective surgery, medication	Genetic services
Examination at birth	Corrective surgery, corrective splints, physical training	Genetic services
Visual inspection at birth	Corrective surgery, physical training	Genetic services
Amniocentesis, prenatal X-ray, ultrasound, maternal blood test, examination at birth	Corrective surgery, prostheses, physical training, special schooling for any mental impairment	Genetic services
Visual inspection at birth	Corrective surgery	Genetic services
Appears in childhood or later; blood and urine tests	Oral medication, special diet, insulin injections	Genetic services
Sweat and blood tests	Treat respiratory and digestive complications	Genetic services
Blood test	Transfusions	Genetic services
Blood test	Clotting factor	Genetic services
Blood test, examination at birth	Medication	Proper prenatal care

(continued on next page)

[c]Last three columns list possible means now known for detection, treatment, and prevention. The techniques may not necessarily be applicable or successful in every case.
[d]Weighing 4 lb 6 oz or less.
Source: *Birth Defects: Tragedy and Hope* by the National Foundation/March of Dimes, 1978.

(table 19.1 Selected Birth Defects, USA continued)

BIRTH DEFECT	TYPE	ANNUAL INCIDENCE[a]	PREVALENCE[b]	CAUSE
Phenylketonuria (PKU)	Metabolic: inability to metabolize a specific amino acid	310	3,100	Hereditary: recessive inheritance
Tay-Sachs disease	Metabolic: inability to metabolize fats in nervous system	30	100	Hereditary: recessive inheritance—most frequent among Ashkenazi Jews
Thalassemia	Blood disease: anemia	70	1,000	Hereditary: incomplete recessive inheritance
Galactosemia	Metabolic: inability to metabolize milk sugar galactose	70	500	Hereditary: recessive inheritance
Erythroblastosis (Rh disease)	Blood disease: destruction of red blood cells	7,000	NA	Hereditary and environmental: Rh– mother has Rh+ child
Turner syndrome	Structural/functional	575	3,100	Chromosomal abnormality
Congenital rubella syndrome	Structural/functional: multiple defects	Varies with occurrence of disease; less than 50	NA	Environmental: maternal infection

TESTS FOR GENETIC ABNORMALITIES

Nuclear Staining

Special staining of nuclear material found within somatic cells can be used to determine the number of X chromosomes. A single dark-stained chromatin mass appears in female cells, but not in normal male cells. The mass, called a *Barr body*, is always one less than the number of X chromosomes. Individuals who are anatomically male but whose cells include one Barr body will therefore have two X chromosomes. Their sex chromosome genotype is XXY. If two Barr bodies are evident in male cells, the genotype is XXXY.

A female with Turner's syndrome would exhibit no Barr bodies in chromosome-staining analysis. Female cells showing two Barr bodies have an XXX genotype.

Amniocentesis

Amniocentesis is a procedure in which a sample of amniotic fluid is withdrawn by means of a hypodermic needle. Amniotic fluid surrounding the fetus contains minerals and nutrients and affords a cushion against physical shock. The fluid also maintains constant, uniform heat distribution for the fetus. During the developmental process, external epithelial cells slough off into the amniotic fluid. A sample of this fluid thus contains epithelial cells. The procedure is shown in Figure 19.10.

The sample cells can be cultured, then processed for microscopic examination. Extra chromosomes, deleted chromosomes, and other chromosomal aberrations may be identified. The purpose of the procedure is to identify the

DETECTION[c]	TREATMENT[c]	PREVENTION[c]
Blood test at birth	Special diet	Carrier identification, genetic services
Blood and tear tests, amniocentesis	None	Carrier identification, genetic services
Blood test	Transfusions	Carrier identification, genetic services
Blood and urine tests, amniocentesis	Special diet	Carrier identification, genetic services
Blood tests	Transfusion: intrauterine or postnatal	Rh vaccine; blood tests to identify women at risk, genetic services
Amniocentesis, chromosome analysis	Corrective surgery, medication	Genetic services
Antibody tests and viral culture	Corrective surgery, prostheses, physical therapy and training	Rubella vaccine

chromosomal aberrations early in the pregnancy for two specific purposes:

1. To alert the parents to the expected abnormalities identified with certain chromosomal defects so that they may be mentally prepared to accept the inevitable, or
2. To identify the chromosomal defect with sufficient lead time to allow parental decision to terminate the pregnancy.

The sex of the fetus can also be determined by amniocentesis (Barr body identification). Although the procedure is considered safe, it would not be performed merely to determine the sex of the fetus.

Blood Tests

Some genetic diseases, particularly those producing an enzyme defect, may be diagnosed in early infancy by blood tests. A missing or defective enzyme will not permit the completion of a specific metabolic pathway. In such cases, some of the incompletely metabolized products accumulate in the blood, where they can be detected.

Children with untreated PKU have a defective liver enzyme. The amino acid phenylalanine (present in many foods as a component of a protein), owing to the defective enzyme, cannot metabolize the amino acid, and therefore it accumulates in the blood, resulting in irreversible nerve damage. Phenylalanine concentration can be determined by blood tests.

Family Pedigree

A family with some inheritable disease recorded in ancestors on either side of the family may wish to know the probability of a recurrence of

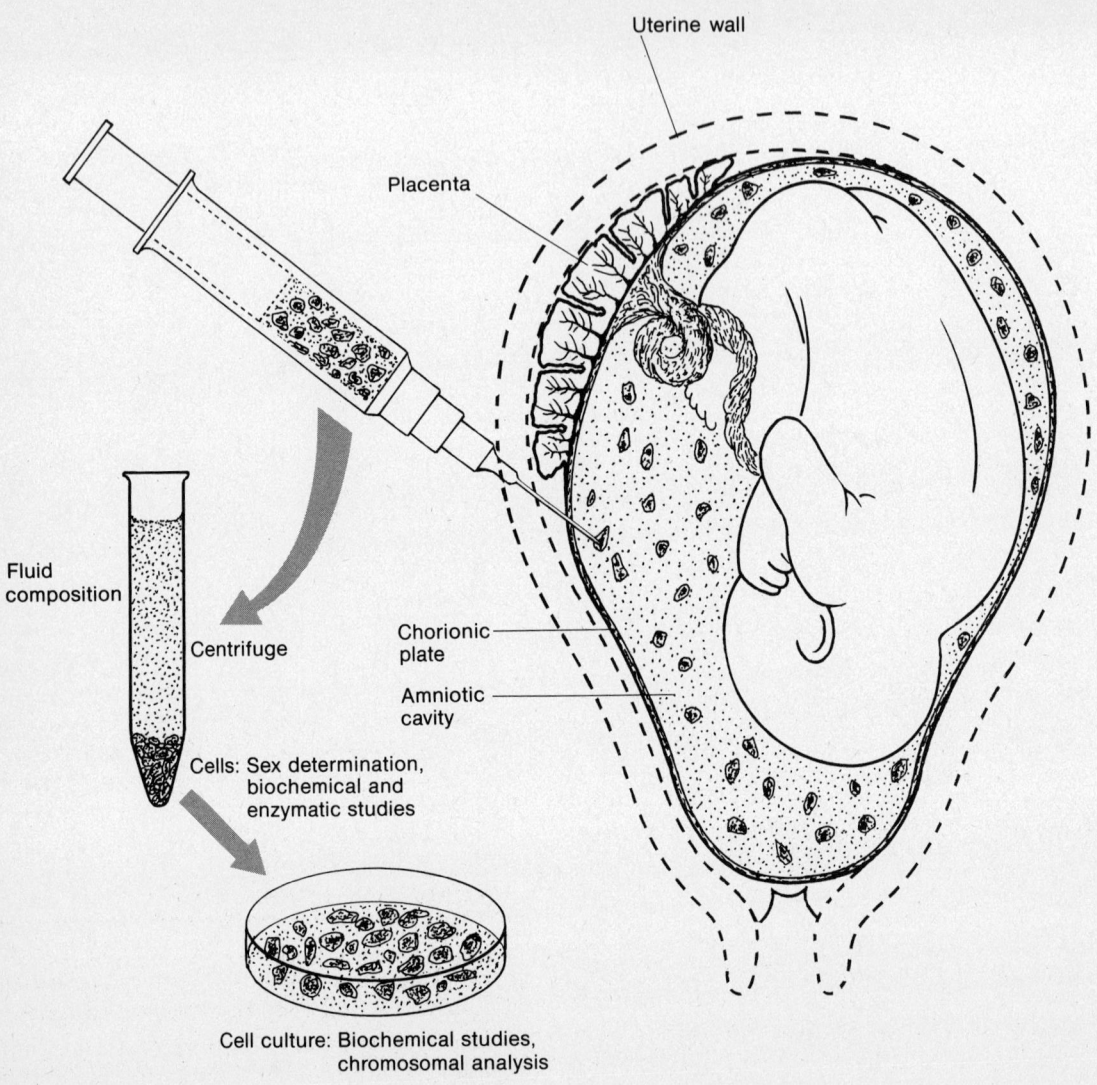

Fig. 19.10 The process of amniocentesis involves removing fetal cells from amniotic fluid and examining them for chromosomal aberrations. The purpose of the procedure is to alert parents to the expected abnormality caused by chromosomal defect so that they may (1) be mentally prepared to accept the inevitable, or (2) decide to terminate the pregnancy.

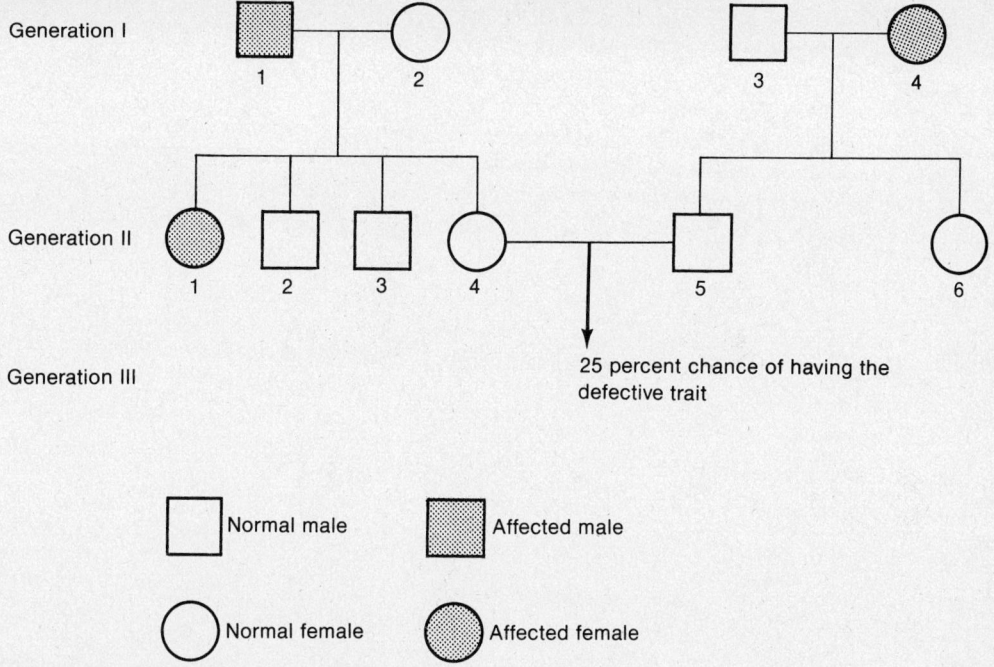

Fig. 19.11 Family pedigree illustrates how a family health history may be used to determine the genotype of some person. If the genotype is known, a prediction may be made concerning probable offspring genotypes and phenotypes. Since II-4 and II-5 are both heterozygous, their offspring in generation III have a 25 percent chance of having the defective trait.

the disease in their children. At least some prediction may be made by what is called a *family pedigree.*

To develop such a pedigree, the family ancestry must be searched as far back as possible to reveal all relatives who had symptoms of the disease in question. A chart is then drawn up as shown in Figure 19.11. If, for example, the disease is known to be inherited as homozygous recessive, then all such individuals may be indicated (shaded figures) and their genotype recorded as homozygous recessive.

Although parents II-4 and II-5 in Figure 19.11 are normal, they must be carriers of the defective gene since each of their parents either had the disease or was a carrier. If any children in generation II show the defect they are homozygous recessive, and thus received a recessive gene from both parents.

On the basis of the genotypes of each of the parents in generation III, Ss × Ss, we may predict that there is a 25 percent risk of transmitting the disease.

SUMMARY AND REVIEW

Genetics is the study of hereditary traits. Parents contribute chromosomes to their offspring. These chromosomes carry genes for all traits.
A. Chromosomes and karyotypes
 1. Chromosomes may be photographed and arranged in order of size and centromere location. Such an arrangement is called a karyotype.
 2. Each parent contributes half the number of chromosomes for each offspring.
 3. A chromosome one parent contributes is identical in structure to one contributed by the other parent. This pair of chromosomes is called a homologous pair.

4. Single-gene defects are those transmitted by only one gene. The gene may be dominant or recessive.
B. Single-gene defects (diseases)
1. Huntington's disease is caused by a dominant gene. The effects of this disease are often delayed until after middle age.
2. Achondroplasia is a defect causing extreme shortening of the extremities. It is the result of a dominant autosomal gene.
3. Sickle cell anemia is caused by a recessive gene. Individuals, predominantly black Americans, who are homozygous recessive for the sickle cell gene possess the defect sickle cell anemia. As its name indicates, the disease causes severe anemia.
4. Tay-Sachs disease is transmitted as an autosomal recessive gene. The disease, found predominantly among persons of Jewish ancestry, is lethal. Children seldom live to age 4.
5. Phenylketonuria (PKU) is a single-gene, autosomal recessive defect. The disease results in the incapability of the individual to metabolize the amino acid phenylalanine.
6. Galactosemia is a disease like PKU except that the sugar galactose cannot be properly metabolized.
C. Sex determination. The sex of an individual depends upon which kind of sperm fertilizes an egg. If an X-bearing sperm fertilizes an egg, the offspring will be a female. If a Y-bearing sperm fertilizes an egg, the offspring will be a male.
D. Sex-linked traits. These are traits induced by genes carried predominantly on the X chromosome. Three sex-linked traits are:
1. red-green color-blindness
2. hemophilia (bleeder's disease)
3. agammaglobullinemia (low immune response)
E. Multifactorial inheritance. Many genetic diseases result from the interaction of many genes. These are less understood than single gene defects.
F. Chromosomal abnormalities. Chromosome numbers that deviate from the normal human number of 46 per cell result in genetic defects.
1. In Down's syndrome, the individual cells have 47 chromosomes rather than the normal 46, owing to nondisjunction of the twenty first pair of chromosomes. The disease is also called mongolism.
2. Structural defects are due to segments of chromosomes being broken and deleted.
G. Sex chromosome defects. Defects owing to nondisjunction of sex chromosomes result in:
1. Turner's syndrome, which results in an egg having no X chromosome or two (XX) X chromosomes. An egg with no X chromosomes fertilized by an X sperm produces an XO female. These individuals exhibit Turner's syndrome.
2. Kleinfelter's syndrome results from an XX egg fertilized by a Y chromosome, producing a XXY male.
H. Tests for genetic defects include:
1. nuclear staining for Barr bodies to determine maleness or femaleness or nondisjunction
2. amniocentesis
3. blood tests
4. family pedigree

APPENDIX A
Basic Chemistry

MAJOR CONCEPTS
ORGANIZED MATTER
ELEMENTS
COMPOUNDS
SOLUTIONS
  Strength of Solutions
  Saturated and Unsaturated Solutions
  Percentage Solutions
  Ratio Solutions
ENERGY
ATOMS AND ATOMIC STRUCTURE
  Atomic Weight
  Molecular Weight
  Periodic Table of Elements
  Molar Solutions
IONS
ACIDS, BASES, AND SALTS
  Acids
  Bases
  pH Scale
  Buffers
  Salts

ORGANIC CHEMISTRY
  Covalent Bonding
  Hydrocarbons
  Alcohols
  Organic Acids
  Aldehydes
  Ketones
  Phenols
BIOLOGICALLY IMPORTANT ORGANIC
  SUBSTANCES
  Carbohydrates
  Lipids
  Proteins
  Amino Acids
  DNA
  RNA
  Nucleotides and Coenzymes
  Vitamins

This appendix is included to serve as refresher material for some students and as a brief introduction to basic chemistry for others. Human physiology is predominately chemical in nature. To explain it fully would certainly require more information than given here, but the discussion will set the stage. Appendix A is divided into organic and inorganic parts.

Chemistry is the study of matter, its properties, characteristics, and interactions. *Matter* is defined as anything that has mass and takes up space.

ORGANIZED MATTER

Matter can be classified as either a gas, a liquid, or a solid. Matter may be further classified into substances of uniform or nonuniform composition. In uniform matter, each part is like every other part of the material. If we examine a kilogram of sugar, each crystal of sugar in this kilogram of sugar would be identical in terms of its components. If we divide a single crystal of sugar into two parts, each half would have identical components. These uniform types of matter are called *substances*. Another type of uniform matter is called a *solution*. If we add a gram of sugar to a liter of water and mix it thoroughly, a drop of this mixture taken either from the top or the bottom of the container would have identical components.

A nonuniform type of matter is called a *mixture*. An example of a mixture might be grains of sand mixed together with crystals of sugar. Some common mixtures include milk, butter, orange juice, and many others.

In summary, matter may be divided into three different classifications: *mixtures, solutions,* and *substances*. Solutions and substances are uniform or homogeneous types of matter; mixtures are nonuniform types of matter.

Substances may be further divided into two categories—*elements* and *compounds*.

ELEMENTS

Elements are pure substances, which cannot be decomposed by chemical change into simpler substances. Common examples of elements include hydrogen, oxygen, and carbon.

There are 92 elements in the Earth's crust. Another dozen or so elements have been artificially produced.

Table A.1 lists some of the more common elements in the human body and the symbols by which they are known. In many instances, the symbol for an element is simply the first letter of the name of the element. For example, H for hydrogen, and C for carbon. Other symbols utilize the first few letters of the name, such as He for helium, Ca for calcium, and Mg for magnesium. Other symbols are derived from Latin or Greek names of elements. For example, the symbol for sodium is Na, from the Latin name *natrium*.

The symbol not only represents the name of an element but also one atom of the element. For example, the symbol Ca stands for one atom of calcium.

Table A.1 The Human Body: Common Elements, Symbols, and Percent by Weight

ELEMENTS	SYMBOLS	PERCENTAGE (BY WEIGHT) IN THE BODY
Oxygen	O	65.0
Carbon	C	18.0
Hydrogen	H	10.0
Nitrogen	N	3.0
Calcium	Ca	2.0
Phosphorus	K	1.0
Potassium	P	0.35
Sulfur	S	0.25
Sodium	Na	0.15
Chlorine	Cl	0.15
Magnesium	Mg	0.05
Iron	Fe	0.004

COMPOUNDS

Compounds are complex substances produced by the chemical union of two or more elements. They can be decomposed into the elements of which they are composed.

Compounds are homogeneous substances and individually show consistent physical and chemical properties. Some common compounds are water (H_2O), table salt (NaCl), and glucose ($C_6H_{12}O_6$). Note that combinations of element symbols are used to indicate individual compounds.

Compounds can be broken down into the elements that compose them. Some compounds are more easily decomposed than others. Decomposition may involve simple heating, or may require an electric current, catalysts, or a combination of these factors.

SOLUTIONS

Solutions are mixtures having two components—the dissolving substance, called a *solvent*, and the substance dissolved, called the *solute*. A solution is not a compound because the proportions of the solvent and solute may vary. A true solution is one in which the solute will not settle out upon standing and the ratio of solute to solvent is precisely the same in all parts of the mixture. An example of a true solution can be prepared by stirring a gram of table salt into a liter of water. A sample of the solution taken from any part of the container would have exactly the same ratio of solute to solvent.

Water is the most universal of solvents; it will dissolve a greater number of solutes than any other substance. Water is so universally accepted as a solvent that we normally assume that a solution has water as the solvent unless otherwise specified.

Strength of Solutions

We often refer to solutions as being diluted or concentrated. These are arbitrary and imprecise terms. *Dilute solutions* are assumed to contain a small amount of solute dissolved in a large quantity of solvent. *Concentrated solutions* contain a large amount of solute dissolved in a small quantity of solvent. When does a solution become concentrated and cease being dilute? This question cannot be answered, and therefore some standards or reference points are needed.

Saturated and Unsaturated Solutions

Sugar can be stirred into a liter of water until no more will dissolve at a given temperature. Any excess sugar added beyond this point will not go into solution, but rather will settle out. This solution is said to be *saturated*, and therefore as concentrated as it can be. The temperature of the water is important because if we raise the temperature, more sugar can be added and dissolved. Any solution that contains less solute than it can hold at a given temperature is called an *unsaturated solution*.

The terms saturated and unsaturated are not wholly satisfactory terms since solutes vary in their ability to be dissolved. For example, more sugar by weight will dissolve in a liter of water at 37°C than will a substance like iodine.

Percentage Solutions

One method of stating the concentration of a solution is to specify how much solute by weight is added to a given volume of solvent. Fifteen g of sugar dissolved in 85 ml of water is a 15 percent solution. Five g of sugar dissolved in 95 ml of water is a 5 percent solution. The percentage figure always represents the weight of the solute. Since 1 ml of water weighs 1 g, the weight of the solute plus the weight of the solvent must equal 100 g.

Ratio Solutions

Another method of specifying concentrations is a *ratio solution*. A solution designated as 1:100 means that 1 g of solute has been dissolved in 100 ml of solvent. A 1:1,000 NaCl solution is 1 g of NaCl dissolved in 1,000 ml of water. As with percentage solutions, the first number in the ratio designates the weight of the solute in grams, and the second number is the volume of the solvent in milliliters. Both percentage and ratio solutions are commonly used in medicine.

Another method of identifying the concentration of a solution is to use the molecular weights of compounds. These solutions are called *molar solutions*. A discussion of this type of solution follows the discussion of molecular weights.

ENERGY

Science has not yet perfected a general definition for the term energy. *Energy* is commonly defined as the ability to do work. Forms of energy include light, heat, and electricity. Whenever a chemical change occurs, there is a conversion of energy into light and heat, or into some other chemical substance.

All living organisms require a source of energy to perform work. The human body requires energy for muscle contraction, synthesis of essential compounds, transfer of nerve impulses, and a host of other functions. The foods we eat are chemically altered by the digestive process to release energy. Some of this energy is released in the form of heat. However, most of it is utilized in the synthesis of other energy-rich compounds. The energy bound up within these compounds is more readily accessible for the body's work than is the food in its original form.

ATOMS AND ATOMIC STRUCTURE

An *atom* is defined as the smallest part of an element capable of exhibiting all the properties of the element.

An Englishman, Jay Thomson, in 1898 suggested that an atom is made up of *negative* and *positive* charges and these were assimilated into a spherical shape. Later, about 1910, Ernest Rutherford argued that the atomic structure was such that the positive charges, called *protons,* were in the center of a solar systemlike arrangement, and the negative charges, called *electrons,* were floating around the outside of this nucleus of positive charges. A few years later, Danish physicist Niels Bohr contended that the protons were indeed located in the nucleus of an atom, but that the electrons, which he assumed were located outside the nucleus, were in a definite pattern. In other words, his contention was that there was a definite number of electrons in different layers or shells around the nucleus.

The electrons revolving around the nucleus are not arranged in a haphazard fashion. It is believed that there is a very precise way in which they are grouped and how they move around the nucleus.

A *neutron* is a particle located in the nucleus of an atom along with protons. It has no electrical charge and has a mass about equal to the mass of a proton.

The parts of an atom with which we are essentially concerned are the protons and electrons. Figure A.1 shows a hydrogen atom. Note that the nucleus contains a single proton, and the orbit surrounding the nucleus contains a single electron. This orbit is known by a series of names, most common of which is *shell*.

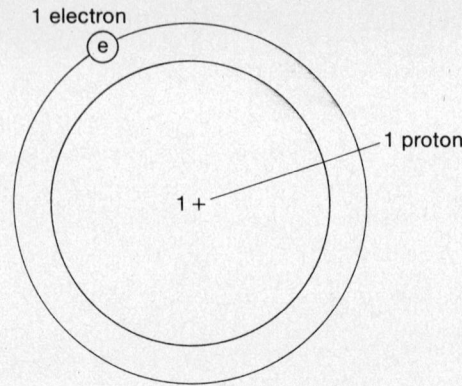

Fig. A.1 Hydrogen atom. The nucleus contains a single proton and the orbit contains a single electron.

The next illustration, Figure A.2, is of the helium atom. Here we find 2 electrons in the shell surrounding the nucleus, which contains 2 protons. All other atoms also have 2 electrons in the first shell. The only element that has less than 2 electrons in the first shell is hydrogen.

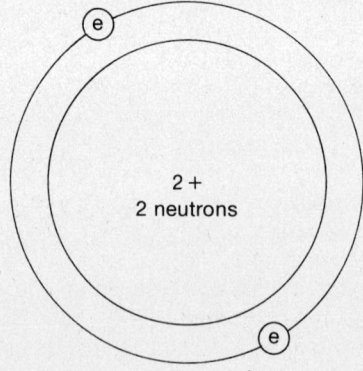

Fig. A.2 Helium atom. The nucleus contains two protons and the orbit contains two electrons.

Figure A.3 illustrates the oxygen atom. The nucleus has 8 protons. There are 2 electrons in the first shell and 6 electrons in a second shell. This makes a total of 8 electrons. It is believed that there can be no more than 2 electrons in the first shell of any atom.

Look again at the examples given for the atomic structure of hydrogen, helium, and oxy-

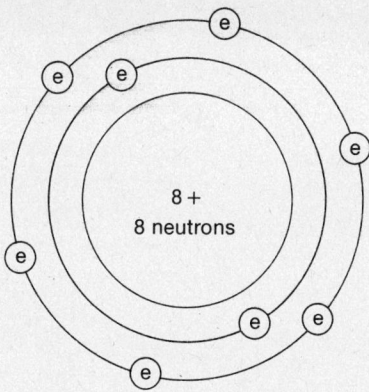

Fig. A.3 Oxygen atom. The nucleus contains eight protons. The first orbit contains two electrons, which fills that orbit. The remaining six electrons overflow into a second orbit.

gen. In each one, note that the number of protons is equal to the number of electrons for each specific element. Hydrogen has 1 electron and 1 proton; helium has 2 electrons and 2 protons; oxygen has 8 electrons and 8 protons. Atoms of all the elements contain equal numbers of electrons and protons. Whenever the number of protons and electrons are the same, we say the particle, in this case, the atom, is neutral.

We may further state that the number of protons any element has is equal to its *atomic number.* The atomic number of hydrogen is 1. The atomic number of helium is 2; and the atomic number of oxygen is 8. If we know the atomic number, then we know two things—the number of protons and (since the number of protons must equal the number of electrons in an atom,) the number of electrons.

The third particle within an atom is the *neutron.* The number of neutrons in the nucleus is not a fixed number for a specific substance. Some hydrogen atoms have no neutrons; some have 1 neutron, and some have 2 neutrons.

Atoms of the same element having different *atomic weights,* due to different numbers of neutrons, are called *isotopes.* Chemically, an atom of hydrogen is the same whether it has 1, 2, or 3 neutrons.

Unstable isotopes of some elements tend to throw off excess neutrons or combinations of neutrons and electrons. These cast-off particles leave the atom with an intense speed and are capable of penetrating such substances as paper, wood, and flesh. These unstable isotopes are said to be *radioactive. Radioactive isotopes* are used in medicine in a variety of ways, in particular, in the destruction of certain cancerous cells.

Atomic Weight

The element carbon has arbitrarily been given the atomic weight 12.00. All other atomic weights are relative to that of carbon. When we say that hydrogen has the atomic weight of 1, we mean that the average weight of a hydrogen atom is one-twelfth as much as the carbon atom. Carbon has an atomic weight of 12 and an atomic number of 6.

Figure A.4 is a diagram of a carbon atom. The carbon atom has a total of 6 protons in the nucleus and has 6 electrons in shells outside the nucleus. If the atomic number of carbon is 12 and the total weight of the atom is composed of protons and neutrons, then there must be 6 neutrons in the nucleus also. Six neutrons plus 6 protons equals 12 as the atomic weight for carbon.

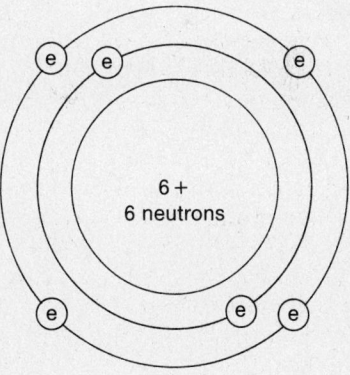

Fig. A.4 Carbon atom. The nucleus has six protons. Two electrons are in the first orbit and four in the second. Note the presence of six neutrons in the nucleus. Atomic weight is determined by adding the number of protons and neutrons. Thus, the atomic weight of carbon is 12.

Chemistry texts show carbon as having an atomic weight of 12.01. This is because some carbon atoms have 2 extra neutrons, thereby making the atomic weight of these atoms 14. Elements having more neutrons than normal in their nuclei are isotopes. There are many more carbon atoms having 12 as the atomic weight rather than 14, but the average comes out 12.01.

Atomic weights have practical uses. If we burn a substance like carbon in the presence of pure oxygen, we find that 3 parts of carbon by weight always combine with 8 parts of oxygen by weight. In other words, if we begin with 3 g of carbon, it would require 8 g of oxygen to completely burn it. The resulting compound would be 11 g of carbon dioxide. Carbon and oxygen can only combine in these proportions in producing carbon dioxide, never in any other proportion. The knowledge that elements combine in precise ratios led to the formulation of the *law of definite proportions*, which was first proposed by John Dalton in 1808. Dalton's Law of Definite Proportions, also known as the *law of definite composition*, states that in a pure compound the numbers of atoms of the different elements combined with each other are always in the same ratio.

Molecular Weight

The *molecular weight* of a compound is found by adding the atomic weights of its components. For example, the molecular weight of carbon dioxide, CO_2, is the sum of the atomic weight of carbon plus two times the atomic weight of oxygen. The sum is 44; therefore 44 is the molecular weight of carbon dioxide.

Periodic Table of Elements

By the early 1900s, most of the presently known natural elements had been discovered. Chemists had also recognized that certain elements have similar chemical properties. For example, the elements fluorine, chlorine, and bromine all react chemically with the element sodium to produce similar chemical compounds called *salts*. The metals lithium, sodium, and potassium also exhibit similar chemical properties. The Russian chemist Mendeleev, in 1869, devised a systematic arrangement of the elements known at that time.

This arrangement is called the *periodic table of elements*, and in its perfected form it provides valuable information for the chemist. Figure A.5 is a modified form of the periodic table. We will not go into the specific details and intricacies of the table, but it will be a source of information as we progress in the study of elements and their chemical activity.

We can now return to the discussion of concentration of solutions. The concepts just presented concerning molecular weights can be used in preparing solutions of known concentration. These are called *molar solutions*.

Molar Solutions

A molar solution is prepared by dissolving 1 g molecular weight (mole) of a substance in enough water to yield 1 liter of solution. A gram molecular weight of any substance is its molecular weight expressed in grams. For example, a 1 molar (1 M) solution of sodium chloride (NaCl) would contain 58.44 g of NaCl plus enough water to provide 1 liter of solution. 58.44 is the molecular weight of NaCl (Na = 22.99, Cl = 34.45).

A 0.5 M solution would be only half as concentrated as a 1 M solution, and would contain 0.5 × 58.44 g (29.22 g) of NaCl, with water added to equal a liter. To prepare a 0.2 M solution of glucose the following procedure is used: Calculate the molecular weight of $C_6H_{12}O_6$ (glucose).

$$C_6 = 12 \times 6 = 72$$
$$H_{12} = 1 \times 12 = 12$$
$$O_6 = 16 \times 6 = 96$$

Molecular weight of glucose = 180

The molecular weight of glucose is 180. Therefore the gram molecular weight of glucose is 180 g. This is also referred to as *one mole* of glucose.

Since we need a 0.2 M solution, it is necessary to multiply 180 g × 0.2, which equals 36 g.

KEY TO ELEMENTS PRESENT IN LIVING MATTER

Primary elements	Secondary elements	Trace elements
1.008 H 1	39.10 K 19	54.94 Mn 25

Atomic weight — Na
Atomic symbol — 11
Atomic number

PERIOD	GROUP IA	IIA	IIIB	IVB	VB	VIB	VIIB	VIII			IB	IIB	IIIA	IVA	VA	VIA	VIIA	INERT GASES COMPLETED
1	1.008 H 1																	4.003 He 2 — 2
2	6.940 Li 3	9.013 Be 4											10.82 B 5	12.011 C 6	14.008 N 7	16.00 O 8	19.00 F 9	20.183 Ne 10 — 2,8
3	22.991 Na 11	24.32 Mg 12											26.98 Al 13	28.09 Si 14	30.975 P 15	32.07 S 16	35.457 Cl 17	39.949 Ar 18 — 2,8
4	39.10 K 19	40.08 Ca 20	44.96 Sc 21	47.90 Ti 22	50.95 V 23	52.01 Cr 24	54.94 Mn 25	55.85 Fe 26	58.94 Co 27	58.71 Ni 28	63.54 Cu 29	65.38 Zn 30	69.72 Ga 31	72.60 Ge 32	74.91 As 33	78.96 Se 34	79.916 Br 35	83.80 Kr 36 — 2,8,18,8
5	85.48 Rb 37	87.63 Sr 38	88.92 Y 39	91.22 Zr 40	92.91 Nb 41	95.95 Mo 42	99 Tc 43	101.1 Ru 44	102.91 Rh 45	106.4 Pd 46	107.88 Ag 47	112.41 Cd 48	114.82 In 49	118.70 Sn 50	121.76 Sb 51	127.61 Te 52	126.91 I 53	131.30 Xe 54 — 2,8,18,18,8
6	132.91 Cs 55	137.36 Ba 56	138.92 La 57	178.50 Hf 72	180.95 Ta 73	183.86 W 74	186.22 Re 75	190.2 Os 76	192.2 Ir 77	195.09 Pt 78	197.0 Au 79	200.61 Hg 80	204.39 Tl 81	207.21 Pb 82	209.00 Bi 83	210 Po 84	210 At 85	222 Rn 86 — 2,8,18,32,18,8
7	223 Fr 87	226 Ra 88	227 Ac 89															

Thirty-six g of glucose added to sufficient water to arrive at a final volume of 1 liter equals a 0.2 M solution of glucose.

IONS

Atoms are neutral particles due to the equal number of protons and electrons. The chemical activity of a specific element depends upon the arrangement of the electrons in the outermost shell of the atom. Atoms having zero or eight electrons in the outermost shell are not chemically active. In the periodic table (Figure A.5) observe Group 0 in the vertical column to the extreme right. All these elements have an outer shell with eight electrons with the exception of helium (He), which has only two electrons. Only two electrons are required to fill the first shell. Therefore we can state that the atoms of all these elements have their outer shell filled. This arrangement of electrons provides for a very stable and nonreactive atom. The elements in Group 0 are called the *inert gases.*

There are no known natural compounds of these elements, and only recently have compounds been synthesized containing them.

Atoms with fewer than eight electrons in the outer shell tend to lose or gain electrons in order to achieve an outer shell of eight.

Since the number of protons is constant, any loss or gain of electrons results in an electrically charged particle called an *ion.* The positive or negative charge of an ion is called its *valence.*

Sodium, in Group 1A of the periodic table, has atomic number 11. The structure of the sodium atom is shown in Figure A.6. This atom tends to lose the single electron from the outer shell, thereby upsetting the electrical balance, leaving the atom with 11 protons and only 10 electrons. The ion produced by the loss of one electron has a valence of +1 and is written Na^+.

The electrical imbalance inherent in ions is the force that attracts and holds different atoms together, forming some types of molecules.

Ordinary table salt is made up of the molecule NaCl, sodium chloride. The formula NaCl indicates that one ion of sodium (Na^+) is joined to one ion of chlorine (Cl^-) to form a molecule.

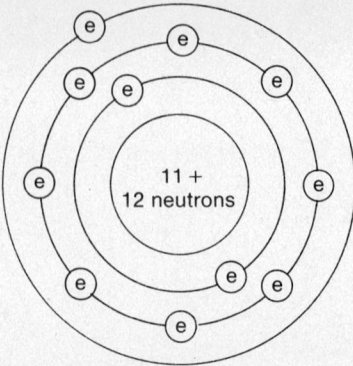

Fig. A.6 Sodium atom. A third orbit is necessary here since the second orbit is filled with eight electrons.

Figure A.7 shows the structure of a chlorine atom. Chlorine needs one additional electron to complete its outer shell. Upon receiving this electron, it becomes Cl^- with 17 protons and 18 electrons. Sodium, on the other hand, has only 1 electron in its outer shell, and it tends to lose this electron, forming Na^+ with 11 protons and 10 electrons. Very simply, then, sodium loses an electron to chlorine, forming complete outer shells in each of the atoms, and the resulting Na^+ and Cl^- are held tightly together by reason of their opposite electrical charges, as illustrated in Figure A.8. Table A.2 shows some common ions and their valences.

When two or more elements chemically combine and react in a molecule as a single ion, the

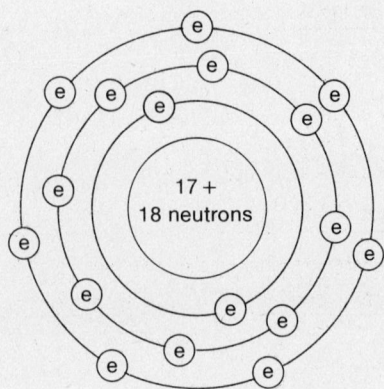

Fig. A.7 Chlorine atom, atomic number 17. Chlorine atoms have seven electrons in the third orbit.

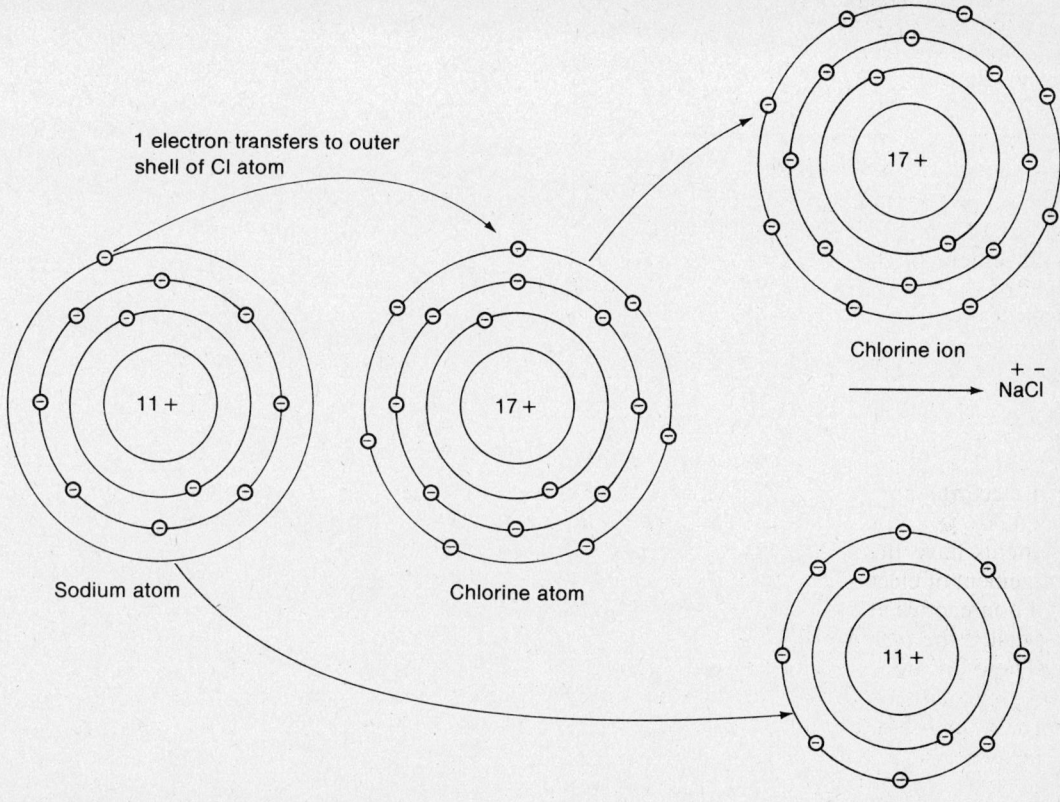

Fig. A.8 Formation of sodium and chloride ions. When chlorine gains an electron and sodium loses one, they are no longer neutral particles (atoms). Because of their unbalanced electrical charges they are now called ions. Opposite charges attract; thus, the positive sodium ion joins with the negative chlorine ion, creating a new particle, a molecule of sodium chloride.

unit is called a *radical*. The sodium hydroxide molecule (NaOH) is a good example of this. In this formula, the OH⁻ fraction of the molecule is called the *hydroxide radical*. The charge on the OH (hydroxide ion) is −1. A correct molecular structure demands that charges must be opposite and equal. The sodium hydroxide molecule (Na⁺OH⁻) satisfies these requirements.

Examples of molecules involving the hydroxide radicals include the following:

$\overset{++}{Ca}(\overset{-}{OH})_2$ calcium hydroxide

$\overset{+}{K}\overset{-}{OH}$ potassium hydroxide

$\overset{++}{Mg}(\overset{-}{OH})_2$ magnesium hydroxide

Other radicals and their valences are listed in Table A.3.

Table A.2 Some Common Ions and Valences

ELEMENT	IONIC CHARGE	ELEMENT	IONIC CHARGE
Sodium	+	Chlorine	−
Magnesium	+ +	Oxygen	− −
Hydrogen	+	Iodine	−
Calcium	+ +		
Potassium	+		

BASIC CHEMISTRY 323

Table A.3 Other Radicals and Valences

RADICAL	NAME	CHARGE	COMMON COMPOUND CONTAINING THE RADICAL
SO_4	Sulfate	--	H_2SO_4 Sulfuric acid
NO_3	Nitrate	-	HNO_3 Nitric acid
NH_4	Ammonium	+	NH_4OH Ammonium hydroxide
PO_4	Phosphate	---	K_3PO_4 Potassium phosphate
CO_3	Carbonate	--	$CaCO_3$ Calcium carbonate

ACIDS, BASES, AND SALTS

Most inorganic compounds (substances that do not contain carbon) can be classified as an acid, a base (alkali), or a salt.

Acids

An *acid* is defined as a substance that will dissociate in solution yielding hydrogen ions. (The term dissociate means the separation of ions.) We have seen that inorganic compounds are composed of positive and negative ions. The positive ion in all inorganic acids is hydrogen (H^+). Table A.4 lists the most common acids. Note the presence of hydrogen in all of them.

The strength of an acid is related to how easily hydrogen ions are separated from the rest of the molecule. For example, hydrochloric acid is considered a strong acid since the H^+ and Cl^- ions separate quite readily in solution. Carbonic acid (H_2CO_3) is a weak acid due to the fact that the H^+ ion does not readily separate from the $\overline{CO_3}$ ion.

Acids have a sour taste, turn red litmus (a dye) blue, react vigorously with most metals, and

Table A.4 Most Common Acids

ACID	FORMULA	IONS PRESENT
Hydrochloric	HCl	$H^+ + Cl$
Sulfuric	H_2SO_4	$H^+ + SO_4$
Nitric	HNO_3	$H^+ + NO_3$
Phosphoric	H_3PO_4	$H^+ + PO_4$
Carbonic	H_2CO_3	$H^+ + CO_3$

react with substances called bases, producing salts and water.

Bases

A *base (alkali)* is defined as a substance that will dissociate yielding free hydroxide (OH^-) ions in solution. The hydroxide ion is negatively charged. Table A.5 lists some common bases.

The strength of a base is dependent upon the degree of the dissociation of the molecule to yield free hydroxide ions. Sodium hydroxide is an example of a strong base. Calcium hydroxide is a weak base. Bases have a bitter taste, feel slippery when a drop is rubbed between the fingers, turn blue litmus red, cause a solution of phenolphthalein to turn a bright red, and react with acids to form salts and water.

Table A.5 Some Common Bases

NAME	FORMULA	IONS PRESENT
Sodium hydroxide	NaOH	$Na^+ + OH^-$
Potassium hydroxide	KOH	$K^+ + OH^-$
Calcium hydroxide	$Ca(OH)_2$	$Ca^{++} + OH^-$
Ammonium hydroxide	NH_4OH	$NH_4^+ + OH^-$

pH Scale

The strengths of acids or bases depend upon the degree to which they dissociate to form free ions in solution. If the dissociation is complete or nearly so the compound is said to be strong, hence the term strong acid or strong base. For those compounds whose dissociation is very limited, the term weak acid or base is used. A rating scale based on a range of 1 to 14 has been devised to express the relative strength of acids and bases. This is called the *pH scale*.

At the midpoint of this range (pH 7), the concentration of hydrogen and hydroxide ions is equal, and the solution is said to be neutral. Pure water has a pH of 7. For each successive integer below or toward one, the hydrogen ion increases by a factor of 10 and for each successive integer above 7 or toward 14, the hydroxide concentration increases by a factor of 10. Thus, a solution of pH 6 indicates an acid 10 times more acid than pH 7; pH 5 is 10 times more acid than pH 6 and 100 times more acid than pH 7. A pH above 7 reflects a similar change in the strength of bases.

Most living organisms can survive only if the pH range of the various fluids in living matter is maintained at a point near pH 7. For example, the pH of blood is maintained at a value very close to 7.35. If the blood pH shifts by 0.2 of a unit, death may result. Living organisms must have a way of protecting themselves from sharp changes in pH. The substances found in the body that help maintain a fairly constant pH are called buffers.

Buffers

Buffers are substances that help maintain the homeostatic balance between acids and bases in living systems. This balance is constantly challenged by the variety of foods eaten, metabolic by-products, and by the level of activity of certain secretory cells.

Buffers are mixtures of a weak acid and a salt of that acid or a weak base and a salt of that base. Under certain conditions, buffers are able to react with free hydrogen or hydroxide ions so that they are no longer free but instead tied up into a neutral molecule.

Salts

A chemical reaction occurs when an acid and a base are poured together. The products of the reaction are molecules of water and a substance called a *salt*. A salt is neither an acid nor a base. It is made up of the positive ion from the base and the negative ion from the acid. For example, if the acid HCl is poured into a solution of NaOH, the positive sodium ion unites with the negative chloride ion, producing sodium chloride (NaCl), a salt. Note in the following chemical equation that water molecules are also produced, the hydrogen atom coming from the acid and hydroxide ion coming from the base.

$$HCl + NaOH \rightarrow NaCl + HOH\,(H_2O)$$

Table A.6 shows several acid-base combinations and the resulting salt.

Salts play a significant role in normal body functions. A constant level of salt concentration, mostly NaCl, is specifically required to maintain osmotic pressures across cell membranes. Profuse sweating and excessive urination cause the loss of an adequate salt concentration level, producing serious physiological disturbances.

Table A.6 Some Acid-Base Combinations and Resulting Salt

ACIDS	BASE	SALT
HCl	KOH	KCl
HCl	Ca(OH)$_2$	CaCl$_2$
H$_2$SO$_4$	NaOH	Na$_2$SO$_4$
H$_2$CO$_3$	Ca(OH)$_2$	CaCO$_3$
HNO$_3$	NaOH	NaNO$_3$

ORGANIC CHEMISTRY

The study of substances containing carbon is called *organic chemistry*. Before 1800, the term organic was applied only to those substances produced by living organisms. In 1828, Friedrich Wohler succeeded in synthesizing *urea*. This was the first artificially synthesized organic compound and marked an end to the theory that only living systems could make organic compounds. The essential difference between inorganic and organic compounds is that organic compounds contain carbon.

Even though there are functional similarities between inorganic and organic compounds, there are some distinct differences, as shown in Table A.7.

Another major difference between organic and inorganic compounds is the way their molecules are held together. Inorganic molecules with

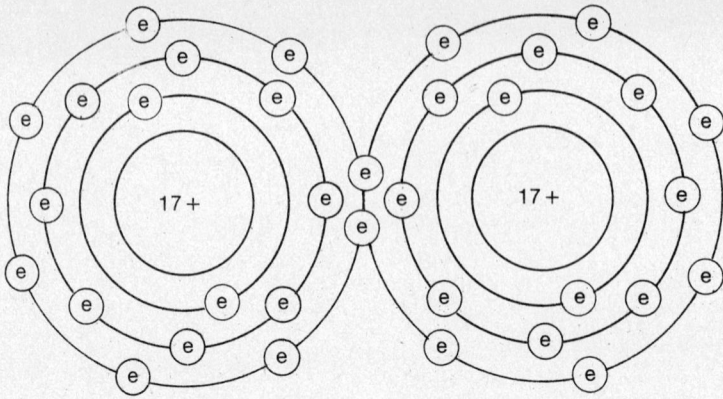

Fig. A.9 Covalent bonding. The two chlorine atoms share electrons in their outer orbit, thus completing the orbit for each with eight electrons.

ionic bonds are held together by the attraction of opposite charges. Organic molecules are formed by *covalent bonds.*

Table A.7 Some Differences Between Inorganic and Organic Substances

INORGANIC	ORGANIC
1. High melting point	1. Low melting point
2. Cannot be distilled	2. Easily distilled
3. Insoluble in liquid hydrocarbons	3. Soluble in liquid hydrocarbons
4. Water soluble	4. Not soluble in water
5. Electrolytic (conduct electricity)	5. Nonelectrolytic

Covalent Bonding

The theory of how organic molecules are held together had its genesis in the study of inorganic substances. The ionic theory of electron transfer cannot explain why gases such as hydrogen (H_2), oxygen (O_2), and chlorine (Cl_2) all exist as two atoms combined into a molecule. When we attempt to apply the electron transfer idea to chlorine gas (Cl_2), it soon becomes obvious that the formation of ions will not explain the formation of the chlorine molecule. A process that explains these molecular configurations without electron transfer is electron-sharing, called *covalence.* Note in Figure A.9 that each chlorine atom shares an electron with the other.

This arrangement satisfies the requirement that the outer shell must be filled. By counting the shared electron, each chlorine atom has eight electrons in its outer shell. A more convenient method of showing the shared electrons is Cl:Cl.

Before leaving gas molecules, observe the covalent bonding in the gases shown in Figure A.10.

Hydrocarbons

We will now apply the concept of covalent bonding to organic compounds. Substances containing only carbon and hydrogen are called *hydrocarbons.* The simplest hydrocarbon is methane gas (CH_4). The molecule is composed of 1 carbon atom and 4 hydrogen atoms held together by covalent bonding. Figure A.11 illustrates the structural formula for the CH_4 molecule.

Figure A.11 illustrates the most commonly used form to show covalent bonds. The – sign represents a single pair of shared electrons.

Carbon is capable of sharing four bonds. In the previous illustration, each of the four bonds is shared with a hydrogen atom. Hydrogen has a

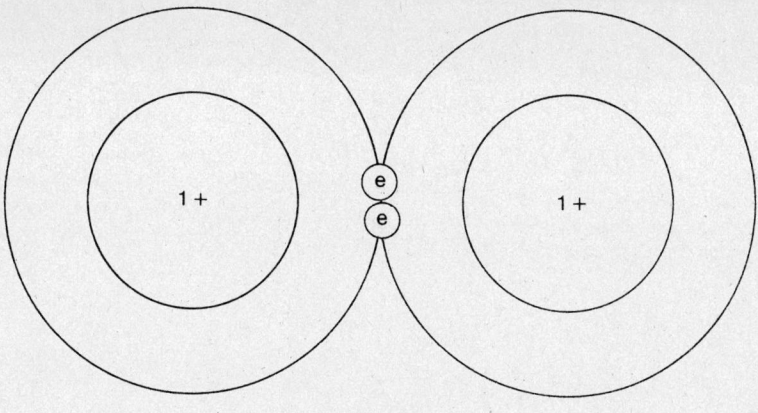

(a) H_2 (hydrogen) H:H

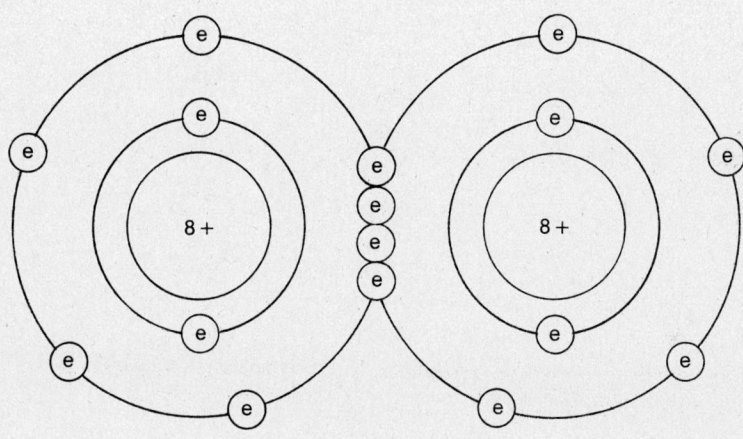

(b) O_2 (oxygen) O::O

Fig. A.10 (a) Covalent bonding for hydrogen produces the H_2 molecule (hydrogen gas). (b) The oxygen molecule is formed by the covalent bonding of two oxygen atoms.

valence of 1. Other elements with a valence of 1 may also bond with carbon, either alone or combined with other elements that have a valence of 1. For example, if one of the hydrogen atoms of CH_4 is replaced with chlorine, the molecule would have the formula CH_3Cl and could be structurally drawn as:

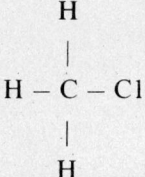

BASIC CHEMISTRY 327

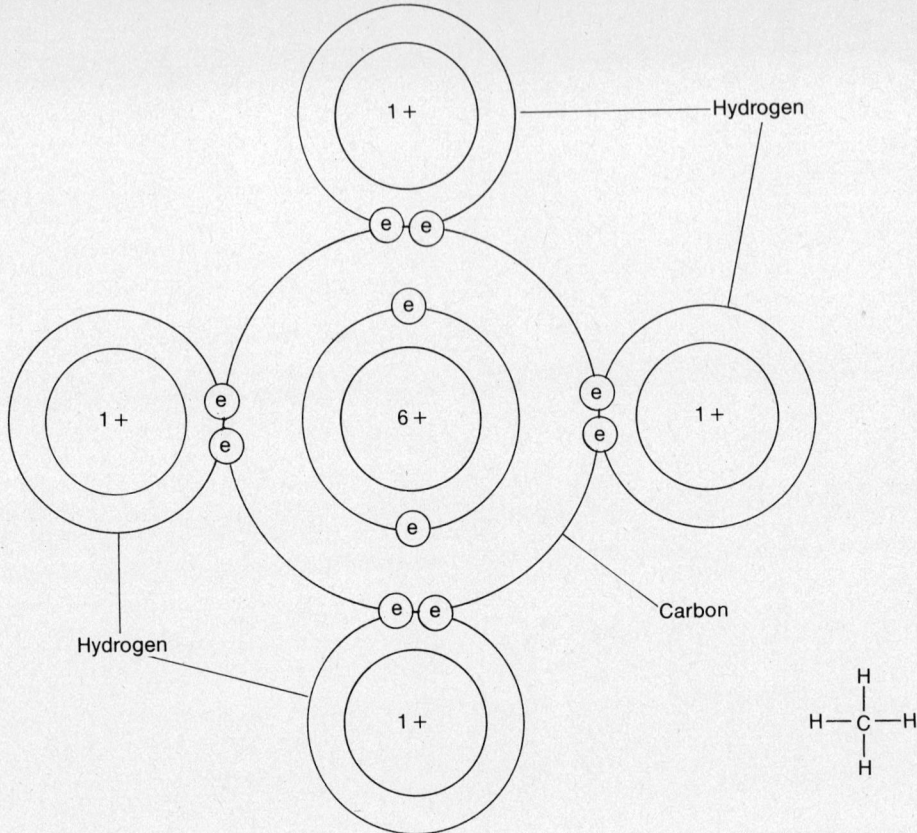

Fig. A.11 The covalent bonds between hydrogen and carbon, which produce the methane molecule, CH_4.

If all four hydrogens are replaced by Cl, carbon tetrachloride (CCl_4) is formed:

$$Cl-\underset{\underset{Cl}{|}}{\overset{\overset{Cl}{|}}{C}}-Cl$$

More than one carbon bond may be shared with some elements, depending on their valence. For instance, oxygen (valence 2) can unite with carbon to form a molecule of carbon dioxide (CO_2):

$$O = C = O$$

Carbon has the capability to share bonds with other carbon atoms, producing chains of carbon atoms. For example, the substance ethane (C_2H_6) has the structural formula:

$$H-\underset{\underset{H}{|}}{\overset{\overset{H}{|}}{C}}-\underset{\underset{H}{|}}{\overset{\overset{H}{|}}{C}}-H$$

Note that each carbon atom is sharing four bonds. Any or all of the hydrogen atoms may be replaced by other atoms, thereby forming new molecules.

Alcohols

The hydroxide radical (OH) has a valence of –1 and therefore may replace one of the hydrogen atoms in the ethane molecule:

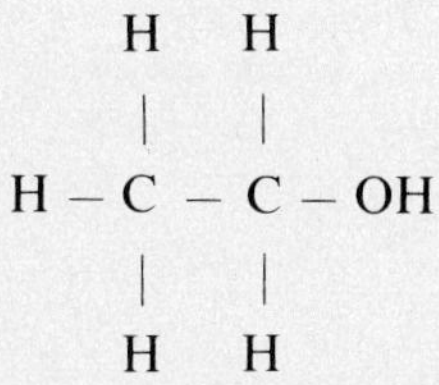

$$\begin{array}{c} H \quad H \\ | \quad\; | \\ H-C-C-OH \\ | \quad\; | \\ H \quad H \end{array}$$

This substance, commonly called *ethyl alcohol*, has the empirical formula C_2H_5OH. If only one OH replaces a single hydrogen atom along a carbon chain consisting of carbon and hydrogen atoms, the substance formed is always an alcohol. Examples of three different alcohols are shown below.

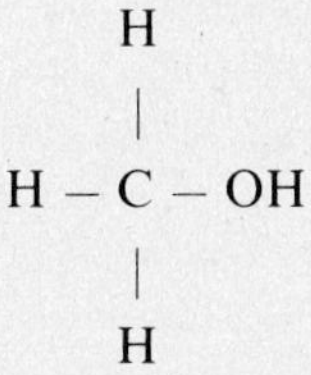

$$\begin{array}{c} H \\ | \\ H-C-OH \\ | \\ H \end{array}$$

Methyl alcohol (methonol)

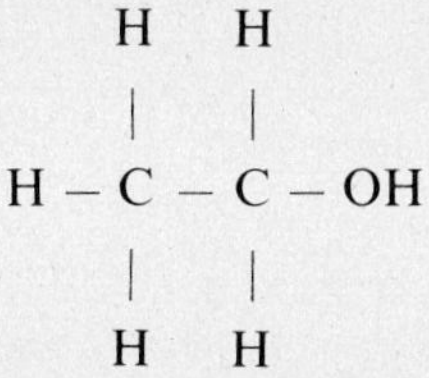

$$\begin{array}{c} H \quad H \\ | \quad\; | \\ H-C-C-OH \\ | \quad\; | \\ H \quad H \end{array}$$

Ethyl alcohol (ethonol)

$$\begin{array}{c} H \quad H \quad H \\ | \quad\; | \quad\; | \\ H-C-C-C-OH \\ | \quad\; | \quad\; | \\ H \quad H \quad H \end{array}$$

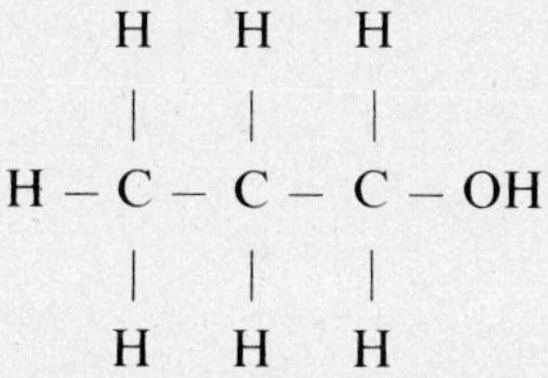

Propyl alcohol (proponol)

The term radical can be applied to organic molecules as well as to inorganic molecules. However, in order to distinguish between organic and inorganic radicals, the term *functional group* is more commonly used to designate organic radicals. For instance, if a hydrogen atom is left off of a methane molecule (CH_4), the methyl group CH_3^- results.

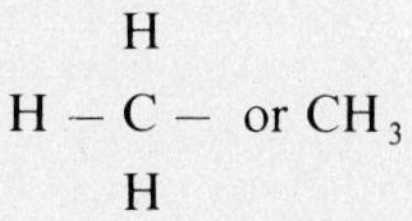

$$\begin{array}{c} H \\ | \\ H-C- \quad \text{or } CH_3 \\ | \\ H \end{array}$$

This leaves the carbon atom with an unshared bond. A functional group is not a complete molecule and must be bonded with another atom or another group.

Organic Acids

The *carboxyl group*, COOH, which is also written

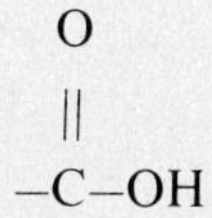

$$\begin{array}{c} O \\ \| \\ -C-OH \end{array}$$

when attached to a carbon chain identifies the substance as an organic acid. Common examples of organic acids include the following:

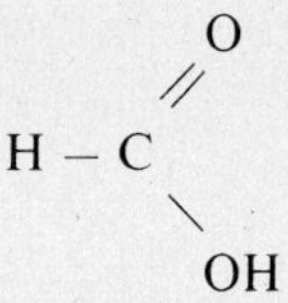

$$H-C\begin{array}{c} \diagup\!\diagup O \\ \diagdown OH \end{array}$$

Formic acid

$$\begin{array}{c} H \\ | \\ H-C-C \\ | \quad\;\; \diagdown \\ H \qquad OH \end{array} \begin{array}{c} O \\ \diagup\!\diagup \\ \\ \end{array}$$

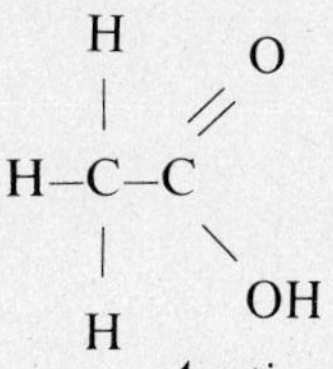

Acetic acid

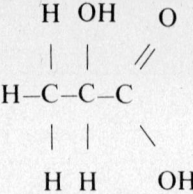

Lactic acid

Note the presence of the carboxyl group in each of these.

Organic acids are components of lipid (fat) molecules and glycerol and play an important role in cellular respiration.

Aldehydes

The characteristic *aldehyde group* is

$$\begin{array}{c} H \\ | \\ -C=O \end{array}$$

or CHO. The simplest aldehyde is formaldehyde,

$$\text{HCHO } (H-\overset{\overset{H}{|}}{C}=O)$$

A 37 percent solution of formaldehyde in water is called *formalin*. Formalin is used commercially as a disinfectant and preservative for biological specimens. Formaldehyde cannot be used as a food preservative because it is toxic to body tissues.

Ketones

The general formula for ketones is

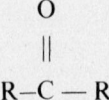

A common ketone is CH_3COCH_3.

$$\begin{array}{c} H \quad O \quad H \\ | \quad \; \| \quad \; | \\ H-C-C-C-H \\ | \qquad \quad | \\ H \qquad \quad H \end{array}$$

Dimethyl ketone (acetone)

Ketones are physiologically important since they are synthesized in the body when sugars are incompletely oxidized, as in diabetes mellitus or in low carbohydrate diets.

Ketones are toxic to body tissues and can cause serious metabolic disturbances if they are not removed.

Until now we have dealt with hydrocarbons as various-sized molecules composed of carbon atoms strung out in a chainlike fashion. During the early nineteenth century, a substance, now called benzene, was found to have the molecular formula C_6H_6. If the carbon atoms were arranged in a chain, it would appear that there would not be enough hydrogen atoms to share all the available bonds. A chainlike molecule having all bonds shared has the formula C_6H_{14}.

One explanation offered was that the molecule was not chainlike but rather a closed circle or hexagon:

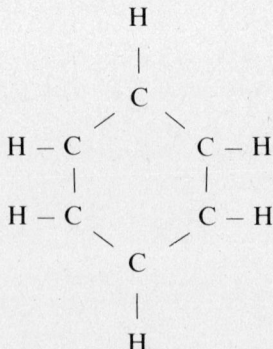

This arrangement still did not account for the presumption that carbon, having a valence of 4, must share four bonds. The hexagonal arrangement allowed only three shared bonds for each carbon. The basic hexagonal arrangement is acceptable today, but it is now assumed that double bonds exist between alternate carbon atoms.

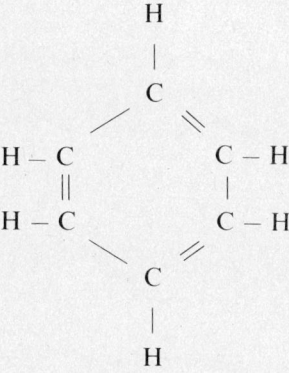

Benzene (C_6H_6)

Since the discovery of benzene and its theoretical molecular structure, thousands of compounds with this basic structure as a nucleus have been discovered or synthesized. The structure is called the *benzene ring*. Hydrocarbons containing the benzene ring are called *cyclic hydrocarbons*. Several medically important compounds are of the benzene ring variety.

Phenols

Phenols are hydrocarbons having an –OH group attached to the basic benzene ring. The name phenol is derived from *phene*, an old name for benzene. The *-ol* suffix designates that it has an alcohol –OH group attached. The benzene ring is so common that a simple hexagon with double bonds included identifies the molecule.

Phenol has an –OH group attached to one carbon in place of a hydrogen.

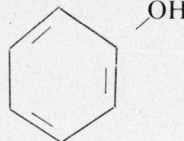

Phenol (C_6H_5OH)

A 5 percent solution of phenol in water is called *carbolic acid*. This was one of the first substances used as a disinfectant. Concentrated phenol is highly toxic and will kill living cells. If placed on the skin, it causes severe deep burns.

BIOLOGICALLY IMPORTANT ORGANIC SUBSTANCES

Although all organic substances found in the body are important, we will concern ourselves here with only three major ones—*carbohydrates*, *proteins*, and *fats*. These substances make up the bulk of the food we eat. They also represent the major substances contained in the body as components of tissues.

Carbohydrates are basically energy storehouses. Proteins are tissue building substances. Fats serve as energy storage protection and insulation against heat loss.

Carbohydrates

The study of carbohydrates is of crucial importance for anyone planning to enter health-related fields. Administering intravenous glucose (a carbohydrate) is a routine hospital activity. The proper care and treatment of the diabetic patient requires a knowledge of glucose metabolism, and certainly students of nutrition will study carbohydrate chemistry.

The word *carbohydrate* originated from experiments in which sugar was heated in a test tube. First the sugar melted, then turned brown and became more viscous. With continued heating at higher temperatures, all that remained in the test tube was a black char identified as almost pure carbon. The vapor given off during heating was found to be water. The term *hydrate of carbon*, or *carbohydrate*, seemed appropriate. We now know that these substances are not mere hydrates. However, the name is still used.

Carbohydrates are substances composed of carbon, hydrogen, and oxygen, with the general ratio of 1 carbon atom, 2 hydrogen atoms, 1 oxygen atom. For example, the common carbohydrate glucose (a simple sugar) has the formula $C_6H_{12}O_6$. Other carbohydrates include sucrose (table sugar), starches, dextrines, cellulose, and glycogen. The body can synthesize some carbohydrates if the basic components are available. The source of these components is plants.

```
        H
        |
        C = O     Aldehyde group
        |
   H — C — OH
        |
   H — C — H
        |
   H — C — OH
        |
   H — C — OH
        |
   H — C — OH
        |
        H

        Glucose
```

```
        H
        |
   H — C — OH      Ketone group
        |
        C = O
        |
   HO — C
        |
   H — C — OH
        |
   H — C — OH
        |
   H — C — OH
        |
        H

        Fructose
```

Fig. A.12 The glucose and fructose molecules. Both molecules have the same formula, $C_6H_{12}O_6$, but their molecular structure is different. Such substances are said to be isomers of each other.

Plants have the capability of synthesizing carbohydrates from carbon dioxide and water in the presence of sunlight and the pigment chlorophyll. The process is known as *photosynthesis*.

Carbohydrates contain several hydroxyl groups (OH) plus either an aldehyde or a ketone group. The simplest carbohydrates are sugars called *monosaccharides*. They may have 2 to 10 carbon atoms and are identified by the suffix *-ose*. A 2-carbon sugar is a *diose*, 3 is a *triose*, 4 a *tetrose*, 5 a *pentose*, 6 a *hexose*, and so on.

All monosaccharides, such as glucose are crystalline solids, soluble in water, and slightly soluble in alcohol. The structural formula would seem to indicate that these substances have both acidic (H^+) and alkaline (OH^-) properties, which they do exhibit in some chemical reactions. However, a solution of sugar has a neutral pH.

Glucose A very common and important body sugar is *glucose* (aldohexose). The term aldohexose refers to the fact that the molecule has an aldehyde group and is composed of a 6- (hex-) carbon chain. The empirical formula for glucose is $C_6H_{12}O_6$. Its structural formula is shown in Figure A.12.

Glucose goes by many names. Others not yet mentioned include *dextrose* and *grape sugar*. In medicine, the term glucose is most commonly used.

Other monosaccharides besides glucose include *fructose*, shown in Figure A.12, and *galactose*. Fructose is found in ripe fruits. Galactose is found in plant seeds and other plant components.

Monosaccharides such as glucose are called simple sugars because they cannot be converted into simpler carbohydrates.

Disaccharides *Disaccharides* are sugars that contain 12 carbon atoms (twice the number of carbon atoms found in a monosaccharide). These substances are important physiologically due to the fact that they make up the bulk of all the sugars we ingest. These include *sucrose* (table sugar), *maltose* (malt sugar), and *lactose* (milk sugar). Disaccharides have the general formula $C_{12}H_{22}O_{11}$. Disaccharides are composed of two monosaccharide molecules minus a water molecule. The chemical process resulting in this combination is called *dehydration synthesis* or *condensation*. When a disaccharide is chemically degraded to the monosaccharide level, the process is called *hydrolysis*. In other words, the hydrolysis of a disaccharide molecule yields two molecules of monosaccharide. (A molecule of water joins in the reaction.)

Polysaccharides The term *polysaccharide* refers to any substance composed of three or more monosaccharide molecules chemically united. Some polysaccharide molecules may be very large molecules made up of hundreds of monosaccharide units.

Polysaccharides include such substances as starch, glycogen, cellulose, and dextrin. From the standpoint of human physiology, the two most important substances are plant starch and glycogen (animal starch).

Plant starch has the general formula $(C_6H_{10}O_5)x$, the x standing for some number of monosaccharide units. There are many kinds of starch—potato, corn, wheat, and so on. Each one has its own peculiar number of monosaccharide units making up its molecules.

Starch is one of the major food substances we ingest. It is hydrolyzed into simple sugars by the digestive process.

Glycogen is often called animal starch. It is a polysaccharide and a storable, energy-rich substance. It is stored in the liver and in muscles whose cells may place extra demands upon energy sources.

Lipids

Lipids are fatty, oily substances and as a group include vegetable and animal fats, waxes, plus other substances scattered throughout the body, such as nerve tissue coverings and cholesterol.

Fats are compounds having a high molecular weight and are composed of carbon, hydrogen, and oxygen. They consist of several large molecules chemically combined by dehydration synthesis. Most fats are composed of three molecules of fatty acids (organic acids) and one molecule of a polyhydroxy (three or more hydroxyl ions) alcohol.

Pure fats composed of the higher fatty acids are odorless and tasteless. Fats with short fatty acid molecules upon hydrolysis yield a noticeable odor. For example, the odor of rancid butter is due to the hydrolysis of butter, yielding butyric acid (an organic acid in butter).

Fats all have a greasy feeling and all have a specific gravity less than 1; therefore they all float on water. They are all insoluble in water and soluble in most organic solvents, such as ether, acetone, and hot alcohol.

Human fat ranges in color from white, translucent, to a yellow color owing to the pigment carotene, which is stored in fat tissue. Carotene pigments are found in carrots and egg yolk.

A very simple test for fats is to rub a bit of the substance to be tested on a piece of unglazed paper. Fats will leave a typical grease spot, making the paper translucent when held up to the light. Lipids are discussed in more detail in Chapter 13.

Proteins

Every living cell in the body contain proteins. Protein molecules are the largest molecules in the body, ranging in molecular weight from about 8,000 to millions. Proteins are made up of smaller units called *amino acids*, which are small molecules of varying molecular weight that can be chemically linked together (dehydration synthesis) to form larger molecules. As a result of these chemical linkages, extremely large molecules can be made.

There are probably no other substances in the body of more critical importance than proteins and the role they have in human physiology.

Amino Acids

Amino acids contain carbon, hydrogen, oxygen, and nitrogen. Some also have traces of sulfur and iron. They are essentially colorless, crystalline substances, soluble in most organic solvents.

Of the 23 known amino acids, 10 are indispensable in the diet. These *essential amino acids* are required for normal growth and maintenance of health. These 10 amino acids (Table A.8) cannot be synthesized by the human body, and thus they must come from foods. If these 10 are present the other 13 can be synthesized by the body.

Figure A.13 illustrates a simple amino acid called *alanine*. Note in the structure of this molecule the presence of two groups that are considered to be functional groups. One group is the NH_2, called the *amino group*. The other is the *carboxyl group*, which we have seen previously as a functional group of organic acids.

Table A.8 Essential Amino Acids

NAME	FORMULA
Threnine	$C_4H_9NO_3$
Valine	$C_5H_{11}NO_2$
Leucine	$C_6H_{13}NO_2$
Isoleucine	$C_6H_{13}NO_2$
Arginine	$C_6H_{14}N_4O_2$
Lysine	$C_6H_{14}N_2O_2$
Methionine	$C_5H_{11}SNO_2$
Phenylalanine	$C_9H_{11}NO_2$
Tryptophan	$C_{11}H_{12}N_2O_2$
Histidine	$C_6H_9N_3O_2$

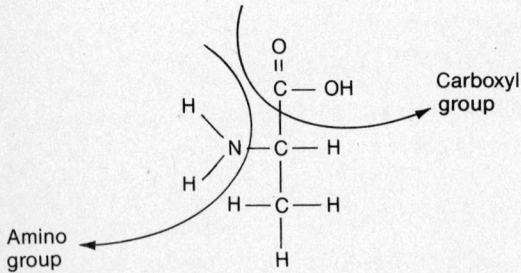

Fig. A.13 The amino acid, alanine. Two functional groups are present, the amino group (NH_2), on the left, and the carboxyl (COOH) group, on the right. These two groups are common to all amino acids.

The major difference between the 23 amino acids is the number of carbon atoms in each molecule. There may be more than one amino functional group attached and one or more carboxyl groups.

Among the things proteins do:

1. Build and repair tissues.
2. Act as buffers in the blood.
3. Supply sulfur and phosphorus to body cells.
4. Act as a source of energy.
5. Serve as emulsifying agents for fats in protoplasm and blood.
6. Provide amino acids for the synthesis of enzymes, hormones, and certain vitamins.

DNA

Although protein metabolism and synthesis are covered in Chapter 14, the discussion here relates protein synthesis with the DNA molecule (*deoxyribonucleic acid*).

The total DNA found within the human body is small, something on the order of 2 ten-trillionths of an ounce. Within this small quantity of material exists a specific code that will direct all the events that occur in an animal from the time the egg is fertilized until the time the animal dies. The DNA molecule will further specifically direct what the adult animal will look like. In the case of a human, the color of eyes, the length of fingers, attached or unattached ear lobes, and so on are determined by DNA within the fertilized egg. Half of the chromosome material comes from each parent. Therefore each parent provides half of the DNA found within a fertilized egg.

Highly purified DNA has been extracted from a wide variety of bacteria, plants, and animals. Upon analysis it has been found, in each case, to consist of a sugar called *deoxyribose*, a *phosphate group* (PO_4), and *nitrogenous bases*.

Four kinds of nitrogenous bases have been found in DNA. Two kinds, called *purines*, are *adenine* and *guanine*. Two others, called *pyrimidines*, are *cytosine* and *thymine*. Figure A.14 illustrates a portion of a DNA molecule.

If a nitrogenous base such as adenine is joined with a phosphate group and the sugar deoxyribose, a three-part molecule called a *nucleotide* is formed. The formation of four different kinds of nucleotides and their precise arrangement forming a chain of molecules represents the formation of a DNA molecule.

The hypothetical arrangement of nucleotides, proposed by Watson and Crick in 1953, suggests that the DNA molecule consists of two sets of nucleotide chains wrapped spirally around each other, with the sugar-phosphate chain on the outside of the helix and the pyrimidines on the inside, these being held together by hydrogen bonds. Figure A.15 represents this configuration.

The outside rails of the spiral ladder represent the sugar-phosphate chains, and the crossbars or the rungs of the ladder represent the nitrogenous bases held together by hydrogen bonds.

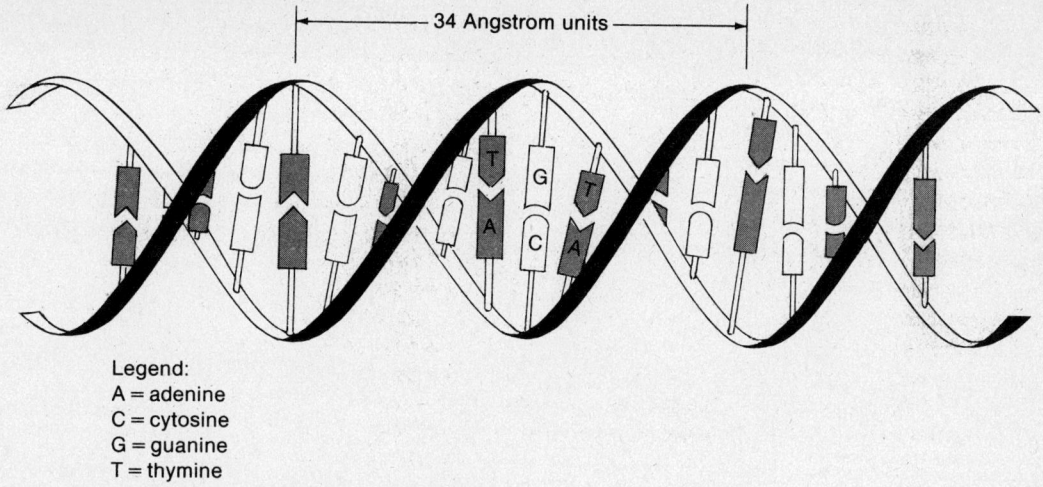

Fig. A.14 A portion of a DNA molecule. This hypothetical arrangement of nucleotides was proposed by Watson and Crick in 1953.

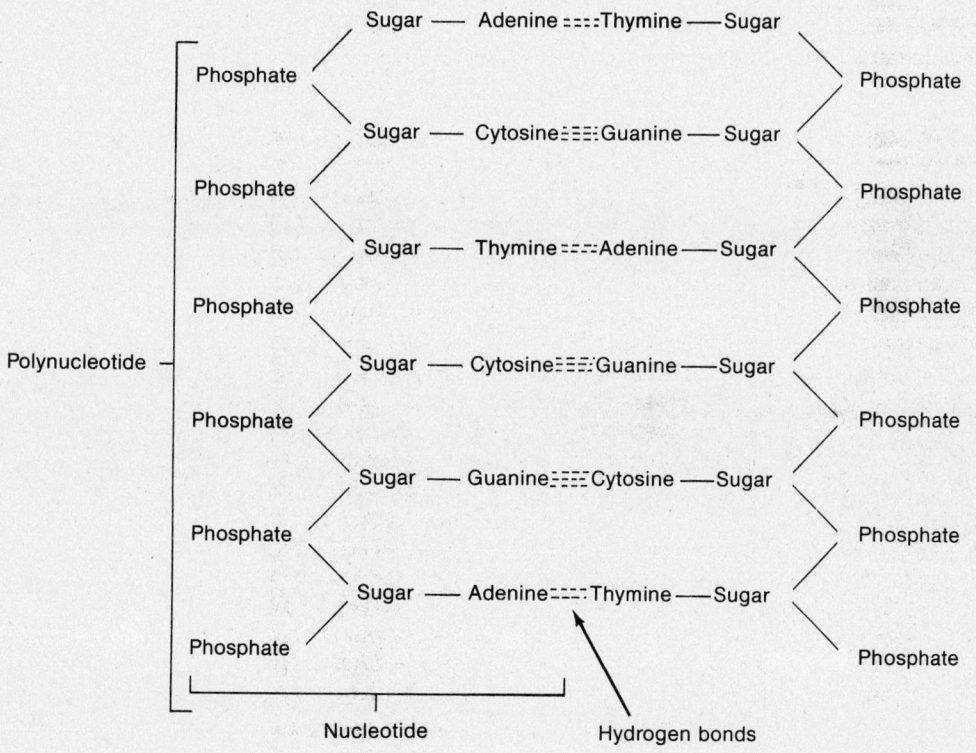

Fig. A.15 The arrangement of the four nucleotide bases with sugar and phosphate molecules.

BASIC CHEMISTRY 335

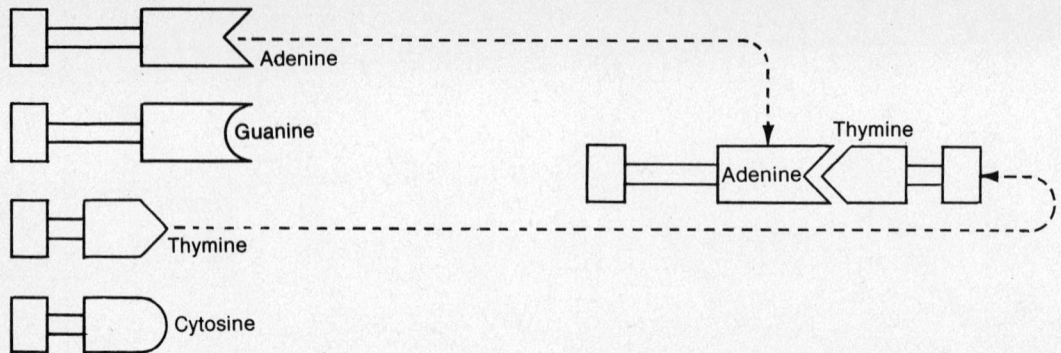

Fig. A.16 Nitrogen base arrangement. Adenine and thymine can unite, owing to their special molecular configuration.

The hydrogen bonds between the purines and the pyrimidines are such that adenine can bond only with thymine, and guanine can bond only with cytosine. Other combinations are not possible because of the size and peculiar configuration of the molecule; they simply will not fit into the hypothetical ladder design.

In Figure A.16 the molecules are drawn as having such a peculiar shape as to prohibit any other arrangement of nitrogenous bases except that which is shown. In every case, then, adenine must be bonded to thymine and guanine bonded to cytosine. If there is an adenine molecule in the chain, there will be a thymine molecule in the chain opposite it.

Living things grow by the process of mitosis—one cell dividing into two cells. The genetic material of which chromosomes are composed is DNA. The process of packaging the chromosome's DNA takes place sometime before cells divide during mitosis. This process is known as *DNA replication.*

It is part of the Watson-Crick theory that when the DNA molecule undergoes replication, the two chains of the molecule separate, possibly just like a zipper opens. Each half of the chain forms a new double chain. Each new chain is made up of half of the old chain plus material taken from cell nucleoplasm. Figure A.17 shows a diagram of a DNA molecule before replication. The sides of the laddershaped molecule, represented by the vertical, parallel lines, are composed of phosphate and sugar molecules. The crossbars are made of the complementary nitrogenous bases. Figure A.17 also shows the chains separating.

The figure shows how free nitrogenous bases from the cytoplasm line up in specific order and become bonded to the chain. Where adenine is located on the chain, a thymine molecule from the cytoplasm will bond with it. After the chains have reformed and regrouped into double chains, phosphate and sugar join the molecule. The process is now complete, and two double chains, in other words, two DNA molecules, now exist where originally there was only one.

Note in Figure A.18 that the order of the nucleotide bases is exactly that of the original molecule of DNA shown in Figure A.17. This specific sequence—AT, CG, TA, CG, TA, GC, AT—may be the code that designates the blueprint that produces a specific protein or some other genetic trait. It is believed that steps along the DNA molecule represent genes. The DNA molecule might be likened to a spiral stairway, with adenine and thymine linked to make a step, and guanine and cytosine linked to make another step. A single gene might be a chunk of DNA 2,000 steps long. Geneticists now think it is the precise order of these steps that gives every gene its particular character.

A single virus may have chromosomes 1/2,000 of an inch long and as many as 170,000 steps. Human DNA might be as much as a yard long and contain six billion steps.

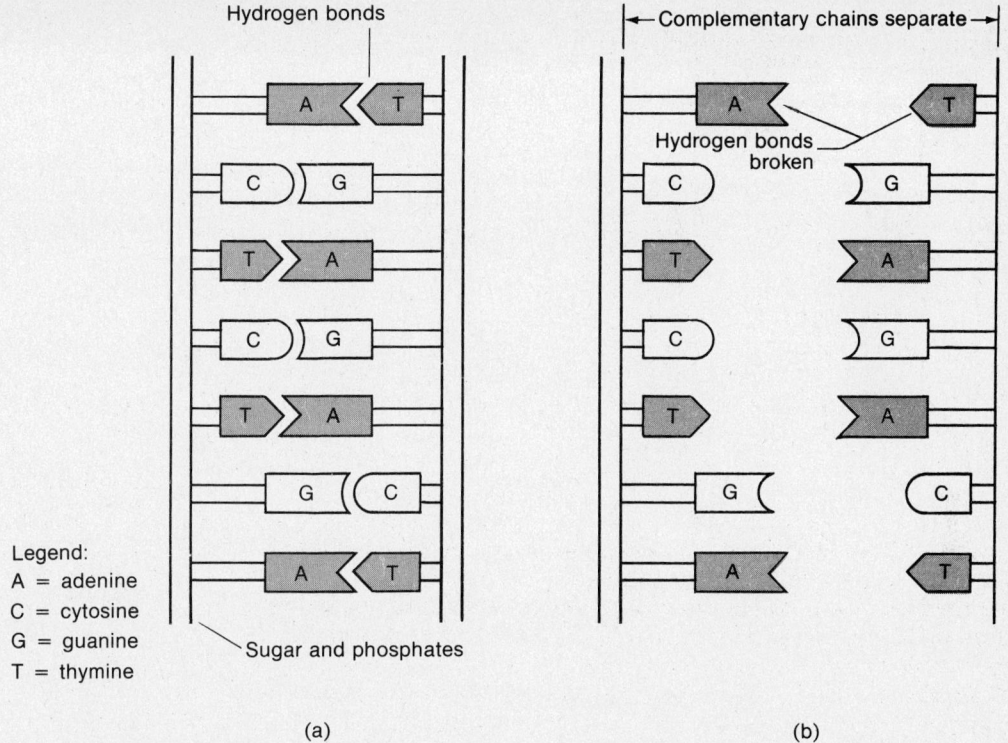

Fig. A.17 (a) A segment of a DNA molecule before replication. (b) The complimentary chains of nucleotides separating.

Any one of these steps out of order could prove to have grave consequences, since the individual could be imperfectly formed. These imperfections then can be passed on to offspring. The precise mechanism whereby a gene directs the formation of inherited traits is not completely understood. Genetic control of protein synthesis is more completely understood, and therefore we will use this process to illustrate how information in the code provided by the DNA molecule is transmitted to other parts of the cell for implementation.

RNA

DNA serves as an instructional code molecule that indirectly determines hereditary traits. The synthesis of specific protein molecules called *enzymes* is no less a hereditary trait than color of eyes or length of fingers.

An intermediate molecule, called *RNA*, (ribonucleic acid) has the function of transmitting the code specified by DNA to cell organelles called *ribosomes*. RNA directs the synthesis of enzyme molecules and other proteins.

RNA is formed in the nucleus of cells as is DNA, but its formation and structure are different from DNA. The basic difference is that RNA is a single strand rather than a double strand of molecules.

Individual nitrogenous bases, phosphates, and sugar molecules may be found in the nucleoplasm unattached to DNA. During RNA synthesis, it is believed that these molecules align themselves along one strand of a DNA molecule. A single DNA strand serves as a pattern or template for the formation of an RNA molecule. In other words, the completed RNA is a copy of a strand of DNA. It is an exact copy

BASIC CHEMISTRY

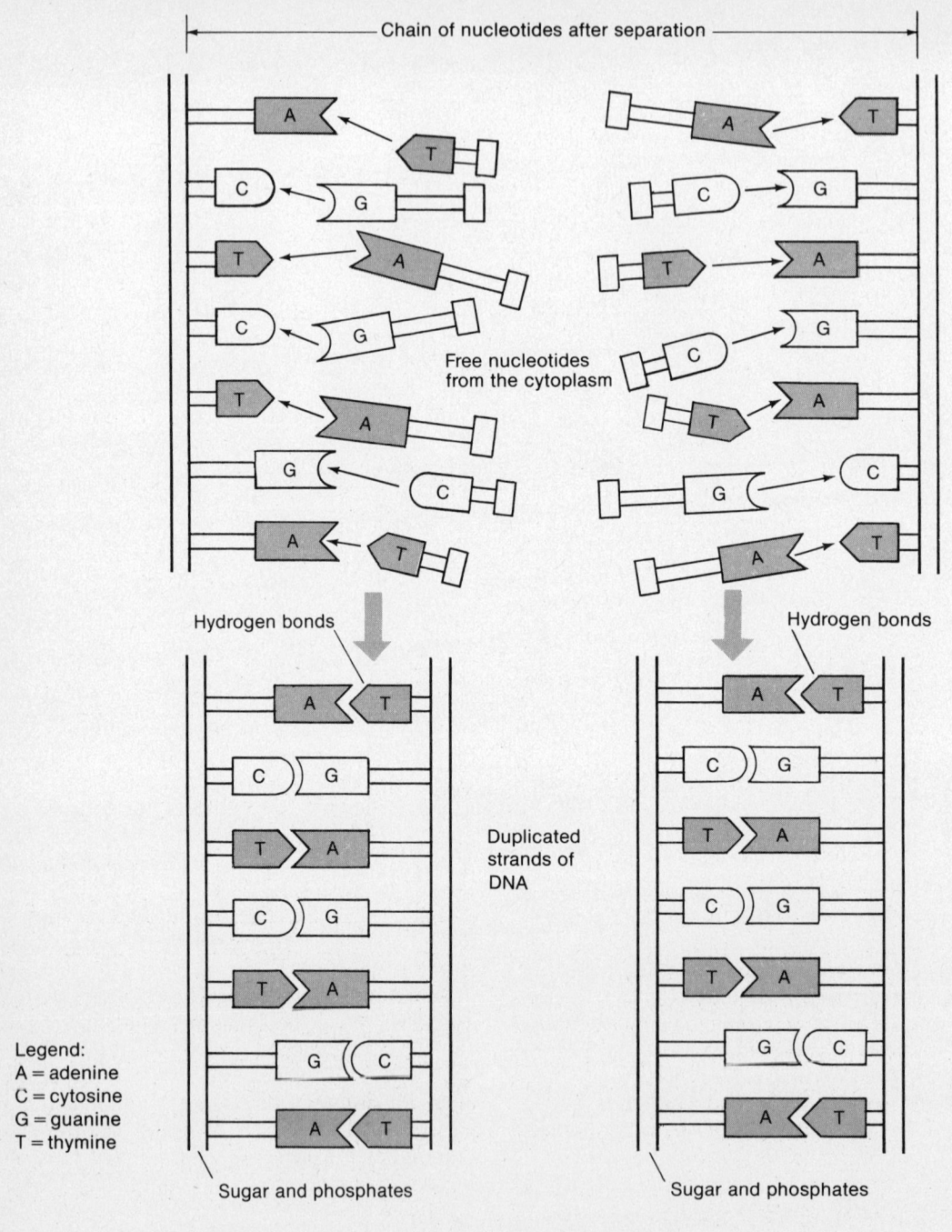

Fig. A.18 The replication process, showing the two identical strands of DNA that are replicas of the original strand.

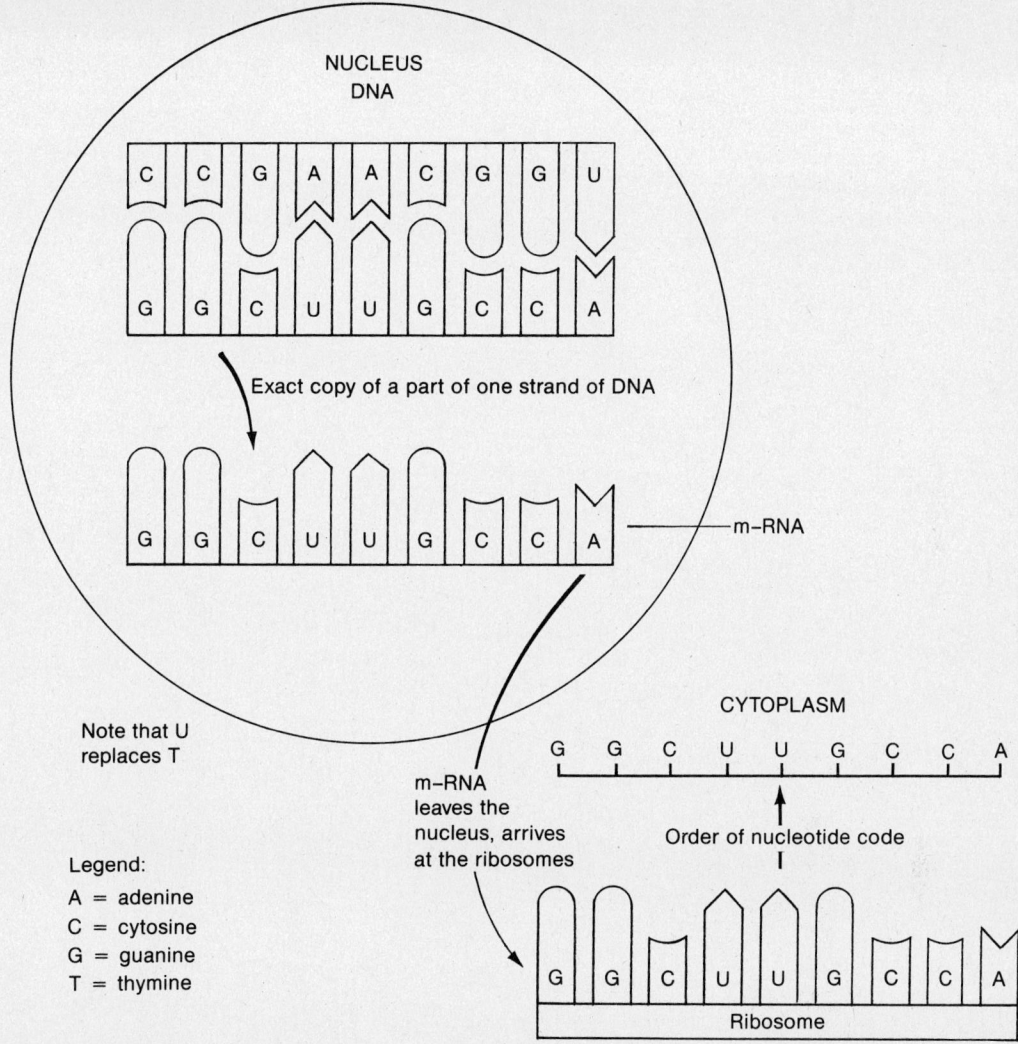

Fig. A.19 The synthesis of mRNA occurs in the nucleus dictated by DNA. mRNA leaves the nucleus and arrives at a cell organelle called the ribosome.

except for one factor—the nitrogenous base *uracil* is substituted for thymine.

Three different kinds of RNA have been identified. *Messenger RNA* (mRNA), formed by the chromosomes, leaves the nucleus and becomes attached to ribosomes located in the cytoplasm. *Transfer RNA* (tRNA), also formed in the nucleus, acts as an amino acid carrier by delivering amino acids to the ribosomes. *Ribosomal RNA* has also been identified, but its precise function in protein synthesis is not clear.

A single molecule of mRNA may be composed of several thousand nucleotides. Figure A.19 illustrates a small piece of such a molecule, its formation, and its arrival at the ribosomes.

We will leave the mRNA at the ribosomes for a moment and look at the formation and activity of tRNA.

BASIC CHEMISTRY **339**

Table A.9 Triplet Codes Contained in tRNA for Specific Amino Acids[a]

AMINO ACID	TRIPLET CODE (A = Adenine, U = Uracil, C = Cytosine, G = Guanine)
Alanine	CCG, CGU
Arginine	CCG
Asparagine	CAU, AAU, AAC
Aspartic Acid	GAU, CAA
Cystine	GUU
Glutamic Acid	GAU, GAA
Glutamine	AAC, AGG
Glycine	GGU
Histidine	CAU, CAC
Isoleucine	AUU
Leucine	AUU, GUU, CUU
Lysine	AAU, AAA
Metnionine	GAU
Phenylalanine	UUU
Proline	CCU, CCC
Serine	CUU, CCU
Threonine	AAC, ACC
Tryptophan	GGU
Tyrosine	AUU
Valine	GUU

[a]Only a few triplet codes have been precisely determined as to their arrangement. This table represents only the nucleotide bases found, not necessarily their order.

Fig. A.21 The process of protein synthesis. Once all the amino acids have found their proper place along the mRNA molecule, chemical (peptide) bonds are formed between adjacent amino acids. A chain of several hundred (in some cases, thousands) of amino acid molecules united by peptide bonds represents a protein molecule.

Transfer RNA, also produced in the nucleus, is a much smaller molecule composed of ribose sugar, phosphates, but only three nitrogenous base molecules. For example, one tRNA molecule may have the order of nitrogenous bases—CCG or AAC or GAU, and so on.

The precise order of the three nitrogenous bases (called a triplet code) permits it to attach to specific amino acids. Free amino acids floating in the cytoplasm are "captured" by one or more of the tRNA molecules and transferred to the ribosomes and mRNA.

Triplet codes contained in tRNA for specific amino acids have been proposed, as shown in Table A.9.

Figure A.20 illustrates three different tRNA molecules and their attachment to the amino acids *alanine,* *glutamine* and *glycine.* Each amino acid in some manner relates to a specific triplet code, as is shown in Table A.9. The code triplets for the 23 different acids are known, but the exact sequence of each triplet has not as yet been worked out for all the amino acids.

To summarize the preceding, messenger RNA molecules synthesized by the chromosomes in the nucleus leave the nucleus and become attached to the ribosomes. Transfer RNA molecules synthesized in the nucleus also enter the cytoplasm, where they attach to specific amino acid molecules.

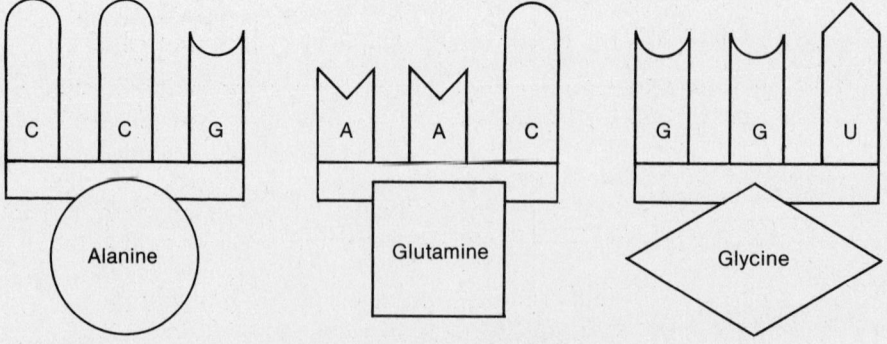

Fig. A.20 Each amino acid forms a loose chemical attachment with a specific tRNA molecule. This figure shows the three amino acids, alanine, glutamine, and glycine, attached to a triplet code.

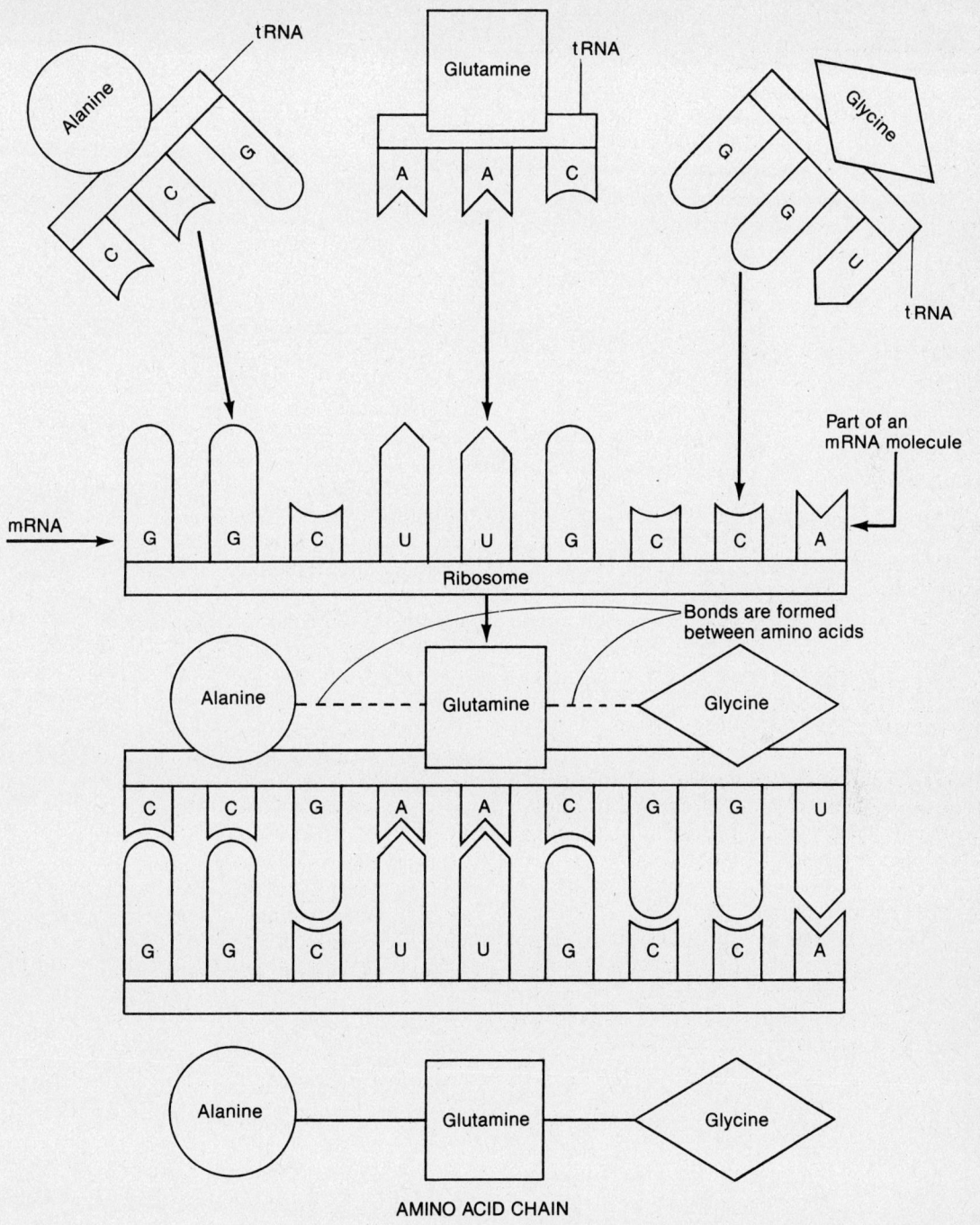

The stage is now set for the synthesis process of combining amino acids in a specified sequence, as shown in Figure A.21. The amino acids alanine, glutamine, and glycine become attached to their complementary tRNA molecules, which carry them to mRNA. Upon arrival at the ribosomes, tRNA finds its place on the mRNA molecule. By finding its place, we mean locating the exact arrangement of nitrogenous bases that are complementary to the triplet code carried by tRNA. For example, note in Figure A.13 that the first three nitrogenous bases of mRNA are GGC. Only tRNA with the code CCG may join the molecule here.

Once all the tRNA molecules have found their proper place along the mRNA molecule, chemical bonds are formed between the amino acids, making a chain. In Figure A.13 the chain of three amino acids, alanine-glutamine-glycine, is formed. Once the chain is formed it peels away from the ribosomes, leaving the "empty" tRNA molecules free to return to the cytoplasm to seek out more specific amino acids and continue the process of synthesizing more chains of amino acids.

The formation of a chain of only three amino acids is used here only for illustrative purposes. Actually, the mRNA molecule is much longer than the one shown, with location sites for thousands of tRNA molecules, each carrying specific amino acids. A chain of several thousand amino acids synthesized in such a manner may represent a single protein molecule.

The human body is composed of many different kinds of protein molecules. Their differences lie in their length (numbers of amino acids) and the sequence of the amino acids. We can therefore assume that there must be at least as many different kinds of mRNA as there are kinds of proteins, since each kind of mRNA can make only one kind of protein.

We receive all our amino acids from protein we eat. The protein of, say, peanuts is different from our muscle protein. Through the process of digestion, peanut protein is broken down to individual amino acids. These amino acids then can be arranged into muscle protein by the mechanism just described, utilizing DNA, mRNA, the ribosomes, and tRNA.

Nucleotides and Coenzymes

Molecules much smaller than DNA or RNA, but related to the nucleic acids, play an important role in metabolism. A *nucleotide* is a nitrogenous base (purine or pyrimidine) joined by a ribose sugar and phosphoric acid. One such substance is *adenosine triphosphate*, abbreviated *ATP*. This is an energy-rich molecule produced by cells to store energy released during cellular respiration. The base adenine is joined by a 5-carbon (ribose) sugar plus three phosphate groups to produce a molecule of ATP.

Other nucleotides include *NAD* (nicotinamide adenine dinucleotide) and *FAD* (flavin adenine dinucleotide). These molecules play an important role in cellular respiration as hydrogen acceptors. (See Chapter 14.) These nucleotides are categorized as *coenzymes*. They are substances that are required along with enzymes to complete specific chemical pathways during cellular respiration.

Vitamins

Vitamins are organic molecules more difficult to categorize than nucleotides. Like nucleotides, they are also coenzymes. They do not belong to any one chemical family. Some are organic acids, some are alcohols, aldehydes, or amino acids. Vitamins are absolutely necessary in many physiological processes. A deficiency of most vitamins results in some impaired physiological process. Chapter 14 discusses specific vitamin deficiencies and related consequences.

SUMMARY AND REVIEW

A. Chemistry is the study of matter, its properties, characteristics, and interactions. It is the study of how matter is organized. Three classifications of matter are:
  1. mixtures (nonuniform matter)
  2. solutions (uniform mixture)
  3. substances (uniform matter)
B. Elements are substances that cannot be decomposed by chemical change into simpler substances.

C. Compounds are produced when elements are chemically combined. Compounds may be chemically degraded back to the elements of which they are composed.
D. Solutions are uniform mixtures of a solute dissolved in a solvent.
  1. The strength of a solution depends upon the ratio of solvent to solute.
  2. A saturated solution is one in which a given volume of solvent has dissolved all of a given solute that it can at a given temperature. An unsaturated solution has less solvent under the same conditions.
  3. Percent solutions are those in which the solute weight and solvent weight are expressed as a part of 100 or as a percent. A 5 percent solution of NaCl would contain 5 g (or some other unit) of NaCl and 95 g of water.
  4. Ratio solutions are expressed as a ratio of solute to solvent.
E. Atoms and atomic structure. Atoms are defined as the smallest part of an element capable of exhibiting all the properties of the element.
  1. Atoms are made up of positive charges (protons), negative charges (electrons), and neutral charges (neutrons).
  2. Atoms are neutral due to equal numbers of protons and electrons.
  3. The atomic weight of an element is a ratio used to aid in determining how much of one substance will combine with another. Atomic weights are shown on the periodic table of elements.
  4. Molecular weight is determined by finding the sum of the atomic weights within a molecule.
  5. Molar solutions are prepared by dissolving 1 mole (gram molecular weight) of a substance in enough water to make a final volume of 1 liter.
F. Ions are charged particles that result when an atom gains or loses electrons. The number of electrons gained or lost in order to fill the outermost orbit determines the combining power of the ion. Radicals are ions composed of two or more elements.
G. Acids, bases, and salts
  1. An acid is a substance that dissociates in solution yielding free hydrogen ions. The greater the dissociation, the stronger the acid.
  2. Bases are substances that dissociate in solution yielding hydroxide (OH^-) ions. The greater the dissociation, the stronger the base.
  3. Salts are products of chemical reactions between an acid and a base.
  4. The pH scale rates the strength of acids and bases on a scale of 1 to 14. Lower numbers represent stronger acids. Higher numbers represent stronger bases. A pH of 7 is neutral.
  5. Buffers are substances that resist changes in pH.
H. Organic chemistry. The chemistry of organic compounds is called organic chemistry. It is also the chemistry of carbon compounds. A type of chemical union called covalent bonding (sharing electrons) is almost unique to organic compounds. Only a few inorganic compounds have covalent bonding.
  1. Compounds of interest include:
     a. hydrocarbons
     b. alcohol
     c. organic acids
     d. aldehydes
     e. ketones
  2. Compounds of biological significance include:
     a. carbohydrates
     b. lipids
     c. proteins
  3. Special treatment of protein synthesis is given in the text. The relationship of amino acids, DNA, and RNA in protein synthesis is discussed.
  4. Nucleotides and coenzymes such as NAD and FAD are critical substances required for the completion of chemical pathways during cellular respiration.
  5. Vitamins are coenzymes required by a variety of physiological processes. A deficit of some vitamins causes deficiency diseases.

APPENDIX B
Some Basic Physical Concepts

MAJOR CONCEPTS
  GASES
    Gas Pressure
    The Siphon
    Evaporation
    Boiling Point
    The Autoclave
  LIQUIDS
    Liquid Pressure
    Pressure Gradient
    Viscosity
    Molecular Forces in Liquids
    Surface Tension
    Pascal's Law
    Archimedes' Principle
  ENERGY
  HEAT
    Surplus Body Heat
    Conduction
    Convection
    Evaporation
  LIGHT
    Ultraviolent, Infrared, and X rays
    Physiological Effects of Light
  ELECTRICITY
    Static Electricity
    Current Electricity
    Dangers of Electricity

This appendix, like Appendix A, may be used as an overview or a general introduction to some physical concepts intimately involved in human anatomy and physiology. Major areas include gases, liquids, solids, and energy.

Physics is the study of the physical aspects of matter. *Matter* is defined as anything that takes up space and has weight. One basic characteristic of matter is that it may be converted into energy. Energy has the capability to do work. First we will look at different forms of matter. These include *gases, liquids,* and *solids.*

We know that all matter is made up of particles called atoms and that atoms combine to form molecules. We also know that molecules of matter are in constant vibratory motion. This motion is due to *kinetic energy.* When heat is applied to any form of matter the kinetic energy is increased, thereby increasing molecular motion. The kinetic energy in any particular form of matter determines whether it is a gas, liquid, or solid. Gas molecules have greater energy levels than do liquids, and liquid molecules have a greater energy level than do solids. Substances with higher energy levels, owing to the higher energy levels and greater molecular motion, have greater spaces between their molecules and tend to have less uniform shape.

GASES

Gases are a normal part of our everyday environment. The influence gases have upon living systems may be as extreme as causing death, owing to the lack of oxygen, or as minimal as discomfort when gas pressures fluctuate. Gases are forms of matter and may exist in either the atomic or molecular form. Atmospheric gases are most commonly found in the molecular configuration. For example, one atom of oxygen is symbolized as O. If two atoms of oxygen are chemically united, a molecule of oxygen is formed, symbolized as O_2. This same principle applies to other gases in the atmosphere. For example, molecular hydrogen is written H_2 and molecular nitrogen N_2.

The essential difference between forms of matter is how closely packed the molecules are. The more tightly packed the molecules are, the higher is the density of a substance. Solids are generally more dense than liquids, and gases are less dense than liquids.

Gases are chemical substances made up of molecules that are widely separated by space. When gas molecules are compressed tightly together and cooled, they form a liquid. Some gases can be compressed and cooled sufficiently to form a solid. An example of this is dry ice, which is solidified carbon dioxide.

The gases most familiar to us are those found in the atmosphere. These include oxygen, carbon dioxide, nitrogen, hydrogen, and a few others of less importance. All gases have weight and take up space and obey the same physical laws affecting liquids and solids. They expand when heated and contract when cooled.

Because of the freedom of movement gas molecules have, they fill and take the shape of any container in which they are placed. If more molecules are forced into a closed container, a greater pressure within the container is created. If a gas is placed in a closed container and heated, the pressure is increased due to the accelerated movement of the molecules since the energy level is increased.

Gas Pressure

Pressure is defined as the force exerted on a unit area. In the case of gases, pressure within a container is created by the weight of the gas plus the force exerted on the inner surface of the container due to the rapid movement of gas molecules and their subsequent collisions with the container walls. As a molecule collides with a surface, a force is created, admittedly very small. But suppose billions of such molecules bombard the surface simultaneously. This cumulative force exerted on an area, say one square centimeter, is called pressure.

When molecules of any form of matter are heated, their movement is speeded up. With their greater speed, the molecules of heated gas bombard surfaces with greater force, hence creating greater pressures.

Gas pressures have a direct influence upon the human body. The most constant pressure is that of the atmospheric gases. The normal pressure of atmospheric gases is approximately 1 kg per square cm at sea level. If you could construct a tube with a cross-sectional area of 1 sq cm and could extend the tube up as far as gases exist, approximately 10 km, the air contained within this tube would weigh 6.8 kg. Since each square cm of body surface area has a force of 6.8 kg pressure upon it, a little simple arithmetic will show that the total force exerted upon the human body by atmospheric pressure alone is tremendous. (Average surface area of an adult body is 44,000 sq cm.) This force would crush the body if counterpressures within the body did not exist. The pressure within cells is fairly constant and cumulatively equalizes the external gaseous pressures. We become aware of unequal pressures when we change altitude. Air pressures decrease as we go higher above sea level and increase as we go lower toward sea level.

Measurement The device most commonly used to measure air pressure is the *barometer*. Two types of barometers are in common use. One, the *mercury barometer*, operates on the principle that a column of liquid is held up by air pressure. A simple mercury barometer can be constructed by filling a glass tube about 80 cm long, closed at one end, with mercury. The open end is then closed and inverted into a small vessel containing mercury. Some of the mercury in the glass tube will run out into the vessel. However, a column of mercury about 76 cm high will remain in the tube. This column is held up to this height by the atmospheric pressure being exerted on the surface of the mercury in the open vessel, as shown in Figure B.1.

If the atmospheric pressure is reduced, more mercury will fall out of the tube, thereby lowering the height of the column. Increasing the pressure will increase the height of the mercury column. The glass tube can be calibrated in inches, centimeters, or millimeters. At sea level the height of the mercury column is 30 in., 76 cm, or 760 mm. Another kind of barometer is the *aneroid* type. An aneroid barometer, as shown in Figure B.2, utilizes the effect that air pressure has upon a flexible material.

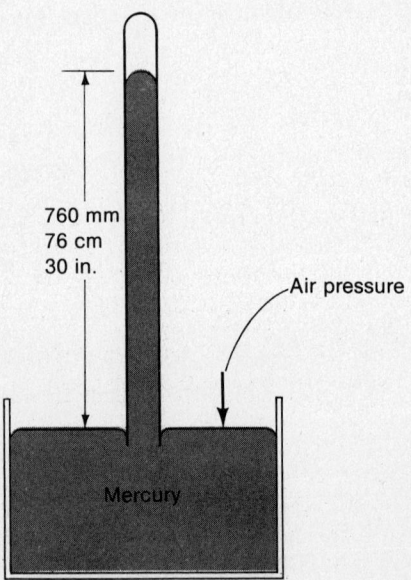

Fig. B.1 Principle of a mercury barometer. Atmospheric air pressure against the mercury in the lower container holds up a column of mercury in the tube. If the atmospheric pressure is reduced, mercury will fall out of the tube, thus lowering the mercury column. Increased atmospheric pressure will lengthen the tube of mercury.

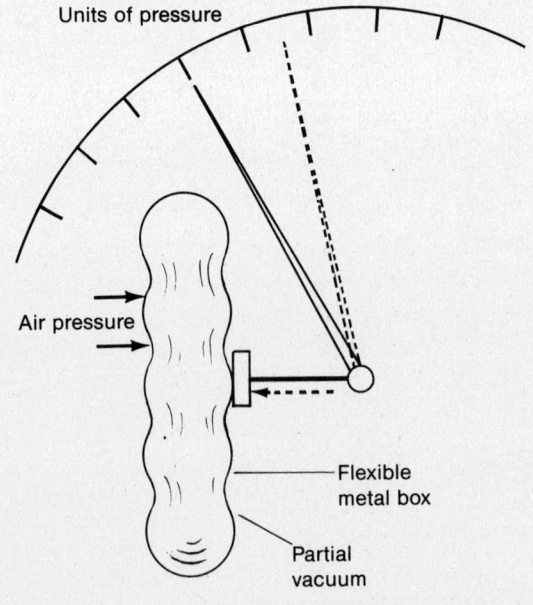

Fig. B.2 Basic principle of the aneroid barometer. Atmospheric pressure pushing against the flexible metal box changes the position of the attached pointer, thus indicating fluctuations in pressure.

A mercury barometer is more sensitive to air pressure fluctuations and therefore more accurate. The aneroid type is most commonly used where precise accuracy is not required, such as in weather forecasting.

Gases exert pressure, they are compressible, and they expand when heated. These properties are so consistent that they can be measured and relationships between volume pressure and temperature can be derived.

Boyle's Law

Suppose that we have a sample of a gas contained within a cylindrical container with a movable piston, as in Figure B.3. If the piston is not moved, the contained gas has a specific volume and exerts a specific pressure on the inner surfaces of the container and the piston. If pressure is applied to the movable piston forcing it downward, the gas would be compressed, therefore reducing the volume. The compression of gases produces heat, which would cause the contained gas to expand. If we wish to show only the relationship between pressure and volume, the temperature would have to remain constant. In other words, as the gas was compressed, the heat, due to increased pressure, would have to be removed.

Pressure-volume relationships were first determined by the Irish scientist Robert Boyle in 1662. The physical law bearing his name, *Boyle's law,* states that if the temperature of a contained gas remains constant, the volume varies inversely as the pressure exerted upon it. Figure B.4 illustrates his law.

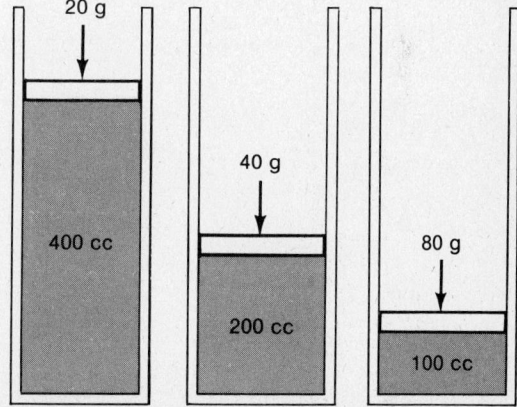

Fig. B.4 Boyle's law. Increased pressure proportionally decreases the volume of an enclosed gas.

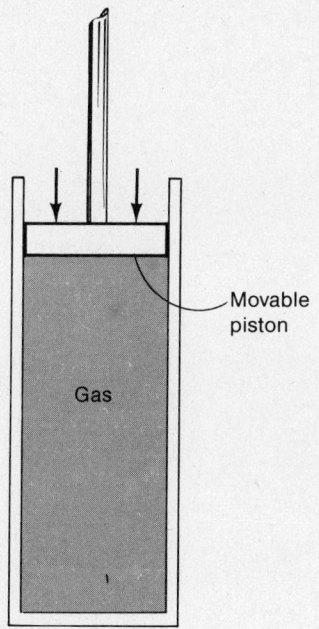

Fig. B.3 Boyle's law. When the piston is moved downward, the enclosed gas is compressed into a smaller space. If the temperature of a contained gas remains constant, the volume varies inversely as the pressure exerted upon it.

An *inverse proportion* is one in which if one factor, in this case, pressure, is increased, the other factor, volume, is decreased. In Figure B.4, note that if the pressure is doubled, the volume is reduced to half its original volume. If the pressure is quadrupled, the volume is reduced to one-fourth its original volume. Gases under extreme pressure do not precisely obey Boyle's law because of other influencing factors. Also, most gases under extreme pressure and cooling become liquids, for which Boyle's law is not applicable. Boyle's law can be expressed in the following formula:

$$\frac{\text{Original volume}}{\text{New volume}} = \frac{\text{new pressure}}{\text{original pressure}}$$

Here is a problem and its solution using this formula. A container having a volume of 1,000 cc is filled with oxygen gas under a pressure of 900 mm Hg. What volume will this gas occupy if it is released into another container of such volume to have a pressure of 760 mm Hg?

Original volume = 1,000 cc
Original pressure = 900 mm Hg
New pressure = 760 mm Hg
New volume = ?

$$\frac{\text{Original volume (1,000 cc)}}{\text{New volume}} = \frac{\text{New pressure (760 mm Hg)}}{\text{Original pressure (900 mm Hg)}}$$

$$\text{New volume} = \frac{1,000 \text{ cc} \times 900 \text{ mm Hg}}{760 \text{ mm Hg}}$$

$$= 1,184.2 \text{ cc}$$

Charles' Law Gases expand when heated. Therefore, if a contained gas is heated, pressure will build up within the container. If we assemble an apparatus similar to that used to illustrate Boyle's law, with a movable piston, the piston will move upward as the gas expands, thereby permitting the pressure to remain constant. As the temperature increases, the volume will also increase. This is a direct proportion.

Dalton's Law The air we breathe is a mixture of gases, mainly oxygen and nitrogen. Each gas in a mixture exerts its own pressure as if it is alone in the container. The total pressure exerted by a mixture of gases is the sum of the pressures exerted by the individual gases.

Atmospheric gases at sea level exert a pressure of 760 mm Hg. If 20 percent of the air we breathe is oxygen, then oxygen exerts a separate pressure of 152 mm Hg (760 × 20 percent). Nitrogen exerts a pressure of 600.4 mm Hg (760 × 79 percent), and carbon dioxide exerts a pressure of 0.3 mm Hg (760 × 0.04 percent).

The transfer of gases across membranes (lungs, cells, capillaries) is accomplished by the pressure of a gas on one side of the membrane being greater than on the other side. Gases maintain their individual (partial) pressures even when dissolved in fluids. For example, we may speak of the partial pressure of oxygen in arterial blood or the partial pressure of carbon dioxide in venous blood. The term partial pressure is usually abbreviated as p; thus the partial pressure of oxygen would be indicated as pO_2 and the partial pressure of carbon dioxide would be indicated as pCO_2.

The Siphon

A siphon is a tube or pipe filled with water used to transfer liquids from one elevation to a lower elevation, as shown in Figure B.5. The direction of liquid flow can be only toward the lower elevation.

The function of a siphon depends upon air pressure, partial vacuums, and the weight of the liquids. Note in Figure B.5 that arm AB of the siphon is longer than arm CD. The atmospheric pressure on the surface of both liquids is so nearly the same that we can ignore this as a factor in moving the liquid from one level to a lower one. The siphon tube must be filled with water and the end of the tube to be placed in container X must be kept submerged in the liquid. The other end of the tube must be lower than the level of the liquid in container X. The liquid will flow into container Y until the level of the liquid in the two containers is equal; if container Y is lower than the bottom of container X, all the liquid will be drained out. Increasing the height of container X will increase the rate of flow into container Y.

The flow of liquids in a siphon is produced in the following manner. The column of fluid AB tends to pull away from the top of the siphon. Also, column CD tends to pull away from the top of the siphon. Since column CD is longer than column AB, it will have a greater downward

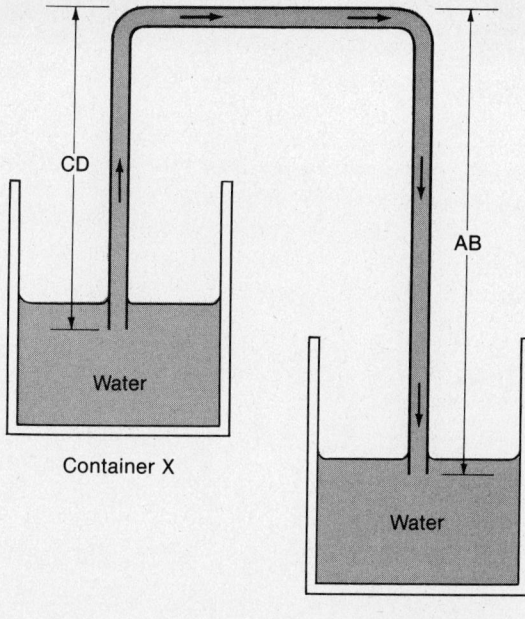

Fig. B.5 A siphon is a tube or pipe filled with a liquid and is used to transfer liquids from one elevation to a lower one.

force. This downward pull tends to form a partial vacuum. If the two containers are the same level, the siphon will cease to function.

Evaporation

In a container of water, such as that shown in Figure B.6, the water molecules. owing to their weight and movement, exert force on the sides and bottom of the containers. The surface of the water in this unclosed container is exposed to the atmosphere.

The surface of a liquid acts as if it were a stretched membrane. It is this phenomenon that permits one to float a needle on a water surface and permits small insects to walk on the surface of water. The membrane effect is produced by the attraction water molecules have for each other. This cohesive force creates a membranelike film on the surface and is responsible for keeping the water molecules from flying out of the container.

The water molecules in a liquid are traveling at various speeds and in many directions. Near the surface, some of the molecules have suffi-cient speed to break away from the surface and escape into the surrounding air. This change of a liquid into a gas is called *evaporation*. Evaporation occurs continuously from the surface of all liquids, but there are several factors responsible for the rate of evaporation.

A higher temperature increases the molecular speed. Therefore a greater number of molecules can escape the surface of a liquid as its temperature rises.

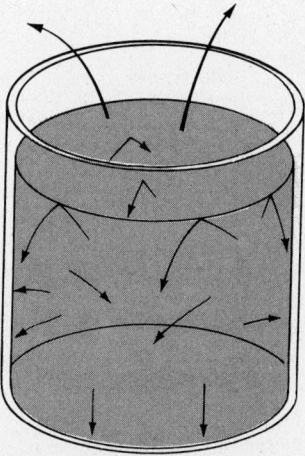

Fig. B.6 Surface tension of water. Some molecules have sufficient energy near the surface to escape (evaporation).

The area of the exposed surface area influences the rate of evaporation. A large surface area permits more rapid evaporation than a small surface area.

As molecules escape the surface of liquids, they may not have sufficient speed to carry them far. They may become concentrated just above the liquid surface and effectively reduce the number of molecules attempting to escape the surface. This accumulation of gas molecules produces a *vapor pressure* that may retard or stop the evaporation process. Circulation of the air above the surface removes some of the gas

SOME BASIC PHYSICAL CONCEPTS **349**

molecules, permitting others to escape more rapidly from the liquid. This is why a jet of air directed onto a liquid causes it to evaporate more rapidly than if it was permitted to stand.

We have all experienced the cooling sensation produced by rubbing a few drops of alcohol on the skin. All evaporation results in a cooling of the liquid and its surroundings. Molecules leaving the liquid are those having the greatest quantity of energy (heat). The remaining molecules have less energy, hence a lower temperature.

Normal body temperature is, in part, maintained by the evaporation of perspiration from the skin. Since the human body has an extensive surface area, caution must be used to prevent excessive evaporation due to skin exposure to the air in a chilly room.

Boiling Point

Air pressure also has an effect upon boiling temperatures. *Boiling point* of a liquid is defined as the temperature at which liquid vapor (a gas) pressure equalizes atmospheric pressure, as shown in Figure B.7. As water is heated, the molecular motion is accelerated until most of the water molecules have gained enough energy and speed to escape the water surface. This is also the point at which water changes from a liquid to a gas. Boiling temperature at sea level is 100°C (212°F).

If the pressure above a liquid is reduced below 760 mm Hg, water molecules can escape at a lower temperature. Water can be made to boil at 4°C if a sufficient vacuum is obtained above the liquid surface. The boiling temperature of a liquid may be increased if increased pressure is put on the surface of the boiling liquid. The increased pressure makes it more difficult for water vapor molecules to escape; therefore a higher temperature is required to give the vapor molecules sufficient energy to escape the water surface. Good examples of the application of pressure to increase the boiling temperature are the pressure cooker and the *autoclave*.

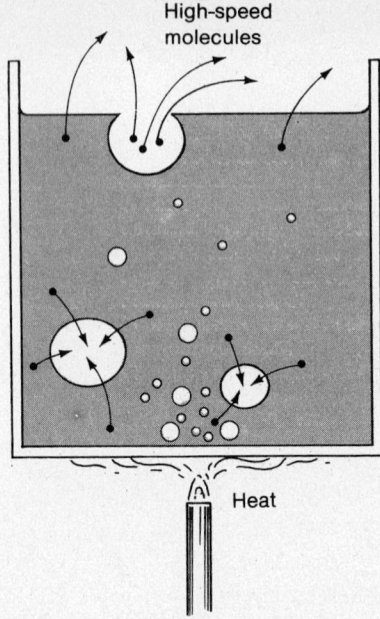

Fig. B.7 When a liquid boils, high speed molecules form bubbles inside the liquid, which rise to the surface, permitting the molecules to escape as a gas.

The Autoclave

Some microorganisms are not killed by mere boiling temperature at sea level. Temperatures exceeding 100°C can be achieved if a container holding water is enclosed and heated so that the vapor which would ordinarily escape is retained. As gas (water vapor) escapes the water surface due to increased heat, it creates a pressure above the liquid, reducing the ease with which molecules may leave the liquid. This increased pressure then retards boiling even though temperatures may exceed 100°C. It has been found that an additional pressure of 760 mm Hg (15 lb) on the surface of water will raise the boiling point of water to 121°C. If the heat is intensified, high boiling points may be achieved. In order to kill some microorganisms, a temperature of 121°C must be maintained for 15 minutes or longer. If all microorganisms are killed, sterilization is said to have been accomplished.

An autoclave is basically a closed container supplied with steam or water. When additional heat is applied to the container, pressure (vapor pressure) builds up in the container, thus raising the boiling point.

LIQUIDS

Liquids and the properties they exhibit play an important role in body functions. The flow of these fluids through vessels or across membranes creates special problems for normal body functions. In addition to fluids in the body, we are faced with fluid properties and their effect in our everyday environment.

Fluids in many respects exhibit properties similar to gases. The major difference lies in the fact that the atoms and molecules of fluids are more dense than those of gases. Another difference is that liquids are practically incompressible.

Liquid Pressure

Liquids create pressure due to the weight they exhibit. *Pressure* is defined as force divided by a unit area. If a container, as shown in Figure B.8, has a volume of 1,000 cc, it has a base (the floor of the container) of 100 sq cm. If we fill this container with water, the entire weight of the water will push down on the base. This total weight is called the force. Because 1,000 cc of water weigh 1,000 g, a force of 1,000 g is pushing against the base. The base has an area of 100 sq cm (10 cm × 10 cm). Since pressure is defined as a force divided by a unit area $(P = F/A)$, we can now put the known units into the formula:

$$P = \frac{1,000 \text{ g}}{100 \text{ sq cm}} = 10 \text{ g sq cm}$$

Another formula for force is $P = A \times H \times D$, where P = pressure, A = area, H = height of the container, and D = density of the fluid. The two formulas for force, F, may now be combined and simplified as follows:

$$F = P \times A$$
Also $$F = A \times H \times D$$
Therefore $$P \times A = A \times H \times D$$

If both sides of the equation are divided by A, then the new formula for pressure would be $P = H \times D$. All that needs to be known now in order to determine the pressure of liquids is the density of the fluid and its height. This relationship is important for many clinical applications. For example, the pressure of fluids flowing from an enema container or intravenous apparatus is the product of fluid density and the height of the fluid volume. An enema container 4 ft above the patient has much greater pressure at the exiting nozzle than a bottle at 2 ft above the nozzle. The height of an intravenous bottle may be adjusted to permit an exact rate of flow per minute from the needle. The rate of flow is increased as the height of the bottle is increased. The distance in height between the upper level of fluid in the bottle and needle exit is called the *fluid head*, as shown in Figure B.9.

Factors other than pressure can affect the rate of flow of a fluid. By rate of flow we mean what volume of a fluid per unit of time flows past a

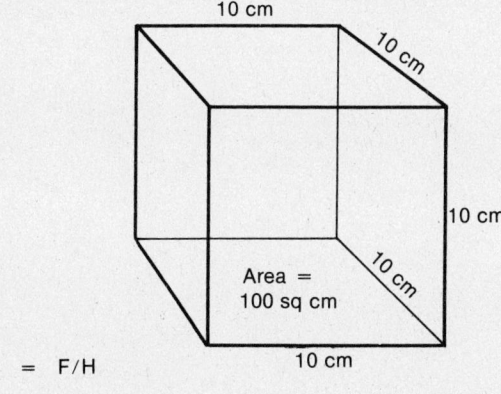

P = F/H

P = $\frac{1,000 \text{ g}}{100 \text{ sq cm}}$

P = 10 g/sq cm
or
P = H x D

P = 10 cm × 1 g/cc = 10 g/sq cm

Fig. B.8 The relationship between force and area, resulting in a unit called pressure.

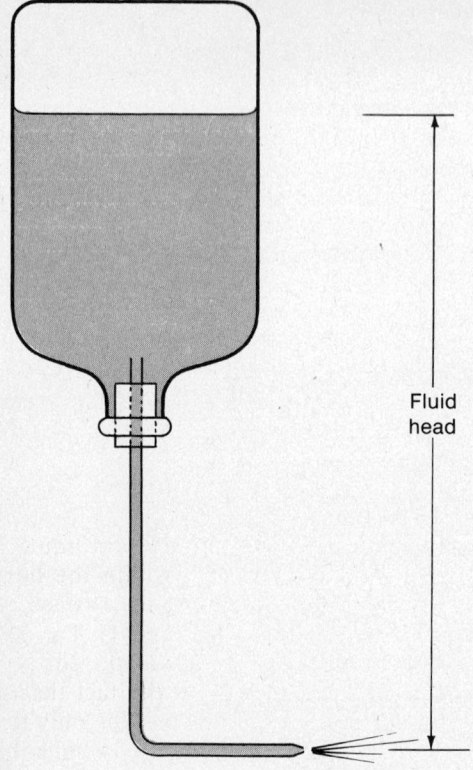

Fig. B.9 Fluid head is the height of a column of fluid above its exit.

given point. Other factors that affect the rate of flow include the diameter of tubes, length of tubes, size of the exiting nozzle, the pressure gradient, and fluid viscosity.

The *diameter of tubes* transferring fluids affects the rate of flow. A smaller diameter tube reduces the rate of flow, owing to friction resulting from fluid molecules flowing along the sides of the tube. The length of a tube may also reduce the rate of flow for the same reason. In the bloodstream, the smallest vessels, called capillaries, have a much reduced rate of flow compared to arteries that feed into them. Some of this reduced rate of flow can be attributed to small vessel diameter and length and the increased friction.

The *diameter of the exiting nozzle* influences the rate of flow. An outlet nozzle of large diameter will permit more fluid to flow per unit of time than will a smaller nozzle; assuming that other factors are constant. But the rate of flow from a smaller outlet may be greater than from a larger outlet if the head of fluid above the smaller outlet is greater.

Pressure Gradient

Pressure gradient is the difference in pressure at two points in a liquid. A liquid standing in a container has no pressure gradient, since the base and all points equidistant from the base have the same pressure. If the container, such as that shown in Figure B.9, is arranged so that the end of the tube is closed, there is also no pressure gradient. If the end of the tube is opened, the liquid will flow out due to gravitational pull. The rate of flow from the tube depends upon the head of the fluid and other factors just discussed. These factors in combinations create the pressure gradient.

Viscosity

Viscosity means resistance to flow. Some liquids have a varied viscosity due to temperature. Sugar syrup has a lesser viscosity when warm. The viscosity of a liquid may influence the pressure gradient, thereby influencing the rate of flow from the container.

A unit called *specific viscosity* is used to compare viscosities of liquids with water. The specific viscosity of water is 1.0. A liquid having a specific viscosity less than 1.0 such as benzene (0.65), has less resistance to flow than water. Olive oil has a specific viscosity of 84. These viscosity relationships have to be determined at the same temperature of 20°C (68° F).

Blood viscosity is influenced by temperature. A decrease in temperature increases the viscosity, and an increase in temperature decreases the viscosity. Factors other than temperature that alter blood viscosity include dehydration and an increase in soluble blood proteins.

The overall effect of increasing blood viscosity is a reduced rate of flow of blood into tissues. A greater stress is placed on the heart in an attempt to move the blood.

Molecular Forces in Liquids

All forms of matter are held together by gravitational forces. Water and other fluid molecules have a mutual attraction for each other. The closer molecules are together, the greater the attracting forces. For example, the force that holds water molecules together is greater than the attraction gas molecules have for each other. Gas molecules are much farther apart than are liquid molecules.

The term *cohesion* is used to describe the force that holds like molecules together. All substances exhibit some degree of cohesion, some more than others. The cohesive forces of mercury are so strong that when mercury is spilled onto a smooth surface the globules tend to form smooth spheres or little ball-like structures. Water behaves in a similar manner when dropped onto an oily surface.

A free falling liquid forms a ball-shaped mass. Raindrops, in the absence of a distorting wind, are round.

The attraction unlike molecules have for each other is called *adhesion*. If water is dropped onto a perfectly clean, oil-free surface, it will spread out on the surface rather than form a ball. In other words, there is a stronger adhesive force between water and the surface than there is cohesive force between water molecules.

Adhesives are a common part of our everyday life in the form of glue, mucilage, rubber cement, and so on. Adhesive tape exhibits the properties of both adhesion and cohesion—it will adhere to skin as well as to itself.

The adhesive properties of unlike molecules produce some phenomena that we must make allowances for—and in some instances can utilize.

The meniscus produced in a graduated cylinder is due to the adhesion of water to the glass sides. This is another example of adhesion being greater than cohesion. If the tube of glass is of very fine bore (small diameter), water will actually climb up the tube against the force of gravity. This phenomenon is called *capillarity*. Capillarity can be at times a useful process; other times it is distracting. It is by capillary action that fuel rises in a wick of a lamp. Water rises from lower soil layers to the surface by capillarity, and absorbant materials such as clothes and bandages rely on capillarity to soak up fluids.

A capillary tube is used to remove a small sample of blood from the fingertip to aid in estimating clotting time.

A specifically designed wicklike material is used as a surgical drain. By capillary action, extraneous fluids may be removed from a wound that could not be drained otherwise.

Adhesion is lessened with higher temperatures. An example of this is the application of solder to mend or hold together metals such as copper. Hot solder has little adhesion. If two pieces of copper with a thin film of molten solder between are allowed to cool, the adhesion is so great that the metal pieces cannot be pulled apart.

Surface Tension

Molecular forces at the surface of a liquid are different from those deeper within the liquid. The action of these forces at the surface produces a stretchable membrane effect. The cohesive force of water molecules at the surface is very great, partly because of the fact that this force is not altered by adhesion. The only thing at the surface that water molecules could adhere to is atmospheric gas.

Some substances such as soaps and detergents lower the surface tension.

The effect that some disinfectants have upon bacteria is due partly to their capability of lowering the surface tension of the bacterial cell wall, thus permitting a more complete covering of bacteria by the disinfectant.

Pascal's Law

The incompressible nature of water (and most other fluids) is utilized in several common devices including hydraulic lifts, chairs, and jacks.

Pascal's law states that pressure applied to any part of a confined liquid is transmitted undiminished to every other part of the container. Figure B.10 illustrates the principle.

The hydraulic jack is also an application of this principle. Figure B.10 illustrates how a very small force can be used to lift heavy weights.

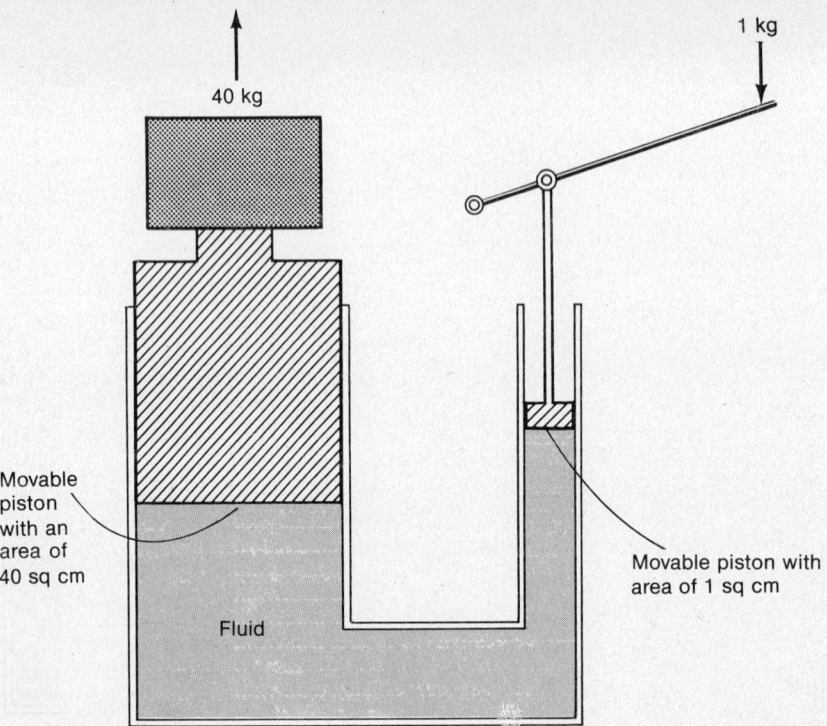

Fig. B.10 Pascal's law states that pressure applied to any part of a confined liquid is transmitted undiminished to every other part of the container. In this figure, if 1 kg of force is applied to 1 sq cm, all this force is transmitted to 40 sq cm in the other container. A total weight of 40 kg could therefore be raised with an initial force of only 1 kg.

Archimedes' Principle

It is common knowledge that objects appear to weigh less in water than in air. This loss in weight is called *buoyant force. Archimedes' principle* shows a specific relationship between a submerged object and its loss in weight. An object immersed in a liquid is buoyed (held) up by a force equal to the weight of the liquid it displaces.

A floating body loses all its weight—it displaces a weight of liquid equal to its entire weight. In order to float, a body must have an average density less than that of the liquid. Figure B.11 illustrates Archimedes' principle.

The buoyant force of liquids due to Archimedes' principle has been applied to several areas of the health field. Patients suffering from muscular disorders or joint conditions that do not permit the full weight of the person upon them gain some relief by being placed in a water bath deep enough to support a large portion of their weight. With weight removed, the patient can more freely exercise the muscle or joint.

ENERGY

The capacity to do work is called *energy*. This definition perhaps seems inadequate since we cannot assign physical attributes. Our concern in this text will be to explore those forms of energy that have the greatest influence upon the body. Theoretical concepts along with their mathematical formulas will be left for other times and other writers.

Two forms of energy have been discussed in other chapters—chemical energy in Appendix A,

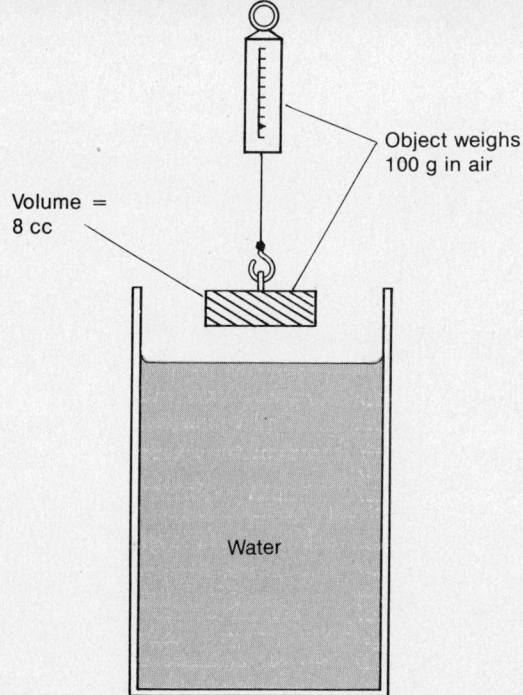
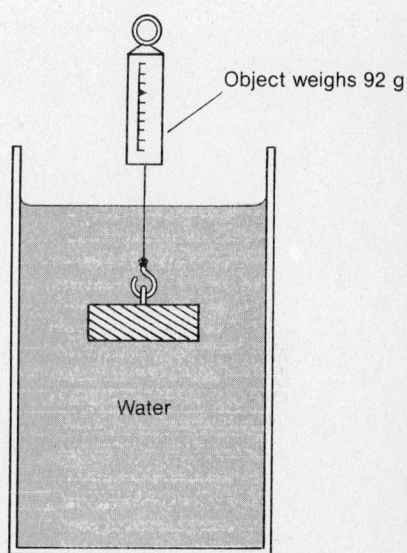

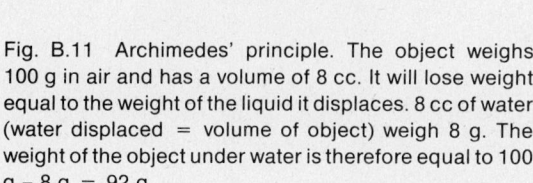

Fig. B.11 Archimedes' principle. The object weighs 100 g in air and has a volume of 8 cc. It will lose weight equal to the weight of the liquid it displaces. 8 cc of water (water displaced = volume of object) weigh 8 g. The weight of the object under water is therefore equal to 100 g − 8 g = 92 g.

and bioelectricity in Chapter 2. Other forms that will be discussed in this chapter include *heat*, *light*, and *electricity*.

HEAT

Heat is the form of energy that provides the vibratory motion of molecules. The different states of matter are a result of different energy levels at the molecular level. Gas molecules have a higher energy level than liquids. Liquid molecules have a higher energy level than solids. When the energy level of a substance is altered (cooled or heated), the state of matter may be changed.

Our concern with heat will be limited to how heat affects the body and its physiological functions.

The human body maintains a temperature very close to 37°C. A change of one degree of temperature is cause for concern. Minute temperature changes generally signal some body malfunction.

Body heat is a by-product of metabolism, the process by which the body releases energy from foods. Energy released from foods is captured by specialized chemical molecules and used to carry on the body's work. Some of the energy released is not captured and is converted into heat. The body utilizes this heat to maintain a constant body temperature. The purpose of body heat is primarily to create an optimal internal environment within which chemical reactions occur.

When too much heat is released, the body must initiate appropriate responses to get rid of the surplus. If insufficient heat is released, appropriate responses are set in motion to conserve heat.

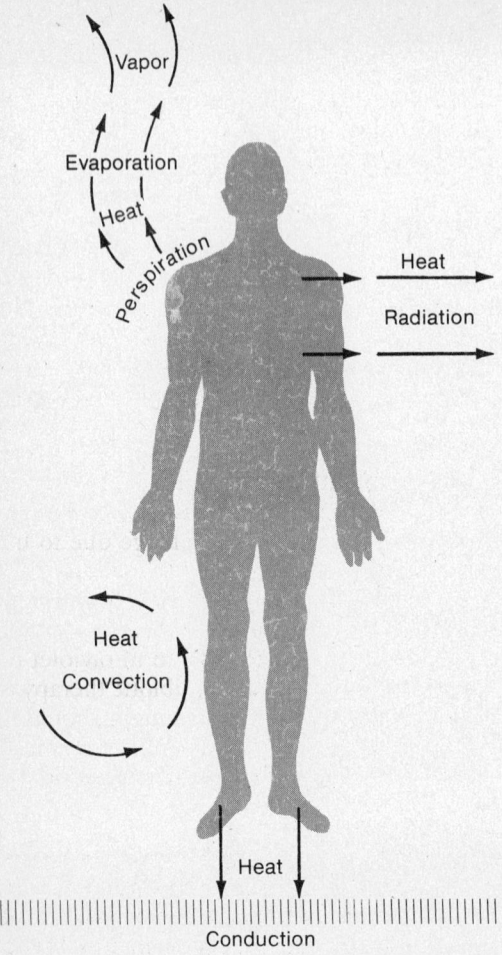

Fig. B.12 Major ways in which heat is lost from the body. These include evaporation, radiation, convection, and conduction.

Surplus Body Heat

Heat is lost from the body mainly by *conduction*, *convection*, and *evaporation* as illustrated in Figure B.12.

Conduction

The process of heat transfer from one object to another containing less heat is called *conduction*. For example, bare feet on a cold floor result in body heat loss. In order to reduce heat loss, body coverings made of insulating materials are used. Substances that reduce conduction include fabric, leather, and synthetics. Many natural fibers (wool, cotton, linen) are better insulators than synthetics (nylon, rayon, and so on).

Convection

Convection is a process similar to conduction except that heat is transferred by air currents rather than by direct contact with another surface. Cold air surrounding an uninsulated body will remove heat from the body. Clinical personnel are trained specifically to protect patients from heat loss due to convection. In extreme cases of heat loss (in temperatures of even 50° to 60°F) a person may develop *hypothermia*. The body temperature drops, requiring prompt action to prevent death. Hypothermia is also caused by conduction, particularly if clothing is wet in cold weather or if a person falls into cold water and is not quickly removed.

Evaporation

Evaporation is the conversion of liquid to the gaseous state accomplished by adding heat energy to sweat, changing it to a vapor (gas). Heat removed from the body in this conversion is a cooling process. Profuse sweating usually accompanies extreme physical activity. The higher metabolic rate of cells, particularly muscles, elevates body heat. One of the mechanisms for the removal of the excess heat is the evaporation of perspiration. (The measurement of the intensity and quantity of heat is discussed in Chapter 1.)

LIGHT

Light is a form of energy that influences the human body in a variety of ways. Light entering the eye and striking the retina stimulates specialized cells that convert the stimulus to nerve impulses. These nerve impulses arriving at the brain are converted into visual images called sight.

Ultraviolet, Infrared, and X Rays

Light, like some other forms of energy, travels in waves. The wavelength of visible light is between

4,000 and 8,000 *angstrom units* (Å). This range will be used here as a reference point. Wavelengths shorter than 4,000 Å to as short as 40 Å are *ultraviolet light*. Although we cannot see ultraviolet light, it can have an effect upon skin (sunburn, tanning). Some individuals are extremely sensitive to ultraviolet rays (*sunpoisoning*). A few antibiotics and other drugs for some reason make some individuals hypersensitive to ultraviolet rays (*phototoxicity*). Ultraviolet light is also essential for the production of vitamin D by the skin.

Light waves in the range of 1 to 40 Å are called *X rays*. These rays will penetrate different body tissues with different velocities. For example, X rays have greater difficulty penetrating bone than softer tissues. This phenomenon, coupled with the fact that X rays alter photographic plates, is an extremely valuable tool in medicine.

On the other end of the spectrum beyond visible light are *infrared* (heat) waves. Beyond these are radio waves, measured in meters.

Sunlight is approximately 50 percent infrared rays, 5 percent ultraviolet rays, and 45 percent visible rays.

Colors used in interiors and the level of light intensity are increasingly becoming a concern of psychologists. We often associate colors with sensation. For example, we say that red is a warm color and blue is a cold color. The colors used for interior walls of hospitals and schools are most often subdued. Sharp flashes of contrasting colors are avoided, presumably to create a quiet atmosphere, since we tend to associate the use of contrasting vivid colors with excitement.

The correct level of light intensity in a room should be influenced by the activity associated with the room. For example, a library would require a different light level than a hallway.

The desirable intensity of light in a room depends upon several variables, such as the type of work being performed, the nature of reflective surfaces, and to some degree upon a person's ability to see.

The rate of radiation of light is commonly measured in units called *candlepower*. One candlepower is the rate of light radiation from a specific kind of candle (called a standard candle) in a specific direction.

The term commonly used to measure the intensity of illumination is the *footcandle*. One footcandle of illumination is the amount of light that falls on a fixed point one foot from a standard candle.

The intensity of illumination decreases proportionately as the distance from the light source increases.

Physiological Effects of Light

Human tissue may be adversely affected by exposure to the extremely long light waves (infrared) and the extremely short (ultraviolet) light waves. Although visible light is suspected to influence some tissues in specific situations, documentation is rather weak. In controlled experiments, mice exposed to concentrated ultraviolet light for extended periods die. The specific metabolic disruption caused by ultraviolet light has not been identified. There are also documented cases of severe retina cell damage due to ultraviolet light.

Recently (since the advent of antibiotics) physicians have recognized that some antibiotics make the skin hypersensitive to ultraviolet rays. Therefore, persons under antibiotic therapy and who are overly sensitive to sunlight should stay out of direct sunlight during and sometimes for several weeks after the medication period.

ELECTRICITY

All matter is held together by electrical forces. This discussion will pursue in general terms the nature of these electrical forces and how they can be utilized to help accomplish work. We will also see that the misuse of these electrical forces may be dangerous.

Atoms, which make up all matter, are neutral. That is, they contain equal numbers of particles called electrons and protons. (See Appendix A.) If a substance has a surplus of electrons, we say it is negatively charged. If a substance has a deficiency of negative charges, we say that it is positively charged. The manner in which these charges behave within and on the surface of substances is called electricity. Two basic types are *static* and *current* electricity.

Static Electricity

Static electricity is caused by the accumulation of an excess of electrons on the surface of substances called *nonconductors* (insulators). Generally, static electricity is considered a nuisance due to the mild shock one may receive after coming into contact with a *grounded* object. The term "ground" refers literally to the earth. The earth acts as a gigantic reservoir of electrical charges. A surplus of electrons on an object in contact with the earth (ground) will leak from the object into the ground. An object with a deficit of electrons will tend to gain electrons from the ground. The mild electrical shock experienced after walking on a nylon carpet and then touching another object (with fewer electrons on the surface) is an example of grounding.

Nonconductors, such as synthetic fibers, rubber, plastic, and glass, do not permit electrons to flow through them. These substances however, permit excesses of electrons to accumulate on their surfaces. The human body may act as a collector of such charges. When these charged objects come in contact with the ground or another object with fewer electrons, a surge of electrons produces the mild shock.

Static electricity may accumulate on a person's clothing or shoes to a sufficiently high concentration to produce a spark when a grounded surface is touched. Hospital personnel may accumulate static electricity on their bodies by simply walking along a corridor. If a patient is then touched without first touching the bed frame or some other grounded object, the patient will experience a mild, although discomforting, shock.

Static electricity poses a far more critical problem in operating rooms where explosive mixtures of oxygen and some anesthetics may exist. A spark of static electricity is sufficient to ignite certain mixtures, causing explosions or fires. Although many anesthetics used today are nonexplosive, hospital personnel should be familiar with the potential dangers of static electricity.

Precautions that hospitals may exercise in operating rooms include:

1. making sure that all equipment, including beds and bedding, is grounded so that the electric charges are drained off before they reach sufficient levels to cause a discharge;
2. providing grounded flooring materials free of waxes or polishes (floor wax acts as an insulator, which allows static charges to build up);
3. requiring personnel to wear special shoes or shoe coverings that will conduct electric charges into a grounded floor;
4. requiring all personnel in an operating room (including the patient) to be grounded by being attached to a flexible conductive cable connected with the ground.

Many hospitals do not provide all these precautionary measures. Nevertheless, static electricity discharge in a potentially explosive environment is a very real danger to the hospital staff and the patient.

Static electricity is minimized in a high-humidity environment. Water vapor is a conductor that helps drain excess electrical charges. There are also substances on the market that can be added to clothes during washing or drying which reduce the accumulation of static charges.

Current Electricity

Current electricity is produced in a much different manner from static electricity. It is not within the scope of this text to pursue the theory and ways in which current electricity is produced. However, it is essential to be aware of some of the dangers and the precautions that must be observed with current electricity.

The major difference between static and current electricity is that current electricity will flow along a conductor. A *conductor* is defined as any substance that will allow the transfer of electricity along its length. Conductors are primarily metallic substances. Insulators permit electrical charges such as static electricity to build up on their surfaces but do not permit the conduction of these charges.

The flow of electric current through a conductor, such as a copper wire, is somewhat analogous to water flowing through a pipe. It is essential that some reservoir of energy, which we call a *generator* or a battery, produces or accumulates a large quantity of electrical charges. These electrical charges traveling along a conductor are an *electric current*.

Electric current flows from the generator or battery along a conductor then back to the source.

Dangers of Electricity

Many of us are almost completely dependent upon electric power to cook meals, heat and light our homes, and run work-saving devices. Electric power is safe if certain precautions are exercised.

Electric wires that have become frayed, or wires from which the insulation has become loose or cracked, are dangerous both from the standpoint of possible electric shock and potential fire hazard.

An electric shock may be only an unpleasant sensation. However, under certain circumstances it might be fatal. If a person is standing on a well-grounded surface and touches a "hot wire" the body becomes a part of the electrical pathway, which may prove fatal.

Moist ground acts as a part of an electrical pathway back to the source of electricity.

Potential danger also exists with some kinds of switches used in combustible environments. You may have observed a small flash of light in a switch when you turn off lights. This flash of heat is capable of igniting fumes of ether, alcohols, and other highly flammable substances.

Great precaution must be exercised around oxygen, not because oxygen is combustible, which it is not, but rather because any combustible substance, such as clothing or bedding, is made almost explosively combustible in a high-oxygen atmosphere. A spark from a switch or even static electricity may cause serious consequences.

There is even a greater danger in producing an open flame near areas utilizing pure oxygen.

Switches and electrical appliances used near combustible gases or oxygen should be explosion-proof types. These appliances have motors and switches that are isolated from the environment. Ether stored in an ordinary household-type refrigerator makes this appliance a potential bomb.

SUMMARY AND REVIEW

Basic physical concepts deal with physical characteristics of matter (gases, liquids, solids) and energy.

A. Gases are substances whose molecules have a higher level of energy than other forms of matter.
 1. Gas pressure is the force exerted upon a specific area.
 2. Gas pressure may be measured with an instrument called a barometer.
 3. Boyle's law states that if the temperature of a contained gas remains constant, the volume varies inversely as the pressure exerted upon it.
 4. Charles' law states that if gases are heated, they expand in a direct proportion to the heat applied.
 5. Dalton's law states that in a mixture of gases, each gas exhibits its own pressure. The total pressure of a mixture of gases in a container is the sum of the individual partial pressures.
 6. The siphon is a device that permits liquids at a specific elevation to be transferred to a lower elevation.
 7. Evaporation is a cooling process that results from matter changing from the liquid state to the gaseous state.
 8. Boiling is the phenomenon resulting in liquid molecules attaining sufficient energy to escape the surface.
 9. The autoclave is a device used to raise the boiling point of water.
B. Liquids
 1. Liquid pressure is a measure of force on each square unit of the base of a container of liquid.
 2. Pressure gradient refers to differences in pressure between two given points.
 3. Viscosity refers to the resistance to flow, usually compared to free-flowing water.
 4. Molecular forces in liquids
    a. Cohesion is the attraction like molecules have for each other.

      b. Adhesion is the attraction unlike molecules have for each other.
  5. Surface tension is the force at liquid surfaces due to cohesive attraction of molecules.
  6. Archimedes' principle states that an object immersed in a liquid loses weight equal to the weight of the liquid displaced.
C. Energy is the capacity to do work.
D. Heat is a form of energy required by the body.
  1. Surplus heat is removed by a variety of processes, which includes evaporation of perspiration.
  2. Conduction is the process of heat removal by direct contact with substances with lower heat energy.
  3. Convection is a type of heat removal by circulating air currents.
E. Light is a form of energy transmitted from the sun and from combustion.
  1. Light travels in extremely short waves measured in units called angstrom units.
  2. Visible light is made up of waves of 4,000 to 8,000 Å.
  3. Ultraviolet rays are invisible, with wavelengths shorter than visible rays. X rays are shorter than ultraviolet rays. Infrared rays have a longer wavelength than visible rays and are also invisible to the eye.
  4. Light affects the body in a variety of ways; it induces vitamin D synthesis in the skin, causes sunburn and tanning, and may cause phototoxicity in conjunction with some antibiotics.
F. Electricity is a form of energy transmitted by the movement of electrons within and on the surface of substances.
  1. Static electricity is due to electrial charges that build up on the surface of nonconductors.
  2. Current electricity is due to the flow of electrons along a conductor, usually a metal. An electrical ground is a conducting pathway to the earth.

APPENDIX C
Discussion of the Color Plates

PLATE 1 - MUSCLES OF THE HEAD AND NECK

The phrase "beauty is only skin deep" might well apply in a casual observation of this drawing. Upon close observation, however, it will soon become evident that beauty is affected by an almost infinite number of facial expressions created by the contraction of head and neck muscles. Few body muscles perform the variety of functions provided by those of the head, face, and neck. Eating, mastication, swallowing, and speech are but a few of these functions. Muscle is *contractile* tissue. All the functions just mentioned, plus many others, are accomplished by the process of muscle shortening. The term *voluntary muscle* is perhaps more appropriate than *skeletal muscle* when referring to face and head muscles. By exercising voluntary control, one can exhibit facial expressions of unlimited variety—rage, joy, sorrow, compassion are but a few. Facial expression and body movement play a significant role in speechless communication. Little is left to be said when the eyebrows are raised, the head tilted slightly to one side (contracting one sternocleido-mastoid muscle) and the shoulders shrugged. This plate also illustrates the integral relationship that exists between bone and muscle. Note here and in Figure 6.12 how the large, thick masseter muscle wraps around the lower jaw, holding it in a sort of sling. Contraction of this muscle exerts more than adequate pressure on the mandible to shread, tear, and chew even the toughest of foods. Two very obvious circular muscles (the *orbicularis oculi*, around the eye, and *orbicularis oris*, around the mouth) contract to close these openings. Variable contraction of these muscles, coordinated with contraction of the sheetlike platysma of the neck, produce some very obvious expressions valuable in human communication. In order to fully appreciate the effect facial muscles contribute to beauty and personality, one need only observe a patient whose facial muscles have been paralyzed. We often use the terms "blank," "expressionless," and "lifeless" to describe such a condition.

PLATE 2 - PRINCIPAL SUPERFICIAL MUSCLES OF THE ANTERIOR (VENTRAL) TRUNK

This view of the torso illustrates the magnificent engineering of muscles that give the body its characteristic shape and the precise location of muscles that permit a wide range of body movement. In Chapter 6, reference is made to the fact that skeletal muscles represent the largest mass of tissue in the body. Note the layer upon layer of muscles, each precisely designed to carry out a specific function. This layering is best illustrated here in the abdominal muscles. These muscles, the external oblique, the rectus abdominis, the internal oblique, and the transverse abdominal, form the anterior wall of the torso. Their major function is to support the viscera, which lies directly posterior to them. The viscera includes the organs of disgestion: stomach, intestines, liver, etc. Another function of these particular muscles is to flex, or bend, the torso forward. These powerful muscles are at least partially responsible for some of the deformities in posture associated with old age. Although it would appear that these layers of muscle would be impenetrable, often there is a separation of the connective tissue that holds together bundles of these muscles. Some of the viscera, namely a segment of the small intestine, may push out through the weakened wall to the exterior. This is called a *hernia*. The massive muscles of the upper chest and shoulders have the dominant functions of locomotion and diversity of motion. The deltoid muscle, for example, lifts the arm. This motion, coupled with the contraction of the pectoralis major and other muscles, permits the lower arm to be rotated with the hand describing a circle. Another well-illustrated feature in this drawing is the connective tissue junction of one muscle to another. These junctions shown as white bands called *aponeuroses* are sheetlike tendons. They are important in that they define certain limits of a muscle's activity.

PLATE 3 - PRINCIPAL SUPERFICIAL MUSCLES OF THE POSTERIOR (DORSAL) TRUNK

Muscles of the back, like those of the chest, are massive and capable of incredible power when contracted. The upper back muscles pull the head backward, lift the shoulders, abduct the arm, and help arch the back. The power these muscles are capable of producing is violently exhibited in Tetanus, a bacteria induced muscular disease. Toxin from the bacterium C. tetani causes many muscles of the body to uncontrollably contract. This includes the masseter muscle, hence the common name for the disease, *lock-jaw*. Because of their bulk and shape the back muscles may, under the influence of tetanus, contract so forcibly that the unfortunate victim's back is broken. Note in this drawing, as in Plate 2, the presence of aponeuroses. One needs to know little Latin to understand the term *gluteus maximus*. What better term could be used to identify the largest and most powerful muscle in the body? This, the major hip muscle, extends the femur during walking, running, swimming, etc. The full power this muscle is capable of producing is like the power of a horse's kick.

PLATE 4 - A SECTION OF SKIN AND THE MUSCULATURE OF THE ARM AND LEG

The specific physical relationship between skin and muscle is illustrated here. The skin is such an important organ that special attention is given to it in Chapter 4. Figures 4.9 and 4.10 illustrate skin components not shown in this plate. The transverse sections of the arm and leg very effectively show the compact—and at first glance, curious—arrangement of some rather large muscles. Muscles that produce special activities are arranged in such a way that they do not hinder other muscle activity. Perhaps you may have heard of someone being "muscle bound." In this case, some muscles are overdeveloped to the point that they do hinder other musclar action. Pumping iron (weight lifting), if not correctly pursued, might lead to this difficulty. The muscular arrangement and size illustrated here is typical of the average person—it might be interesting to speculate about the appearance of a transverse section of the appendages of a Mr. Universe.

Two excellently illustrated features of muscle-bone relationship should be pointed out in these drawings. First, note the white tendons that lead from the fingers and toes up to arm and leg muscles. The tendon attachments to bones at these locations are called *insertions*, the bone to which they join will be moved when the muscle contracts. The criss-crossed bands of connective tissue at the wrist and ankle are *ligaments*. These tough bands bind many small bones into a compact package providing intricate and delicate movements of the hand and foot. Here again, one must be impressed with the massive bulk of muscle compared to other kinds of tissue. What is not shown in Plate 4 is the extensive network of blood vessels required to carry nourishment and oxygen to muscle cells and to carry waste products away. Muscle is not only the most abundant tissue in the body, muscle cells are among the most demanding for nutrition.

The higher rate of respiration of muscle cells during vigorous exercise often requires that other less important functions (at the time) be halted or at least reduced. For example, blood may be shunted away from the digestive system during exercise, thus raising the blood pressure at muscle. Higher blood pressure speeds nutrients and oxygen to muscle cells.

PLATE 5 - POSITION OF ORGANS OF THE TRUNK

This drawing illustrates how trunk organs would appear if the torso was sliced in transverse section. The close physical relationship of the lungs and heart is shown in the upper section. The physiological interrelationship of these organs is discussed in Chapters 10 and 11 (Heart and Blood) and Chapter 17 (Respiration). Although the heart and lungs fit rather snugly in the thoracic cavity, they are separated by synovial membranes. In fact, each lung is enclosed separately. In Chapter 17 (Figure 17.2) particularly observe the pleura. In this plate the parietal pleura is clinging to the thoracic cavity wall (on the rib side). The visceral pleura clings tightly to the lung. The relative size of different organs making up the viscera is well illustrated here. It is one thing to be told (or to read) that the liver is the largest single organ in the body. It is another thing to actually observe the liver in a dissected animal or as illustrated here. The abdominal cavity, with the diaphragm as a cap, contains, in addition to the liver, all the organs of digestion. The stomach, small intestine, and large intestine are firmly packed into the cavity. In addition, the excretory and reproductive organs are tucked around and between other viscera. The abdominal cavity is certainly distensible, as demonstrated in pregnancy. It doesn't tax the imagination much to realize that a 6- or 7-pound fetus requires substantial space in the uterus, which bulges into the abdominal cavity. The protective function provided by skeletal components (Chapter 6) is also illustrated here. Notice how the ribs encircle the contents of the thoracic cavity. Vertebrae of the spinal column, shown in each section, not only protect the spinal cord but also cumulatively provide a semi-rigid rod for the erect body. This plate emphasizes so many body systems and their interrelationships that frequent reference to it may prove very helpful.

PLATE 6 - SPINAL CORD, NERVE PATHWAYS, AND ASSOCIATED TISSUES

This plate portrays a panoramic view of the spinal cord, its covering (the *menenges*), and spinal nerve fibers. Although this view represents only a small part of the nervous system, the complexity of the entire system, both anatomically and physiologically, presents a challenge to even the most ambitious student. The structures shown here are all involved in a process known as a *reflex*. This plate, along with Figures 7.9 and 7.10 in Chapter 7, illustrates the precise nature of nerve pathways involved in reflex acts. Although the representation here may at first glance appear to be complicated, the fact is that this shows one of the simplest nerve pathways. An obvious feature of the nerve segments shown here is the bilateral symmetry of the spinal cord. Notice how nerves enter and leave both sides. Sensory neurons enter the spinal cord via posterior (dorsal) root ganglia. Motor neurons exit via anterior (ventral) roots. If we follow the motor neurons leaving the spinal cord on the left side of the illustration (the anatomical right side), we see that they terminate in *synaptic knobs*. Remember the neuromuscular junction discussed in Chapter 7? These synaptic knobs reach into muscle and other tissue where they secrete transmitter substances (acetylcholine is one). These substances initiate muscle contraction, glandular secretion, and a host of other functions. This plate shows in excellent detail how well the spinal cord is protected. The bony ring (vertebrae) provides a secure lodging for delicate cord tissue. In addition, the three-layered meninges, a close-adhering tough membrane, provides additional protection. The autonomic nervous system, you may recall, is divided into two parts, the sympathetic and the parasympathetic systems. Observe, in this plate, the location of sympathetic ganglia lying along and very near the vertebral column. The synaptic knobs are shown here overlapping the vertebrae. Their termination is muscle or gland tissue; the synaptic knobs do not enter vertebrae.

PLATE 7 - CIRCULATION—THE HEART AND BLOOD VESSELS

This is an illustration of the major components of the circulatory system. Blood is squeezed out of the heart (left ventricle) into muscular tubes called arteries. Arteries sub-divide into smaller, less muscular arterioles, then into capillaries that have walls only one cell thick. Here nutrients seep out around tissue cells, then into the cells by diffusion, osmosis, and active transport. Wastes, by similar processes, seep into capillaries that unite to form venules. Venules drain into larger vessels called veins which return blood to the heart. The amazing durability of heart muscle cannot be overemphasized. It begins to pump blood very early in fetal life and continues for an average of 70+ years. An average heartbeat of 72 times per second pumps approximately 5 liters of blood, under considerable pressure, into the aorta each minute. This is the normal resting rate. Vigorous exercise may increase this volume per minute sixfold. Textbooks emphasize that heart muscle actually rests longer than it contracts, nevertheless, one cannot help but be amazed at the durability of this totally essential organ. Recall from the discussion in Chapter 9 that the right side of the heart (atrium and ventricle) collects and pumps impure (low-oxygen) blood to lung capillaries. Pure (oxygenated) blood, still under pressure from the right ventricle, enters the left side of the heart to be pumped to the tissues. Observe the sectioned artery and note the layered muscles necessary to withstand high pressures. Elastic tissue here permits the artery to bulge under pressure, then return to its normal size once the pressure is released by cardiac diastol. Nearby, observe the sectional vein. Much of the pressure in the aorta is dissipated along the tortuous routes leading to tissues. Veinous pressure is much less than arterial pressure. Thus, less muscular support is needed around veins. Another major difference between arteries and veins, as shown here, is the presence of valves in veins. Since blood pressure in veins is very low, without one-way valves blood would tend to flow backward toward the tissues. The pull of gravity when one stands is greater than venous pressure. Valves in veins, along with some help from muscle contraction, keep blood moving toward the heart. Notice the minute arterioles and venules in the muscle layers of artery and vein walls. Cells that make up this tissue, like any other tissue, must receive nutrients and oxygen. Only the innermost layer, the *tunica intima*, is capable of taking some nutrition from blood flowing in the artery.

PLATE 8 - DIGESTIVE SYSTEM

Here in a single wide-screen view one can observe most of the organs of digestion. Cutaway segments reveal the internal structure of the stomach and intestines. Food is propelled through the tubular pathway by rhythmic muscular contractions called *peristalsis*. Each segment of the route provides its own special treatment of the ingested material. The teeth shred, crush, and tear food into small particles. The stomach continues the mixing process, making foods more soluble, and in addition secretes enzymes to begin the process of hydrolysis. The small intestine secretes enzymes, plus it accepts enzymes from the pancreas and bile from the liver. The ultimate end products, mostly small molecules, are then absorbed into the bloodstream. Undigested substances continue into the large intestine, eventually to be discarded through the bowel. Each segment of the digestive tract (alimentary canal or gastrointestinal tract) is treated separately and in detail in Chapter 13. An interesting phenomenon occurs along the tract. Food is not permitted to rush uninhibited along the route. Sphincters (circular muscles) act as gates that temporarily halt or impede foods so that mixing may be more complete or so that enzyme action can take place. The villi of the small intestine also need time to absorb the physically and chemically modified foodstuff. Make special note of the fingerlike villi protruding into the lumen of the small intestine. They are precisely engineered to absorb nutrients which immediately enter capillaries of the blood stream. Most of the absorbed nutrients are then carried directly to the liver via the hepatic portal blood system. The liver is capable of detoxifying the absorbed nutrients and, along with pancreatic hormones (insulin and glucagon), regulates the concentration of some substances (glucose in particular) before they enter the inferior vena cava, then ultimately are carried to body tissues. Notice the circular and longitudinal muscle fibers shown in the large intestine. These fibers (found also in the small intestine) when rhythmically contracted, produce peristaltic motion to propel substances along the tract. Although one is tempted to extol each body system as being the most important, medical science has found ways, if necessary, to completely by-pass many body systems. Prepared nutrients in proper concentrations may be introduced directly into the bloodstream. Perhaps this procedure should be looked upon as only a minor miracle since you and I spent, on the average, nine months as a fetus, receiving nutrients from maternal blood.

PLATE 9 - KIDNEY

The drawing of kidneys and associated tissues shown here dramatically illustrates the anatomical relationship of the kidneys and major blood vessels. Observe that the kidneys are not in the main line of blood circulation. The renal artery arises as an offshoot from the descending aorta. In Chapter 15, it was pointed out that about 1200 milliliters of blood enters both kidneys per minute. Since the total cardiac output per minute is 5 to 6 liters, it is obvious that only about 20 percent of the total blood volume leaving the heart passes into the renal arteries to be purified by kidney structures. The major purifying unit of the kidney is the nephron. Only one (of several million) is illustrated here at the upper right of the drawing. Excess substances, such as water and electrolytes, and nitrogen wastes (urea) are removed from blood. This mixture of substances, called urine, is drained by a system of ducts and tubes into the urinary bladder. Purified blood leaving nephrons enters renal veins which empty into the inferior vena cava. It is interesting to speculate about how this pathway evolved. Purified, clean blood is emptied into a dirty stream. Blood in the inferior vena cava, returning to the heart from tissues, contains impurities dervived from cellular respiration.

Whatever the reason for this design, the kidneys (by constantly sampling aortic blood) perform adequately to reduce the concentrations of potentially harmful substances in the blood. Normal kidney function perfectly describes homeostasis. The internal environment of the body is stabalized by nephron activity. The full impact of this statement may not be significant until you realize that every drop of sweat and every drop of urine comes from blood. When kidneys fail, excess fluid, electrolytes, and wastes rapidly accumulate in the blood, creating stress for many body organs. The ultimate consequence of untreated kidney failure is death.

PLATE 10 - RESPIRATORY SYSTEM

The respiratory apparatus, shown here, once again points out the intimate anatomical relationship of the circulatory system with body systems (see Plates 8 and 9 for other examples of this relationship). One cannot help being impressed with the efficiency exhibited by this system. Although these tissues are not actually involved in the physical process of breathing (the diaphragm and intercostal muscles accomplish this), they condition inhaled air and direct oxygen contained in it to blood capillaries. Observe the elaborate design of the conchae of the nose. These shelves of bone covered with mucous membranes provide an extensive surface area over which fresh air flows. This air is warmed and filtered by the highly vascular membranes with their sticky surfaces. Also observe the rigid cartilage framework on the anterior surface of the trachea. This prevents the trachea from collapsing when we inhale. The posterior port of the trachea, lacking this cartilage support, closely adheres to the esophagus. The detail of the entire respiratory system is discussed in Chapter 17. Here, also, you will find a discussion of this system's role in body homeostasis. The ultimate respiratory membrane is the alveolus shown on the left, surrounded by blood capillaries. This is where the action is, so as to speak.

Oxygen diffuses across alveolar membranes into blood capillaries and carbon dioxide accumulated from the tissues leaves blood capillaries and diffuses into the alveoli. A respiratory membrane must be thin, moist, and have extenisve surface area. The alveoli meet these standards exceedingly well. It has been estimated that the total alveolar surface is equivalent to a sheet of elastic tissue one cell thick that could cover an area of about 60 square meters (sufficient to cover the end zone of a football field). A fourth requirement of respiratory tissues is that they must be located very close to the circulatory system. The only thing that separates blood capillaries and alveolar tissue is a thin layer of fluid. The total distance between the two probably doesn't exceed one micron. The process of oxygen and carbon dioxide transfer across these tissues is purely physical. Gases follow pressure gradients, high pressure diffuses to lower pressure. You may wish to refer once again to Chapter 17 and the discussion of partial pressure. This is the driving force of gases across respiratory membranes.

PLATE 11 - MALE REPRODUCTIVE SYSTEM

The basic function of the male reproductive system shown here is to produce the male sex cells, sperm, and to provide an access by which these cells can be introduced into the female vagina. Of all animals, mammals' reproductive systems provide the surest way to place the sperm in the near vicinity of an egg. Sperm originates (by meiosis) from cells that line *seminiferous tubules* of the testes. Smooth muscle contractions of the *ductus deferens* propel sperm toward the urethra. Here the sperm is joined by alkaline secretions from seminal vesicles and prostate gland, producing a mixture called *semen*. Sexual stimulation produces several effects. The penis becomes engorged with blood, making it rigid and erect and hence capable of penetrating the female vaginal opening. When the male orgasm is reached, a rhythmic contraction of muscle in the ejaculatory ducts propels semen toward the tip of the penis. As related in Chapter 18, the male reproductive system has yet another function, the production and secretion of the male sex hormone, testosterone. Cells located in the spaces between the seminiferous tubules (interstitial cells of Leydig) produce testosterone. Because testosterone is a hormone, it diffuses directly into the blood stream. Its influence in producing male characteristics is discussed on pages 290-291. Observe the sperm cells in the drawing. The tails, or *flagella,* although fully formed as shown here, have no motile function until the sperm reaches the area of the seminal vesicles. It has been determined that the low pH of the seminiferous tubules and ductus deferens in some manner inhibits flagellar movement. The alkaline secretion emptied from the seminal vesicles and prostate gland either removes the inhibitor or perhaps stimulates the flagella to begin a whiplike motion. Here, all of a sudden, we find wandering cells free of other tissue and capable of independent, though brief, existence. A single ejaculation releases millions of these microscopic cells. This is perhaps, an area in which man has not evolved beyond some lower animal forms. Why do you suppose millions of sperm are released simultaneously when only one is required to join an egg in fertilization? Part of the answer is that the union of the egg and a sperm is a random, almost accidental, event. Another factor is the extremely small size of the egg. Although it is larger than sperm cells (about the size of a pin head), one can imagine the odds of a single sperm finding this tiny cell in the female uterus. This process is discussed further in Plate 12.

PLATE 12 - FEMALE REPRODUCTIVE SYSTEM

The female reproductive system illustrated here plays the major role in all human reproduction. This system is not only responsible for producing a sex cell, the egg, but must nourish, protect, and finally release another human being into the world. Within a framework of modern technology and efficiency, one must seriously question what appears to be wasteful loss of materials and energy in the human female reproductive cycles. On the other hand, we must admit, it works. In fact, it works so well that for many, many centuries, mankind has sought ways to inhibit the process. These contraceptive attempts include inhibiting the formation and release of an egg, instituting barriers to keep the sperm away from the egg, stimulating tissues so that they will not nourish the fertilized egg, and abstaining from sexual intercourse. If these methods fail (or are not used) and an unwanted fetus develops, mankind has invented abortion to prevent birth. The progression of events that leads to the development of a fetus is directed by the ebb and flow of hormones. you may recall from the discussion in Chapter 18 on human reproduction and the discussion in Chapter 12 on the endocrine system that hormone action directs the female cycle just as surely as a concert master directs a difficult musical passage. When the major hormones cease to be released, the process stops. Let us observe for a moment some of the potential hazards to the production of a fetus. The ovary, shown in this plate as an almond-shaped structure, normally after puberty prepares and releases a mature egg (ova) each month for the next 30-35 years. This tiny cell (with only 23 chromosomes), forcibly ejected from the ovary, is sucked up by the octopuslike fimbriae at the ends of the uterine tubes (also called fallopian tubes or ovaducts). Rhythmic contractions of smooth muscles that line the uterine tubes propel the egg in the direction of the uterus. If sperm is present here at the proper time, fertilization may occur. If sperm is not present, the egg continues its journey, only to die. The stage is then re-set for the process to begin all over again.

If fertilization results, the cell (now with 48 chromosomes—23 from each sex cell) immediately begins the process of mitosis. The uterine tube continues to propel the accumulating mass of cells toward the uterus. What follows is explained in the discussion of Plate 13.

PLATE 13 - CONCEPTION AND IMPLANTATION

This illustration depicts a single sperm penetrating an ovum and the series of events which follow. Although millions of sperm are present in each ejaculate, only one sperm unites with the ovum. Other than the fact that the single ovum may present a very small target for a smaller number of sperm, there are perhaps other factors that require the great numbers normally present. When the ovum is ejected from the ovary, it carries with it a coat of follicle cells. This coat and the plasma membrane of the ovum must be penetrated by a sperm in order for it to unite its nucleus with ovum nuclein. Sperm heads contain the enzyme *hyaluronidase,* which is capable of digesting the cementing substances holding the layers of cells attached to the ovum. If the sperm count falls below about 35 million per ml, there is apparently insufficient hyaluronidase for even a single sperm to penetrate the ovum.

After a sperm has penetrated the egg, mitosis begins immediately, first resulting in 2 diploid cells then 4, 8, 16, and so forth until a mass of cells are accumulated. This mass of cells *(blastocyst)* after one to three days reaches the *endometrium.* By this time, as a result of hormone influences, the endometrium is prepared to receive the blastocyst. The endometrium shown here is a highly vascular, spongy, thick tissue. The blastocyst, by enzyme action, digests a segment of the endometrium and implants itself, thus initiating pregnancy. As the embryo grows, it and the attending membranes bulge into the uterine cavity. The developing embryo, for an average of nine months, will receive its nutrition and oxygen from blood capillaries in the uterine tissue. Fetal respiratory wastes will diffuse into these capillaries to be carried away by the mother's venous blood system. This process is discussed in more detail on page 295.

Glossary

abdominal cavity (ab-dom'i-nal) the part of the body located between the diaphragm and the pelvis

abdomino-pelvic cavity (ab-dom'i-no pel'vik) cavity containing organs inferior to the diaphragm

abductor (ab-duk'tor) any muscle that functions without a movement away from the midline of the body; the act of turning outward

accommodation (ah-kom-mo-da'shun) a change in the shape of the eye lens so that vision is more acute; an adjustment of the eye lens for various distances; focusing

acetabulum (as-e-tab'u-lum) the rounded cavity of the oscoxae that receives the head of the femus

acetylcholine (as-e-til-ko'lēn) an ester of choline released from certain nerve endings thought to play a role in transmission of nerve impulses at synapses and myoneural junctions; easily destroyed by the enzyme cholinesterase

achalasia (ak-ah-la'ze-ah) failure to relax muscles, such as sphincters

Achilles reflex (ah-kil'ēz) reflex of the Achilles tendon which inhibits further contraction

achondroplasia (ah-kon-dro-pla'se-ah) defect in the formation of cartilage at the epiphyses of long bones, producing a form of dwarfism

acid (as'id) any substance releasing hydrogen ions and neutralizing basic substances; a pH less than 7

acidosis (as-i-do'sis) a serious disorder of body chemistry in which the normal alkaline substances of the blood are reduced in amount

acne (ak'ne) a skin disease in which the sebaceous (oil) glands become inflamed; a common type often forms elevations called papules (pimples)

acoustic nerve (ah-koos'tik) nerve pertaining to sound or the sense of hearing

acromegaly (ak-ro-meg'ah-le) enlargement of the extremities and jaw owing to hypersecretion of somatotrophic or growth hormone from the anterior pituitary after full growth has been achieved

acromion process (ah-kro'me-on) the flattened projection that extends from the shoulder blades (scapula) to form the top of the shoulder itself

actin (ak'tin) a muscle protein responsible for the shortening of the muscle when it contracts. Forms a combination with the protein myosin when muscle contracts

action potential the measurable electrical charges associated with conduction of a nerve impulse or contraction of a muscle

active transport using carriers, energy (ATP), and enzymes of cell to cause a substance to cross a membrane

acute renal shutdown (ah-kut' re'nal) kidney failure resulting in death

Addison's disease (ad'i-sonz) one due to deficiency in the secretion of adrenocortical hormones

adductor (ah-duk'tor) muscle that functions with a movement toward the body; the act of turning inward

adenine (ad'e-nin) solid substance of the uric acid group derived from nucleic acids

adenosine diphosphate (ADP) (ah'de-no-sen di-fos'fat) a compound of adenosine containing two phosphoric acid groups; this enzyme is produced during muscle contraction

adenosine triphosphate (AT) (ah'de-no-sen tri-fos'fat) the major source of cellular energy, found in all cells; composed of a nitrogenous base (adenine), ribose sugar, and three phosphoric acid radicals

adhesion (ad-he'zhun) the adherence or sticking together of two surfaces

adrenocorticotrophic hormone (ACTH) (ad-ren-o-kor-te-ko-trof'ik) hormone produced by the anterior pituitary gland, which influences the adrenal glands

adrenal cortex (ad-re′nal kor′tex) part of adrenal gland that secretes steroids that affect carbohydrate and protein metabolism

adrenal gland (ad-re′nal) hormone-producing gland located near the kidney

adrenal medulla (ad-re′nal me-dul′ah) part of the adrenal gland that secretes two hormones that affect blood pressure and gastrointestinal movement

adrenalin (ad-ren′ah-lin) proprietary name for epinephrine, the active secretion of the medulla of the adrenal gland

adrenergic neurons (ad-ren-er′jik nu′ronz) term used to describe nerve fibers that release norepinephrine (noradrenalin or sympathin) at their endings when stimulated

adrenogenital syndrome (ad-ren-o-jen′i-tal sin′drōm) symptoms due to oversecretion of the adrenal cortical hormones

afferent arteriole (af′er-ent ar-te′re-ol) the blood vessel that brings blood to the kidney glomerulus

agammaglobullinemia (a-gam-ah-glob-u-li-nē′me-ah) rare disease characterized by the virtual absence of gamma globulin from the blood plasma with resulting loss of the ability to produce immune antibodies

agglutinin (a-glu′ti-nin) an antibody in the red blood plasma capable of causing clumping of specific antigens on red blood cells

alanine (al′ah-nēn) an amino acid

albumin (al-bu′min) one of a group of simple proteins in both animal and vegetable cells or fluids

aldehyde (al′de-hid) a compound containing the CHO group; derived by oxidation of a primary alcohol

aldosterone (al-do-ster′on) a hormone produced by the adrenal cortex that regulates electrolytes

alimentary canal (al-e-men′tar-e) a continuous passageway from the mouth, where food is taken in, to the anus, where waste products are discharged; the process of digestion takes place here

alkalosis (al-kah-lo′sis) increased bicarbonate content of the blood due to excess of alkalies or withdrawal of acid or chlorides from the blood

allergen (al′er-jen) a substance that causes sensitivity; something that induces allergy

allergy (al′er-jē) a condition characterized by reaction of body tissues to specific antigens without production of immunity

alpha cells (al′fah) acidophil cells of the hypophysis and the pancreas

alveoli (al-ve′o-li) a cluster of air sacs at the end of the bronchial trees; sockets for the teeth or any small hollow or cavity

ameboid (ah-me′boid) having the appearance and the characteristics of an ameba

amino acid (ah-mē′no) the building blocks of protein

amino group (ah-mē′no grup) the NH^2 group that characterizes the amines

amniocentesis (am-ne-o-sen-te′sis) perforation of the abdominal wall and the uterine wall in order to remove amniotic fluid for examination

amphiarthrosis (am-fe-ar-thro′sis) a form of articulation midway between diathrosis and synarthrosis, in which the articulating bony surfaces are separated by an elastic substance to which both are attached, so that the mobility is slight but may be exerted in all directions

ampulla (am-pul′lah) a saclike dilation of a tube or duct

amylase (am′i-las) a ferment enzyme of saliva, pancreatic juice, and intestinal juice that hydrolyzes starch

anal sphincter (a′nal sfink′ter) sphincter that closes the anus

anaphylaxis (an-ah-fi-lak′sis) a severe reaction caused by extreme sensitivity to a foreign protein or other substance

androgen (an′dro-jen) a substance stimulating or producing male characteristics

anemia (ah-nēm′e-ah) a decrease in certain elements of the blood, especially red cells and hemoglobin

aneurysm (an′u-rizm) a saclike enlargement of a blood vessel caused by a weakening of the wall

angina pectoria (an-ji′na pek′tor-is) heart muscle deprived of blood flow with severe pain referred to the chest and left arm

angstrom (ang′strem) international unit of wavelength measurement

anion (an′i-on) a negatively charged ion or radical

annular ligament (an′u-lar lig′ah-ment) a circular band or sheet of strong fibrous connective tissue enclosing the head of the radius and holding footplate of stapes

anoxia (an-ok′se-ah) a lack of oxygen

anterior (an-te′re-or) before or in front of; the ventral (belly) portion

anterior pituitary (an-te′re-or pi-tu′i-tar-e) dried, defatted, powdered anterior lobe of pituitary gland of domestic animals

antibody (an′ti-bod-e) a protein (usually a flobulin) developed against a specific substance (antigen), whether the antigen is foreign or not

anticoagulant (an-ti-ko-ag′u-lant) a substance that prevents the clotting of blood

antidiuretic hormone (ADH) (an-ti-di-u-ret′ik hor′mon) hormone that inhibits urine formation

aorta (a-or′tah) the largest artery

apnea (ap′ne-ah) a temporary cessation of breathing

aponeurosis (ap-o-nu-ro′sis) a sheetlike layer of connective tissue connecting a muscle to the part that it moves, or acting as a sheath enclosing a muscle

appendicitis (a-pen-di-si′tis) inflammation of the vermiform appendix

appendicular (ap-en-dik′u-lar) that part of the skeleton which forms the framework for the extremities

aqueous humor (a′kwe-us hu′mor) watery fluid in the anterior and posterior chambers of the eye

arachnoid (ar-ak′noid) the middle of the three coverings (meninges) of the brain

arteriole (ar-te′re-ol) the smallest artery, one that branches into microscopic capillaries

arteriosclerosis (ar-te-re-o-skle-ro′sis) general term referring to many conditions where there is thickening, hardening, and loss of elasticity of the walls of blood vessels

artery (ar′ter-e) any vessel carrying blood away from the heart to the tissues. Usually refers to the larger vessels

arthritis (ar-thri′tis) inflammation of a joint, usually accompanied by pain and frequently structural changes

articular cartilage (ar-tik′u-lar kar′til-aj) cartilage that joins together a joint so as to permit motion between parts

ascending tract (ah-send′ing trakt) the chief bundle of fibers located inside the central nervous system that conducts impulses up the cord

asthma (az′mah) intermittent severe difficulty of breathing accompanied by wheezing and cough; due to spasms or swelling of the bronchioles

astigmatism (ah-stig′ma-tizm) defective curvature of refractive surfaces of the eye; as a result a ray of light is not focused sharply on the retina, but is spread over a diffuse area

atelectasis (at-e-lek′tah-sis) a condition in which lungs of a fetus remain unexpanded at birth

atherosclerosis (ath-e-ro-skle-ro′sis) a form of arteriosclerosis in which there are localized accumulations of lipid-containing material within or beneath the surfaces of blood vessels

atlas (at′las) the first cervical vertebra by which the head articulates with the occipital bone. Named for Atlas, who was supposed to have supported the world on his shoulders

atom (at′om) any one of the ultimate units of an element that can exist and still have the properties of the element; the particles that together form a molecule in a compound

atomic number number of protons in nucleus of an atom

atomic weight ratio of weight of an atom of an element to one-twelfth the weight of the standard reference carbon atom

atrioventricular node (A-V node) (a-tre-o-ven-trik′u-lar nod) a tangled mass of Purkinje fibers located in lower part of interatrial septum from which atrioventricular bundle arises

atrium, left (a′tre-um) upper chamber of the heart, which receives oxygenated red blood from the lungs through the pulmonary vein

autoclave (aw′to-klāv) apparatus for sterilization by steam pressure, usually at 250°F for a specified length of time

autoimmunity (aw′to-im-mu-ni-te) a condition in which antibodies are produced against a person's own tissue

automatic nervous system (aw-to-ma′tik) the involuntary or self-regulating portion of the peripheral nervous system

axial (ak′se-al) that part of the skeleton which includes the bony framework of the head and trunk

axis (ak′sis) a line running through the center of a body, or about which a part revolves. The second cervical vertebra that has the dens (odontoid process) about which the atlas rotates

axon (ak′son) a neuron process that conducts impulses away from (efferent) the cell body

bacteria (bak-ter'e-ah) microscopic plants including some that are disease producers and others valued for fermentation

bacterial skin disease (bak-ter'e-al) disease of the skin caused by plantlike organisms

barometer (ba-rom'et-er) an instrument for measuring atmospheric pressure

Barr body (bahr) sex chromatin found in normal females but not males

basal ganglia (ba'sal gang'le-ah) masses of gray matter within the lower part of the forebrain that aid in maintaining muscle coordination and steadiness of muscle contraction

base (bās) a nonacid, a proton acceptor, characterized by excess of OH ion and pH greater than 7

basement membrane (bas'ment mem'bran) a thin layer of solid substance underlying the epithelium of mucous surfaces

basilic (bah-sil'ik) large vein on inner side of biceps

basophile (ba'so-fil) a white blood cell characterized by a pale nucleus and large densely basophilic granuoles

benign (be-nin') mild; not malignant

Claude Bernard French physiologist, 1813–1878

beta cells (ba'tah) insulin-secreting cells of the pancreatic islets of Langerhans

bile (bil) a secretion of the liver

bile salts (bil sawltz) alkali salts of bile

bilirubin (bil-i-roo'bin) the orange or yellow pigment in bile; derived from breakdown of heme

bioelectricity (bi-o-e-lek-tris'i-te) electricity produced in the body due to shift in ions

bipolar neuron (bi-po'lar nu'ron) nerve cell bearing an axon and a dendrite

bladder calculi (blad'er kal'ku-le) stones formed within the urinary bladder

bolus (bo'lus) a mass of food ready for swallowing

Bordetella pertussis (bor-de-tel'lah per-tus'sis) cause of whooping cough

Bowman's capsule (bo'manz kap'sel) the capsule containing the glomeruli of the kidney; the cuplike depressions on the expanded ends of the renal tubules or nephrons that surround the capillary tufts

Boyle's law (boilz) the volume of a given mass of gas at any temperature varies inversely with the pressure applied

brachial (bra'ke-al) pertaining to the arm (the part between the shoulder and the elbow)

brachial artery (bra'ke-al art'er-e) main artery of arm

brachiocephalic vein (brak-e-o-se-fal'ik vān) vein located in the arm or head

bronchial (brong'ke-al) pertaining to the bronchi or bronchioles

bronchiole tube (brong'ke-ol tūb) smaller division of the bronchi

bronchitis (brong-ki'tis) inflammation of bronchial mucous membrane

Brunner's glands (brun'erz) glands in the duodenal submucosa secreting intestinal juice

buffer (buf'er) a substance preserving pH upon additions of acids or bases; reduces the change in hydrogen ion concentration

bulbourethral glands (bul-bo-u-re'thral) two small glands whose secretions form part of the seminal fluid

calcaneus (kal-kā'ne-us) the largest and strongest tarsal bone, located in the heel of the foot

calcitonin (kal-si-to'nin) a hormone secreted by the thyroid gland that aids in regulating calcium-phosphorous metabolism

calcium (kal'se-um) silver-white metallic element that takes part in blood coagulation, formation of teeth and bones, and the functions of muscles, nerves, and heart

callus (kal'us) circumscribed area of the horny layer of the skin that has thickened; osseous material formed between the ends of a fractured bone

calorie (kal'o-re) a unit of heat. The small calorie is the amount of heat required to raise 1 g of water 1°C. The large calorie, or kilocalorie, is the one used in nutrition and metabolic studies, and is the amount of heat necessary to raise 1 kg of water 1°C.

cancellous bone (kan'se-lus) having a latticework or reticular structure

capillaries (kap'i-lar-ēz) minute blood vessels that connect the arterioles with venules to carry blood; minute lymphatic ducts. Capillary networks allow passage of oxygen and nutrients from the blood to the tissues, and of wastes from the tissues into the blood

capitate (kap′i-tāt)   head-shaped; having a rounded extremity

capitulum (kah-pit′u-lum)   a small rounded end of a bone that articulates with another bone

capsule (kap′sul)   a sheath or continuous enclosure around an organ or structure

carbaminohemoglobin   (kar-bam-i-no-he-mo-glo′bin)   the compound formed by the combination of carbon dioxide with the amine groups of the globin molecule of hemoglobin

carbohydrate (kar-bo-hi′drāt)   a starch, sugar, cellulose, or gum; a compound containing carbon, hydrogen, and oxygen

carbonic acid (kar-bon′ik)   an acid formed from carbon dioxide dissolved in water

carbonic anhydrase (kar-bon′ik an-hi′dras)   an enzyme catalyzing the reaction of carbon dioxide and water to form carbonic acid

carboxyl group (kar-bok′sil)   the COOH group of organic carboxyl acids

cardiac muscle (kar′de-ak)   muscle of the heart

cardiac output (kar′de-ak)   amount of blood ejected from one ventricle per minute

cardiac sphincter (kar′de-ak sfingk′ter)   plain muscle about the esophagus at cardiac opening into the stomach

cardiovascular disease (stroke) (kar-de-o-vas′ku-lar)   the total set of symptoms resulting from a cerebral vascular disorder

carotene (kar′o-tēn′)   a yellow, crystalline pigment obtained from yellow vegetables; stored and converted to vitamin A in the liver

carotid sinus (kah-rot′id si′nus)   a dilated area of the common carotid artery that is richly supplied with sensory nerve endings of the sinus branch of the vagus nerve

carpals (kar′pals)   the eight bones of the wrist

carrier (kar′e-er)   a large molecule transporting substances across cell membranes in active transport

cartilage (kar′ti-lij)   a form of connective tissue usually with no blood or nerve supply of its own; it is firm, elastic, and a semiopaque bluish-white or gray; cells lie in cavities called lacunae

catalyst (kat′ah-list)   a substance that alters or speeds up the rates of chemical reactions without being itself altered in the process

cataract (kat′ah-rakt)   an opacity of the eye lens

cation (kat′i-on)   an ion carrying a positive charge

caudal (kaw′dal)   pertaining to any taillike structure; inferior in position

cecum (se′kum)   a blind pouch at the junction of the small intestines with the ascending colon, and to which the ileum is attached

cell (sel)   a small enclosed or partly enclosed cavity; a mass of protoplasm containing a nucleus or nuclear material

Celsius scale (sel′se-us)   relating to the thermometer scale on which the interval between the freezing and boiling points of water is divided into 100 degrees, with 0 degrees representing freezing point and 100 degrees representing boiling point

centigram (sen′ti-gram)   one hundredth of a gram

centimeter (sen′ti-me-ter)   2.5 cm equals 1 inch; 1 cm equals 1/100 meter

central fissure (sen′tral fish′ūr)   groove separating the frontal and parietal lobes

central nervous system (sen′tral ner′vus sis′tem)   the control center for the entire nervous system, consisting of the brain and spinal cord

centromere (sen′tro-mer)   the structure at the junction of the two arms of a chromosome

cephalic (se-fal′ik)   cranial, pertaining to the head; superior in position

cerebellum (ser-e-bel′um)   the part of the hindbrain that lies below the occipital part of the cerebrum on each side; concerned with voluntary muscle movement

cerebral palsy (ser′e-bral pawl′ze)   group of motor disorders caused by damage to the motor areas of the brain during fetal life, birth, or infancy

cerebrospinal fluid (ser-e-bro-spi′nal floo′id)   fluid formed in the ventricles of the brain that cushions the organ against shock

cerebrum (ser′e-brum)   the upper, largest part of the brain; contains the motor and sensory areas and the association areas concerned with the higher mental faculties

cervical (ser′vi-kal)   relating to the neck or any cervix, including the cervix of the uterus

cervical vertebra (ser′vi-kal ver′te-bre)   relating to the vertebrae of the neck

cervix (ser′viks)   any neck or constricted portion of an organ, especially the lower cylindrical part of the uterus

chemoreceptor (ke-mo-re-sep′tor)   a sense organ or sensory nerve ending that is stimulated by a chemical substance or change

chemotaxis (kem-o-tak'sis) attraction and repulsion of living protoplasm to a chemical stimulus

chloride shift (klo'rid) the movement of chloride ions into red blood cells; bicarbonate ions are formed and leave the cell during the reaction of carbon dioxide and water in the cell

cholecystokinin (ko-le-sis-to-kin'in) a hormonelike chemical secreted by the duodenal mucosa that stimulates contraction of the gallbladder

cholesterol (ko-les'ter-ol) an organic, fatlike compound found in many parts of the body

cholinergic neurons (kol-in-er'jik nu'ronz) nerve endings of the nervous system that liberate acetylcholine at a synapse

chondrocyte (kon'dro-sit) a cartilage cell

chordae tendineae (kor'de ten'din-e) small tendinous cords that connect the edge of an atrioventricular valve to a papillary muscle and prevent valve reversal

chromatid (kro'mah-tid) one of the two bodies resulting from longitudinal separation of duplicated chromosomes

chromosomes (kro'mo-sōms) small, rod-shaped bodies that stain deeply and appear in the nucleus at the time of cell division containing the hereditary factors

chronic renal insufficiency (kron'ik ren'al in-su-fish'en-see) reduced capacity of the kidney to perform its functions

chyme (kim) the semifluid mixture of partly digested food and digestive secretions found in the stomach and small intestine during digestion of a meal

cilia (sil'e-ah) hairs or hairlike processes; may refer to eyelashes or to microscopic extensions of the cell protoplasm. Singular: cilium (sil'e-um)

ciliary body (sil'e-er-e) flattened, ring-shaped muscle, with a hole, that is the size of the outer edge of the iris that alters the shape of the lens

circle of Willis (ser'k'l of wil'is) an arterial circle on the base of the brain composed of anterior cerebral, anterior communicating, internal carotid, posterior communicating, and posterior cerebral arteries

clitoris (kli'to-ris) female erectile genital organ that is homologous to the penis of the male

Clostridium perfringens (klo-strid'e-um per-frin'gens) a cause of gas gangrene

Clostridium tetani (klo-strid'e-um tet'a-ni) the causative organism of tetanus or lockjaw. Produces a powerful exotoxin, a portion of which affects nerve tissue

coccygeal (kok-sij'e-al) pertaining to the coccyx or tailbone, the last four fused spinal bones

coccyx (kok'siks) the tail bone, the last four fused spinal bones

cochlea (kok'le-ah) anything having a spiral form; part of the inner ear

coenzyme (ko en' zim) a nonprotein substance that activates enzymes or chemical compounds

cohesion (ko-he'zhun) the property of matter to cohere, unite, or stick together

colon (ko'lon) the portion of the large intestine from the end of the ileum to the rectum

colostomy (ko-los'to-me) the surgical creation of a new opening from the colon to the body surface

columnar epithelium (ko-lum'nar ep-i-the'le-um) tissue composed of cells shaped like tubes or cylinders

compact bone (kom-pakt') the hard or dense part of bones containing subunits of structure called osteons or Haversian systems

conception (kon-sep'shun) the point in time when a sperm unites with an ovum to initiate formation of a new individual

conditioned response (kon-dish'und re-spons') response acquired as result of training and repetition

conduction deafness (kon-duk'shun def'nes) resulting from failure of the ossicles to transmit the sound waves to the cochlea

condyle (kon'dil) a rounded projection on a bone, usually for articulation with another bone

congenital torticollis (kon-jen'i-tal tor-tik-ol'is) stiff neck caused by spasmodic contraction of neck muscles, drawing the head to one side with chin pointing to the other side

conjunctiva (kon-junk-ti'vah) the thin, delicate membrane that lines the eyelids and is reflected over the front of the eyeball

conjunctivitis (kon-junk-ti-vi'tis) inflammation of the conjunctiva

constipation (kon-sti-pa'shun) difficult defecation; often infrequent passage of unduly hard or dry fecal material

contractile tissue (kon-trak'til tish'u) tissue having the ability to contract or shorten

convection (kon-vek'shun) transference of heat in liquids or gases by means of currents

cornea (kor'ne-ah) the transparent front part of the eyeball; the forward continuation of the outer coat (sclera)

corocoid process (kor'o-koid) projection of the anterior surface of the scapulae to which the muscles are attached

coronal plane (kor'o-nal plān) vertical plane that separates the body into anterior and posterior parts

coronal suture (kor'o-nal su'chur) immovable joint of the skull between the frontal bone and two parietal bones

coronary artery (kor'o-na-re ar'ter-e) one of a pair of arteries that supplies blood to the myocardium of the heart

corpora cavernosa (kor'poh-ra kav'ern-o-sah) two columns of erectile tissue on dorsum of the penis

corpus luteum (kor'pus lu'te-um) a yellow mass formed in the ovarian follicle which produces the hormone progesterone

corpus spongiosum (kor'pus spun-je-o'sum) erectile tissue surrounding the urethra

cortex (kor'tex) an outer portion or layer of an organ or structure, also called gray matter

cortisone (kor'ti-sōn) a hormone isolated from the cortex of the adrenal gland important for its regulatory action in metabolism

cranial cavity (kra'ne-al kav'i-te) the portion of the skull that encloses the brain, consisting of eight bones

cranial nerve (kra'ne-al) one of the twelve pairs of nerves arising from the brain and making its exit through a foramen of the cranium

crest (krest) a ridge or long prominence, as on bone

cretinism (kre'tin-izm) severe congenital thyroid deficiency leading to physical and mental retardation

croup (krōōp) disease characterized by suffocative and difficult breathing, laryngeal spasm, and sometimes by the formation of a membrane

crypt of Lieberkühn (kript of le'ber-kenz) a tubular gland of the intestine that secretes intestinal juice

cryptorchidism (krip-tor'ki-dizm) a failure of the testes to descend normally; a defect in the normal development and descent of the testes into the scrotum

cuboidal epithelium (ku'boi-dal ep-e-the'e-um) tissue composed of cells shaped like cubes

cuneiform bone (ku-ne'e-form) those of the tarsus, internal, middle, and external

Cushing's syndrome (koosh'ingz sin'drom) a syndrome resulting from hypersecretion of the adrenal cortex

cutaneous membranes (ku-ta'ne-us mem'brānz) membranes of the skin

cystitis (sis-ti'tis) inflammation of the urinary bladder

cytogenetics (si-to-je-net'iks) study of cytology in relation to genetics, especially chromosomal behavior in mitosis and meiosis

cytokinesis (si-to-ki-ne'sis) division of cytoplasm in latter stages of mitosis

cytoplasm (si'to-plazm) the cellular substance between the membrane and the nucleus of the cell

deaminization (de-am-i-ni-za'shun) a chemical decomposition whereby substances like amino acids and alkaloids lose their amino groups and form ammonia

deciduous teeth (de-sid'u-us) temporary or milk teeth; those comprising the first set, which are shed

decubitus ulcer (de-ku'bi-tus ul'ser) an open lesion upon the skin or mucous membrane of the body with loss of substance and necrosis of the tissue caused by pressure of a lying-down position

dehydration (de-hi-dra'shun) a condition due to excessive water loss from the body or its parts

dendrite (den'drit) a nerve fiber that conducts impulses to the cell body

dentine (den'tēn) the osseous tissues of a tooth enclosing the pulp cavity

deoxyribose (de-ok-se-ri'bōs) phosphoric ester of a pentose present in nucleic acid

depolarization (de-po-lar-i-za'shun) loss of polarity or the polarized state

dermis (der'mis) the sensitive vascular inner layer of the skin

descending tracks (de-send'ing tt05aks) bundles of fibers that carry impulses down the cord

detoxification (de-tok-si-fi-ka'shun) to remove the toxic quality of a substance

dextrose (deks′trōs) another name for glucose, a simple sugar (monosaccharide)
diabetes insipidus (di-ah-be′tēz in-sip′id-us) polyuria due to vasopressin deficiency
diabetes mellitus (di-ah-be′tēz mel-li′tus) a disease in which sugar is not "burned" in the tissues for transformation into energy because of insufficient insulin
diabetogenic (di-ah-bet-o-jen′ik) causing diabetes
diaphysis (di-af′i-sis) the shaft of a long bone
diarrhea (di-ah-re′ah) frequent passage of watery bowel movements
diarthrosis (di-ar-thro′sis) an articulation in which opposing bones move freely, as in a hinge
diastolic pressure (di-ah-stol′ik) pressure existing during the relaxation phase between heart beats
differential permeability (dif-er-en′shal per-me-ah-bil′i-te) ability of membrane to differentiate between liquids or gases passing through it
diffusion (di-fu′zhun) eventual even mixing of solutes and solvents as a result of motion of molecules
dipeptidase (di-pep′ti-dās) an enzyme that hydrolyzes dipeptides to amino acids
diptheria (dif-the′re-ah) acute infectious disease characterized by the formation of a false membrane on any mucous surface and occasionally the skin
diploid number (dip′loid) cell having twice the number of chromosomes present in the egg or sperm of a given species
disaccharide (di-sak′ah-rid) a sugar composed of two simple sugar molecules
distal (dis′tal) farthest from the center of the body or point of attachment; opposite of proximal
diuretic (di-u-ret′ik) any agent that increases the secretion of urine
diverticula (di-ver-tik′ūlah) sacs or pouches in the walls of a canal or organ
diverticulitis (di-ver-tik-u-li′tis) an inflammation of the sacs that form as a result of weakness in the muscle wall of a tubular organ
diverticulosis (di-ver-tik-u-lo′sis) diverticula of the colon without inflammation or symptoms
Down's syndrome (dounz sin′drōm) 21 trisomy or mongolism, a variety of congenital, moderate to severe mental retardation; characteristics are sloping forehead, oriental appearance, and dwarfed physique
ductus arteriosus (duk′tus ar-te-re-o′sus) a channel of communication between main pulmonary artery of fetus and aorta
duodenum (du-o-de′num) the first or proximal part of the small bowel (about 12 inches in length)
dura matter (du′rah) the outermost and toughest of the three coverings of the brain and spinal cord
dwarfism (dwarf′izm) an abnormal state that produces small, short, or disproportioned persons
dyspnea (disp-ne′a) difficult or labored breathing
edema (e-de′mah) swelling due to increase of extracellular fluid volume; dropsy
ejaculatory duct (e-jak′u-la-to-re dukt) terminal portion of the seminal duct
electrocardiogram (EKG) (e-lek-tro-kar′de-o-gram) the tracing of the electric current produced by heart-muscle activity (contraction); the record produced by an electrocardiograph
electrolyte balance (e-lek′tro-līt) condition in which electrolytes are maintained in suitable concentrations for maintenance of fluid and osmotic environment
electrolyte (e-lek′tro-lit) a solution that conducts electricity by means of ions that are positively or negatively charged
electron (e-lek′tron) a minute body or charge of negative electricity; a component of atoms
embryology (em-bre-ol′o-je) the study of the development of the embryo
emphysema (em-fi-se′mah) a swelling or inflation of air passages with resulting stagnation of air in parts or the lungs; loss of elasticity in the alveoli
enamel (en-am′el) the hard, white, dense substance forming a covering for the crown of the teeth
endocartitis (en-do-kar-di′tis) inflammation of the endocardium
endocardium (en-do-kar′de-um) the membrane that lines the heart chambers and assists in forming the heart valves
endochondral (en-do-kon′dral) formation of bone in cartilage
endocrine gland (en′do-krin) an organ or structure which secretes a hormone that is absorbed into the blood or lymph; a ductless gland

endolymph (en'do-limf) fluid of the membranous canals of the ear

endometrium (en-do-me'tre-um) the mucous membrane lining the inner surface of the uterus

endoplasmic reticulum (en-do-plas'mik re-tik'u-lum) the series of tubular structures found in the cytoplasm of cells; transports substances through cells

endosteum (en-dos'teūm) membrane that lines the medullary cavity of bones

endothelium (en-do-the'le-um) the layer of cells that lines blood and lymph vessels, the heart, and the serous body cavities

enterogasterone (en-ter-o-gas'trōn) a hormone secreted by the intestinal mucosa

enzyme (en'zīm) a substance that causes chemical changes; an organic catalyst, usually a protein

eosinophil (e-o-sin'o-fil) a cell or cellular structure that stains readily with the acid stain

epidermis (ep-i-der'mis) the outer epithelial layer of the skin, which contains no blood vessels and which rests on the dermis

epididymis (ep-i-did'i-mis) a long convoluted tubule or duct resting on the testis and conveying sperm to the vas deferens

epigastric (ep-i-gas'trik) the upper-middle section of the abdominal cavity, just below the breastbone

epiglottis (ep-i-glot'tis) a leaf-shaped structure located over the superior end of the larynx. Aids in closing the larynx during swallowing

epilepsy (epi-lep-si) disorder characterized by short, periodic attacks of motor, sensory, and/or psychological malfunction

epinephrine (ep-i-nef'rin) a hormone produced by the adrenal medulla; also produced synthetically

epiphyseal plate (ep-i-fiz'e-al) separates a parent bone from a secondary bone-forming (ossification) center in the developing infant

epiphysis (ep-i'fiz-is) one of the two ends of a long bone, separated from the diaphysis by a growth line

epithelial tissue (ep-i-the'le-al tish'u) the tissue that forms the outer part of the skin, lines blood vessels, hollow organs, and passages that lead to the outside of the body

erythroblastosis fetalis (e-rith-ro-blas-to'sis fe-tal'is) a congenital disorder in which an Rh-negative mother transmits antibodies against the Rh protein to an Rh-positive baby

erythrocyte (e-rith'ro-sît) a red blood cell

erythropoietin (e-rith-ro-poi'e-tin) a substance produced by the kidney that stimulates production of red blood cells

Escherichia coli (esh-er-i'ke-a co'le) the colon bacillus; a short, plump, gram-negative, nonsporeforming bacillus almost constantly present in the alimentary canal of humans and some animals

esophagus (e-sof'ah-gus) the tubular passage extending from the pharynx to the stomach

essential amino acid group of organic compounds that are converted to protein and cannot be synthesized by the body but must be obtained from food

estrogen (es'tro-jen) any substance that induces estrogenic activity or stimulates the development of secondary female characteristics

eupnea (ūp-ne'a) normal breathing

eustachian tube (u-sta'ke-an) a tubelike structure connecting the middle ear cavity and the throat that equalizes air pressure

evagination (e-vaj-i-na'shun) emergence from a sheath; protrusion of an organ or part

exocrine gland (ek'so-krin) gland whose secretion reaches an epithelial surface either directly or through a duct

extensor (eks-ten'sor) any muscle that functions in straightening or stretching

external auditory meatus (eks-ter'nal aw'di-to-re me-a'tus) canal in the temporal bone that leads to the middle ear

extracellular fluid (eks-trah-sel'u-lar) fluid outside a cell or cells

Fahrenheit scale (far'en-hît) relating to a thermometer scale on which the boiling point of water is 212 degrees and the freezing point is 32 degrees above its zero point

fallopian tubes (fal-lo'pe-an) ducts connected to the uterus carrying the ova from the ovaries into the uterine cavity and providing a place for the fertilization of the ovum by the spermatozoan

fascia (fash'e-ah) a fibrous membrane covering supporting and separating muscles

feces (fe'sēz) material discharged from the bowel, which is made up of bacteria, secretions, and food residue

feedback control (fēd'bak) detection of the nature of an output and using that to control the process producing the output. May be negative (inhibitory) or positive (stimulating)

femoral artery (fem-or-al art'er-e) begins at external iliac artery and terminates behind the knee as the popliteal artery on inner side of femur

fertilization (fer'ti-li-za'shun) the impregnation of an ovum by a sperm

fibrin (fi'brin) a white or yellowish insoluble fibrous protein formed when blood clots

fibroblast (fi'bro-blast) a flat, long, connective tissue cell that forms the fibrous tissues of the body

fibroelastic cartilage (fi-bro-e-las'tik kar'ti-lij) connective tissue containing both white, nonelastic, collagenous fibers and yellow elastic fibers

fibrosis (fi-bro'sis) abnormal deposition or increase in fibrous tissue in a body part, organ, or tissue

fibula (fib'u-lah) the smaller long bone of the lower leg, running parallel to the tibia

filtration (fil-tra'shun) the passage of a liquid through a filter or a membrane that acts as a filter

flagella (flah-jel'ah) microscopic, whiplike processes or threadlike appendages that enable certain bacteria to move rapidly. Singular: flagellum (flah-jel'um)

flat bone thin, flat bone composed of two more or less parallel plates of compact bone enclosing a layer of spongy bone

flexor (flek'sor) any muscle that functions in bending, decreasing the angle between two parts

follicle stimulating hormone (FSH) (fol'li-k'l) hormone secreted by the anterior lobe of hypophysis that stimulates development of the ovarian follicles

fontanelle (fon-tah-nel') a soft area in a baby's skull; a membrane-covered spot where bone formation has not yet occurred

foramen (fo-ra'men) a natural opening or passageway; a general term especially for a passage into or through a bone. Plural: foramina (fo-ram'i-nah)

foramen magnum (fo-ra'men mag'num) passage or opening that pierces the occipital bone through which passes the spinal cord from the brain

foramen ovale (fo-ra'men o-val'e) opening at lower posterior of septum in fetus between two cardiac atria

fossa (fos'ah) a furrow or shallow depression

fovea centralis (fo've-ah sen-tra'lis) a fossa, or cup; depression in the retina containing the cone cells

fracture, greenstick (frak'tūr) bone is partially bent and partially broken as when a green stick breaks

fracture, simple bone is broken, but there is no external wound

frontal (frun'tal) the bone of the skull forming the anterior part of the cranium of the forehead, the upper portion of the orbits, and most of the anterior part of the cranial floor

fundus (fun'dus) the part of a hollow organ farthest from the opening

galactose (gah-lak'tōs) milk sugar, a simple hexose sugar found in milk

galactosemia (gah-lak-to-se'me-ah) galactose in the blood

gallstone (gawl'stōn) concretion formed in the gallbladder or bile ducts

gamete (gam'ēt) a male or female reproductive cell, the spermatozoan or ovum

ganglionic chain (gang-gle-on'ik) a mass of nerve cell bodies outside of the brain and spinal cord

gas gangrene (gang'grēn) gangrene in a wound infected by a gas bacillus

gastric juice (gas'trik) the digestive juice of the glands of the stomach

gastrin (gas'trin) a hormone that stimulates secretion of the glands in the cardiac end of the stomach

gastroenteritis (gas-tro-en-ter-i'tis) inflammation of the stomach and intestine

gene (jēn) one of the biological units of heredity, an ultramicroscopic, self-reproducing DNA particle located in a definite position on a particular chromosome

genotype (jen'o-tip) the basic hereditary combination of genes of an organism

gestation (jes-ta'shun) the period of intrauterine fetal development

giantism (ji'ant-izm) abnormal state producing large, tall, or disproportioned person

glans penis (glanz pe'nis) bulbous end of the penis

glaucoma (glaw-ko'mah) an eye disorder in which there is increased pressure due to an excess of fluid within the eye

glenoid fossa (gle'noid fos'ah) the mandibular fossa, which receives the capitulum of the mandible

glomerular filtrate (glo-mer'u-lar fil'trāt) fluid that passes from the blood through the capillary walls of the glomeruli of the kidney

glomerular nephritis (glo-mer'u-lar ne-fri'tis) inflammation of the renal glomeruli

glomerulus (glo-mer'u-lus) a rounded mass of nerves or blood vessels, especially the microscopic tuft of capillaries that is surrounded by the expanded part of each kidney tubule

glottis (glot'is) the opening between the vocal cords in the larynx

glucagon (gloo'kah-gon) a hyperglycemic glycogenolytic factor secreted by the alpha cells of the pancreas, stimulating breakdown of glycogen and the release of glucose by the liver

glucocorticoid (gloo'ko-kor'ti-koid) a general classification of adrenal cortical hormones that are primarily active in protecting against stress and in affecting protein and CHO metabolism

glucose (gloo'kōs) most important carbohydrate in body metabolism, formed during digestion from di- and poly-saccharides

glutamine (gloo'tah-min) monoamide present in the juices of many plants, essential in hydrolysis of proteins

glycine (gli'sēn) aminoacetic acid derived from gelatin and many proteins

glycogen (gli'ko-jen) a complex polysaccharide; "animal starch," stored in liver and muscle

glycogenesis (gli-ko-jen'e-sis) the formation of glycogen as occurs after the eating of a carbohydrate meal

glycogenolysis (gli-ko-je-nol'i-sis) conversion of glycogen into dextrose in the liver

glycolysis (gli-kol'i-sis) hydrolysis of sugar by a ferment in the body

goblet cell (gob'let) a one-celled mucus-secreting gland found in epithelia of the respiratory and digestive systems

Golgi body (gol'je bod'e) specialized portion of the endoplasmic reticulum believed to collect and package products of cell synthesis

gonadotropin (gon-ah-do-tro'pin) a sex gland–stimulating hormone

gonorrhea (gon-o-re'ah) a contagious veneral disease characterized by discharge of pus from mucous membranes, chiefly of the urethra and other parts of the genitourinary system

greater trochanter (tro-kan'ter) projection that serves as a point of attachment for some of the muscles of the thigh and buttock

growth hormone pituitary hormone promoting normal growth

guanine (gwan'in) organic compound found in the liver, pancreas, and muscle

gyrus (ji'rus) one of the tortuous elevations (convulsions) of the cerebral cortex region of the brain. Plural: gyri (ji'ri)

haploid number (hap'loid) having half the normal number of chromosomes characteristic of the species; sperm and ova are haploid cells

hard palate (pal'at) anterior part supported by the maxillary and palatine bones

haustra (haws'trah) the sacculations or pouches of the colon

haversian canal (ha-ver'shan) circular cavities that run longitudinally through the dense bone

haversian system (ha-ver'shan) system consisting of a haversian canal with its surrounding lamellae, lacunae, osteocytes, and canaliculi

hematocrit (he-mat'o-krit) the volume percentage of red blood cells in whole blood

hematology (hem-ah-tol'o-jē) science of the blood

hematoma (hem-ah-to-mah) a tumor or swelling filled with blood

heme (hēm) the iron-containing red pigment that, with globin, forms hemoglobin

hemodialysis (he-mo-di-al'i-sis) removal of chemical substances from the blood by passing the blood through tubes made of semipermeable membranes

hemoglobin (he-mo-glo'bin) the oxygen-carrying colored compound in the red blood cells

hemophilia (he-mo-fil'e-ah) a hereditary blood disorder in which there is deficient production of certain factors involved in blood clotting, resulting in bleeding into joints and deep tissues

Henle's loop (hen'lēz) U-shaped portion of a renal tubule lying between the proximal and distal portions

heparin (hep'ah-rin) a complex acid compound, found most abundantly in the liver, that prevents blood clotting

hepatic flexure (ha-pat'ik flek'sher) the bend on right side forming junction of the ascending with the transverse colon

hepatic vein (he-pat'ik) the three vessels returning blood from the liver and discharging into the inferior vena cava

Hering-Breuer reflex reflex inhibition of inspiration resulting from stimulating of pressoreceptors by inflation of the lungs

herpes simplex (her'pēz sim'plex) a skin disease in which small blisters appear, often in clusters

heterozygous (het-er-o-zi'gus) having unlike genes

hiatus hernia (hi-a'tus her'ne-ah) protrusion of the stomach upward into the mediastinal cavity through the esophageal hiatus of the diaphragm

hilus (hi'lus) an area, depression, or pit where blood vessels and nerves enter or leave the organ

Hodgkin's disease (hoj'kinz) a disease-producing enlargement of lymphoid tissue, spleen, and liver, with invasion of other tissues

homeostasis (ho-me-o-sta'sis) a consistency and uniformity of the internal body environment that maintains normal body function, stability of body fluids and their constituents

humerus (hu'mer-us) longest and largest bone of the arm

hyaline cartilage (hi'ah-lin kar'ti-lij) connective tissue that is clear, translucent, or glassy

hydrocephaly (hi-dro-sef'ah-le) a disorder characterized by gross enlargement of the cranium and atrophy of the brain due to an accumulation of excessive amounts of fluid in and around the brain

hydrolysis (hi-drol'i-sis) any reaction in which water is one of the reactants; the combination of water with a salt to produce an acid and a base

hydrostatic pressure (hi-dro-stat'ik) pressure exerted by liquids

hymen (hi'men) fold of mucous membrane that partially covers the entrance to the vagina

hyperkalemia (hi-per-kah-le'me-ah) excessive amount of potassium in blood plasma

hypercapnia (hi-per-kap'ne-ah) abnormal amount of carbon dioxide in the blood

hyperglycemia (hi-per-gli-se'me-ah) an abnormal increase in the amount of sugar (glucose) in the blood

hyperkalemia (hi-per-kah-le'me-ah) excessive amount of potassium in blood plasma

hyperparathyroidism (hi-per-par-ah-thi'roid-izm) condition due to increased activity of the parathyroid glands

hyperthyroidism (hi-per-thi'roid-izm) a condition caused by excessive secretion of the thyroid glands, which overstimulates the basal metabolism, causing an increased demand for food to prevent oxidation of body tissues

hyperventilation (hi-per-ven-ti-la'-shun) increased exchange of air by increasing both rate and depth of breathing

hypervolemia (hi-per-vo-le'me-ah) plethora of blood

hypocalcemia (hi-po-kal-se'me-ah) abnormally low blood calcium

hypokalemia (hi-po-ka-le'me-ah) low blood potassium levels

hypoparathyroidism (hi-po-par-ah-thi'roid-izm) insufficient secretion of the parathyroid glands

hypothalamus (hi-po-thal'ah-mus) a part of the forebrain, near the third ventricle, containing groups of nerve cells that control temperature, sleep, water balance, and other chemical and visceral activities

hypothermia (hi-po-ther'me-ah) having a body temperature below normal

hypothyroidism (hi-po-thi'roid-izm) underactivity of the thyroid

hypovolemia (hi-po-vo-le'me-ah) diminished blood supply

hypoxia (hi-pok'se-ah) lack of an adequate amount of oxygen

ileocecal valve (il-e-o-se'kal) sphincter muscles that guard the aperture of the ileum at the cecum, where the small intestines open into the ascending colon

ileum (il′e-um) the last or distal part of the small intestine, ending at the cecum of the large intestine

iliac region (il′e-ak) pertaining to the bone called the ileum (upper oscoxae); the two regions at each side of the lower abdomen

ilium (il′e-um) the upper wing-shaped portion of the oscoxae (hipbone)

immunity (i-mu′ni-te) protected from getting a given disease

impetigo (im-pe-ti′go) inflammatory skin disease marked by isolated pustules that become crusted and rupture

implantation (im-plan-ta′shun) embedding of the blastocyst in the uterine lining

incus (ing′kus) the middle of the three ossicles of the ear, called the anvil

infraspinous fossa (in-frah-spi′nus fos′ah) groove below the scapular spine

infundibulum (in-fun-dib′u-lum) tube connecting the frontal sinus with the middle nasal meatus

inguinal canal (in′gwi-nal) canal carrying the spermatic cord in the male and the round ligament in the female

insulin (in′su-lin) hormone secreted by the beta cells of the islets of Langerhans of the pancreas

integument (in-teg′u-ment) a covering, especially the skin

intercostal muscles (in-ter-kos′tal) muscles between the ribs

internuncial (in-ter-nun′she-al) a connector between two other items, as neurons

interstitial cells of Leydig (in-ter-stish′al sels of li′dig) cells producing testosterone of the testes

interstitial fluid (in-ter-stish′al) fluid lying between the vessels and cells or between cells

interstitial space (in-ter-stish′al) pertaining to spaces or structures between the functioning active tissues of any part or organ

intrapleural (in-trah-ploor′al) within the pleural cavity

intrinsic factor (in-trin′sik) substance that increases absorption of vitamin B complex

invagination (in-vaj-i-na′shun) the pushing of the wall of a cavity into the cavity itself

iris (i′ris) colored or pigmented part of the eye

ischium (is′ke-um) lower portion of the innominate or hipbone

islets of Langerhans (i′lets of lang′er-hanz) groups of specialized cells scattered throughout the pancreas

jejunum (je-joo′num) the second portion of the small intestine; the part of the small intestine between the duodenum and the ileum

ketone (ke′tōn) any substance having the carbonyl group in its molecule

ketosis (ke-to′sis) accumulation of ketone bodies in blood or body

Krebs cycle the citric acid cycle; a series of energy-yielding steps in the catabolism of carbohydrates

Kupffer's cell (koop′ferz) a fixed phagocytic cell found in the sinusoids of the liver

kwashiorkor (kwash-e-or′kor) disease resulting from a deficiency of protein in infancy or early childhood

labia majora (la′be-ah ma′jor-ah) two folds of cellular adipose tissue lying on either side of the vaginal opening forming the lateral borders of the vulva

labia minor (la′be-ah mi′nor) two thin folds of integument that lie within the labia majora and enclose the vestibule

lacrimal glands (lak′ri-mal) the glands of the eye that secrete tears

lactic acid (lak′tik as′id) a colorless syrupy liquid ($C_3H_6O_3$) formed in milk, sauerkraut, and in certain types of pickles by the fermentation of the sugars by microorganisms; it is formed in muscles during activity by the breakdown of glycogen (glycolysis)

lacuna (lah-ku′nah) a small, hollow space, such as that found in bones, in which lie the osteoblasts. Plural: lacunae (lah-ku′ne)

lambdoidal suture (lam′doid-al su′chur) immovable joint of the skull between the parietal bones and the occipital bone

lamella (lah-mel′ah) circularly arranged layers, as in the elastic layers in a large artery; a ring of bony tissue around a haversian canal

lateral malleolus (lat′er-al ma-le′o-lus) the protuberance on the lower extremity of the fibula

lesser trochanter (tro-kan′ter) projection that serves as a point of attachment for some of the muscles of the thigh and buttock

leukemia (lu-ke′me-ah) a cancerous disorder of the blood-forming organs characterized by a marked increase in the number of white blood cells plus the presence of many immature cells in the circulating blood

leukocyte (lu′ko-sît) a white blood cell

leukopenia (lu-ko-pe′ne-ah) a decrease of the number of white blood cells, below 5,000 per cubic millimeter

levator (le-va′tor) a muscle that raises a part; opposed to depressor

lipase (li′pās) a fat-splitting enzyme found in the blood, pancreatic secretion, and tissues

lipid (lip′id) a fat or fatlike substance not soluble in water

long bone a bone greater than its width, with a diaphysis and two epiphyses. More-or-less curved for greater strength, and composed of more compact bone than spongy bone

longitudinal fissure (lon-ji-tu′di-nal fish′ūr) most prominent deep downfold of the brain, completely separating the cerebrum into right and left halves

lumen (lu′men) the space within an artery, vein, intestine, or tube

luteinizing hormone (LH) (lu′te-in-i-zing) hormone produced by the anterior lobe of hyphophysis that induces ovulation and the formation of the corpus luteum. Also stimulates development of interstitial cells of the testes

lymph node (limf nōd) a rounded body consisting of accumulations of lymphatic tissue, found at intervals in the course of lymphatic vessels

lymphatic system (lim-fat′ik) system including all structures involved in the conveyance of lymph from the tissues to the blood

lymphocyte (lim′fo-s′it) a white blood cell that is formed in lymph nodes instead of in bone marrow

lysosome (li′so-s′om) cell organelle concerned with digestion of large molecules

lysozyme (li′so-zim) a substance present in tears, saliva, and other body fluids that has antibacterial activity

malar (ma′lar) commonly referred to as the cheekbone, forming the prominence of the cheeks and part of the outer wall and floor of the orbits

malignant (mah-lig′nant) referring to disorders that tend to become worse and cause death

malleus (mal′e-us) the largest of the auditory ossicles; also called the hammer

manubrium (mah-nu′bre-um) the upper bone of the sternum articulating with the clavicle and first pair of costal cartilages

marrow, red (mar′o) soft tissue in cancellous tissue of bone concerned with the production of blood cells and hemoglobin

marrow, yellow (mar′o) soft tissue in the medullary canal of long bones consisting principally of fat cells

mastoid process (mas′toid) rounded projection of the temporal bone behind the external auditory meatus serving as a point of attachment for several neck muscles

mastoiditis (mas-toi-di′tis) inflammation of the bony cells in the mastoid of the skull

matrix (ma′triks) intercellular substance of cartilage; formative portion of a tooth or nail; the uterus

medulla (me-dul′lah) inner or central portion of an organ; the medulla oblongata of the brain stem

meiosis (mi-o′sis) a form of cell division that reduces chromosome number to haploid. Occurs in formation of sex cells

melanin (mel′ah-nin) the dark pigment found in some parts of the body, such as the skin, the middle coat of the eye, and certain tissues in the brain

melanocyte (mel′ah-no-sît) a cell of the stratum germinativum that forms melanin

melatonin (mel-ah-to′nin) a hormone that acts on the pigment cells of the skin

membrane (mem′brān) a thin, soft layer of tissue lining a tube or cavity, or covering or separating one part from another

meninges (me-nin′jēz) the three membranes that cover the brain and spinal cord

meningitis (men-in-ji'tis) inflammation of one or more of the three membranes that cover the brain and spinal cord

menopause (men'o-pawz) the termination of the menstrual cycle

menstruation (men-stroo-a'shun) periodic discharge of a bloody fluid from the uterus occurring at intervals during the life of a woman from puberty to menopause

metacarpal (met-ah-kar'pal) the part of the hand between the wrist and fingers; the five elongated bones in the hand

microcephaly (mî-krō-sef'ah-le) mental retardation caused when the fontanels of the skull close earlier than normal, preventing the brain from growing

micturition (mik-tu-rish'un) the act of expelling urine from the bladder; urination

midsaggital plane (mid-saj'i-tal) vertical plane separating body into right and left symmetrical halves

mineralocorticoid (min-er-al-o-kor'ti-koid) a biologically active principle of the adrenal cortex affecting the retention or excretion of sodium or potassium

mitochondria (mit-o-kon'dre-ah) very small rod-shaped structures or granules in the cytoplasm of cells that are responsible for oxidative reactions to release energy from the food materials

mitosis (mi-to'sis) indirect cell division by which the two daughter cells receive identical complements of chromosomes characteristics of the somatic cells of the species

mitral, bicuspid (mi'tral) valve between the left atrium and ventricle of the heart

molar (mo'lar) a grinding or back tooth, one of three on each side of the jaw

molar solution a solution in which one gram molecular weight of a substance is present in each liter of the solution

molecular weight weight of a molecule obtained by adding the weights of its constituent atoms, expressed in grams

monocyte (mon'o-sît) a large mononuclear leukocyte

monosaccharide (mon-o-sak'ah-rîd) a simple sugar that cannot be further decomposed by hydrolysis

motor neuron neuron that transmits a stimulus to muscle tissue

mucosa (mu-ko'sah) a mucus-producing lining membrane found in spaces connected with the outside, such as the alimentary and respiratory tract; mucous membrane

mucous membrane (mu'kus) mucus lining cavities and canals communicating with the air and kept moist by secretion of mucus

mucus (mu'kus) the thick fluid secretion of the mucous glands and mucous membrane

multipolar neuron nerve cell with one axon and many dendrites

muscle fatigue the reduced capacity of a muscle to perform work

myasthenia gravis (mi-as-the'ne-ah gra'vis) a disease characterized by chronic muscular fatigue brought on by the slightest exertion

myelin (mi'e-lin) a fatty substance forming a sheath or covering around many nerve fibers. Speeds impulse conduction

myelinated nerve fiber elongated process of a nerve cell possessing a myelin sheath

myocardial infarction (mi-o-kar'de-al) development of stoppage in cardiac muscle usually resulting from coronary thrombosis

myofibril (mi-o-fi'bril) a tiny fibril, found in muscular tissue, running parallel to the cellular long axis from one cell to another

myofilament (mi-o-fil'ah-ment) thin, threadlike component of a myofibril

myoneural junction (mi-o-nu'ral) a junction of the motor nerve fibers

myopia (mi-o'pe-ah) nearsightedness

myosin (mi'o-sin) a muscle protein acting as an enzyme to aid in initiating muscle contraction

myxedema (mik-se-de'mah) mental and physical sluggishness as the result of atrophy of the thyroid in the adult

nasolacrimal duct canal that transports the lacrimal secretion into the nose

nephron (nef'ron) the microscopic functional of kidney tissue, consisting of a glomerulus with its capsule, convoluted tubules, and Henle's loop, plus the collecting tubule

nerve deafness condition resulting from damage to the cochlea and/or the auditory pathways to the brain

neurilemma (nu-ri-lem'mah) a very thin membrane wrapping the nerve fibers of the peripheral nervous system

neuroglia (nu-rog'le-ah) pertaining to the glial or nonnervous cells of the nervous system

neurohypophysis (nu-ro-hi-pof'i-sis) posterior portion of the pituitary gland

neuron (nu'ron) the cell serving as the unit of structure and function of the nervous system; it is excitable and conductile

neutrophil (nu'tro-fil) a leukocyte that stains easily with neutral dyes

Nissl bodies chromophil substance in the form of granules found in the cell bodies and dendrites of neurons

nodes of Ranvier nonmyelinated gaps between the segments of sheath created by the Schwann cells

nondisjunction condition in which one or more pairs of homologous chromosomes fail to separate following synapsis

nonmyelinated fiber elongated process of a nerve cell not possessing a myelin sheath

norepinephrine (nor-ep-i-nef'rin) one of the "fight-or-flight" hormones, related to epinephrine; both are put to use in emergency situations

nucleolus (nu-kle'o-lus) a tiny globule located within the nucleus of a cell

nucleotide (nu'kle-o-tid) a unit or compound formed of a nitrogenous base, a five-carbon sugar, and a phosphoric acid radical

occipital (ok-sip'i-tal) relating to the back part of the head (the occiput)

occlusion (o-kloo'zhun) the state of being closed, blocked, or obstructed

olecranon fossa (o-lek'rah-non) posterior depression that receives the olecranon of the ulna when the forearm is extended

olecranon process large process of the ulna projecting behind the elbow joint and forming the bony prominence of the elbow

organ of Corti an elongated spiral structure running the entire length of the cochlea in the floor of the cochlear duct and resting on the basilar membrane

organelle (or-gan-el') a tiny specific particle of living material present in most cells and serving a specific function

osmosis (oz-mo'sis) the passage of a pure solvent, such as water, from a solution of lesser concentration to one of greater concentration through a semipermeable membrane

osmotic pressure (os-mot'ik) pressure developing when two solutions of different concentrations are separated by a membrane permeable to the solvent

ossification (os-i-fi-ka'shun) formation of bone substance

osteitis fibrosa (os-te-i'tis) a condition resulting from overactivity of the parathyroid glands with resulting disturbances in calcium and phosphorus metabolism. Characterized by decalcification and softening of bone

osteoblast (os'te-o-blast) immature or primitive bone

osteoclast (os'te-o-klast) giant, multinuclear cell found in depressions on the surface of a bone causing entire resorption of bone substance

osteocyte (os'te-o-sit) bone cell

osteomalacia (os-te-o-mah-la'she-ah) a disease marked by increasing softness of the bones so that they become flexible and brittle and cause deformities. Occurs chiefly in adults

osteomyelitis (os-te-o-mi-e-li'tis) inflammation of bone by a pus-producing organism. The condition may be a localized inflammation or it may spread through the marrow and bone tissue to the periosteum

otitis externa (o-ti'tis) inflammation of the outer ear

otitis media inflammation of the ear, most often the middle ear

oval window inner part of the ear that contains a membrane which vibrates and conducts waves to the internal ear

ovary (o'vah-re) one of two glands in the female producing the reproductive cell, the ovum, and two known hormones

oxygen debt (ok'si-jen) the amount of oxygen required after muscular activity for the removal of lactic acid and other metabolic products that accumulate when the supply of oxygen is below the needs of the organism

oxytocin (ok-se-to'sin) a pituitary hormone that stimulates the uterus to contract. It also acts on the mammary gland to force out milk

pancreatic duct (pan-kre-at′ik) narrow tubular vessel that conveys pancreatic juice to the duodenum

pancreatic juice a clear, viscid, alkaline digestive juice of the pancreas poured into the duodenum

pancreatitis (pan-kre-ah-ti′tis) inflammation of the pancreas

pancreozymin (pan′kre-o-zi-min) a hormone extracted from the duodenal mucosa that stimulates the secretion of pancreatic juice

parasympathetic (par-ah-sim-pah-thet′ik) pertaining to the portion of the autonomic nervous system that controls normal body functions; the craniosacral division

parathormone (par-ah-thor′mōn) an extract from fresh or frozen parathyroid glands of domestic animals that contains the active principle of these glands

parathyroid gland (par-ah-thi′roid) located near the thyroid gland in the neck; any of the four small glands embedded in the capsule covering the thyroid gland

parathyroid hormone hormone that regulates calcium and phosphorus metabolism. Deficiency results in tetany

parietal (pah-ri′e-tal) relating to the walls of a space or cavity

parietal peritoneum (pah-ri′e-tal per-i-to-ne′um) the serous membrane that lines the walls of the abdominopelvic cavity

parietal pleura (pah-ri′e-tal ploor′ah) the outer layer of serous membrane attached to the walls of the pleural cavity

Parkinson's disease (syndrome) progressive malfunction of the cerebral nuclei of the cerebrum characterized by tremors and rigidity of muscles

patent ductus arteriosis (pa′tent duk′tus ar-te-re-o′-sis) persistence after birth of the foramen ovale. A treatable form of congenital heart defect

pelvis (pel′vis) any basinlike structure; an oblong trough; the lower portion of the trunk of the body bounded by the sacrum and coccyx at the back, the two hipbones at the sides and front, and the tissues of the pelvic floor at the outlet

pelvis, false portion of the pelvis above the iliopectineal line

penicillinase (pen-i-sil′i-nās) a substance produced by bacteria that inactivates most but not all penicillins. Used to treat allergic reactions to penicillin

pepsinogen (pep-sin′o-jen) the zymogen or antecedent of pepsin existing in the form of granules in the chief cells of gastric glands

perforated ulcer an ulcer that permeates the entire thickness of the part, as the foot or intestine

pericardium (per-i-kar′de-um) the serous membrane that lines the sac enclosing the heart, plus the reflection that attaches itself to the heart itself

perichondrium (per-i-kon′dre-um) a membrane that covers the surface of cartilage

periosteum (per-e-os′te-um) the special fibrous connective tissue membrane covering the bones of the body; the surface tissue that plays an important part in the repair of bone fractures and other injuries

peristalsis (per-i-stal′sis) a rhythmic wavelike motion of the muscle tissue of the alimentary canal that moves the food through the digestive tube

peritoneal membrane (per-i-to-ne′al) serous membrane over the viscera and lining the abdominal cavity

peritonitis (per-i-to-ni′tis) inflammation of the peritoneum, the membranous coat lining the abdominal cavity

pernicious anemia (per-nish′us ah-ne′me-ah) a chronic anemia characterized by progressive decrease in red blood corpuscles, muscular weakness, and gastrointestinal and neural disturbances

phagocytosis (fag-o-si-to′sis) engulfing of particles by cells

phenotype (fe′no-tîp) physical appearance or makeup of an individual

phenylketonuria (PKU) (fen-il-ke-to-nu′re-ah) a disorder due to faulty chemistry of an amino acid (phenylalanine) required for growth and normal nitrogen balance, resulting in mental deficiency

pinocytosis (pi-no-si-to′sis) intake of solutions by cells through "sinking in" of the cell membrane

Pitocin (pi-to′sin) trademark for an aqueous solution containing the oxytocic fraction of the posterior pituitary gland

pituitary dwarfism (pi-tu′i-tār-e) abnormally small, short, or disproportioned condition caused by malfunction of the pituitary

placenta (plah-sen′tah) the structure attached to the inner uterine wall through which the fetus gets its nourishment and excretes its wastes

plasma cell (plaz′mah) one capable of producing antibodies in response to antigenic challenge

platelet (plāt′let) round or oval disk playing an important role in blood coagulation, hemostatis, and blood thrombus formation

pleura (ploor′ah) serous membrane that enfolds the lungs and lines the walls of the chest and diaphragm

pleurisy (ploor′i-se) inflammation of the serous membrane covering the lungs and lining the chest cavity

polycythemia (pol-e-si-the′me-ah) excess of red blood cells

polymorphonuclear granulocyte (pol-e-mor-fo-nu′kle-ar) white blood cell that possesses a nucleus composed of two or more lobes or parts

polysaccharide (pol-e-sak′ah-rĭd) one of a group of carbohydrates, which upon hydrolysis yields more than two molecules of simple sugar

portal system (por′tal) the portal vein and its branches by which blood is collected from abdominal viscera and conveyed to the sinusoids of the liver from which it passes through the hepatic veins to the inferior vena cava

postganglionic neuron (pōst-gang-gle-on′ik nu′ron) beyond or past a ganglion or synapse

postsynaptic neuron (pōst-si-nap′tik nu′ron) dendritic portion of a neuron located after the synapse

preganglionic neuron (pre-gang-gle-on′ik nu′ron) in front of or before a ganglion or synapse

prepuce (pre′pūs) foreskin or fold of skin over the glans penis

pressoreceptor (pres-o-re-sep′tor) sensory nerve ending such as those in the aorta and carotid sinus which are stimulated by changes in blood pressure

progesterone (pro-jes′te-ron) a hormone produced in the female sex glands that assists in the normal development of pregnancy

prolactin (pro-lak′tin) hormone derived from the anterior pituitary lobe that stimulates lactation

pronation (pro-na′shun) the act of lying prone or face downward; the act of turning hand so that palm faces downward or backward

proprioceptor center (pro-pre-o-sep′tor) receiving stimulations within the body tissues, especially in muscles, tendons, and in the inner ear

prostate gland (pros′tāt) gland surrounding male bladder neck and urethra

prothrombin (pro-throm′bin) chemical substance existing in circulating blood that interacts with calcium salts to produce thrombin

provitamin (pro-vi′tah-min) substance that can be transformed in the body to an active vitamin

pubic symphysis (pu′bik sim′fi-sis) joint between the two coxal bones

puerperal fever (pu-er′per-al) fever following childbirth resulting from absorption of septic products into blood and tissues

pulmonary circulation (pul′mo-ner-e) the venous blood received into the right atrium passes through the tricuspid valve into the right ventricle, then into the pulmonary artery to take up oxygen from inspired air

pulmonary edema (pul′mo-ner-e e-de′mah) abnormal accumulation of fluid in the lungs

pulmonary semilunar valve (pul′mo-ner-e sem-e-lu′nar) valve between the heart and the pulmonary artery

pulse pressure the difference between the systolic and diastolic pressures

Purkinje fibers (pur-kin′je) atypical muscle fibers lying beneath endocardium of heart which constitute the impulse-conducting system of the heart

pus (pus) liquid product of inflammation containing leukocytes or their remains and debris of dead cells

pyelonephritis (pi-e-lo-ne-fri′tis) an inflammation involving the kidney pelvis and the tissues of the kidney itself, usually due to an infection with microorganisms

pyloric antrum (pi-lor′ik an′trum) bulge in the pyloric portion of the stomach along the greater curvature on distention

pyloric sphincter (pi-lor′ik sfingk′ter) a thickening of the muscular wall around the pyloric orifice

refraction (re-frak′shun) the bending of a ray of light as it passes from one medium into another of different density

renal calculi (re′nal kal′ku-li) abnormal concretions within the kidney

renal papilla (re′nal pah-pil′ah) apex of a pyramid in the kidney

renal pyramid (re'nal pir'ah-mid) one of 8 to 18 cone-shaped structures comprising medulla of the kidney

renin (ren'in) protein formed in an ischemic kidney that acts as an enzyme

rennin (ren'-in) a coagulating enzyme found in the stomach of ruminants that curdles milk

respiratory acidosis (re-spi'rah-to-re as-i-do'sis) acidosis secondary to pulmonary insufficiency, resulting in retention of carbon dioxide

rete testis (re'te test'is) network of tubules that receive sperm from the seminiferous tubules

retina (ret'i-nah) the third and innermost coat of the eye; contains visual receptors (rods and cones)

retinene (ret'i-nēn) a derivative of vitamin A that is found in rhodopsin

Rh factor factor, discovered in the rhesus monkey, that may result in serious transfusion reactions

rhodopsin (ro-dop'sin) a visual pigment found in rod cells; "visual purple"

ribosome (ri'bo-sōm) dense aggregations of RNA and protein, usually attached to the endoplasmic reticulun; site of protein synthesis

rickets (rik'ets) disease of metabolism affecting children, characterized by ineffective nutrition, and often resulting in deformities

rugae (roo'je) folds as of mucous membrane found in the lining of the stomach and elsewhere in the body

sacroiliac joint (sa-kro-il'e-ak) the articulation between the hipbone and sacrum

sacrum (sa'krum) the triangular bone formed of five united vertebrae, wedged between the two innominate bones. It forms the base of the vertebral column

sagittal plane (saj'i-tal) like an arrow or arrowhead; a suture in the skull; a plane dividing the body into right and left portions

sagittal suture (saj'i-tal su'chur) arrow-shaped joint in the skull

salmonella (sal-mo-nel'ah) a genus of bacteria belonging to the family Enterobactericeae. Several species are pathogenic, some producing mild gastroenteritis, others producing a severe and often fatal food poisoning

sarcolemma (sar-ko-lem'ah) cell membrane of a skeletal muscle fiber

sarcomere (sar'ko-mēr) the portion of a myofibril lying between two adjacent dark lines

Schlemm's canal (schlemz kah-nal) ring of veins that lies at the junction of the sclera and cornea and drains the humor from the anterior chamber

sclera (skle'rah) the outer coat of the eyeball; the "white of the eye"

scotopsin (sko-top'sin) a protein substance found in rhodopsin

scrotum (skro'tum) a sac suspended between the thighs of the male that contains the testes

sebaceous glands (se-ba'shus) glands that secrete a fatty (oily) fluid

secretin (se-kre'tin) a hormone produced by the stomach and small intestine that stimulates other digestive organs to produce their digestive juices

secretion (se-kre'shun) the process of producing a new substance from materials in the blood; the new substance produced by glandular activity using materials in the blood

semen (se'men) the thick whitish secretion from the male reproductive organs; a combination of male germ cells and secretions from the several glands of the reproductive system

semilunar valve (sem-e-lu'nar) valve between heart and the aorta and valve between heart and the pulmonary artery

seminal vesicle (sem'in-al ves'i-k'l) one of the two saclike structures in the male lying behind the bladder that secrete a thick viscous fluid that forms a part of the semen

sensory cell (sen'so-re) sensory neuron; a cell which when stimulated gives rise to nerve impulses that are conveyed to the central nervous system

sensory neuron (sen'so-re nu'ron) an afferent neuron that conveys impulses that give rise to sensations

septicemia (sep-ti-se'me-ah) blood poisoning: the presence of bacterial toxins in the blood

septum (sep'tum) a wall dividing two cavities

serous membrane (se'rus) membrane producing a secretion having a watery nature

serum (se'rum) the watery portion of the blood remaining after coagulation

sickle cell anemia hereditary, chronic anemia characterized by presence of large numbers of crescent-shaped red blood cells in the blood

sinoatrial node (S-A node) (si-no-a′tre-al nōd) node at junction of superior vena cava with right cardiac

sinus (si′nus) a hollow in a bone or other tissue; a channel for blood; any cavity having a narrow opening

sodium pump (so′de-um) active transport of sodium

soft palate (pal′at) posterior muscular, membranous fold partly separating the mouth and pharynx

somatic reflex (so-mat′ik re′fleks) involuntary response induced by stimulation of somatic sensory nerve endings

somatotrophic hormone (STH) (so-mah-to-trof′ik hor′mōn) hormone produced by the anterior lobe of the hypopysis that regulates growth of the body

specific gravity (spe-sif′ik grav′i-te) weight of a substance compared to that of an equal volume of pure water, with water assigned a value of 1

spermatid (sper′mah-tid) cell arising by division of the secondary spermatocyte to become a spermatozoan

spermatocyte (sper′mah-to-sît) cell originating from a spermatogonium that forms by division the spermatids

spermatogenesis (sper-mah-to-jen′e-sis) the formation and development of the spermatozoa

spermatogonia (sper-mah-to-go′ne-a) large unspecialized germ cells which in spermatogenesis give use to a primary spermatocyte

spermatozoon (sper-mah-to-zo′on) the male sex cell, produced in the testes

sphincter (sfingk′ter) a band of circularly arranged muscle that narrows an opening when it contracts

sphincter of Oddi (sfingk′ter of odē) contracted region in common bile duct

sphygmomanometer (sfig-mo-mah-nom′e-ter) an instrument for measuring arterial blood pressure

spinal tract (spi′nal) a bundle of nerve fibers in the central nervous system that carries particular kinds of motor or sensory impulses

splenic flexure (splen′ik flek′sher) bend at junction of transverse with descending colon

splenomegaly (sple-no-meg′ah-le) enlargement of the spleen

squamous suture (skwa′mus su′chur) immovable joint of the skull between the parietal bones and the temporal bones

stapes (sta′pēz) the innermost of the ossicles of the ear, also called the stirrup

stratum corneum (stra′tum kor′ne-um) outermost horny layer of the epidermis

stratum germinativum (stra′tum jer-mi-na′ti-vum) innermost layer of epidermis; a row of columnar cells which divide to replace the rest of the epidermis as it wears away

stratum granulosum (stra′tum gran-u-lo′sum) a layer of cells containing deeply staining granules of keratohyalin found in the epidermis of the skin

stratum lucidum (stra′tum lu-sid′um) a translucent layer of the epidermis lying between the stratum corneum and stratum granulosum

substrate (sub′strāt) the substance an enzyme acts on

sucrase (su′krās) an enzyme in the intestinal juice that splits cane sugar into glucose and fructose, which are absorbed into the portal circulation

sulcus (sul′kus) a groove or depression between parts, especially a fissure between the convolutions of the brain; any furrow, as in the teeth, the bones, or in the lung surfaces. Plural: sulci

supinator (su-pi-na′tor) a muscle producing the motion of supination of the forearm

supraspinous (su-prah-spi′nus) a groove above the spine of the scapula

surface tension the phenomenon whereby liquid droplets tend to assume the smallest area for their volume. The surface acts as though it had a skin on it as a result of cohesion among the surface water molecules

surfactant (sur-fak′tant) a lipid-protein substance that reduces surface tension in the lung alveoli and thus reduces chances of alveolar collapse

suture (su′chur) a type of joint, especially in the skull, where bone surfaces are closely united

synapse (sin′aps) the region where parts of two neurons are anatomically related so that impulses are transmitted from one neuron to another; also called the synaptic junction

synarthrosis (sin-ar-thro′sis) a type of joint in which the skeletal elements are united by a continuous intervening substance. Movement is absent or limited, and a joint cavity is lacking

synovial membrane (si-no′ve-al) membrane producing thick fluid in the space between bones of a freely moving joint

systolic blood pressure (sis-tol′ik) greatest force caused by the contraction of the left ventricle of the heart

Tay-Sachs disease (ta-saks′) infantile form of family idiocy characterized by a disorder of lipid metabolism

tendon (ten′dun) fibrous connective tissue serving for the attachment of muscles to bones and other parts

tensor (ten′sor) any muscle that stretches or pulls on a part to make it tense

testosterone (tes-tos′te-ron) the hormone produced by the male sex glands

tetanus (tet′ah-nus) a sustained contraction of a muscle; a disease caused by a bacterium, characterized by a sustained contraction of jaw muscles ("lockjaw")

thalamus (thal′ah-mus) the part of the brain at each side of the third ventricle that acts as the chief relay center for sensory impulses to the cerebral cortex. It includes two large masses of gray matter

thoracic cavity (tho-ras′ik) space lying above the disphragm and enclosed within the walls of the thorax

thoracic vertebra (tho-ras′ik ver′te-brah) vertebra relating to the chest portion of the body

threshold of stimulation the least or minimal stimulus that will give rise to a sensation or bring about a response such as a muscle contraction

thrombin (throm′bin) enzyme formed in shed blood from prothrombin that reacts with soluble fibrinogen forming the basis of a blood clot

thrombocyte (throm′bo-sīt) a particular of protoplasm found in the circulating blood; a blood platelet believed to play a part in the process of blood clotting

thromboplastin (throm-bo-plas′tin) a substance released by the injured tissues that triggers the clotting mechanism

thymus gland (thi′mus) an unpaired organ located in the cavity above the heart important in the development of immune response in newborn

thyroglobulin (thi-ro-glob′u-lin) an iodine-containing protein secreted by the thyroid gland and stored within its colloid substance

thyroid gland (thi′roid) a gland of internal secretion located in the base of the neck on both sides of the lower part of the larynx and upper part of trachea

thyroid storm fulminating increase in all the signs and symptoms of intoxication due to excessive thyroid secretion

thyrotropic stimulating hormone (TSH) (thi-ro-trop′ik) hormone produced by anterior lobe of hypophysis that regulates development and functioning of the thyroid gland

thyroxine (thi-roks′in) the hormone produced by the thyroid gland. It increases the metabolic rate and is needed for normal growth

tinea capitis (tin′e-ah cap′i-tis) a fungus skin disease of the scalp, also called scalp ringworm

tonus (to′nus) partial, steady contraction of muscle that determines tonacity or firmness

trapezium (trah-pe′ze-um) the first bone in the distal row of carpal bones lying between the navicular and first metacarpal bones

treppe (trep′e) an increase in strength of muscular contraction when a muscle is stimulated maximally and repeatedly

tricuspid (tri-kus′pid) a valve in the right side of the heart that closes when the ventricle begins pumping

trigone (tri′gōn) triangular space at the base of the bladder

trochlea (trok′le-ah) pulleylike surface that articulates with the ulna

trypsin (trip′sin) an enzyme in the gastric juice that splits proteins into amino acids

trypsinogen (trip-sin′o-jen) the inactive form of trypsin found in pancreatic juice

tunica albuginea (tu′ni-kah al-bu-jin′e-ah) white fibrous coat of the eye, testicle, ovary, or spleen

tympanum (tim′pah-num) the cavity of the middle ear

ulcer (ul′ser)   an open sore or lesion in the skin or a mucous membrane-lined organ such as the stomach, mouth, or intestine
ulcerative colitis (ul′ser-a-tiv ko-li′tis)   ulceration of mucosa of colon
umbilical region (um-bil′i-kal)   referring to the section of the abdominal cavity around the navel in the central part of the abdomen
uracil (u′rah-sil)   a base found combined with D-ribose and phosphoric acid in yeast nucleic acid
urea (u-re′ah)   a nitrogen waste product excreted in the urine; an end-product of protein metabolism
ureter (u-re′ter)   in the urinary system, the two ureters conduct the secretion from the kidneys to the urinary bladder
urethra (u-re′thrah)   the excretory tube for the bladder
urinalysis (u-ri-nal′i-sis)   analysis of the urine
uterus (u′ter-us)   a muscular, pear-shaped organ in the female pelvis within which the fetus grows

vagina (vah-ji′nah)   the lower part of the birth canal, which opens to the outside of the body
vasa recta (va′sah rek′tah)   tubules that become straight prior to entering the mediastrinum testis
vas deferens (vas def′er-enz)   the excretory duct of the testis
vasoconstriction (vas-o-kon-strik′shun)   a decrease in the diameter of blood vessels, especially the constriction of the smallest arteries (arterioles), resulting in a decrease of blood to a part
vasodilation (vas-o-di-la′shun)   an increase in the diameter of blood vessels, especially the dilation of the smallest arteries (arterioles), resulting in an increase in the blood to a part
vasopressin (vas-o-pres′in)   a hormone formed in the hypothalamus. It has an antidiuretic and a pressor effect, elevating blood pressure
vena cava (ve′nah ca′vah)   principal vein; one branch drains the upper portion of the body and the other branch drains the lower portion of the body
ventral (ven′tral)   pertaining to the anterior or front side of the body; opposite of dorsal
venule (ven′ūl)   a small vein that gathers blood from capillary beds

vermiform appendix (ver′mi-form)   a worm-shaped process projecting from the cecum, whose mucous membrane also lines the appendix
vestibular apparatus (ves-tib′u-lar ap-ah-ra′tus)   the equilibrium structures of the inner ear and their nerves
villi (vil′i)   tiny, fingerlike projections in the mucosa which lines the small intestine
visceral peritoneum (vis′er-al per-i-to-ne′um)   the serous membrane that covers some of the organs of the digestive system and constitutes their serosa
visceral pleura (vis′er-al ploor′ah)   the inner layer of serous membrane that covers the lungs
vital capacity   volume of air that can be forcibly exhaled after full inspiration
vital signs   essential to or contributing to life
vitamin (vi′tah-min)   an essential organic substance, not serving as a source of body energy, but working with enzymes to control body function
vitreous humor (vit′re-us)   the glassy body of the eyeball found between the lens and retina
Volkmann's canal   small passageway found in bone through which blood vessels pass from the periosteum; connects with the blood vessels of Haversian canals or the marrow cavity
vulva (vul′vah)   female genitalia consisting of the labia majora, labia minora, clitoris, vestibule of the vagina, and bulbs of the vestibule
vulvovaginitis (vul-vo-vaj-i-ni′tis)   inflammation of the vulva and vagina

zygomatic arch (zi-go-mat′ik)   the formation on each side of the cheeks of the zygomatic process of each malar bone articulating with the zygomatic process of the temporal bone
zygomatic bone   commonly referred to as the cheekbone, forming the prominence of the cheeks and part of the outer wall and floor of the orbits
zygomatic process   a thin projection from the temporal bone bounding its squamous portion; a part of the malar bone helping to form the zygoma
zygote (zi′got)   the fertilized ovum; the cell formed by the union of the spermatozoon and the egg cell

Bibliography

CHAPTER 1: INTRODUCTION

Barnum, M. R., and Pruitt, M. L. 1974. *Use of the metric system.* Unpublished.

Conant, J. B. 1951. *Science and common sense.* New Haven, Conn.: Yale University Press.

Donovan, F. 1970. *Prepare now for a metric future.* New York: Weybright & Talley.

Fulton, J. F., ed. 1966. *Selected readings in the history of physiology.* Springfield, Ill.: Charles C. Thomas.

Ritchie-Calder, (Lord). 1970. Conversion to the metric system. *Scientific American* 223 (July):17–25.

Singer, C. 1957. *A short history of anatomy and physiology from the Greeks to Harvey,* 2d ed. New York: Dover.

CHAPTER 2: OVERVIEW OF ANATOMY AND PHYSIOLOGY

Landau, B. R. 1976. *Essential human anatomy and physiology.* Glenview, Ill.: Scott, Foresman.

Segerberg, O. J. 1974. *The immortality factor.* New York: Dutton.

Tortora, G. J., and Anagnostakos, N. P. 1975. *Principles of anatomy and physiology.* San Francisco: Canfield Press.

CHAPTER 3: CELL STRUCTURES AND FUNCTIONS

Brachet, J. 1961. The living cell. *Scientific American* 205:50–61. offprint 90.

Goldstein, L., ed. 1966. *Cell biology.* Dubuque, Iowa: Brown.

Kennedy, D., ed. 1965. The living cell. *Readings from Scientific American.* San Francisco: Freeman.

Kleinsmith, L. J. 1972. Molecular mechanisms for the regulation of cell function. *Bioscience* 22(6): 343–48.

Mazia, D. 1961. How cells divide. *Scientific American* 205(September):74–82. Offprint 131.

Solomon, A. K. 1962. Pumps in the living cell. *Scientific American* 207(August):100–108. Offprint 131.

Swanson, C. P. 1964. *The cell.* Englewood Cliffs, N.J.: Prentice-Hall.

CHAPTER 4: TISSUES, MEMBRANES, SKIN

Bloom, W., and Fawcett, D. W. 1968. *A textbook of histology,* 9th ed. Philadelphia: Saunders.

Ham, A. W. 1974. *Histology.* Philadelphia: Lippincott.

Kavchak-Keys, M. A. 1977. Treating decubitus ulcers using four proven steps. *Nursing 77* 7(September):58–61.

Porter, K., and Bonneville, M. A. 1968. *An introduction to the fine structure of cells and tissues.* Philadelphia: Lea & Febiger.

Roach, L. B. 1977. Color changes in dark skin. *Nursing 77* 7(January):48–51.

Spratt, N. T., Jr. 1964. *Introduction to cell differentiation.* New York: Van Nostrand Reinhold.

CHAPTER 5: THE SKELETAL SYSTEM

Anderson, W. A. 1971. *Pathology*, 6th ed. St. Louis: Mosby.
Beeson, P. B., and McDermott, W. 1975. *Cecil-Loeb textbook of medicine.* Philadelphia: Saunders.
Beck, W. S. 1971. *Human design: molecular, cellular and systematic physiology.* New York: Harcourt Brace Jovanovich.
Griffin, D. R., and Hovick, A. 1970. *Animal structure and function.* New York: Holt, Rinehart & Winston.
Walker, P. 1977. Bone marrow transplant. *Nursing 77* 7(January):24–25.

CHAPTER 6: THE MUSCULAR SYSTEM

Basmajian, J. V. 1967. *Muscles alive.* Baltimore: Williams & Wilkins.
Brunner, L. S., and Suddarth, D. S. 1975. *Textbook of medical-surgical nursing*, 3d ed. Philadelphia: Lippincott.
Farago, P., and Lagnado, J. 1972. *Life in action.* New York: Vintage.
Gray, J. 1968. *Animal locomotion.* New York: Norton.
Huxley, H. E. 1965. The mechanisms of muscular contraction. *Scientific American* 213(December):18–27.
Katz, B. 1966. *Muscle, nerve and synapse.* New York: McGraw-Hill.
Margaria, R. 1972. The sources of muscular energy. *Scientific American* 226(May):84–91.
Merton, P. A. 1972. How we control the contraction of our muscles. *Scientific American* 226(May):30–37.

CHAPTER 7: THE NERVOUS SYSTEM

Axelrod, J. 1974. Neurotransmitters. *Scientific American* 230(September):58–66.
Baker, P. F. 1966. The nerve axon. *Scientific American* 214(March):74–82.
Beck, W. S. 1971. *Human design: molecular, cellular and systematic physiology.* New York: Harcourt Brace Jovanovich.
Heimer, L. 1971. Pathways in the brain. *Scientific American* 225(July):48–57.
Katz, B. 1966. *Muscle, nerve and synapse.* New York: McGraw-Hill.
Schmidt-Nielson, K. 1970. *Animal physiology*, 3d ed. Englewood Cliffs, N.J.: Prentice-Hall.
Stent, G. S. 1972. Cellular communications. *Scientific American* 227(March):42–51.
Zacke, S. 1964. *The motor endplate.* Philadelphia: Saunders.

CHAPTER 8: COORDINATION: THE SENSES

Amoore, J. E. et al. 1964. The stereochemical theory of odor. *Scientific American* 210(February):42–49.
Case, J. 1966. *Sensory mechanisms.* New York: Macmillan.
Langley, L. L. et al. 1974. *Dynamic anatomy and physiology*, 4th ed. New York: McGraw-Hill.
Miller, W. H. et al. 1961. How cells receive stimuli. *Scientific American* 205(April):222–38.
Neisser, A. 1968. The processes of vision. *Scientific American* 219(September):204–208.
McVan, B. 1977. Odors: what the nose knows. *Nursing 77* 7(April):46–49.
Vander, A. J. et al. 1970. *Human physiology: the mechanisms of body function.* New York: McGraw-Hill.
Werblin, F. S. 1973. The control of sensitivity in the retina. *Scientific American* 228(January):70–79.
Young, R. W. 1970. Visual cells. *Scientific American* 223(April):80–84.

CHAPTER 9: THE HEART AND BLOOD VESSELS

Adolph, E. F. 1967. The heart's pacemaker. *Scientific American* 216(March):32–37. Offprint 1067.
Berne, R. M., and Levy, M. N. 1972. *Cardiovascular physiology*, 2d ed. St. Louis: Mosby.
Brunner, L. et al. 1970. *Textbook of medical-surgical nursing.* Philadelphia: Lippincott.
D'Alonzo, C. A. 1962. *Heart disease, blood pressure and strokes.* Houston: Gulf.
Kilgour, F. G. 1952. William Harvey. *Scientific American* 185(June):56–60.

Guyton, A. C. 1967. *Functions of the human body.* Philadelphia: Saunders.
———. 1971. *Textbook of medical physiology.* Philadelphia: Saunders.
Longmore, D. 1971. *The heart.* New York: McGraw-Hill.
Spain, D. M. 1966. Atherosclerosis. *Scientific American* 215(February):48–50.
Sweetwood, H. 1977. Patients with pacemakers. *Nursing 77* 7(March):44–51.
Wiggers, C. J. 1957. The heart. *Scientific American* 196(May):74–87.
Wood, J. E. 1968. The venous system. *Scientific American* 218(January):86–96.
Zweifach, B. 1959. The microcirculation of the blood. *Scientific American* 200(January):54–60.

CHAPTER 10: BLOOD

Bevelander, G. 1970. *Essentials of histology,* 6th ed. St. Louis: Mosby.
Calin, A. 1977. Immunology: a facet of aging. *Geriatrics* 32(September):54–55.
Desotell, S. 1977. A brighter future for leukemia patients. *Nursing 77* 7:18–21.
Perutz, M. F. 1964. The hemoglobin molecule. *Scientific American* 211(May):64–76.
Ponder, E. 1957. The red blood cell. *Scientific American* 196(January):95–98.
Snively, W., and Thuerbach, J. 1969. *Sea of life.* New York: McKay.
Zucker, M. B. 1961. Blood platelets. *Scientific American* 204(February):58–64.

CHAPTER 11: THE LYMPHATIC SYSTEM

Abramoff, P., and La Via, M. F. 1970. *Biology of the immune response.* New York: McGraw-Hill.
Cooper, M. C., and Lawton, A. R. 1974. The development of the immune system. *Scientific American* 231(November):58–70.
Glasser, R. J. 1977. How the body works against itself. *Nursing 77* 7(September):38–43.
Hoyle, G. 1970. How muscle is turned on and off. *Scientific American.* 222(April):84–93.

Jerne, N. K. 1973. The immune system. *Scientific American* 229(July):52–60.
Lerner, R. A., and Dixon, F. J. 1973. The human lymphocyte as an experimental animal. *Scientific American* 228(June):82–91.
Mayerson, H. S. 1963. The lymphatic system. *Scientific American* 208(June):80–90.
Nester, E. W.; Roberts, C. E.; Pearsall, N. N.; McCarthy, B. J. 1973. *Microbiology, molecules, microbes and man.* New York: Holt, Rinehart & Winston.

CHAPTER 12: THE ENDOCRINE SYSTEM

Foster, G. V. 1968. Calcitonin (thyrocalcitonin). *New England Journal of Medicine* 279(August):349–60.
Greene, R. 1970. *Human hormones.* New York: McGraw-Hill.
Guyton, A. C. 1974. *Functions of the human body,* 4th ed. Philadelphia: Saunders.
Netter, F. 1965. *The endocrine system and selected metabolic diseases.* New York: Ciba Pharmaceutical.
Selye, H. 1956. *The stress of life.* New York: McGraw-Hill.
Whalen, R. E. 1967. *Hormones and behavior.* New York: Van Nostrand Reinhold.
Williams, R. 1968. *Textbook of endocrinology.* Philadelphia: Saunders.

CHAPTER 13: THE DIGESTIVE SYSTEM

Anthony, C. P., and Kolthoff, N. J. 1975. *Textbook of anatomy and physiology,* 9th ed. St. Louis: Mosby.
Best, C. H., and Taylor, V. B. 1958. *The living body,* 4th ed. New York: Holt, Rinehart & Winston.
Davenport, H. W. 1971. *Physiology of the digestive tract,* 3d ed. Chicago: Yearbook Medical Publishers.
———. 1972. Why the stomach does not digest itself. *Scientific American* 226(January):86–93.
Goss, C. M. 1973. *Gray's anatomy,* 29th ed. Philadelphia: Lea & Febiger.
Selkurt, E. E. 1975. *Basic physiology for the health sciences.* Boston: Little, Brown.
Vander, A. J., Sherman, J. H., and Luciano, D. S. 1975. *Human physiology: the mechanisms of body functions,* 2d ed. New York: McGraw-Hill.

CHAPTER 14: NUTRITION AND METABOLISM

Barnum, M. R.; Gillespie, R. J.; Greer, A. J.; and Peardon, L. M. 1969. *Introductory biology principles.* Beverly Hills: Glencoe Press.
Diehl, H. S. 1973. *Textbook of healthful living,* 9th ed. New York: McGraw-Hill.
Goodhart, R. S., and Shils, M. E. 1973. *Modern nutrition in health and disease,* 5th ed. Philadelphia: Lea & Febiger.
Robinson, C. H. 1973. *Fundamentals of normal nutrition,* 2d ed. New York: Macmillan.
Schutte, K. H. 1964. *The biology of the trace elements: their role in nutrition.* Philadelphia: Lippincott.
Williams, S. R. 1977. *Nutrition and diet therapy,* 3d ed. St. Louis: Mosby.

CHAPTER 15: THE EXCRETORY SYSTEM

Beeson, P. B., and McDermott, W. 1975. *Cecil-Loeb textbook of medicine.* Philadelphia: Saunders.
DeCoursey, R. M. 1974. *The human organisms,* 4th ed. New York: McGraw-Hill.
Easton, Dexter. 1974. *Mechanisms of body functions.* Englewood Cliffs, N.J.: Prentice-Hall.
Ganong, W. F. 1973. *Review of medical physiology,* 6th ed. Los Altos, Calif.: Lange Medical Publications.
Juliani, L. 1977. Kidney transplant: your role in aftercare. *Nursing 77* 7(October):46–53.
Smith, H. W. 1953. The kidney. *Scientific American* 188(January):40–48. offprint 37.
Winton, F. R., and Bayliss, L. E. 1969. *Human physiology,* 6th ed. Revised and edited by O. C. J. Lippold and F. R. Winton. Baltimore: Williams & Wilkins.

CHAPTER 16: BODY FLUIDS AND ELECTROLYTES

Brooks, S. M. 1973. *Basic facts of body water and ions,* 6th ed. New York: Springer.
Goldberger, E. 1970. *A primer of water, electrolyte and acid-base syndromes.* Philadelphia: Lea & Febiger.
Howells, E. M. 1977. Managing fluids and electrolytes in surgical patients. *Geriatrics* 32(May):100–101.
Metheny, N., and Snively, W. 1967. *Nurses handbook of fluid balance.* Philadelphia: Lippincott.
Moyer, C. 1952. *Fluid balance.* Chicago: Yearbook Medical Publishers.
Weisberg, H. F. 1962. *Water, electrolyte and acid-base balance.* Baltimore: Williams & Wilkins.

CHAPTER 17: THE RESPIRATORY SYSTEM

Birren, J. E. 1961. *Handbook of aging and the individual.* Chicago: University of Chicago Press.
Boyd, W. 1970. *A textbook of pathology.* Philadelphia: Lea & Febiger.
Brunner, L., and Suddarth, D. 1970. *Textbook of medical-surgical nursing.* Philadelphia: Lippincott.
Clements, J. A. 1962. Surface tension in the lungs. *Scientific American.* 207(December):120–24.
Comroe, J. H., Jr. 1966. The lung. *Scientific American* 214(February):56–68.
Fenn, W. O. 1960. The mechanisms of breathing. *Scientific American* 202(January):138–48.
Guyton, A. 1976. *Textbook of medical physiology.* Philadelphia: Saunders.
Lynne-Davies, P. 1977. Influence of age on the respiratory system. *Geriatrics* 32(August):57–60.
Rossier, R. H. 1960. *Respiration.* St. Louis: Mosby.
Snively, W. D., and Beshear, D. R. 1972. *Textbook of pathophysiology.* Philadelphia: Lippincott.
Sweetwood, H. 1977. Acute respiratory insufficiency. *Nursing 77.* 7(December):24–31.

CHAPTER 18: HUMAN REPRODUCTION

Basmajian, J. V. 1975. *Grant's method of anatomy,* 9th ed. Baltimore: Williams & Wilkins.
Bonner, J. T. 1963. *Morphogenesis.* New York: Atheneum.
Csapo, A. 1958. Progesterone. *Scientific American* 198(April):4–46, offprint 163.
Demarest, R. J., and Sciarra, J. J. 1969. *Conception, birth and contraception: a visual presentation.* New York: Blakiston.
Ham, A. W. 1974. *Histology.* Philadelphia: Lippincott.
Jacob, S. W., and Francone, C. A. 1970. *Structure and functions in man.* Philadelphia: Saunders.

Katchadorian, H., and Lunde, D. T. 1972. *Fundamentals of human sexuality.* New York: Holt, Rinehart & Winston.

Masters, W., and Johnson, V. 1966. *Human sexual response.* Boston: Little, Brown.

Snively, W. D., and Beshear, D. R. 1972. *Textbook of pathophysiology.* Philadelphia: Lippincott.

Tyler, A. 1954. Fertilization and antibodies. *Scientific American* 190(June):70–75.

CHAPTER 19: MEDICAL GENETICS

Bartalos, M. 1968. *Genetics in medical practice.* Philadelphia: Lippincott.

Federman, D. D. 1967. *Abnormal sexual development.* Philadelphia: Saunders.

McKusick, V. A. 1964. *Human genetics.* Englewood Cliffs, N.J.: Prentice-Hall.

Saxen, L., and Rapola, J. 1969. *Congenital defects.* New York: Holt, Rinehart & Winston.

Stempfel, R. S., and Thomkins, G. M. 1966. *The metabolic basis of inherited disease.* New York: McGraw-Hill.

Votano, J. R., Gorecki, M., and Rich, A. 1977. Sickel hemoglobin aggregation. *Science* 96 (June 10): 1216–19.

APPENDIX A: BASIC CHEMISTRY

Barry, J. M., and Barry, E. M. 1969. *An introduction to the structure of biological molecules.* Englewood Cliffs, N.J.: Prentice-Hall.

Benfey, O. T. 1964. *From vital force to structural formulas.* Boston: Houghton Mifflin.

Brown, T. L., and LeMay, E. H. 1977. *Chemistry: the central science.* Englewood Cliffs, N.J.: Prentice-Hall.

Eblin, L. P. 1968. *Chemistry: a survey of fundamentals.* New York: Harcourt Brace Jovanovich.

Grillot, G. F. 1974. *A chemical background for the paramedical sciences.* Reading, Mass.: Addison-Wesley.

Hoffman, K. B. *Chemistry for the applied sciences.* Englewood Cliffs, N.J.: Prentice-Hall.

Holum, J. R. 1975. *Elements of general and biological chemistry.* New York: Wiley.

Sackheim, G. I., and Schultz, R. M. 1969. *Chemistry for the health sciences.* New York: Macmillan.

APPENDIX B: BASIC PHYSICAL CONCEPTS

Flitter, H. 1972. *An introduction to physics in nursing,* 6th ed. St. Louis: Mosby.

Index

Boldface numbers refer to illustrations

A

Abdomen
  arteries of, 155, **156**
  surgical regions, 13, 16, **15**
Abdomino-pelvic cavity, 16, 39, **17**
Abdominal reflex, 106
Abducens nerve, 118, **117**
Abductor, 85
ABO factors, 175-76, 177, 178
Absorption in intestines, 221
Accelerator nerves of heart, **144**
Acetabulum, 60, **62**
Acetic acid molecule, **329**
Acetoacetic acid, 205
Acetone, 205, **330**
Acetyl coenzyme A, 233, 234
Acetylcholine, 80, 95, 106-7, 121, 147
Achalasia, 228
Achilles reflex, 106
Achondroplasia, 303
Acids, 324-25
  organic, 329-30. *See also* Amino acids, Fatty acids, *names of specific acids*
Acid-base balance, 167, 251-52, 266
Acid-base combinations, 325
Acidosis, 200, 266, 277
Acne, 43-44
Acoustic nerve, 118, 135, **117, 133**
Acromegaly, 71-72, 195
ACTH. *See* Adrenocorticotropic hormone
Actin, 78, **78, 81**
Action potential, 19
Active transport, 17-18, 260, **18**
Addison's disease, 187, 201
Adductor, 85
Adenine, 334, 336
Adenohypophysis, 191, 194-96, **193**
Adenosine diphosphate (ADP), 81, 233, **234, 235**
Adenosine triphosphate (ATP), 17, 28, 81-82, 197, 342
  systhesis, 233-35, **234, 235**
ADH. *See* Antidiuretic hormone
Adhesion, 353

Adipose tissue, 37
ADP. *See* Adenosine diphosphate
Adrenal cortex, 195, 196, 199, 200-201, 261, **200**
Adrenal gland, 195, 196, 199-201, 206, **119, 200**
Adrenal medulla, 120, 195, 199, **200**
Adrenalin, 120, 121, 195, 199
Adrenergic neurons, 121
Adrenocorticotropic hormone (ACTH), 195, 196
Adrenogenital syndrome, 199
Aerobacter aerogenes, 255
Aerobic respiration, 81-82
African diet, fiber in, 225
Agammaglobulinemia, 305, 306
Aging, effects on
  blood, 180
  bone, 37, 74
  cardiovascular system, 163-64
  digestive system, 230
  endocrine function, 206
  immune system, 188
  kidneys, 258
  muscles, 96
  nervous system, 122
  reproductive system, 297
  respiratory system, 283
  senses, 127, 134, 136
  skin, 44
Agglutinins and agglutinogens, 175
Agranulocytes, **171**
Air flow obstruction, 278
Air volumes, respiratory, 281
Alanine, 333, 340, **334**
Albumin, 253, 254
Alcohol, 194, 216, 329
Aldehydes, **330**
Aldohexose. *See* Glucose
Aldosterone, 195, 200, 249-50, 261
Algae, 24
Alimentary canal. *See* Digestive system
Alkali, 324. *See also* Base
Alkalosis, 200, 266, 277
All-or-nothing law of muscular contraction, 83

Allergy, 44, 187-88
Alpha cells, 202, 204
Altitude and polycythemia, 180
Alveoli, 269, 273-74, **270, 273, 279**
Alzheimer's presenile psychosis, 306
Ameba, 30
Ameboid, 30
Amino acids
  chemistry, 333-34
  decomposition, 195, 201
  essential, 333, 334
  in protein digestion, 215, 227, 237
  in triplet codes, 340
Amino groups, 227, 237, **334**
Ammonium and ammonium hydroxide, 324
Amniocentesis, 310-11, **312**
Amniotic cavity, **312**
Amphiarthroses, 72
Ampulla of ear, 134
Ampulla of Vater, **226**
Amylase, 218, 219, 220, 223
Amyloidosis, 306
Anaerobic respiration, 81, 235
Anal reflex, 106
Anal sphincter, 225
Anaphylaxis, 188
Anatomy, 12, 21-22
Androgens, 201, 290
Anemia, 169-70, 178-79, 187, 188
  and oxygen transport, 278, 279
  pernicious, 169-70
  sickle cell, 179, 303, 308-9
Anesthesia and hyperthermia, 84
Aneurysm, 149-50, 163
Angina pectoris, 161
Animal starch. *See* Glycogen
Anions, 262, 265
Ankle bones, 66, **55, 65**
Annular ligament, 58
Anoxia, 278
Antagonistic muscles, 82-83, **83**
Anterior aspect, 13, **12, 14**

Anterior descending artery, 152, **139**
Anterior pituitary, 191, 194–96, **193**
Anterior tibial syndrome, 92
Antibiotics and ultraviolet sensitivity, 357
Antibodies, 44, 175, 185, 186, 195
Anticoagulants, 174
Antidiuretic hormone (ADH) 191, 194, 195, 251, 261
Antigen, 44, 175, 186
Antigen-antibody reaction, 44, 186
Antiinflammatory agents, 195
Antral pump, 213
Anus, 208
Anvil of ear, 70, 131, **133, 135**
Aorta, 140, 152, 161, **139, 141, 154, 156, 200**
 fetal, **160**
 and kidneys, **243**
Aortic arch, **139, 153, 156**
Apnea, 277
Aponeurosis, 82, **82, 83**
Appendicitis, 223, 225, 229
Appendicular skeleton
 lower, 60, 63–66, **62, 63, 64, 65**
 upper, 54, 57–60, **57, 58, 59, 60**
Appendix, vermiform, 223, **217, 224**
Appetite, 116
Aqueous humor, 125, **125, 132**
Arachnoid, 109, **110**
Archimedes' principle, 354, **355**
Arginine, 334, 340
Arteries, 139, 148, 149–50, 152–55
 of abdomen and chest, 155, **154, 156**
 diseases and disorders, 161–64
 of head, shoulder, and arm, 152–53, 155, **152, 153**
 of leg and foot, 155, **157**
Arterioles, 151, **151**
 of nephron, 245, 246, **247**
Arteriosclerosis, 162–63, 255
Arm
 arteries, 152–53, **153**
 bones, 58–60, **59**
 muscles, 85, 89, **83, 90, 91**
 veins, 159, **158**
Arthritis, 74, 187
Articular cartilage, 52, **40, 51**
Articular cavities, 72
Articulations, 72. *See also* Joints
Artificial environment in tissue culture, 31
Asparagine, 340
Aspartic acid, 340
Aspects of the body, 12–13, **12, 13**

Asthma, 278
Astigmatism, 127
Astrocyte, 102
ATP. *See* Adenosine triphosphate
Atelectasis, 278
Atherosclerosis, 163
Athlete's foot, 44
Atlas, 70
Atom(s), 318–21, **318, 319, 322**
 structure, 318–19
Atomic number 319
Atomic weight, 319–20
Atrial septum defect, 161
Atrioventricular (A-V) node, 145–46, **144**
Auditory canal, 131, **133**
Auditory meatus, 66, **67**
Auditory nerve, 118, 135, **117, 133**
Autoclave, 350–51
Autoimmunity, 95, 187–88, 255
Autonomic functions, control of, 116
Autonomic nervous system, 118–21, 146, **119**
Autosomal gene defects, 303
A-V (atrioventricular) node, 145–46, **144**
Axial skeleton, 54, 66–70, **67, 68, 69**
Axial vein, 159
Axillary arteries, 152, 153, 155, **153, 156**
Axillary vein, 159, **158**
Axis, 70
Axon, 99, 102, **27, 99, 101, 103, 107**
 in myoneural junction, **80**

B

Bacteria, 24, 179, 224, 229
Ball-and-socket joint, **83**
Banting, Frederick G., 202
Barometers, 346–47, **346**
Barr body, 310
Basal metabolism test (BMT), 197
Bases, 324–25
Base bicarbonate, 266
Basement membrane, 35, **35**
Basilar artery, 153, 155
Basilar membrane, 134, **134**
Basilic vein, 159, **158**
Basophil, 167, 171, **166, 171**
Bedsores, 44
Benzene, 330–31, **330, 331**
Bernard, Claude, 2
Best, Charles H., 202
Beta cells, 202, 205
Bicarbonate, 251, 261, 266
Biceps
 brachii, 89, **83, 90, 91**
 femoris, 92, **93, 94**
 reflex, 106
Biconcave lenses, 129, **130**
Biconvex lenses, 129, **130**

Bicuspid valve, 140, **141**
Bile, 171, 218, 220–221
Bile duct, 220, 221, **203, 218, 226**
Bilirubin, 171, 220, 221
Bioelectricity, 18–19
Bipolar cells, **126**
Birth canal, 291. *See also* Vagina
Birth defects, 306, 308–11
Birthweight, low, 308–9
Black skin color changes, 41
Blacks and sickle cell anemia, 303
Bladder. *See* Gallbladder, Urinary bladder
Blister, 39
Blockage, digestive, 228
Blood, 38, 165–81
 and aging, 180
 analysis, 179
 and breathing rate, 116
 calcium levels, 195, 198, 199, 256
 cells, 167, **27, 166**. *See also* Erythrocytes, Leukocytes
 circulation patterns, 151–61, **152**
 Clotting, 172–75, **173**
 components, 167–74, **166, 167, 171**
 count, 168, 172
 diseases and abnormalities, 178–80
 flow resistance, 142
 groups, 175–78, **176**
 and liver, 225
 oxygen levels, 277
 sugar levels, 202, 227
 summary, 180–81
 transfusion, 174, 175
 viscosity, 352
 volume, 162
Blood pressure, 115, 146, 150, 246. *See also* Hypertension
 measurement, 142–43, **143**
Body
 cavities, 16, **17**
 locations, 12–16, **12, 13, 14, 15**
 odors, 136
 planes, 12, 13, **12, 14**
 position and inner ear, 134–35
 temperature, 84, 116, 235, 355–56, **356**
Body fluids, 259–61, 267
 and body weight, 260
 flow direction, 265
 functions, 260
 regulation, 260–61
 summary, 267
Bohr, Niels, 318
Boiling point, 350, **350**
Bolus, 209, **211**
Bone(s) 49–75, **50, 51, 53, 54**. *See also* Skeletal,
 and aging, 74
 cells. *See* Osteocytes
 classification, 51
 composition, 49–50, **50, 51**

Bones. *(continued)*
  diseases, 73–74
  formation, 52–53
  growth, 52–54, 70–72, **54**
  individual, 54–70
  length, and height, 53
  markings, 52
  marrow, 50–51, 169, 172, 185–86, **50, 51**
  morphology, 52
Bordetella pertussis, 282
Bowman's capsule, 187, 246–47, **245, 247, 248**
Boyle, Robert, 347
Boyle's law, 347–48, **347**
Brachial artery, 152, 155, **153, 155, 156**
  in blood pressure measurement, 142–43
Brachial vein, 159, **158**
Brachialis, 89, **90**
Brachiocephalic artery, **152, 153, 154**
Brachiocephalic vein, 159, **158**
Brachioradialis, 89, **90**
Brain, 111–16, **111, 112, 113, 114, 117**
  areas, **114**
  damage, 121, 174–75
  fissures, 111, 114, **111, 114, 117**
  folds, 111, **111, 114**
  furrows, 111, **111, 117**
  geography, **111**
  hemispheres, 115, **112**
  hormone, 112
  lobes, 113, **114, 117**
  and nerve impulses, 102
  and reflexes, 104–5, **105**
  tumors, 112, 116
  ventricles, 111, **113**
Breastbone, 55, 57, 70, 71
Breath holding, 276
Breathing, 269–78. *See also* Respiratory system
  abnormal, 270
  apparatus, 269
  mechamisms, 269–73
  regulation, 116, 274–77
  types, 277–78
Bronchi, 269, **119, 270, 271, 273**
Bronchial artery, **156**
Bronchial tubes, 269
Bronchioles, 269, **270, 273**
Bronchitis, 282
Brow, 66
Brunner's glands, 218
Buccinator, 87, **86**
Buffers, 325
Bulbae fascia, 125
Bulbourethral glands, 287, **286**
Bundle branch, **144**
Bundle of His, 145, **144**
Buoyant force, 354
Burkitt, Denis P., 225
Burns, 45

C

Caffeine, 194, 251
Calcaneus, 66, **56, 65, 94**
Calcitonin, 195, 198, 199
Calcium, 70–71, 239, 316
  absorption, 256
  in blood, 173, 174, 195, 198, 199
  in bone, 37, 50, 73
  as electrolyte, 261, 264–65
  level and aging, 74
  milliequivalents, 262
  in muscle contraction, 80–81
Calcium hydroxide, 324
Calculi, 254, 256, **256**. *See also* Gallstones
Callus, 72
Caloric value of foods, 235
Calorie, 9, 235
Canals of Schlemm, 125
Canaliculi, 50, **50**
Cancer, 31, 180, 187, 188, 225, 297
Candida albicans, 296
Candlepower, 357
Canine teeth, **209**
Cannon, Walter B., 2
Capillaries, 41–42, 139, 151, **151**
Capillary bed, **151**
Capillarity, 353
Capitate carpal, 60, **61**
Capsule, kidney, 244, **244**
Carbaminohemoglobin, 280
Carbohydrates, 331–33
  caloric value, 235
  digestion, 221, **223**
  digestion, 221, **223**
  metabolism, 227, 233–35
Carbolic acid, 331
Carbon, 316, 319, 325, 326–28, **319**
Carbon dioxide, 2–3, 328
  and breathing, 277
  and cardiovascular system, 146, 151
  transport, 269, 279–81, **280**
Carbon monoxide, 136, 282
Carbon tetrachloride, **328**
Carbonate, 324
Carbonic acid, 266, 280, 324
Carbonic anhydrase, 280
Carboxyl group, 329, 333, **334**
Carboxypeptidase, 219, 220
Cardiac muscle, 37, 77, 139, **38, 77**
Cardiac acceleratory center, 146
Cardiac inhibitory center, 146
Cardiac output, 141–142
Cardiac sphincter, 212, 228, **211, 213**
Cardiovascular system, 138–64. *See also* Arteries, Blood, Heart, Veins
  and aging, 163–64
  circulation patterns, 151–61, 162
  diseases, 121, 161–64
  structures, 139–51
  and vagus nerve, 118

Carotene, 40
Carotid arteries, 153, 155, **141, 152, 153, 154**
Carotid sinus, 146
Carpals, 60, **91**
Carpus, **55, 56**
Cartilage, 36–37, 49, **48, 49, 53**
Casts in urine, 254
Catalyst, 237
  enzymes as, 20–21
Cataract, 127, 136
Cation, 262
Caudal aspect, 12, **12, 14**
Cecum, 223, **217, 224**
Celiac artery, 155, **156**
Cell(s), 12, 23–32. *See also* Erythrocyte, Leukocyte, Osteocyte, Neuron
  activities, 24, 30–32
  blood, 167, **166**
  body, 27
  bone, 37
  cartilage, 49, **48**
  connective, 36–37, **37**
  epithelial, 35–36, **34, 35**
  fat, **37**
  fluids, 28–29
  growth rate, 26
  membrane, 26–27, **25**
  movement, 30
  muscle, 37–38, **38**
  nerve, 38
  reproduction of, 30
  size, 24, 26
  specialization, 36
  structure, 23–30, **25**
  summary, 32
  tendon, 37
  theory, 24, 30
  types, 26, **27**
  wall, **25**
Cell-cluster hypothesis, 275, **276**
Cells of Hensen, **134**
Cellular organelles, 24, 26
Cellular respiration, 269, **234**
Celsius, Anders, 7
Celsius scale of temperature measurement, 7–9, **9**
Cementum, tooth, 208
Centigrade *see* Celsius
Centriole, **25**
Centromere, 300, **300**
Cephalic vein, 159, **158**
Cerebellum, 113, 114–115, 117
Cerebral cortex, 112
Cerebral palsy, 121–22

Cerebrospinal fluid, 109, 111
Cerebrum, 111–114, **111, 114**
Cervical curvature, 70
Cervical nerve, 109, **119**
Cervical vertebrae, 70, **55, 56, 69**
Cervix, 291, 297, **292**
Charles' law, 348
Cheeks, 209, **209**
Cheek bone, 66, 70, **67, 210**
Chemistry, 315–43
   inorganic, 318–25
   organic, 325–42
   summary, 342–43
Chemoreceptors, 277
Chemotaxis, 172
Chest arteries, 155, **154**
Chest breathers, 272
Childbirth, oxytocin in, 194
Chlamydia, 132
Chloride ion, 265, 322, 323
Chloride shift, **280**
Chlorine, 239, 316
   atom, **322, 323, 326**
Choking, 212, 269, 278
Cholecystokinin, 214, 220–21
Cholesterol, 163, 200, 220, 221, 227
Cholinergic neuron, 121
Cholinesterase, 80, 95, 107, 121
Chondrocyte, 49, **48**
Chordae tendineae, 140, **141**
Chorea, Huntington's, 303
Chorionic gonadotropin, 290
Chorionic plate, **312**
Choroid layer, 125, **125**
Chromatid, 300, **300**
Chromation, **25**
Chromosome, 300–302, **300**
   abnormalities, 305, 307
   number, 32, 289, 290, 295, 300
Chronic obstructive pulmonary disease (COPD), 283
Chyme, 213
Chymotrypsin, 219, 220
Chymotrypsinogen, 220
Cilia, 28, 34, 35, 36, 269
Ciliary body, 127, **125, 132**
Ciliated border, **25**
Circle of Willis, 153
Circulation, fetal, 159, 161, **160**
Circulatory shock, 161–62
Circulatory system. *See* Cardiovascular system
Circumcision, 289
Circumflex artery, 152, **139**
Citric acid, 233, 234, 235

Clavicle, 58, **55, 57**
Cleft lip and palate, 305, 308–9
Clitoris, 292, **292**
Cloning, 36
Clostridium botulinum, 230
Clostridium perfringens, 95, 230
Clostridium tetani, 95–96
Clot retraction, 174. *See also* Blood clotting
Clubfoot, 308–9
Clumping in blood matching, **176**
Coccygeal nerve, 109, **119**
Coccyx, 70, **55, 56, 63, 69**
Cochlea, 133–34, **133, 134, 135**
Cochlear nerve, **134**
Coenzymes, 237, 238, 342
Cohesion, 353
Cold, common, 282
Colic flexures, 223, **217**
Colitis, 229
Collagen, 37
   fiber, **37, 38**
Collarbone, 58, **55, 57**
Collecting ducts, renal, 244, 246, **245, 248, 250**
Colon, 223–25, 243, **217, 224**
Colors, psychological effects of, 357
Color-blindness, 304–5
Colostomy, 229
Columnar cells, **27**
Columnar epithelium, 36, **34, 35**
Commensual light reflex, 106
Compounds, 316–17
Computers in blood analysis, 179
Computerized medical care, 19
Concentrated solutions, 317
Concentration gradient, 17
Conception, 295
Conditioned response, 105
Conduction of heat, 356, **356**
Conductor, electric, 358
Condyle, 52
Cones of retina, 127, **126**
Conjunctiva, 127
Conjunctivitis, 127, 131, 132
Connective tissue, 36–37, **27, 35, 37**
Constipation, 224
Contractile tissue, 37–38
Convection, 356, **356**
Convolutions of the brain, 111, **111**
COPD, 283
Cornea, 125, 127, **125**
Corneal reflex, 106
Coronal plane, 13, **12, 14**
Coronal suture, 66, **67, 68**
Coronary arteries, 152, **139, 156**
   diseases, 161–62
Coronary risk factors, 163
Coronary vein, **139, 158**
Corpora cavernosa, 287, **286**
Corpus luteum, 195, 293
Corpus spongiosum, 287, **286**

Cortex. *See also* Adrenal cortex
   of brain, 111, 112–13, **113**
   of kidney, 244, **244**
Corticospinal tract, 108, **109**
Cortisol, 195, 200–201
Cortisone, 195, 200–201
Corynebacterium diphtheriae, 132, 282
Costal cartilage, 70, **71**
Counter-current mechanism, 249, **250**
Covalent bonding, 326, **326, 327**
Cowper's gland, 287, **286**
Cranial aspect, 12, **12, 14**
Cranial cavity, 16, **17**
Cranial nerves, 116–18, **117**
Cranium, 66, **55, 56**
Creatine phosphokinase, 84
Crest of bones, 52
Cretinism, 196–97, 306
Crick, Francis, 334
Cross-bridges of myosin filaments, 80–81, **81**
Cross-section, body, 13, **14**
Croup, 282
Crypts of Lieberkuhn, 218
Cryptorchidism, 287
Cubital vein, 159
Cuboid bone, 66, **65**
Cuboid epithelium, 35, 36, **34**
Cuneate tract, 108, **109**
Cuneiform bones, 66, **65**
Current electricity, 357, 358–59
Curvature, spinal, 70
Cushing's syndrome, 201
Cutaneous membrane, 39
Cyanosis detection in dark skin, 41, 180
Cystic duct, **218, 226**
Cystic fibrosis, 306, 308–9
Cystine, 340
Cystitis, 256
Cytogenetics, 305, 307
Cytokinesis, 31
Cytoplasm, 28, 29, **25**
Cytosine, 334, 336

D

Dalton, John, 320
Dalton's law of partial pressures, 273–74, 348
Deafness, 135
Deaminization, 205, 227, 237
Deciduous teeth, 208
Decomposition, chemical; enzymes in, 20–21
Decubitus ulcers, 44
Deep reflexes, 106
Defecation, 225
Dehydration, 43, 249
   synthesis, 332
Delta cells, 202
Deltoid, **90**

Dendrite, 99, 102, **27, 99, 105, 107**
Density, 9
Dentine, 208, **208**
Deodorants, 136
Deoxyribonucleic acid. *See* DNA
Deoxyribose, 334
Depressor, 85
Dermatitis, 44
Dermatophytoses, 44
Dermis, 39, **41**
Detoxification of blood, 225, 227
Detrusor, 254, **254**
Dextrose, 332
Diabetes insipidus, 191, 194, 251
Diabetes mellitus, 112, 194, 202, 204–5, 306, 308–9
 and arteriosclerosis, 163
 and eye disease, 129, 131
Dialysis, 256–57
Diapedesis, 30
Diapers and pyelonephritis, 255–56
Diaphragm, 270–72, **17, 270, 271, 272**
Diaphysis, 52, **51**
Diarrhea, 224
Diarthroses, 72
Diastolic blood pressure, 143
Diet, 163, 215, 225, 235
Dietary goals for the United States, 236
Differential permeability, 26
Diffusion, 16, 260
 across alveoli, 273–74
Digestive enzymes, 35, 202, 214, 218, 219, 223
Digestive hormones, 214
Digestive system, 207–31, **218**
 and aging, 230
 and autonomic nervous system, 118, 120, **119**
 components, 208–27
 diseases and disorders, 215, 227–29
 functions, 208, 210, 212–13, 219–21, 224–25, 226
Digital arteries, 153, **153, 156, 157**
Digitalis, 251
Dilute solutions, 317
Diose, 332
Dipeptidase, 219, 220, 221
Diphtheria, 132, 282
Diplococcus pneumoniae, 283
Diploid chromosome number, 32, 290
Disaccharides, 221, 223, 332
Disc, intervertebral, **108**
Diseases and disorders
 blood, 178–80
 cardiovascular, 161–64
 digestive, 215, 227–29
 ear, 133, 135–36
 eye, 129–31
 genetic, 303–11
 muscle, 92, 95–96
 of nervous system, 121–22

Diseases, *(continued)*
 nutritional, 239–41
 renal, 255–58
 of reproductive system, 296–97
 respiratory, 277–79, 282–83
 of reticuloendothelial system, 188
 of skeletal system, 66, 71–72, 73–74
 skin, 43–45
Distal, 13, **13**, 14
Diuretics, 251
Diverticula, 225, 228–29
Diverticulitis, 228–29
Diverticulosis, 228, 230
DNA (deoxyribonucleic acid), 28, 300, 334–37
 in autoimmune disease, 187
 composition, 334
 and heredity, 336–37
 molecule, **335, 336, 337**
 replication, 336
Dominance, genetic, 301
Dorsal aspect, 13, **12, 14**
Dorsal metatarsal arteries, **157**
Dorsal venous arch, **158**
Dorsalis pedis artery, 155, **157**
Down's syndrome, 307, 308–9
Drugs in pregnancy, 296
Drug-induced nutritional deficiency, 240
Drug sensitivity, 188
Ducts of Schlemm, 131, **132**
Ductless glands. *See* Endocrine
Ductus arteriosus, 148, 159, 161, **160**
Ductus deferens, 286, 287, **286, 288**
Ductus venosus, **160**
Duodenum, 214, 217, **203, 213, 217, 218, 224, 226**
Dura mater, 109, **110**
Dwarfism, 71, 194, 303
Dyspnea, 278

E

E. coli. *See Escherichia coli*
Ear, 131, 133–36, **133, 134, 135**
 and aging, 134, 136
 bones, 70
 disorders and diseases, 133, 135–36
 external, 131, **133**
 inner, 131, 133–35, **135**
 middle, 118, 131, 133, **133**
Eardrum, 131, **133**
Ectoderm, 295
Edema, 183, 249, 252–53
 causes, 184
 pulmonary, 162
Edward's syndrome, 307
Effective filtration pressure, 246–47, **247**
Effector, **104**
Efferent ducts, 287, **288**
Egg, 32, 195, 291

Ejaculatory duct, 287
EKG (Electrocardiograph) 148, **149**
Elastic cartilage, **48**
Elastic fiber, **37**
Elbow, 58, 60, **60**
Electricity, 357–59
Electrocardiograph (EKG) 148, **149**
Electrolytes, 260, 261–67
 balance, 262
 in fluid compartments, 262–63
 functions, 261–66
Electron, 318, 319
 acceptors, 233
Elements, 316
 periodic table, 320, **321**
Embryonic pelvis, **62**
Embryonic skeleton formation, 52–53
Emphysema, 278, 282, 283, **279**
Emulsification, 220
Endocarditis, 163
Endocardium, 139
Endochondral bone formation, 52, **53**
Endocrine system, 36, 190–206, 219
 and aging, 206
 gland locations, **192**
 summary, 206
Endoderm, 295
Endolymph, 134–35
Endometrium 291, 292, **292**
Endoneurium, **110**
Endoplasmic reticulum, 27–28, **25, 26**
Endosteum, 52, **51**
Endothelium, 149, **150**
Energy, 20, 317–18, 345, 354–56
 and ATP, **235**
 in muscle contraction, 81–82
 and nutrients, 233
Enterogasterone, 213, 214
Enterogastric reflex, 213–14
Enterokinase, 218
Enzyme(s), 19–21, 237, **21**
 digestive, 35, 202, 214, 218, 219, 223
 synthesis, 197, 337
Eosinophil, 167, 171, **166, 171**
Epidermis, 39–41, **41**
Epididymis, 287, **286, 288**
Epigastric arteries, **154**
Epigastric region, 13, 16, **15**
Epiglottis, 210, 269, **196, 211**
Epilepsy, 121–22
Epinephrine, 120, 121, 195, 199
Epineurium, **110**
Epiphyseal cartilage, 49
Epiphyseal line, **54, 65**

Epiphyseal plate, 53, **51**
Epiphysis, 52, **51**
Epithelial tissue, 35–36, **34, 35**
Ergot, 230
Erythroblastosis, 177, 179, 310–11
Erythrocyte, 51, 167, 168–71, **166**
Erythrocyte count, 168
    and anemia, 178
    in polycythemia, 180
Erythropoietin, 169
Escherichia coli, 74, 112, 255–56
Esophagus, 155, 210, 212, 269, **198, 211, 213, 217, 218, 224**
Estrogen, 196, 201, **293**
    and males, 291
    secretion, 195, 291, 292
Ethane molecule, **328**
Ethyl alcohol molecule, **329**
Eupnea, 277
Eureka phenomenon, 155
Eustachian tube, 131, 133, **133**
Evagination, 27, **25**
Evaporation, 349, 350, 358, **358**
Excretion as blood function, 167
Excretory system, 242–58. *See also* Kidney, Renal
Exercise and atherosclerosis, 163
Exercise and breathing rate, 116, 274–75
Exhalation. *See* Expiration
Exocrine glands, 36, 191, 219
Expiration, 270, **271, 272**
    in asthma, 278
Expiratory reserve volume, 281
Extensor, 85, **83**
    carpi radialis brevis, **90**
    carpi radialis longus, 89, **91**
    carpi ulnaris, 89, **91**
    digitorum, 89, **91**
Extracellular compartment, 260
Extracellular fluid, 252, 261, 266
Eye, 125–32, **125, 128, 130, 132**
    and aging, 127, 136
    anatomy, 125, 127
    cavity, 66
    diseases, 129–31
    motion, 118
    muscles, 128, **128**

F

Face
    bones, 66, 70, **67**
    muscles, 85, 87, **86**
    nerves controlling, 118
Facial nerve, 118, **117**
Fahrenheit scale of temperature measurement, 7, **9**
Fallopian tubes, 291, **292**
False pelvis, **62, 63**
Family pedigree, 311, 313, **313**
Fanconi's syndrome, 306
Farsightedness, 129, **130**
Fascia, 83
Fasciculus, spinal. *See* Spinal tract
Fat(s), 253, 333
    and atherosclerosis, 163
    caloric value, 235
    cell, **37**
    digestion, 214, 219, 220–21, **220**
    metabolism, 205, 236
Fatty acids, 227, 236, 333
Feces, 221, 224–25
Feedback control, 2–3, 169, 195, 293. *See also* Homeostasis
Female reproductive system, 291–95, **292**
Femoral artery, 155
Femoral vein, 155, 159, **158**
Femur, 63, **51, 55, 56, 64, 94**
    fractured, **73**
Fertilization, 295
Fetus, **312**
    circulation in, 159, 161, **160**
    development of, 295–96
    erythrocyte formation in, 169
    skeleton of, 49, 60, **62**
Fever blisters, 44
Fiber
    in nutrition, 225
    in skeletal muscle, 82
Fibrin and fibrinogen, 173, 174
Fibroblast, 72, **37**
Fibroelastic cartilage, 49, **48**
Fibrosis, 278, **279**
Fibula, 65, **55, 56, 64, 94**
Filtration, 18
    of blood, 225
Fissure of Rolondo, 111
Flagella, 28
Flavin adenine dinucleotide (FAD), 342
Flexor, 85, **83**
    carpi radialis, 89, **91**
    carpi ulnaris, **91**
    digitorum longus, 92, **94**
    hallucis longus, 92, **94**
Floating ribs, 70
Flow rate, 351–52
Fluid. *See also* Body fluid
    balance, 167, 252
    head, 351, **352**
Fluorides and tooth decay, 208
Flushing, 42
Follicles, ovarian, 291
Follicle stimulating hormone (FSH), 195, 196, 290, 292
Fontanel, 66, **68**
Food poisoning, 230
Foot
    arteries, 155, **157**
    bones, **65**
Footcandle, 357
Foramen, 52
    ovale 159, 161, **160**
Foreskin, 289
Formaldehyde, **330**
Formalin, 330
Formic acid molecule, **329**
Fossa, 52
Fovea centralis, 127 **125**
Fractures, 72, **73**
Frontal bone, 66, **67, 68**
Frontal plane, **12**
Frontalis, 87, **86**
Fructose, 332, **332**
FSH. *See* follicle stimulating hormone
Functional group, 329

G

Gag reflex, 106
Galactose, 332
Galactosemia, 240, 304, 310–11
Gallbladder, 208, 220, **119, 203, 217, 218, 226**
Gallstones, 221, 230. *See also* Calculi
Gametes, 32
Ganglia, 111, **104, 112**
Ganglion cells, **126**
Ganglionic chain, 118
Gangrene, 95
Gas(es), 345–51
    atmospheric, 345
    concentration, and respiration, 273–74
    exchange, 273–74, 281, **275, 280, 281**
    inert, 322
    pressure, 274, 345–47
    measurement, 346–47
    volume relationships, 347–48
Gaseous transport, 278–82
Gastroenteritis, 229
Gastric arteries, 155
Gastric juice, 214
Gastrin, 214, 215–16
Gastrocnemius, 92, **94**
Gastrointestinal tract. *See* Digestive system
Gel, 184
Gene(s), 300–302
    and DNA, 336
    faulty, 301–2
    locus, 300
Genetics, human, 299–314

Genetic defects, 301–313
　Chromosomal, 305–7
　late-appearing, 306
　multifactorial, 305
　probability of, 301–2, **302**
　sex-linked, 304–5
　single-gene, 303–4
　tests for, 309–13
Genicular arteries, **157**
Genotype, 301
Gestation, 295
GH. *See* Growth hormone
Giantism, 71, 194–95
Gland, 36. *See also* Endocrine system, *names of individual glands*
Glandular epithelium, 36
Glans penis, 289, **286**
Glaucoma, 125, 131
Globin, 170
Globulins, 186
Glomerular filtrate, 247, 249
Glomerular nephritis, 187, 255
Glomerulus, 187, 245–46, **245, 247**
Glossopharyngeal nerve, 118, **117**
Glottis, 269
Glucagon, 195, 202, 203, 204, 227
Glucocorticoids, 195, 200–201, 205, 240
Gluconeogenesis, 205, 240
Glucose, 227, 233, 332, **332**
　and insulin, 195, 202, 204
　and osmotic pressure, 262
　oxidation, 81
　in urine, 253, 254
Glutamic acid, 340
Glutamine, 340
Gluteus maximus, 92, **93**
Glycerol, 236
Glycine, 340
Glycogen, 195, 204, 227, 333
Glycogenesis and glycogenolysis, 227
Glycolysis, 233, **234, 236**
Goblet cells, 36, **34**
Goiter, 197–98, 306
Golgi, Camillo, 30
Golgi body, 30, **25, 99**
Gonadal arteries, 155
Gonadotropic hormones, 293
Gonadotropins, 196
Gonoccal ophthalmia neonatorum, 132
Gonorrhea, 296
Gout, 306
Gracile tract, 108, **109**
Gracilis, 92, **93, 94**
Grand mal seizure, 122
Granulocytes, **171**
Grape sugar, 332
Grave's disease, 197–98
Gravity and the heart, 142

Gray matter
　of brain, 111, **112**
　of spinal cord, 107, **104, 105, 108**
Growth hormone (GH), 71–72, 194–95
Guanine, 334, 336
Gums, **208**
Gyri of brain, 111, **111**

H

H-zone, 79, **78, 79**
Haemophilus aegyptius, 132
Hair, 43, 44
　cells of ear, 134, **134**
　follicle, **42**
　nasal, 269
Hairshaft, **42**
Hamate carpal, 60, **61**
Hammer of ear, 70, 131, **133, 135**
Hand
　arteries, 152–53, **153, 156**
　bones, 60, **61**
　infant, **62**
　muscles, 85, 89, **91**
　rotation, 58
Haploid chromosome number, 32
Hardening of the arteries, 162–63, 255
Hashimoto's disease, 187
Haustra, 224, **217, 224**
Haversian canal, 49, **50**
Haversian system 49–50, **50**
Hay fever, 187
Head
　arteries, 155, **152**
　of bones, 52
　muscles, 85, 87, **86**
Health, 2
Hearing, 113, 114. *See also* Ear
Heart, 139–48, **139, 141, 156**
　anatomy, 139–40, **139, 141**
　and autonomic nervous system, 120, **119**
　diseases and disorders, 148, 161–64, 308–9
　fetal, 159, 161, **160**
　function, 140–43
　physiology, 143–47, **144**
　sounds, 147–48, 162
　valves, 140, 141, **140**
Heartbeat, 115, 140, 143–47
Heartburn, 216
Heat, 355
Height, and bone length, 53
Helium atom, **318**
Hematoblast, 185–86
Hematocrit, 142, 168, 178–79, **168**
Hematology, 178
Hematoma, 72
Heme, 170, **170**
Hemeoxygen complex, 170

Hemodialysis, 256–57, **257**
Hemoglobin, 167, 170–71, 178, **170**
　in anemia, 178
　in gaseous transport, 280, **280**
　in skin color, 40
Hemphilia, 174, 305, 308–9
Hemorrhoids, 225
Heparin, 174
Hepatic system. *See also* Liver
　artery, 155
　duct, **218, 226**
　flexure, 223, **217**
　portal system, 159
　portal vein, 159, 223, 225
　vein, 159, 225, **158**
Hepatitis, 229
Hering-Breuer reflex, 275–76
Herpes simplex, 44
Heterozygous genes, 301
Hexose, 332
Hiatal hernia, 228, **228**
Hilus, 244, **243**
Hinge joint, **83**
Hip dislocation, congenital, 306
Hipbone, 60, **93**
Histidine, 334, 340
Histoplasmosis, 179
Hodgkin's disease, 188
Homeostasis, 2–3, 169, 250, **169**
Homologous chromosome pairs, **301**
Homozygous genes, 301
Hooke, Robert, 24
Hormone(s), 167, 190–206
　and bone growth, 70–72
　diffusion, 191
　digestive, 214
　growth, 71–72, 194–95
　production, 36, 293
　sex, 199, 200, 289–91, 292–93
Humerus, 58, **55, 56, 57, 58, 60, 91**
Huntington's disease, 303, 306
Hyaline cartilage, 49, 72, **48, 49**
Hyaline membrane disease, 273
Hydrocarbons, 326–28, 331
Hydrocephalus, 109, 305, 308–9
Hydrocephaly, 66
Hydrochloric acid, 214, 215, 324
Hydrogen, 316
　atom, **318, 327**
　ion, 251–52, 324, **252**
Hydrolysis, 220
Hydrostatic pressure, 18, 246, 265
Hydroxide, 323, 324
Hymen, 291
Hyoid bone, 70, **196**
Hyperbaric chamber, 279

Hypercalcemia, 256, 265
Hypercapnia, 277
Hyperglycemia, 204
Hyperkalemia, 264
Hyperopia, 129, **130**
Hyperparathyroidism, 199
Hyperplastic thyroid, 197
Hyperpnea, 277
Hypertension, 163-64
Hyperthermia, 84
Hyperthyroidism, 197-98
Hyperventilation, 266, 277
Hypervolemia, 261
Hypocalcemia, 265
Hypochrondriac regions, 13, 16, **15**
Hypogastric artery, 155
Hypogastric region, 13, 16, **15**
Hypoglossal nerve, 118, **117**
Hypoglycemic factors, 202, 204
Hypokalemia, 264
Hypoparathyroidism, 199
Hypophysis, 112, 191, 194-96, 261, **113, 117, 192, 193**
Hypopnea, 277
Hypothalamus, 111, **113, 193**
 functions, 116, 191
 hormones, 194, 196
Hypothermia, 356
Hypothyroidism, 163, 196-97
Hypovolemia, 161-62, 261
Hypoxia, 169, 278-79

I

Ileocecal junction, 223
Ileocecal sphincter, 223
Ileocecal valve, 217, 223, **224**
Ileum, 217, 228, **217**
Iliac arteries, 155, **156, 160**
Iliac regions, 13, 16, **15**
Iliac veins, 159, **158**
Ilium, 60, **55, 56, 62, 63**
Image and light in vision, 129-30
Impetigo, 43
Implantation of zygote, 295
Immune response, 186
Immune system. See also Lymphatic system
 and aging, 188
 and hormones, 195, 201
 and kidney transplants, 257
 and thymus, 186
Immunity, 186-88
Incisors, **209**
Incus, 70, 131, **133, 135**

Indican, 254
Indochondral bone formation, 53
Inert gases, 322
Infants, circulatory changes in, 159, 161
Infant digestion, 216, 219
Infantile amaurotic idiocy, 303
Inferior aspect, 12, **12, 14**
Influenza, 282
Infrared light, 358
Infundibular stalk, 191
Inguinal canal, 287
Inhalation. See Inspiration
Innoculation, 187
Innominate, 60
 artery, 153, **141**
Inorganic compounds, and organic, 325-26
Inspiration, 270-72, **271, 272**
Inspiratory reserve volume, 281
Insulin, 195, 202, 203, 204
Integument, 39
Interatrial defect, 148
Intercostal arteries, 155, **154**
Intercostal muscles, 70, 272, **272**
Intercostal space, 70, **71**
Interstitial cells of Leydig, 289-90, **288**
Interstitial fluid, 29, 183, 184, 260, **184**
 electrolytes, 262, 263
 renal, 249, 250
Interstitial space, 28-29
Interventricular defect, 148
Intestinal cell, **222**
Intestines, 217-25
 absorption in, 221-23
 large, 223-25, **218**
 obstruction of, 228
 small, 215, 217-18, 219-23, **224**
Intracellular compartment, 260
Intracellular fluid, 28, 252, 262, 263
Intramembranous bone formation, 52-53, 54
Intrapleural cavity, **271**
Intrapleural pressure, 272
Intrinsic factor, 170, 239
Invagination, 27
Inverse proportion, 347
Involuntary muscle, 37, **38**
Iodine, 197, 239
Ions, 322-23, 262, 265, **323**. See also Electrolytes
Iris, 120, 127, **125, 132**
Iron, 170, 171, 179, 239, 316
Ischemic phase of menstruation, 294, **294**
Ischium, 60, **55, 56, 62, 63**
Islets of Langerhans, 202, 204
Isoleucine, 334, 340
Isomers, 332
Isotopes, 319

J

Jaundice detection in dark skin, 41
Jaw bones, 70, 208, **55, 67**
Jejunum, 217, **217**
Jews and Tay-Sachs disease, 303
Joints, 39, 72, **40**
Jugular vein, 159, **158**

K

Karyotype, 300, **301**
Keratinization, 40
Ketoglutarid acid, 233, 234, 235
Ketone, 254, 330-31, **330**
Ketone bodies, 205
Ketosis, 205
Kidney(s), 242-58, **119, 200, 243**. See also Renal system
 and aging, 258
 arteries, 155, **159**
 diseases, 187, 255-58
 function, 246-55, 260-61, **169**
 structure, 243-46, **244**
 summary, 258
 transplants, 257-58
Kinetic energy, 345
Kleinfelter's syndrome, 307
Knee joint, **40, 41, 65**
Kneecap, 63-64, **55, 93**
Kreb's cycle, 233-34, **236**
Kupffer cells, 185, **185**
Kwashiorkor, 240

L

Labia majora and minora, 292, **292**
Lacrimal canaliculi, **129**
Lacrimal ducts, **129**
Lacrimal glands, 128-29, **129**
Lacrimal sac, 129, **129**
Lactation, 35, 194, 195, 196
Lactic acid, 81, 82, 234, 235, 265, **330**
Lactose, 332
Lacunae, 49-50, **50**
Lambdoidal suture, 66, **67, 68**
Lamellae, 50, **50**
Larynx, 118, 210, 269, 282, **211, 270, 271**
Lateral aspect 13, **13, 14**
Latissimus dorsi, 83
Law of definite proportions, 320
Left brain hemisphere, 115, **112**
Leg
 arteries, 155, **156, 157**
 bones, 63-65, **64**
 muscles, 92, **93, 94**
 veins, 155, 159
Length, measurement of, 3-4, **4, 5**
Lens of eye, 127, 132, **125**
Leptospirosis, 132

Lethal genes, 303
Leucine, 334, 340
Leukemia, 180, 188
Leukocytes, 30, 51, 167, 171–72, **166, 171**
Leukopenia, 180, 188
Levator, 85
  scapulae, 89, **88**
LH. *See* Luteinizing hormone
Life, 24
Ligaments, 36
Light, 356–57
  in vision, 120–30, **130**
Liminal threshold, 83
Lipase, 219, 220
Lipids, 227, 333
Liquid(s), 351–54
  volume measurement, 7, **7**
Liver, 185, **185,** 203. *See also* Hepatic system
  and autonomic nervous system, 120, **119**
  in blood clotting, 174
  and blood sugar levels, 204
  blood vessels of, 155, 159
  and digestion, 208, 220, 223, 225–27, **217, 218, 226**
  and erythrocyte formation, 169
  fetal, **160**
  and iron metabolism, 171
Lobes of the brain, 111, 114, **114**
Lobules, liver 225
Lock-and-key theory of enzyme specificity, 21
Loop of Henle, 246, **245, 248, 250**
Lower respiratory tract, 282–83
Lowwi, Otto, 106
Lumbar curvature, 70
Lumbar nerves, 109, **119**
Lumbar regions, 13, 16, **15**
Lumbar vertebrae, 70, **55, 56, 69**
Lumbodorsal fascia, **83**
Lumen of small intestine, 218, **226**
Lunate carpal, 60, **61**
Lung(s), 269, 270–72, 243, **271**
  capacity, 281
  collapsed, 278
  fetal, 159
  and gas exchange, 281
Lupus erythematosis, 187
Luteinizing hormone (LH) 195, 196, 290, 293
Lymph channels, 183–84, **183, 184**
Lymph ducts, 183
Lymph movement, 183
Lymph nodes, 184, 185, 188, **183**
Lymph valve, 183, **184**
Lymphatic system, 182–89
Lymphocytes, 167, 172, 185, 201, **166, 171**
Lymphoid tissue, 184, **183**
Lyase, 220

Lysine, 334, 340
Lysis, 179
Lysosome, 29, **25**
Lysozyme, 129, 209

M

McGovern, George, 236
McLeod, John J., 202
Macrophage, **37**
Magnesium 262, 265, 316
Malar. *See* Zygomatic bone
Malaria resistance and sickle cell anemia, 179, 303
Male reproductive system, 286–91, **286, 288, 289**
Male sex hormones, 199, 289–91
Malleolar arteries, **157**
Malleus, 70, 131, **133, 135**
Maltose, 209, 210, 332
Mammary gland and hormones, 194, 195, 196
Mammillary body, **117**
Mandible, 70, **55, 67**
Manganese, 239
Manubrium 70, **57, 71**
Marginal artery, 152, **139**
Marrow, 50–51, 169, 172, 185–86, **50, 51**
  cavity, **50**
Masculinization, 199
Masseter, 87
Mast cell, **37**
Mastication, 87, 118
Mastoid process, 66, **67**
Mastoiditis, 66, 133, 135
Matrix of cartilage, 49, **48**
Matter, 316, 345
Maturation failure of erythrocytes, 169–70
Maxilla, 70, **67**
Measurement, 3–10
  English and metric systems compared, 3, 5, **4**
Meatus of bones, 52
Medial aspect, 13, **13, 14**
Medial canthus, 129, **129**
Mediastinum, 139
Medication and digestive bacteria, 229
Medulla
  adrenal, 120, 195, 199, **200**
  of brain, 115–16, 275, **113, 117, 119**
  of kidney, **244**
Medullary cavity, 52, **51**
Meiosis, 32, **290**
Melanin, 40
Melanocyte, 40
Melatonin, 205
Membranes, 39
Mendeleev, Dmitri, 320
Meninges, 109, **108, 110, 113**

Meningitis, 109, 122
Meniscus, 7, **7**
Menopause, 293, 297
Menstrual phase of menstruation, 294–95, **294**
Menstruation, 293–95, **294**
Mesenteric arteries, 155, **156**
Mesenteric veins, **158**
Mesentery, 155
Mesoderm, 295
Messenger RNA, 339, **339**
Metabolic pathways in cellular respiration, **234**
Metabolism, 232–41. *See also* Carbohydrate metabolism, Fat metabolism, Protein metabolism
  thyroxin in, 195, 196, 197
Metacarpals, 60, **61, 91**
Metacarpus, **55, 56**
Metatarsals, 66, **65**
Metatarsus, **55**
Methane, 326, **328**
Methionine, 334, 340
Methyl alcohol molecule, **329**
Metric system, 3–10, **4, 5, 6, 8**
  compared to English, 3, 5, **4**
Microaneurysm, 129, 131
Microcephaly, 66
Microglia, 102
Microvilli, 217–18, 221, 222
Micturition, 254–55
Midbrain, **113, 119**
Midsaggital plane, 13
Milk
  digestion, 219
  production, 35, 194, 195, 196
Milk teeth, 208
Milliequivalents (mEq), 262
Minerals in human nutrition, 239
Mineral oil and nutritional deficiency, 240
Mineralocorticoids, 195, 200
Minimal stimulus, 83
Mitochondria, 28, **25, 26, 79, 99**
  in Kreb's cycle, 233
Mitosis, 30–31, 295–96
Mitral valve, 140, **141**
Mixtures, 316
Moisturized air and respiratory function, 269
Molars, 208, **209, 210**
Molar solutions, 317, 320–21
Mole, 320
Molecular forces in liquids, 353
Molecular weight, 320
Mongolism. *See* Down's syndrome

Monocytes, 167, 172, **166, 171**
Monosaccharides, 221, 223, 332
Morula, 295
Motor end-plates, 79
Motor pathways, 108
Motor unit, 80
Movement, neural control of, 118
Mouth, 208, 209–12, 269, **209, 210, 211, 270**
Mouth breathing, 269
Mucosa, digestive, 212, 214, 219
Mucous membranes, 39, 269
Mucus, 45, 46, 209, 218, 225
Multifactorial inheritance, 305
Multiple sclerosis, 187
Muscle, 76–97
  and aging, 96
  attachment, 82–83, **83**
  cells, 78, **27, 77, 78**
  classification, 85–92, **86, 88, 90, 91, 93, 94**
  contraction, 78–82
  diseases, 92, 95–96
  eye, 128, **128**
  fatigue, 84
  fibers, **78, 79**
  relaxation, 80
  summary, 96–97
  tissue structure, 37–38, 77–79, **38, 77, 78**
Muscular dystrophy, 95, 306, 308–9
Muscular insertion, 82, **83**
Muscular origin, 83, **83**
Musculophrenic artery, **154**
Myasthenia gravis, 95
Mycobacterium tuberculosis, 283
Myelin, 100
  sheath, **99, 101, 110**
Myocardial infarction, 161–62
Myofibril, 77–78, **78, 80**
Myofilament, 78
Myoneural junction, 80, **80**
Myopia, 129, **130**
Myosin, 78, **78, 81**
Myositis ossificans, 92
Myxedema, 197

N

Nares, 269, **270**
Nasal bones, 66, 70, **67**
Nasal ducts, **129**
Nasal passageways, **270**
Nasolacrimal duct, 129, **129**
Nasopharynx, **270**
Navicular bone, 66, **65**

Nearsightedness, 129, **130**
Neck muscles, 85, 87, 118, **86**
Neck vertebrae, 70
Neisseria gonorrheae, 74, 132, 296
Neisseria meningitis, 122
Nephritis, glomerular, 187, 255
Nephron, 244–46, 255, **245**
Nerve, 108, **110**. *See also* Neuron
  cranial, 116, 118, **117**
  end-plate, 79–80
  fiber, 108, **79, 110**
    regeneration, 100–101, **101**
  impulse, 79–80, 102–3, 107, **103, 107**
  myelinated and nonmyelinated, 100, **101**
  pathways, 79, 103–8, 115, 120–21
  spinal, 109
  tissue, 38
  tracts, 108–9, **109**
Nervous system, 98–123. *See also* Senses
  and aging, 122
  autonomic, 118–121
  central, 99
  diseases, 121–22
  peripheral, 99, 109
  summary, 118–21
Neurilemma, 101–2, **99, 101, 110**
Neuroglia, 102
Neurohypophysis, 191, 194, 195, 261, **193**
Neuron, 38, 99–102, 103, 104, **27, 99, 100, 103**
  of autonomic system, 118–19, 120–21
  bipolar, 100, **100**
  internuncial, 103, 104, **105**
  motor, 103, 104, **104, 105**
  multipolar, 100
  sensory, 103, **104, 105**
  unipolar, 99–100, **100**
Neutron, 318, 319
Neutrophil, 167, 171, 172, **166, 172**
Nicotinamide adenine dinucleotide (NAD), 342
Night blindness, 128
Nissl bodies, 99
Nitrate, 324
Nitric acid, 324
Nitrogen, 237, 316
Nitrogen base in DNA, 334, **336**
Nitroglycerine in angina, 161
Node of Ranvier, 100, **99, 101**
Noise and stress, 133
Noradrenalin, 121, 147, 195, 199, 251
Norepinephrine, 121, 147, 195, 199, 251
Nose, 269, **270**
Nostrils, 269, **270**
Nuclear membrane, 28, **25**
Nuclear staining, 310

Nucleolus, 29, **25, 99**
Nucleoplasm, 28
Nucleotide, 334, 342
  bases, **335**
Nucleus, 28, **25, 27**
  and mitosis, 30–31
  of neuron, **99**
  Schwann cell, **99, 101**
Nutrient
  artery, **51**
  foramina, 60, **59**
  storage, 227
Nutrition, 167, 232–41

O

Obesity, 163, 204, 240–41, 260
Obstruction, intestinal, 228
Occipital bone, 66, **67, 78**
Occipitalis muscle, 87, **86**
Occulomotor nerve, 118, **117**
Odor detection, 118
Oil secretion, 35
Olfactory nerve, 116, 118, **117**
Oligodendrocyte, 102
Operating room, static electricity in, 358
Ophthalmia neonatorum, gonococcal, 132
Optic chiasm, **117**
Optic nerve, 118, **117, 125**
Orbicularis oculi, 87, **86**
Orbicularis oris, 87, **86**
Orbit, atomic, 318, **318, 319, 322, 323**
Orbit of skull, 125
Orbital cavity, 66
Organ, 12, 24, 34
  of Corti, 134, **134**
Organ system. *See* System
Organelle, 24, 26
Organic acids, 329–30
Organic chemistry, 325–42
Organic compounds, 325–26
Organic radicals, 329
Organism, 24, 34
Osmosis, 16–17, 260, **17**
Osmotic pressure, 17, 261–63, 265–66, **17**
Ossification, 52, 53, **61**
  centers, 53, **53**
Osteitis fibrosa, 73
Osteoblast, 52, 53, 70
Osteoclast, 52, 53–54, 70
Osteocyte, 37, 49, 70, **50**
Osteogenesis imperfecta tarda, 306
Osteomalacia, 74
Osteomyelitis, 74
Otitis, 135–36
Oval window of ear, 134, **135**
Ovarian follicles, 195, 196
Ovary, 196, 291, **192, 292**
Ovulation, 292–93, 295

Ovum, 32, 195, 291
Oxaloacetic acid, 235
Oxygen, 81, 197, 316
    atom, **319, 327**
    and breathing rate, 277
    debt, 82, 235
    and erythrocyte production, 169
    and heart rate, 146
    supply to heart, 151, 159, **169**
    transport, 170–71, 269, 279, **280**
Oxytocin, 194, 195

P

Pacemaker, artificial, 145, **145**
Pain receptors, 43, 112
Palate, 209, 210, **209, 211, 270**
Palmar arches, **156**
Palmar arteries, 153, 155, **153**
Palmaris longus, 89, **91**
Pancreas, 202–5, 206, **203**
    in digestion, 208, 215, 218–19, 223, **217, 218, 226**
    exocrine function, 219
Pancreatic duct, 202, 219, **203, 226**
Pancreatic enzymes, 219, 220
Pancreatic islets, 120, 192, 195, 202, **203**
Pancreatic juice, 218, 219
Pancreatitis, 112, 229
Pancreozymin, 214, 219
Pap smear test, 297
Papilla
    duodenal, 213
    renal, 244
    of tongue, 209
Papillary muscles, **141**
Paralysis, periodic, 306
Parasitic worms, 229
Parasympathetic nervous system, 118, 120, 121, **119**
Parathormone, 195, 198–99
Parathyroid, 70–71, 73, 195, 198–99, **192, 198**
Parietal bone, 66, **67, 68**
Parkinson's disease, 122
Parotid duct, **210**
Parotid gland, 209, **210**
Partial pressures of gases, 273–74, 348
Pascal's law, 353, **354**
Patau's syndrome, 307
Patella, 63–64, **55, 93**
Patellar ligament, **93**
Patellar reflex, 106
Pelvic canal, 60, **63**
Pelvic cavity, 16
Pelvic girdle, 60, **61, 63**
Pectoralis minor, 89, **88, 90**
Pelvis, 60, **62, 63**
    of kidney, 244
Penicillinase, 20

Penis, 286, 287, 289, **286**
Pentose, 332
Pepsin, 214, 215–16, 219
Pepsinogen, 214
Peptides, 214, 215
Percentage solutions, 317
Pericardium, 39, 139
Perichondrium, 49, **48**
Perineurium, **110**
Periodic table of elements, 320, **321**
Periosteum, 52, **50, 51, 208**
Peristalsis, 212, 224, **211**
Peritoneal dialysis, 257
Peritoneum, 39
Peritonitis, 229
Peritubular capillaries, 246, 248, **250**
Permanent teeth, 208
Peroneal artery, 155, **157**
Peroneal vein, 155, **158**
Peroneus brevis, 92, **94**
Peroneus longus, 92, **94**
Pertussis, 282
Petit mal seizure, 122
pH scale, 324–25
Phagocytic cells, 183, 184
Phagocytic vesicle, **25**
Phagocytosis, 29, 30, 172, 185, **31**
Phalanges, **55, 56**
    of feet, 66, **65**
    of hands, 60, **61, 91**
Pharyngeal reflex, 106
Pharyngitis, 282
Pharynx, 118, 269, **196, 198, 270**
    in swallowing, 209, 210, **209, 211**
Phenols, 331, **331**
Phenotype, 302
Phenylalanine, 334, 340
    and phenylketonuria, 239, 311, 313
Phenylketonuria (PKU), 239–40, 304, 308–9
    tests for, 311, 313
Phonocardiograph, 148
Phosphate, 265, 324, 334
Phosphoric acid, 324
Phosphorus, 239, 265, 316
    and bone, 37, 50, 70–71, 73
Photosynthesis, 332
Phototoxicity, 357
Physics, basic, 344–60
Physiology, 12, 22
Pia mater, 109, **110**
Pineal gland, 205, **192**
Pinocytic vesicle, **25**
Pinocytosis, 30
Pisiform carpal, 60, **61**
Pitocin, 194, 195
Pituitary dwarfs, 71
Pituitary gland, 112, 191, 194–96, 261, **113, 117, 192, 193**
PKU. *See* Phenylketonuria
Placenta, 159, 161, 293, 295, **160, 312**
Plantar arteries, 155, **157**

Plasma, 29, 167, 227, 260
    in blood clotting, 173
    cells, 185, 186
    composition, 249, 262–63
Platelets, 51, 167, **166, 171**
    and blood clotting, 172–73
Platysma, 87, **86**
Pleura, 39, 270, 272, **271**
Pleural membranes, **270, 271**
Pleurisy, 272–73, 282–83
Pneumonia, 278, 282, 283, **279**
Poison ivy and oak, 44
Polarization in muscle contraction, 80
Polarization and nerve impulse transmission, 102–3, **103**
Polys (polymorphonuclear granulocytes), 171, **171**
Polycystic kidney, 306
Polycythemia, 180
Polydactyly, 308–9
Polymorphonuclear granulocytes (polys), 171, **171**
Polynucleotide, **335**
Polypeptide hydrolysis, 220
Polysaccharides, 50, 219, 333
Pons, 111, **113, 117**
    and breathing, 276–77
Popliteal artery, 155, **157**
Popliteal vein, 155, **158**
Portal system, 159
Portal vein, **158, 160**
Portal vessel, 223
Posterior aspect, 13, **12, 14**
Posterior descending artery, 152, **139**
Posterior pituitary, 191, 194, 195, 261, **193**
Postural reflexes, 106
Potassium, 261, 264, 316
    function, 102, 239
    milliequivalents, 262
Potassium hydroxide, 324
Potential osmotic pressure (P.O.P.), 262
Potential space, 184
Pott's disease, 74
Pregnancy, 295
    drugs during, 296
    mineral drain in, 52, 74
    and Rh factor, 176–77
Prematurity, 308–9
Premolars, **209**
Prepuce, 289, 292
Pressoreceptors, 146
Pressure in circulatory system, 141
Pressure gradient, 279, 352
Pressure receptors, 42–43

Progesterone, 196, 291, 293, 295
Prolactin, 195, 196, 293
Proliferative phase of menstruation, 294, **294**
Proline, 340
Pronation, 58
Pronator, 85
 teres, 89, **90, 91**
Proprioceptor centers, 277
Propyl alcohol molecule, **329**
Prostate cancer, 297
Prostate gland, 255, 287, **286**
Proteases, 215
Protection as blood function, 167
Protective reflexes, 119–20
Protein, 184, 265, 333
 in bone, 50
 caloric value, 235
 deficiency, 95, 240
 digestion, 214, 215, 219–20, **215**
 functions, 334
 metabolism, 205, 227, 237
 synthesis, 195, 320, 342, **341**
Protein Bound Iodine Test (PBI), 197
Proteus, 74
Proteus vulgaris, 255
Prothrombin, 173, 174
Proton, 318, 319
Protozoa, 24, 229
Provitamin, 238
Proximal aspect, 13, **13, 14**
Pseudomonas, 74, 179
Pseudomonas aeruginosa, 135–36, 255
Pseudostratified columnar epithelium, 36, **34**
Psychosomatic illness, 216
Pubic arch, **63**
Pubic symphysis, 49, 60, **62, 63**
Pubis, 60, **55, 62**
 female, **292**
Puerperal fever, 297
Pulmonary artery, 140, 151, 161, **139, 141, 156, 160**
Pulmonary circulation, 139–40, 151, **152**
Pulmonary trunk, **139**
Pulmonary veins, 140, 151, **139, 141, 158**
Pulp cavity, 208
Pulse pressure, 143
Pupil, 127, **119, 125**
Pupillary reflex, 106
Purines, 334
Purkinje fibers, 145, **144**
Pus, 30, 172, 254
Pyelonephritis, 255–56

Pyloric antrum, 212, **213**
Pyloric pump, 213
Pyloric sphincter, 212, 213, **213**
Pyloric stenosis, 228
Pyloric valve, **213**
Pylorus, **217**
Pyramids, renal, 244, **244**
Pyrimidines, 334
Pyruvic acid, 233, 234, 235

Q

Quadriceps femoris, **93**

R

Race and sickle cell anemia, 179
Radial artery, 152–53, 155, **153, 156**
Radial vein, 159
Radiation of heat, **358**
Radioactive iodine uptake test, 197
Radioactive isotopes, 319
Radicals, 322–24
 organic, 329
Radius, 58, **55, 56, 59, 60, 61, 91**
Rashes, 44
Ratio solutions, 317
Reasoning and brain geography, 113–14
Rebound phenomenon, 115
Receptor, **104, 107**
Recessive genes, 301
Rectum, 223–25, **217, 218, 224**
 female, **292**
 male, **286**
Rectus femoris, 92, **93**
Red blood cells. *See* Erythrocytes
Reed-Sternberg cells, 188
Reflex(es), 104–6, 118, 119–20, **105**
 arc, 103–4, **104, 105**
 protective, 119–20
 somatic, 106, 118, 120
 visceral, 106, 118, 120
Refraction, 127
Release factors, 194
Releasing hormones, 116
Renal system. *See* Kidney
Renal arteries, 155, **156, 243, 245**
Renal column, 244, **244**
Renal diseases, 255–58
Renal papilla, **244**
Renal pelvis, **244**
Renal veins, 246, **158, 243**
Rennin, 216, 219
Reproductive system, 285–98
 and aging, 297
 diseases, 296–97
 female, 291–95, **292**
 fetal, 287
 hormones, 196
 male, 286–91, **286**
 summary, 297–98

Residual volume, 281
Respiration as blood function, 167
Respiration, cellular, 269, **234**
Respiratory center of medulla, 275
Respiratory distress syndrome, 273
Respiratory system, 268–84, **270**. *See also* Breathing
 and aging, 283
 diseases of, 272–73, 277, 278, 282–83, **279**
 summary, 283–84
 and vagus nerve, 118
Rete testis, 287, **288**
Reticular fibers, **37**
Reticuloendothelial system, 185–88
Reticulospinal tract, 108, **109**
Reticulum cells, 185
Retina, 125–27, **125, 126**
 detachment of, 131
Retinene, 127, **128**
Rh factor, 176–78, 179
Rheumatic fever, 162, 187
Rheumatoid arthritis, 187
Rhinovirus, 282
Rhodopsin, 127–28, **128**
Rhomboideus major, 89, **88**
Rhomboideus minor, **88**
Ribs, 70, **55, 56, 57, 71**
 in breathing, **272**
Ribonucleic acid. *See* RNA
Ribosomes, 28, 337, **25, 26**
Rickets, 73–74, 306
Right brain hemisphere, 115, **112**
Ringworm, 44
Risorius, **86**
RNA (ribonucleic acid), 337, 339–42, **339**
Rods of retina, 127–28, **126**
Rotator, 85
Rubella syndrome, 310–11
Rubeola, 132
Rubrospinal tract, 108, **109**
Rugae of stomach, 212
Rutherford, Ernest, 318

S

S-A (sinoatrial) node, 144–46, **144**
Sacral artery, **156**
Sacral nerves, 109, **119**
Sacroiliac joint, 60, **62, 63**
Sacrum, 70, **55, 56, 63, 69**
Sagittal plane, 13, **14**
Sagittal sinus veins, **158**
Sagittal suture, 66, **68**
Saliva, 35, 309–10,
Salivary amylase, 209
Salivary glands, 120, 208, 209, **119, 210, 218**
Salmonella, 74
Salt, 320, 324, 325
 in diet, 263–64

Saphenous veins, 155, 159, **158**
Sarcolemma, 78, **80**
Sarcomere, 78, **78**, **79**
Sarcoplasmic reticulum, 80, **80**
Sartorius, 92, **93**
Saturated solutions, 317
Scaphoid carpal, 60, **61**
Scapula, 54, **55, 56, 57, 83**
Schleiden, Matthias J., 24, 30
Schwann, Theodor, 24, 30
Schwann cell, 100, **101**
Sciatic notch, **62, 63**
Scientific method, 3
Sclera, 125, 127, **125**
Scotopsin, 127, **128**
Scrotum, 287, **286**
Sebaceous glands, 43, **42**
Secretin, 214
Secretion and epithelial cells, 35
Secretory phase of menstruation, 294, **294**
Selective permeability, 29
Sella turcica, 191, **193**
Semen, 286
Semicircular canals, 134–35, **133, 135**
Semilunar valves, 140, **141**
Semimembranosus, 92, **93, 94**
Seminal vesicle, 287, **286**
Seminiferous tubules, 287, 289, **288**
Semitendinosus, 92, **94**
Senility, 122
Senses, 114, 124–37. *See also* Ear, Eye, Smell, Taste
  and aging, 136
  summary, 136–37
Sense organs, 125
Sensory cells in ear, 134
Sensory nerves, 41–43
Sensory pathways, 108
Septa of testes, 287, **288**
Septum, interventricular, 139, 144, 161, **141, 144**
Serine, 340
Serous membranes, 39
Serous secretions, 209
Serratus anterior, 89, **88**
Serum, 174, 175
Sesamoid bone, 64
Sex
  and body fluid level, 252, 253
  chromosome defects, 307
  determination, 304, **304**
  and disease, 163, 215, 228, 255–56
Sex hormones, 199, 200, 289–91, 292–93
Sex-linked traits, 304–5
Shaking palsy, 122
Shell of atoms, 318
Shin bone, 64, **55, 56, 64**
Shinsplints, 92
Shoulder
  arteries of, **153**
  muscles of, 85, 89, **88**
  nerve controlling, 118
  veins in, 159, **158**
Shoulder blade, 54, **55, 56, 57, 83**
Sickle cell anemia, 179, 303, 308–9
Single-celled forms of life, 24
Single-gene defects, 303–4
Sinoatrial (S-A) node, 144–46, **144**
Sinus of bones, 52
Siphon, 348–49, **349**
Skeletal muscle, 37, 77, **27, 38, 77, 79**
Skeletal system, 47–75. *See also* Bone, Cartilage
  and aging, 74
  articulation, 72 **57**
  components, 54–70, **55, 56**
  diseases and disorders, 71–72, 73–74, **73**
  summary, 74–75
  tissues, 48–54, **48, 49, 50, 51**
Skeleton, **55, 56**
Skin, 39–45, **41, 42, 43**
  accessory structures, 43
  and aging, 44
  color, 40
  diseases, 43–45
  functions, 39, 243
  structure, 39–41
  summary, 46
Skin fold test of obesity, 240
Skull, 66, **67, 68,** 113
Sleep regulation, 116
Smell, sense of, 113, 114, 136
Smoking and heart disease, 163
Smooth muscle, 37, 77, **27, 38, 77**
Sodium, 195, 200, 239, 263, 316
  atom, **322**
  as electrolyte, 261, 263–64
  ion, 80, 102, 322, **323**
  milliequivalents, 262
  pump, 17–18, **18**
  regulation of, 248–49, **250**
Sodium chloride, 262, 263–64
Sodium hydroxide, 324
Soleus, 92, **94**
Solutions, 316, 317
Solvent, 317
Somatastatin, 112
Somatic reflexes, 118, 120
Somatropic hormone. *See* Growth hormone
Sound perception, 118, 131–36
Specific gravity, 9–10
Specific viscosity, 352
Sperm, 32, 196, 286, 289, **288, 290, 291, 301**
Sperm duct. *See* Vas deferens
Spermatid, 289, **290**
Spermatocyte, 289, **290**
Spermatogenesis, 289, **290**
Spermatogonia, 289, **288**
Spermatozoan *see* Sperm
Sphenoid bone, 66, **67**
Sphincter, 85, 212
  of Oddi, 221, **226**
  of urinary bladder, **254**
Sphygmomanometer, 142–43, **143**
Spina bifida, 305, 308–9
Spinal accessory nerve, 118, **117**
Spinal cord, 102, 104, 107–11, **104, 105, 108**
Spinal nerves, 109, **108**
Spinal tracts, 107–8, **109**
Spinocerebellar tract, 108, **109**
Spinothalamic tract, 108, **109**
Spine. *See also* Vertebral column
  of bones, 52
Spinous process, 52
Spleen, 169, 185, 188
Splenic artery, 155, **156**
Splenic flexure, 223, **217**
Splenic vein, **158**
Splenis capitis, 87
Splenomegaly, 180
Split-brain theory, 115
Squamous epithelium, 35, **34**
Squamous suture, 66, **67, 68**
Stapes, 70, 131, **133, 135**
Staphylococcus, 43, 135, 179, 282
Staphylococcus aureus, 74, 132, 163, 230
Starch, 333. *See also* Carbohydrate, Glycogen
  conversion, 209, 210, 219, 220
  digestion, 223
Static electricity, 357, 358
Stem cells, 201
Stenosis, 161
Sterilization, 350–51
Sternomastoid, 87, **86**
Sternum, 70, **55, 57, 71**
Steroids, 200
Stirrups, 70, 131, **133, 135**
Stethoscope, 148
Stomach, 212–17, **211, 217, 218, 224**
  anatomy, 212, **213**
  functions, 212–14
  secretions, 212, 214–16
  volume, 212
Stomach breathers, 272
Storage vesicles, **80**
Straight sinus vein, **158**
Strain gauge, 142

Stratified epithelium, 34, 35–36, **34**
Strata of epidermis, 40–41
Streptococcus, 74, 135, 162, 179, 187, 255, 282, 297
Streptococcus viridans, 163
Stress, 216
 and disease, 95, 163, 215
 and hormones, 195, 199, 206
 and noise, 133
Striated muscle tissue, 37, 77, **27, 38, 77, 79**
Stroke, 121
Subclavian arteries, 152, 153, **141, 152, 153, 154, 156**
Subclavian veins, 159, 183, **158**
Subliminal stimulus, 83, 103, **103**
Sublingual gland, 209, **210**
Submandibular gland, 209, **210**
Substances, 316
Substrates, enzyme, 13–21, **21**
Succinic acid, 234
Sucrase, 19–20
Sucrose, 332
Sugar, 201, 331–33
 digestion, 219
 level in blood, 202
 and tooth decay, 208
Sulci of brain, 111, **111, 117**
Sulfate, 324
Sulfur, 239, 316
Sulfuric acid, 324
Sun poisoning, 357
Suntan, 40
Superior aspect, 12, **12, 14**
Supination, 58
Supinator, 85, 89, **90**
Suprarenal gland. *See* Adrenal gland
Surface tension, 349, 353, **349**
Surfactant, 273
Surgical regions, 13, 16, **15**
Surgical risk, 241, 252, 260
Suspensory ligaments of eye, 127, **125**
Sutures, skull, 72, **67, 68**
Swallowing, 209, 210, 212, **211**
Sweat, 35, 41–42, 43, 136, 358
 secretion, 35
Sweat glands, 43, 120, **42**
Sympathetic nervous system, 118–21, **119**
Synapse, 104, 106–7, **107**
Synarthroses, 72
Synovial cavity, **40, 41**
Synovial membrane, 39, **40**
Syphilis, 122, 179, 296, 308–9
Synthesis, chemical, enzymes in, **21**
System, 2, 12, 24

Systemic circulation, 140, 151, 152–59, **152**
Systolic blood pressure, 142–43, **143**

T

Taeniae coli, **217**
Talus, 65, 66, **65**
Tarsals, 66, **55, 65**
Tarsus, **55**
Taste, 113, 114, 118, 136, 209
Taste buds, 136, 209
Tay-Sachs disease, 303, 306, 310–11
Tears, 35, 128–29
Tectorial membrane, **134**
Teeth, 208, 209, **208, 209**
Temperature. *See also* Body temperature
 and enzyme function, 20
 Fahrenheit and Celsius equivalents, 8–9
 measurement, 7–9, **9**
Temperature receptors, 41–42
Temperature regulation, 167
Temporal bones, 66, **67**
Temporalis, 87, **86**
Tendons, 36, 82, **37, 82**
Tensor, 85
 fasciae latae, 92, **93**
Testes, 286, 287, 289, **192, 286, 288**
Testosterone, 196, 201, 289–90
Tetanus, 95–96
Tetany, 83, 84, 96, **84**
Tetracycline deactivation, 240
Tetralogy of Fallot, 161
Tetrose, 332
Thalamus, 111, 116, **113**
Thalassemia, 310–11
Thomson, Jay, 318
Thoracic arteries, **154, 156**
Thoracic cavity, 16, 39, 269–70, **17**
Thoracic nerves, 109, **119**
Thoracic vertebrae, 70, **55, 56, 69**
Threonine, 334, 340
Threshold of stimulation, 103
Throat. *See* Pharynx
Thrombin, 173, 174
Thrombocytes, 51
Thromboplastin, 173
Thymine, 334, 336
Thymus, 186, 195, 201
Thyroglobulin, 197
Thyroid cartilage, **196**
Thyroid gland, 195, 196–98, 206, **192, 196, 198**
Thyroid storm, 198
Thyrotropic stimulating hormone (TSH), 195–96
Thyrotropin, 197–98
Thyroxin, 195, 196–97
Tibia, 64, **56, 64, 94**
Tibial arteries, 155, **157**

Tibial veins, 155, **158**
Tibialis posterior, 92, **94**
Tidal volume, 274, 281
Tinea capitis, 44
Tissue, 12, 24, 34–39
 culture, 31
 summary, 45–56
Titer, 175
Tongue, 118, 209, 210, **210, 211, 270**
Tonsils, 185, 186
Tonus, 84
Tonticollis, 92, 95
Touch receptors, 42–43
Trace substances, 237
Trachea, 269, **119, 196, 198, 270, 271, 273**
Trachoma, 132
Transfer RNA, 339–42
Transferrin, 180
Transmitter substance, 106–7
Transplants, 127, 201, 257–58
Transverse plane, 13, **14**
Transverse sinus vein, **158**
Trapezium, 60, **61**
Trapezius, 89, **88**
Trapezoid carpal, 60, **61**
Treponema pallidum, 122, 296
Treppe, 83–84
Triangular carpal, 60, **61**
Triceps brachii, 89, **83, 90, 91**
Triceps reflex, 106
Trichomonas vaginalis, 296, 297
Tricuspid valve, 140, **141**
Trigeminal nerve, 118, **117**
Triglycerides and insulin, 204
Trigone, 254, **254**
Triose, 332
Triplet codes, 340, **340**
Trisomy conditions, 307
Trochlea, 128, **128**
Trochlear nerve, 118, **117**
Tropic hormones, 195–96
Trunk, arteries of, 155, **156**
Trypsin, 215, 219–20
Trypsinogen, 219–20
Tryptophan, 334, 340
TSH. *See* Thyrotropic stimulating hormone
Tuberculosis, 74, 282, 283
Tuberous sclerosis, 306
Tubular reabsorption and secretion, 247–48, **248**
Tubules of testes, 287, 289, **288**
Tumors, 229, 256, 297. *See also* Cancer
Tunica adventitia, **150**
Tunica albuginea, 287, 289, **288**
Tunica intima, **150**
Tunica media, **150**
Turner's syndrome, 307, 310–11
Tympanum, 131, **133**
Tyrosine, 239, 340

U

Ulcer, 215
Ulna, 58, **55, 56,** 59, 60, 61, **83, 91**
Ulner artery, 152–53, 155, **153, 156**
Ulnar vein, 159
Ultraviolet light, 74, 356, 357
Umbilical arteries, **160**
Umbilical cord, 161, **160**
Umbilical region, 13, 16, **15**
Umbilical vein, **160**
Underweight, 241
Unsaturated solutions, 317
Upper respiratory tract, 282
Uracil, 339
Urea, 227, 237, 262, 325
Ureter, 244, 254, **200, 243, 254**
Urethra, 254, 255, **243, 254**
  female, **292**
  male, **286**
Urinalysis, 253
Urinary bladder, 244, 254–55, **243, 254**
  and autonomic nervous system, 120, **119**
  female, **292**
  male, **286**
Urinary system anatomy, 243–46, **243**
Urination, 254–55
Urine, 243, 246, 247, 253–54
Uterus, 194, 195, 291, 297, **292, 312**
Uvula, **209**

V

Vacuole, 28, **25, 31**
Vagina, 291, **292**
Vagus nerve, 115, 118, **117, 144**
Valence, 322
  of common ions, 323
  of radicals, 324
Valine, 334, 340
Valves in veins, 150, **150**
Valvular heart disease, 162
Vapor pressure, 349–50
Varicose veins, 150–51
Vas deferens, 286, 287, **286, 288**
Vasa recta, 249, **245, 250**
Vasopressin. *See* Antidiuretic hormone
Vasoconstriction, 42, 115
Vasodilation, 42
Vastus muscles, 92, **93**
Vegetative responses, 116
Veins, 139, 150–51, 155, 159, **150, 158**
Venae cavae, 141, 159, **139, 141, 158, 160, 200, 243**
Venereal diseases, 296–297.
Ventral aspect, 13, **12, 14**
Ventricles of the brain, 111, **113**
Ventricular septal defect, 161

Venules, 151, **37, 151,** 246
Vertebrae, **55, 56, 108**
Vertebral arteries, 153, 155, **152, 153**
Vertebral canal, 17
Vertebral column, 70, **17, 69**
Vestibular apparatus, 134–35, **133, 135**
Villi, 217–18, 221, **222**
Viruses in digestive disease, 229
Viscera, 116, 156
Visceral muscle, 37, 77, **27, 38, 77**
Visceral reflexes, 106, 118, 120
Viscosity, 352
Vision, 125–31,
  and aging, 127, 136
  chemistry of, 127–28
  low light, 127–28, **128**
  neural control of, 113, 114, 118
Visual purple, 127
Vital capacity of lungs, 281
Vital signs, 24
Vitamin(s), 169–70, 237–38, 342
  A, 127, 128
  B_{12}, 169
  D, 72, 73–74, 256
  K, 174
Vitreous humor, 125, **125**
Vocal cords, 210
Voice box. *See* Larynx
Volar arches, **153**
Volar digital vein, **158**
Volkmann's canals, 49, **50**
Volume measurement, 6–7, **6, 7, 8**
Voluntary muscle, 37, 77, **27, 38, 77, 79**
Vomiting, 216–17
Vulva, 292
Vulvovaginitis, 296

W

Warts, 44
Water, 239, 317
  absorption, 195, 224
  balance, 191, 194, 201
  level, and sodium, 263
  regulation, 249, 250–51
Watson, James, 334
Watson-Crick theory, 336
Weight. *See also* Body weight; Molecular weight
  gain, and blood flow, 142
  measurement, 4–6, **6**
White blood cells. *See* Leukocytes
White matter
  of brain, 111, **112**
  of spinal cord, 107, **104, 108**
Whooping cough, 282
Wilson's disease, 306
Wisdom tooth, 208
Wohler, Friedrich, 325
Wrist bones, 60, 91, **61**
Wryneck, 92, 95

X

X rays, 121–22, 357

Z

Z-line, 78, 79, **78, 79**
Zygomatic arch, 66, **67, 210**
Zygomatic bone, 66, 70, **67**
Zygomaticus major, 87, **86**
Zygote, 295